Formal Methods
for Open Object-based
Distributed Systems

IFIP – The International Federation for Information Processing

IFIP was founded in 1960 under the auspices of UNESCO, following the First World Computer Congress held in Paris the previous year. An umbrella organization for societies working in information processing, IFIP's aim is two-fold: to support information processing within its member countries and to encourage technology transfer to developing nations. As its mission statement clearly states,

> IFIP's mission is to be the leading, truly international, apolitical organization which encourages and assists in the development, exploitation and application of information technology for the benefit of all people.

IFIP is a non-profitmaking organization, run almost solely by 2500 volunteers. It operates through a number of technical committees, which organize events and publications. IFIP's events range from an international congress to local seminars, but the most important are:

- the IFIP World Computer Congress, held every second year;
- open conferences;
- working conferences.

The flagship event is the IFIP World Computer Congress, at which both invited and contributed papers are presented. Contributed papers are rigorously refereed and the rejection rate is high.

As with the Congress, participation in the open conferences is open to all and papers may be invited or submitted. Again, submitted papers are stringently refereed.

The working conferences are structured differently. They are usually run by a working group and attendance is small and by invitation only. Their purpose is to create an atmosphere conducive to innovation and development. Refereeing is less rigorous and papers are subjected to extensive group discussion.

Publications arising from IFIP events vary. The papers presented at the IFIP World Computer Congress and at open conferences are published as conference proceedings, while the results of the working conferences are often published as collections of selected and edited papers.

Any national society whose primary activity is in information may apply to become a full member of IFIP, although full membership is restricted to one society per country. Full members are entitled to vote at the annual General Assembly, National societies preferring a less committed involvement may apply for associate or corresponding membership. Associate members enjoy the same benefits as full members, but without voting rights. Corresponding members are not represented in IFIP bodies. Affiliated membership is open to non-national societies, and individual and honorary membership schemes are also offered.

Formal Methods
for Open Object-based
Distributed Systems

Volume 2

**IFIP TC6 WG6.1 International Workshop
on Formal Methods for Open Object-based
Distributed Systems (FMOODS '97),
21–23 July 1997, Canterbury, Kent, UK**

Edited by

Howard Bowman

and

John Derrick

*Computing Laboratory
University of Kent,
Canterbury,
UK*

Published by Chapman & Hall on behalf of the
International Federation for Information Processing (IFIP)

CHAPMAN & HALL

London · Weinheim · New York · Tokyo · Melbourne · Madras

Published by Chapman & Hall, 2–6 Boundary Row, London SE1 8HN, UK

Chapman & Hall, 2–6 Boundary Row, London SE1 8HN, UK

Chapman & Hall GmbH, Pappelallee 3, 69469 Weinheim, Germany

Chapman & Hall USA, 115 Fifth Avenue, New York, NY 10003, USA

Chapman & Hall Japan, ITP-Japan, Kyowa Building, 3F, 2-2-1 Hirakawacho, Chiyoda-ku, Tokyo 102, Japan

Chapman & Hall Australia, 102 Dodds Street, South Melbourne, Victoria 3205, Australia

Chapman & Hall India, R. Seshadri, 32 Second Main Road, CIT East, Madras 600 035, India

First edition 1997

© 1997 IFIP

Printed in Great Britain by Athenæum Press Ltd, Gateshead, Tyne & Wear

ISBN 0 412 82040 4

A catalogue record for this book is available from the British Library

⊗ Printed on permanent acid-free text paper, manufactured in accordance with ANSI/NISO Z39.48-1992 and ANSI/NISO Z39.48-1984 (Permanence of Paper).

CONTENTS

PREFACE

This volume contains the proceedings of the Second IFIP WG 6.1 International Conference on Formal Methods for Open Object-based Distributed Systems (FMOODS'97). The conference was held on the campus of the University of Kent, Canterbury, UK on 21st-23rd July, 1997. The event was the second meeting of a new conference series, initiated in Paris in March 1996. The objective of FMOODS is to represent work at the convergence of three important and related fields: formal methods, distributed systems and object-based technology. This convergence is representative of some of the latest advances in the field of distributed systems (for example, the ODP reference model and the work of the OMG) and provides links between a number of important communities (for example, FORTE/PSTV, ICODP, ECOOP et cetera).

The papers presented at the event reflect the scope of the conference, which provided an excellent forum to debate a number of important issues. Our invited speakers at FMOODS'97 were drawn from the UK, the USA and France. Robin Milner of the University of Cambridge is well known for his work on CCS and latterly the Pi-Calculus which was the subject of his invited talk. Jeannette Wing of Carnegie-Mellon University has made important contributions in the field of object-orientation, and presented some of her latest work on subtyping in OO systems. Elie Najm (ENST) and Jean-Bernard Stefani (CNET) have both made major contributions to the application of formal methods to distributed systems, and in their invited talk they discussed the current state of the art in this area.

The technical papers in this proceedings comprise 20 regular papers together with 8 short papers. The conference was composed of a number of sessions covering the topics of: Mobility and Pi-Calculus; Concurrent OO Specification and Programming; Actors; Distributed Systems; OO Requirements Analysis and Design; Formal Specification, and Subtyping and Inheritance.

The conference enjoyed the continuing support of IFIP, and thanks are due to Otto Spaniol and Harry Rudin for their efforts in this respect. We are also grateful to the Engineering and Physical Sciences Research Council in the UK, who provided generous support which enabled a number of postgraduate students to attend the meeting. In addition, we would like to thank all programme committee members and external referees for all their efforts in reviewing and shaping the final programme. The members of the local organisation committee also worked hard during all stages of the conference preparation, and our thanks go to them. Finally, thanks go to all authors for their high quality submissions and Aileen Parlane of Chapman & Hall for her work preparing this volume.

Howard Bowman and *John Derrick*
Canterbury, July, 1997.

Conference Co-Chairs

Howard Bowman and John Derrick (University of Kent at Canterbury, UK)

Programme Committee

Gul Agha (U. of Illinois, Urbana, USA), Patrick Bellot (ENST, Paris, France), Gregor Bochmann (U. Montreal, Canada), Howard Bowman (UKC, Kent, UK), Ed Brinksma (U. Twente, Netherlands), John Derrick (UKC, Kent, UK), Michel Diaz (LAAS-CNRS, Toulouse, France), Kokichi Futatsugi (Jaist, Ishikawa, Japan), Reinhard Gotzhein (U. Kaiserslautern, Germany), Haim Kilov (IBM T.J. Watson Research Center, USA), Guy Leduc (U. of Liege, Belgium), Luigi Logrippo (U. of Ottawa, Canada), Jan de Meer (GMD Fokus, Germany), Elie Najm (ENST, Paris, France), Oscar Nierstrasz (U. of Bern, Switzerland), Claudia Linnhoff-Popien (RWTH Aachen, Germany), Kerry Raymond (DSTC, Brisbane, Australia), Omar Rafiq (U. of Pau, France), Gerd Schuermann (GMD Fokus, Germany), Jacob Slonim (IBM, Toronto, Canada), Jean-Bernard Stefani (FT/CNET, Paris, France), Ben Strulo (British Telecom Research, Ipswich, UK), Sebastiano Trigila (F. Ugo Bordoni, Roma, Italy), Juan Quemada (ETSI Telecomunicacion, Madrid, Spain), and Akinori Yonezawa (U. of Tokyo, Japan).

Invited Speakers

Robin Milner (University of Cambridge, UK), Jeannette Wing (Carnegie-Mellon University, USA), Elie Najm (ENST, France) and Jean-Bernard Stefani (CNET, France).

Local Arrangements Chair

Olga Fernandes (University of Kent)

Organization Committee

Eerke Boiten, Charles Briscoe-Smith, Geraldina Fernandes, Olga Fernandes, Erik Poll, Helena Rodrigues and Maarten Steen (all University of Kent).

LIST OF REFEREES

Daniel Amyot
Pierre Azima
Patrick Bellot
Eerke Boiten
Howard Bowman
Stephen Crawley
Juergen Dittrich
Kazi Farooqui
Yuen Fay (Vivien) Fong
Reinhard Gotzhein
Philippe Hunel
Barry Kitson
Lacayrelle Laurent
Giovanny Lucero
Zoran Milosevic
Elie Najm
Abdelkrim Nimour
Juan A. de la Puente
Pedro R. D'Argenio
Dunia Ramazani
Frank Roessler
Jean-Guy Schneider
Jacob Slonim
Ben Strulo
Tayfun Umman
Nalini Venkatasubramanian
Mario Winkler
Jan de Meer

Laurent Andriantsiferana
Herman Balsters
Andrew Berry
Rob Booth
Leo Cacciari
Serge Demeyer
Jawad Drissi
Arnaud Fevrier
Kokichi Futatsugi
George Yanbing Guo
Jadwiga Indulska
Eckhart Koerner
Guy Leduc
Markus Lumpe
Achour Mostefaoui
Robb Nebbe
Owezarski Philippe
Franz Puntigam
Omar Rafiq
Kerry Raymond
Theo C. Ruys
Gerd Schuermann
Maarten W.A. Steen
Carolyn Talcott
Hasan Ural
Francois Vernadat
Pete Young

Mark Astley
Veronique Baudin
Gregor v. Bochmann
Marc Born
Jay Che
John Derrick
Rachida Dssouli
Stefan Fischer
Brahim Ghribi
Lex Heerink
Haim Kilov
Rom Langerak
Luigi Logrippo
Bernd Meyer
Nancy Moussa
Oscar Nierstrasz
Erik Poll
Juan Quemada
Andry Rakotonirainy
Shangping Ren
Carlos Sanchez-Tarnawie
Richard Sinnott
Jean-Bernard Stefani
Simon Thompson
Carlos Varela
Takuo Watanabe
Job Zwiers

PART ONE

Mobility and Pi-Calculus

1

The Pi Calculus and its Applications

Robin Milner
University of Cambridge, UK

The pi calculus [MPW92, Mil93] was defined by Milner, Parrow and Walker as a "Calculus of Mobile Processes", extending work by Engberg and Nielsen [EN86]. It provides an underlying formal model for interactive systems which can change their configuration on the fly; this spans a large spectrum from mobile telephone networks to Java-like languages. The calculus aims to be a model for interactive behaviour as basic as is the lambda calculus for sequential computation. In fact, the lambda calculus can be modelled straightforwardly within it, and thus sequential computation can be seen as a special case of interaction.

The pi calculus is very simple; in my talk I shall presume no previous knowledge of it, but I shall not need to spend long in describing its primitive constructions. I shall focus on how it can be applied; in particular, how it admits a pleasant type system in which "type" can be understood to mean "pattern of interaction". In particular, I shall show how properties like "each mobile agent (ambulance? ..) will never be connected to more than one transmitter station at a time" are statically checkable types.

If time allows, I shall briefly discuss the language PICT (for distributed interactive systems) based upon pi calculus by Pierce and Turner [PT97], and the application of pi calculus to model authentication protocols by Abadi and Gordon [AG97].

REFERENCES

[AG97]   M. Abadi and A. Gordon. A calculus for cryptographic protocols: the spi calculus. Technical Report 414, Computer Laboratory, University of Cambridge, UK, 1997.

[EN86]   U. Engberg and M. Nielsen. A calculus of communicating systems with label-passing. Technical Report DAIMI PB-208, Computer

Science Department, University of Aarhus, Denmark, 1986.

[Mil93]   R. Milner. The polyadic pi-calculus: a tutorial. In F.L. Bauer, W. Brauer, and H. Schwichtenberg, editors, *Logic and Algebra of Specification*, pages 203–246. Springer Verlag, 1993.

[MPW92] R. Milner, J. Parrow, and D. Walker. A calculus of mobile processes, Parts I and II. *Information and Computation*, 100:1–77, 1992.

[PT97]   B. Pierce and D. Turner. Pict: a programming language based on the pi calculus. Technical report, Computer Science Department, Indiana University, USA, 1997.

2

A Calculus of Object Bindings[†]

E. Najm, A. Nimour
École Nationale Supérieure des Télécommunications
46, rue Barrault, 75 634 PARIS CEDEX 13, France
{najm, nimour}@res.enst.fr

Abstract

We introduce COB (Calculus of Object Bindings) which features objects with dynamically changing service offers, and which embodies a type discipline that guarantees that no object may experience an unexpected service request at run-time. COB is compliant with the ODP computational model: objects run in parallel and interact by exchanging messages. The originality of COB lies in the distinction that is made between two kinds of interfaces: public and private. A public interface can be known by (and thus can react to messages coming from) more than one object at a time. In contrast, the private interfaces may be used only by pairs of objects, a client and a server, and their interaction is private without any possibility for other objects to disturb their "protocol" of interaction. Thus the services available on a private interface may change depending on the way the two partners have agreed to interact. This leads to a new interface typing system which handles dynamic service offers on interfaces. An equivalence, a compatibility and a subtyping relation are defined on interface types, inspired from the well known simulation and bisimulation relations. In spite of the non-uniform service availability on the (private) interfaces a safety theorem guarantees that, for well typed configurations, no "service not specified" errors occur at run-time.

Keywords

Process calculi, Actors, typing and subtyping, type compatibility, bisimulation, ODP

[†]Work partially supported by France Telecom/CNET

1   INTRODUCTION

We define COB, a Calculus of Objects Binding which is compliant with the ODP Computational Model (see [Najm-Stefani, 1995]). The main contribution brought by COB is a new type system which allows to safely compose objects having dynamically evolving interfaces. We consider collections of objects running in parallel and communicating with each other by exchanging messages. Each object can be addressed by one of its interfaces. We introduce the notion of interface mode. The mode of an interface can be public or private. More than one object can have the reference of a public interface and so can invoke the services offered by this interface*. In contrast, a private interface can be known (invoked) only by one object at a time. The services available at a public interface are uniform: a service available once will always be available in the future. In contrast, the set of processable messages at a private interface may change thus making a service that was previously processable possibly not available.

The way of exchanging messages differs also when using a public or a private interface. The invocation of a public interface is asynchronous: the invoker creates a message that can be absorbed by the interface in a subsequent step. In contrast, the communication on a private interface is by *rendez-vous* between a sender and the receiver.

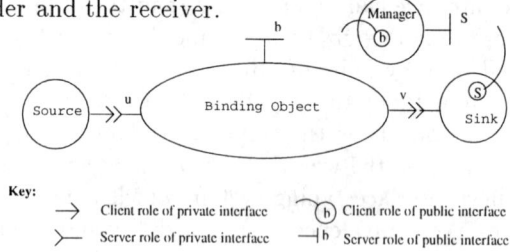

Figure 1 An example of an object configuration in COB

The invocation of a public interface corresponds actually to the ODP implicit binding where the invoker is not related to the receiver by a binding object and thus emits an asynchronous message that may , in turn, interact with the remote receiver. In contrast, the private invocation models a synchronous interaction of an (basic) object with a co-located binding object. The communication between an object and a binding object is private by nature, the private interfaces of the COB typing system allows to express and to check this privacy. The COB communication paradigm corresponds to the ODP signal basic operation. The other operations (announcements and interrogations) can be written in term of signals.

The non-uniform service availability (on the private interfaces) leads us to develop a new typing approach that allows to describe the evolution of the

---

*Without loss of generality, we consider only one way invocations. Indeed, invocations with return values can be modelled in terms of one way invocations.

type of an interface. This is the main contribution of COB which extends the ODP-CM type system defined in the ODP standard.

The remainder of the paper is structured as follows: section two is an informal introduction to COB, section three discusses the type language used in COB, section four deals with static semantics of COB while section five is devoted to the operational semantics. The safety property of well typed COB objects is presented in section six and in section seven we discuss related work and conclude.

2 THE COB CALCULUS

The COB language is designed to describe object systems compliant with the ODP computational model. The objects run in parallel and each object can invoke the services offered by other objects. The services are made available at the interfaces of the objects. The syntax of a service offer is as follows:

$$?u[m(\ldots) > B_1] \qquad \text{(Ex-1)}$$

The behavior of this object is to wait for the message m on the interface u and then process m as specified in B_1. The invocation of service m at interface u is written:

$$!u.m(\ldots) > B_2 \qquad \text{(Ex-2)}$$

This is the invocation of the service m of the interface u.

We will so distinguish between two roles for a given interface:

- client role: an object that has the client role of an interface can invoke the services offered by this interface.
- server role: an object that have the server role of an interface can accept messages (and thus it offers services) on this interface.

For instance, the behavior expression of the example Ex-1 describes an object which has the client role on u whereas the one of example Ex-2 has the client role. An object can have both roles for the same interface.

Another important characteristic of the interfaces is their mode: public or private. A public interface is an open interface that offers its services to any other object who knows that interface. More than one object is allowed to know this interface. This means that the client role of a public interface can be given to any object without restriction. Since many objects can independently invoke the services available on a public interface these services must be uniform i.e. must not change over time. Thus it is possible to duplicate the server role of a public interface as long as we are sure that its services are uniform. On the other hand, the private interfaces are used by pairs of objects to interact in a private way without any possibility for other objects to disturb their "protocol" of interaction. So the services available on a private interface may change depending only on the way the two partners want to interact. This privacy is ensured by forbidding the duplication of the roles of a private interface. This does not mean that the private interfaces are static. An object can pass to another object the server role (respectively the client role) of a private interface, but, in this, case it does not possess the role anymore and thus will not be able to receive (respectively to emit) on this interface.

An object can communicate to the other objects the roles of its interfaces by invoking their methods with the interface roles as arguments. For instance:

$$!u.m(v^!) > B$$

is the behavior of sending on u the client role of v to the receiver of the message. If v has been declared private, there can be no invocation of v in B. In the second case:

$$!u.m(v^?) > B$$

we describe the behavior of sending the server role of v. In the same way, if v has been declared private, there can be no service offer on v in B. An object can also send both capacities:

$$!u.m(v^{!?}) > B$$

When offering a service on an interface, an object must declare the type of the interfaces that it is expecting to receive. An interface type is a triplet (μ, ρ, X), noted X^{ρ^μ} where:

- μ is the mode of the interface: public (noted $\square$) or private (noted $\triangle$).
- ρ is the role of the interface: client (noted !), server (noted ?) or both (noted !?).
- X is the behavior type of the interface. It describes the succession of messages that can be handled by the interface.

Thus, a service offer can be written: $?u[m(w:X^{\rho^\mu}) > B]$

Let us now consider the behavior of a configuration of objects that contain a client and a server ready to interact on the same interface:

$$\underbrace{!u.m(v^!) > B_1}_{\text{Emitter}} \mid \overbrace{?u[m(w:X^{!^\triangle}) > B_2]}^{\text{Receiver}}$$

This configuration can evolve to a new one by passing the client role of the private interface v on u. The new behavior expression is thus:

$$B_1 \mid B_2[v/w]$$

in this new configuration the emitter has passed to the receiver its capacity to emit on v and cannot use it anymore: B_1 should not contain actions of the form $!v.m'(\ldots)$. On the other hand, B_2 has now the capacity to invoke the services of v or to pass it to another object (in which case it looses this capacity). It is important to note that B_2 cannot receive messages on v because it just received the client role and not the server role. In the case where v is a public interface, the emitter can still invoke the methods of v after passing to the receiver the capability to emit on v.

In COB the behavior of objects can be named. This is achieved by the following declaration:

$$A[u:X_1^{?^\square}, v:X_n^{!^\triangle}] = B_A$$

An object can then be instantiated by providing the actual parameters.
The object creation is another feature of COB. It is different from the object instantiation in the sense that, in a creation, the creator object has its proper continuation that runs in parallel with the created object. For instance, the expression:

$$\textbf{create } A[w_1^?, w_2^!] > B$$

evolves to a new one where A and B run in parallel:

$$B_A[w_1/u, w_2/v] \mid B$$

B has now lost the capacity to emit on w_2 because it has been passed to A. But it still has the capacity to receive on w_1 because w_1 is public.

An object can create a new interface at any time as follows:

$$\textbf{new } u:X^\mu \textbf{ in } B$$

the behavior of this expression is to first create a new interface u (having mode μ and behaving according to the behavior type X) and then to behave like B. A newly created interface has both roles (client and server).
In all the examples we presented until now only one message was handled by reception actions. In the general case, more that one service may be available on an interface and a reception action is written:

$$?u[m_1(\tilde{v}_1) > B_1, \ldots, m_n(\tilde{v}_n) > B_n]$$

Finally, it is possible for an object to offer services on more that one interface at the same time. This is written :

$$?u_1[\ldots] + \ldots + ?u_n[\ldots]$$

3 THE TYPE LANGUAGE

In our approach a *behavior type* describes the succession of messages that can be processed by an interface. Behavior types are dynamic: at a given state, a behavior type specifies the set of messages than can be handled by the interface and for each message the continuation state. For example, the interface of a one place buffer can be described by a two-state type. In the initial state it handles *put* messages and evolves to a state where it handles *get* messages.

$$
\begin{array}{rcl}
I & ::= & u \colon X^{\rho^{\mu}} \\
Dcl & ::= & A[\tilde{I}] = B \\
B & ::= & a > B \;\Big|\; \overset{n}{\underset{i=1}{\Sigma}} Recep_i \;\Big|\; A[\tilde{u}^{\tilde{\rho}}] \;\Big|\; \mathbf{new}\ I\ \mathbf{in}\ B \;\Big|\; 0 \\
a & ::= & !u.m(\tilde{v}^{\tilde{\rho}}) \;\Big|\; \mathbf{create}\ A[\tilde{u}^{\tilde{\rho}}] \\
Recep & ::= & ?u[m_1(\tilde{I}_1) = B_1, \ldots, m_n(\tilde{I}_n) = B_n] \\
C & ::= & B \;\Big|\; C|C \;\Big|\; \mathbf{new}\ I\ \mathbf{in}\ C
\end{array}
$$

Terminals	Meaning
A	behavior name
u	interface names
X	interface behavior type
$m, m_1, \ldots, m_n$	method names
ρ	interface role
μ	interface mode

Table 1 Summary of the COB Syntax

3.1   Syntax

Let *Meth* be a set of method (or message) names ranged over $m, m', m_i, \ldots$ and *TVar* a set of behavior type variable names ranged over $x, x', x_i, \ldots, y, \ldots, z, \ldots$ We build the set *ITypes* of interface behavior types ranged over by $X, \ldots, Y, \ldots, Z, \ldots$ as follows: an interface behavior type is a pair, noted $E \triangleright x$, where x is a type variable and E a finite collection of equations of the form $x_i = e_i$ where each x_i is a type variable that appears once in the left-hand side of an equation and e_i is an expression of the form ($\tilde{x}_i$ denotes a finite list: $x_i^1, x_i^2, \ldots$):

$$
e ::= \overset{n}{\underset{i=1}{\Sigma}} m_i(\tilde{x}_i); x_i'
$$

The empty summation ($n = 0$) denotes the type constant ε.

Here are some examples of type definitions:

- $\{x = \varepsilon\} \triangleright x$ is the type of an interface that does not expect to handle any message.
- an interface type of a one position buffer that can handle indefinitely the message *put* followed by the message *get* is :
  $\{x = put(z); y, \; y = get(); x\} \triangleright x$, for some type variable z.
- an interface type of a boolean object is:

$$
\begin{array}{rcl}
\{Bool & = & not(); Bool \\
& + & and(Bool); Bool \\
& + & or(Bool); Bool\} \triangleright Bool
\end{array}
$$

this type handles, at any time, one of the three methods: *not* (with no arguments), *or* and *and* (with one argument of type *Bool*). After accepting one such method, the type of an interface of type *Bool* is again *Bool*.

In our model we consider only deterministic types, i.e., if there is a choice between two messages of a behavior type expression, the names of the messages must be different. More precisely, for every type equation of the form $\sum_{i=1}^{n} m_i(\tilde{x}_i); x_i'$, we have $m_i \neq m_j$ for every pair, (i, j), of distinct indices.

3.2 Type behavior

Since we want to deal with non-uniform service availability, we have to specify how types evolve. A type evolves by performing type actions. A type action is a method signature: $m(X_1, \ldots, X_n)$. After performing an action a type behaves as another type as shown in the following rule.

$$\frac{(x = \sum_{i=1}^{n} m_i(\tilde{x}_i); x_i') \in E}{E \vartriangleright x \xrightarrow{m_j(E \vartriangleright \tilde{x}_j)} E \vartriangleright x_j'} \quad 1 \leq j \leq n$$

The notation $m_j(E \vartriangleright \tilde{x}_j)$ denotes the type action $m_j(E \vartriangleright x_j^1, E \vartriangleright x_j^2, \ldots)$.

We define $Succ(X)$ as the set of all the types that X can evolve to in one step:

Definition 1 $Succ(X) = \{Y \in \text{ITypes}, \exists m, \tilde{X} \text{ such that } X \xrightarrow{m(\tilde{X})} Y\}$

3.3 Behavior types equality

Bisimulation is a behavioral equality relation over interface behavior types. Two interface behavior types are bisimilar if they can perform the same set of action types and for each action the resulting types are bisimilar. This definition is very close to the definition of process bisimulation in CCS (see [Milner, 1989]). The principal difference is that in our case the actions need to be bisimilar too.

Definition 2 *(Bisimulation)*
A binary relation β over types is a bisimulation if $(X, Y) \in \beta$ implies:

i) $X \xrightarrow{m(\tilde{X})} X' \Rightarrow \exists Y', \tilde{Y} \text{ / } Y \xrightarrow{m(\tilde{Y})} Y' \text{ and } (X', Y') \in \beta, \forall i \ (Y_i, X_i) \in \beta$

ii) $Y \xrightarrow{m(\tilde{Y})} Y' \Rightarrow \exists X', \tilde{X} \text{ / } X \xrightarrow{m(\tilde{X})} X' \text{ and } (X', Y') \in \beta, \forall i \ (X_i, Y_i) \in \beta$

Proposition 1 *A bisimulation relation is an equivalence relation.*

The type equality is then defined, as for CCS, using the bisimulation relation.

Definition 3 *(Type equivalence)*
Two types X, Y are equivalent, noted $X \sim Y$, iff $(X, Y) \in \beta$ for some bisimulation β.

Definition 4
Let $\mathcal{F}$ be the function over binary relations over types defined as follows:
$(X, Y) \in \mathcal{F}(\mathcal{R})$ *iff*

i) $X \xrightarrow{m(\tilde{X})} X' \Rightarrow \exists Y', \tilde{Y} ~/~ Y \xrightarrow{m(\tilde{Y})} Y'$ *and* $X' \mathcal{R} Y', \forall i ~ X_i \mathcal{R} Y_i$

ii) $Y \xrightarrow{m(\tilde{Y})} Y' \Rightarrow \exists X', \tilde{X} ~/~ X \xrightarrow{m(\tilde{X})} X'$ *and* $X' \mathcal{R} Y', \forall i ~ Y_i \mathcal{R} X_i$

Proposition 2
- $\mathcal{F}$ *is monotonic:* $\mathcal{R}_1 \subseteq \mathcal{R}_2 \Rightarrow \mathcal{F}(\mathcal{R}_1) \subseteq \mathcal{F}(\mathcal{R}_2)$
- β *is a bisimulation iff* $\beta \subseteq \mathcal{F}(\beta)$

Theorem 1 *The type equivalence relation ($\sim$) is the largest fixed-point of $\mathcal{F}$*

3.4   Compatibility

Informally, a type X is said to be compatible with a type Y if a client of type X can interact safely with a server of type Y. So if X can perform an action, Y will be able to perform this action too.

Definition 5 *(Compatibility relation)*
A binary relation, $\mathcal{C}$, over types is a compatibility relation if $(X, Y) \in \mathcal{C}$ implies: $X \xrightarrow{m(\tilde{X})} X' \Rightarrow \exists Y', \tilde{Y} ~/~ Y \xrightarrow{m(\tilde{Y})} Y'$ *and* $(X', Y') \in \mathcal{C}, \forall i ~ (Y_i, X_i) \in \mathcal{C}$
Note how, in the previous definition, the types X_i and Y_i are contravariant.

Definition 6 *(Behavior type compatibility)*
A behavior type X is compatible with a behavior type Y, noted $X \angle Y$, iff $(X, Y) \in \mathcal{C}$ for some compatibility relation $\mathcal{C}$

We are not interested here by the algorithmic concerns, but it is easy to see that the type equivalence ($\sim$) as well as type compatibility are computable. This is due to the fact that the number of variable names (and so of possible types) is finite for a particular program text.

4 STATIC SEMANTICS

4.1 Introduction

As seen in section 2, an interface type is a triplet (μ, c, X), noted $X^{c^{\mu}}$ where

- μ is the mode of the interface: public ($\square$) or private ($\triangle$).
- c is the set of capacities of the interface: $c \subseteq \{!, ?\}$
- X is the behavior type of the interface.

The meta-variables $T, T', \ldots, T_i, \ldots$ will range over interface types. A context Γ is a set of bindings of the form: $x : T$ or $A : (T_1, \ldots, T_n)$. The static semantics is given using the judgments given in table 2.

judgment	meaning
$\Gamma \vdash u : T$	in the context Γ, the interface u has type T
$\Gamma \vdash A : (T_1, \ldots, T_n)$	in the context Γ, the object A has type $(T_1, \ldots, T_n)$
$\Gamma \vdash B$	in the context Γ, the behavior B is well typed
$\Gamma \vdash C$	in the context Γ, the configuration C is well typed
$\Gamma \vdash D \Rightarrow \Gamma'$	in the context Γ, the declaration D yields context Γ'

Table 2 Judgments of the typing rules

The context extension, noted $\Gamma, u:T$, is defined such that $\Gamma, u:T \vdash u:T$.

In order to formalize the notion of interface role passing we need a partial function, $\ominus$, which substracts a role, i.e., an element of $\{!, ?\}$, from an interface type. For an interface type $T = (\mu, c, X)$, $T \ominus r$ is defined when $r \in c$. $\ominus$ is given by:

$$(\triangle, c, X) \ominus \{!\} = (\triangle, c \setminus \{!\}, X)$$
$$(\triangle, c, X) \ominus \{?\} = (\triangle, c \setminus \{?\}, X)$$
$$(\square, c, X) \ominus \{!\} = (\square, c, X)$$
$$(\square, c, X) \ominus \{?\} = (\square, c, X)$$

Passing any of the capacities of a public interface does not change the role of this interface, but we have to check anyway that the interface already had this role. The definition of $\ominus$ is quite intuitive, but in our typing rules we are also going to use the opposite of this function : the $\oplus$ function defined by:

$$T \oplus c = T', \text{ where } T' \ominus c = T$$

We extend $\oplus$ to contexts as follows:

- $(\Gamma, u:T) \oplus u^!$ denotes the context $\Gamma, u:(T \oplus \{!\})$
- $(\Gamma, u:T) \oplus u^?$ denotes the context $\Gamma, u:(T \oplus \{?\})$
- $(\Gamma, u:T) \oplus u^{!?}$ denotes the context $\Gamma, u:(T \oplus \{!\} \oplus \{?\})$

4.2   Subtyping

We define a subtyping relation over interface types, noted $\preceq$, based on the behavior types compatibility (see definition in section 3.4). For the sake of simplifying the presentation, we deal with a subset of COB where only client roles can be sent along with method invocation. However, the subtyping relation can be easily extended to the whole COB calculus.

Definition 7 *Interface Subtyping*
An interface type $X_1{}^{c_1{}^{\mu_1}}$ *is a subtype of* $X_2{}^{c_2{}^{\mu_2}}$ *if* $\mu_1 = \mu_2$, $c_2 \subseteq c_1$ *and*

i) $c_2 = \{!\} \Rightarrow X_1 \measuredangle X_2$

ii) $c_2 = \{?\} \Rightarrow X_2 \measuredangle X_1$

iii) $c_2 = \{!, ?\} \Rightarrow X \sim Y$

This subtyping relation allows more flexibility in the language. It allows formal interfaces to be instantiated by actual interfaces which types are subtypes of the declared types. This relation ensures that the declared type of the client role of an interface will always be compatible with the declared type of its corresponding server role.

4.3   Typing Rules

The basic idea underlying our typing rules is to guaranty that each object use the interfaces in a way compatible with their declared behavior type. Our rules ensure also that there is no duplication of the roles of the private interfaces.

$$\frac{\Gamma \vdash B, \Gamma \vdash u : Y^{!\mu} \qquad \Gamma \vdash v_1 : X_1{}^{c_1{}^{\mu_1}}, \ldots, \Gamma \vdash v_n : X_n{}^{c_n{}^{\mu_n}} \qquad X \xrightarrow{m(X_1, \ldots, X_n)} Y, \; \Gamma \oplus \tilde{v}^{\tilde{\rho}} \text{ defined}}{\Gamma, u : X^{!\mu} \oplus \tilde{v}^{\tilde{\rho}} \vdash !u.m(\tilde{v}^{\tilde{\rho}}) > B}$$

This rule could be read like this : If $!u.m(\tilde{v}^{\tilde{\rho}}) > B$ is well-typed in a context where u has the client role of an interface of type X and where the interfaces v_i have the capacities $\{\rho_i\}$ then B is well-typed in a context where the type of u has evolved to Y and where the capacities interfaces v_i have been updated.

$$\frac{\Gamma, u : Y_1{}^{?\mu}, \tilde{v}_1 : \tilde{X}_1^{\tilde{c}_1^{\tilde{\mu}_1}} \vdash B_1, \ldots, \Gamma, u : Y_n{}^{?\mu}, \tilde{v}_n : \tilde{X}_n^{\tilde{c}_n^{\tilde{\mu}_n}} \vdash B_n \qquad X \xrightarrow{m_1(\tilde{X}_1)} Y_1, \ldots, X \xrightarrow{m_n(\tilde{X}_n)} Y_n, \; Succ(X) = \{Y_1, \ldots, Y_n\}}{\Gamma, u : X^{?\mu} \vdash ?u[m_1(\tilde{v}_1 : \tilde{X}_1^{\tilde{c}_1^{\tilde{\mu}_1}}) = B_1 \ldots m_n(\tilde{v}_n : \tilde{X}_n^{\tilde{c}_n^{\tilde{\mu}_n}}) = B_n]}$$

In this rule it is important that $Succ(X) = \{Y_1, \ldots, Y_n\}$. This ensures that all the messages that can be processed by an interface of type X are handled by the reception action.

$$\frac{\Gamma, u : X^{!?^\mu} \vdash C}{\Gamma \vdash \textbf{new } u : X^\mu \textbf{ in } C}$$

A newly created interface has both roles: client and server.

$$\frac{\Gamma \vdash Recep_i \ldots \Gamma \vdash Recep_n}{\Gamma \vdash \overset{n}{\underset{i=1}{\Sigma}} Recep_i}$$

The context is propagated as it is in all the branches of a choice.

$$\frac{\Gamma \vdash (A[\tilde{u} : \tilde{X}^{\tilde{\rho}^{\tilde{\mu}}}] = B) \Rightarrow \Gamma, \Gamma \vdash B}{\Gamma \vdash (A[\tilde{u} : \tilde{X}^{\tilde{\rho}^{\tilde{\mu}}}] = B) \Rightarrow \Gamma, \tilde{u} : \tilde{X}^{\rho^{\tilde{\mu}}}, A : (\tilde{X}^{\rho^{\tilde{\mu}}})}$$

The first condition is here to allow recursive definition. It ensures that the behavior B is typed in a context that already contains the bindings needed for the instantiation of A.

$$\frac{\Gamma \vdash \tilde{u} : \tilde{X}^{c^{\tilde{\mu}}}, \Gamma \vdash A : (\tilde{T}), \tilde{X}^{c^{\tilde{\mu}}} \preceq \tilde{T}}{\Gamma \vdash A[\tilde{u}^{\tilde{c}}]}$$

An interface having a subtype of another can replace it in an object instantiation.

$$\frac{\Gamma \vdash B, \Gamma \vdash \tilde{u} : \tilde{X}^{\tilde{c}^{\tilde{\mu}}}, \Gamma \vdash A : (\tilde{T}), \tilde{X}^{\tilde{c}^{\tilde{\mu}}} \preceq \tilde{T}, \Gamma \oplus \tilde{u}^{\tilde{c}} \text{ defined}}{\Gamma \oplus \tilde{u}^{\tilde{c}} \vdash \textbf{create } A[\tilde{u}^{\tilde{c}}] > B}$$

This is the same case as the precedent except that here we must be careful about how B is going to use the interfaces $\tilde{u}$.

$$\frac{\Gamma_1 \vdash C_1, \Gamma_2 \vdash C_2, \Gamma_1 \oplus \Gamma_2 \text{ defined}}{\Gamma_1 \oplus \Gamma_2 \vdash C_1 | C_2}$$

Here again we must be sure that there is no duplication of the roles of a client interface.

5 OPERATIONAL SEMANTICS

We present the operational semantics of configurations in two steps. We first define a structural congruence relation and then we give a reduction relation that specifies how the configurations evolve. The COB language is statically typed so will omit all the information about types in out dynamic semantics. To distinguish between the interactions on public and private interfaces, the receiving and emitting actions will be labeled, respectively by, $\Box$ and $\triangle$.

5.1　Structural Congruence

The first structural rules state that the choice operator between receptions is commutative and associative:

$$Recep_1 + Recep_2 \equiv Recep_2 + Recep_1$$

$$(Recep_1 + Recep_2) + Recep_3 \equiv Recep_1 + (Recep_2 + Recep_3)$$

The parallel operator is also commutative and associative and 0 is its neutral element:

$$C_1 \mid C_2 \equiv C_2 \mid C_1 \,, \ (C_1 \mid C_2) \mid C_3 \equiv C_1 \mid (C_2 \mid C_3) \,, \ C \mid 0 \equiv C$$

The order of the introduction of the interfaces in meaningless:

$$\textbf{new } u_1 \textbf{ in } (\textbf{new } u_2 \textbf{ in } C) \equiv \textbf{new } u_2 \textbf{ in } (\textbf{new } u_1 \textbf{ in } C)$$

The two last rules are the pi-calculus scope extrusion and the alpha-conversion:

$$(\textbf{new } u \textbf{ in } C_1) \mid C_2 \equiv \textbf{new } u \textbf{ in } (C_1 \mid C_2) \quad \text{if } u \text{ is not free in } C_2$$

$$\textbf{new } u \textbf{ in } C_1 \equiv \textbf{new } v \textbf{ in } C_2 \text{ if } C_1[w/u] = C_2[w/v] \text{ for some interface name } w$$

5.2　Reduction Rules

We define now the reduction rules that specify how a configuration can evolve by making a single and atomic step. The evolution of configurations may generate messages. The syntax of a message is similar to the syntax of the method invocation except that the message has no continuation. To avoid any ambiguity messages will be written between brackets: $[!u.m(\tilde{v})]$

The reduction relation is defined by the following rules:

$$?^{\triangle} u[m(\tilde{u}) > B + \Sigma m_i(\tilde{u}_i) > B_i] \mid !^{\triangle} u.m(\tilde{v}) > B' \longrightarrow B[\tilde{v}/\tilde{u}] \mid B'$$

The synchronization on a private interface is by *rendez-vous*.

$$!^{\square} u.m(\tilde{v}) > B \longrightarrow B \mid [!u.m(\tilde{v})]$$

The invocation of a public interface generates a message ($[!^{\square} u.m(\tilde{v})]$) whose behavior is to synchronize with this interface.

$$?^{\square} u[m(\tilde{u}) > B + \Sigma m_i(\tilde{u}_i) > B_i] \mid [!u.m(\tilde{v})] \longrightarrow B[\tilde{v}/\tilde{u}]$$

A message is absorbed by the appropriate (public) interface and then vanishes.

$$\frac{C_1 \longrightarrow C_1'}{C \mid C_1 \longrightarrow C \mid C_1'}$$

This rule states that if a subconfiguration can evolve to a new one then the whole configuration can evolve too.

$$\frac{C \longrightarrow C'}{\textbf{new } u \textbf{ in } C \longrightarrow \textbf{new } u \textbf{ in } C'}$$

The reduction rule concerning the interface creation operator (equivalent to the restriction operator of the *pi*-calculus) is straightforward. In a labeled-transition semantics it would have been more complicated.

$$\frac{C_1 \equiv C_1' \longrightarrow C_2' \equiv C_2}{C_1 \longrightarrow C_2}$$

This rule states that configurations that are equivalent (according to $\equiv$) behave equally.

5.3   Run-time safety

The static and dynamic semantics we have defined ensure the the run-time safety of well-typed COB programs. The executions of the COB programs are free of "non-specified service" errors. We formally define run-time errors using the following rules:

$$\frac{m \neq m_i \, , \; \forall i}{?^{\triangle}u[\Sigma m_i(\tilde{u}_i) > B_i] \mid !^{\triangle}u.m(\tilde{v}) > B' \longrightarrow error}$$

A configuration where the server on a private interface cannot accept a message m while the client of this interface is trying to invoke this service reduces to *error*.

$$\frac{k \neq l}{?^{\triangle}u[m(u^1,\dots,u^k) > B + \Sigma m_i(\tilde{u}_i) > B_i] \mid !^{\triangle}u.m(v^1,\dots,v^l) > B' \longrightarrow error}$$

Similarly, a configuration where the server is ready to process the message m, but where there is a parameters number mismatch reduces to *error*.

$$\frac{m \neq m_i \, , \; \forall i}{?^{\square}u[\Sigma m_i(\tilde{u}_i) > B_i] \mid [!^{\square}u.m(\tilde{v})] \longrightarrow error}$$

A configuration where a message addressed to a public interface cannot be processed reduces to *error*.

$$\frac{k \neq l}{?^{\square}u[m(u^1,\dots,u^k) > B + \Sigma m_i(\tilde{u}_i) > B_i] \mid [!u.m(v^1,\dots,v^l) > B'] \longrightarrow error}$$

As for in the private interface case, a configuration where there is a parameters number mismatch reduces to *error*.

$$\frac{C_1 \longrightarrow error}{C_1 \mid C_2 \longrightarrow error}$$

If a subconfiguration reduces to *error* the entire configuration does so.

Before presenting our main result, let us introduce the properties of COB configurations that will help us the prove our run-time safety theorem.

Lemma 1 *In a well-typed configuration, only one object can be client on a private interface and symmetrically only one object can hold the server role for a private interface.*

Outline of the proof: this follows directly from the use of the $\oplus$ function in the typing rule of the parallel composition.

The second lemma states that a well-typed configuration always reduces to another well-typed configuration.

Lemma 2 *If $\Gamma \vdash C$ and $C \longrightarrow C'$ then there exits a typing context Γ' such that $\Gamma' \vdash C'$*

Outline of the proof: the proof is by induction on the structure of C and the length of SOS rules used in the reduction.

Our main theorem states that a well-typed configuration never reduces to *error*.

Theorem 2
$$\Gamma \vdash C \Rightarrow C \not\longrightarrow error$$

Outline of the proof: this follows immediately from the preceding lemma. In well-typed configurations all the instances of an interface have the same behavior type and thus can handle the same messages.

6   CONCLUSION AND RELATED WORK

We defined a language compliant with the ODP computational model. COB extends the ODP type system by introducing the notion of private interface. We also allow the services on a private interface to be non-uniform and in spite of this the COB typing system guarantees the absence of run-time errors. This typing system can be adapted to a wide variety of languages that have the same communication paradigm.

Type systems for concurrent object oriented languages is an active research topic. Many authors have tackled this issue in the realm of the actors [Agha et al., 1993] and the π-calculus [Milner et al., 1992] paradigms. Concerning the latter, a wide variety of typing systems has been proposed that deal the problem of channel typing. The simplest one [Milner, 1992] just checks the arity of the channels. This type system has been extended such that it can handle polymorphism and type inference [Gay, 1993, Vasconcelos-Honda, 1993, Turner, 1995] and subtyping [Pierce et al., 1995, Pierce et al., 1995]. All these typing systems do not handle dynamic service behavior.

The importance of distinguishing public from private interfaces have been identified by [Nierstrasz, 1995], but, however, without given it a formal treatment. [Nierstrasz, 1995] has also introduced the concept of non-uniform service availability and has used traces to specify the constraints on the ordering of the messages that can be handled by a channel (an interface). Our work extends [Nierstrasz, 1995] in two ways: by formally introducing the concept of privacy of interfaces and by allowing message types to include parameter types.

The work reported in [Puntigam, 1996] is close to the one discussed in the present paper: the author provides a typing system which handles the dynamic behavior of concurrent configurations of actors. In contrast with [Puntigam, 1996] where the type language is based on traces (where the possibility of having recursion and non terminating behavior are not treated thoroughly), in COB, we explicitly deal with recursive types, both in the type language and the type system. We also use the private/public qualifier for interfaces which, we believe, greatly simplifies the type system and the operational semantics.

Takeuchi and al. [Takeuchi et al., 1994] define a typed process calculus called $\mathcal{L}$ with the notion of *session*: "a semantically atomic chain of communication actions" between two processes. In COB a session is represented by a communication between two objects using a private interface. But unlike COB, the session channels are static and the roles of the partners of a session cannot be passed. In COB, a object can pass its client or server role of a private interface and so delegate to another object the continuation of a "session".

The technical treatment of the contexts in the static semantic of COB has been inspired from [Kobayashi et al., 1995]. In this version of the π-calculus the authors use the linear capabilities of some special channel to ensure that there are used (at most) once. We use a similar mechanism to ensure that there is no duplication of the roles of private interfaces. "Linear channels" as defined in [Kobayashi et al., 1995] can be defined very simply in COB $(linear = m(\ldots); \varepsilon)$.

Acknowledgments: The authors would like to thank Jean-Bernard Stefani and Roland Groz, from CNET, with whom they had inspiring discussions on topics related to the present paper.

REFERENCES

[Agha et al., 1993] G. A. Agha, I. A. Mason, S. F. Smith and C. L. Talcott, *A Foundation for Actor Computation*, J. Functional Programming 1 (1), 1993.

[Gay, 1993] Simon J. Gay. *A sort inference algorithm for the polyadic π-calculus*. Twentieth ACM Symposium on Principles of Programming Languages, January 1993.

[Honda, 1993] Kohei Honda. *Types for Dyadic Interaction.* CONCUR'93, LNCS 612, Springer-Verlag.

[Kobayashi et al., 1995] Naoki Kobayashi, Benjamin C. Pierce, David N. Turner. *Linearity and the Pi-Calculus*. Technical report, Department of Information Science, University of Tokyo and Computer Laboratory, University of Cambridge, 1995.

[Milner, 1989] Robin Milner. *Communication and Concurrency.* Prentice-Hall, 1989.

[Milner, 1992] Robin Milner. *The polyadic π-calculus: a tutorial.* Technical Report ECS-LFCS-91-180, Laboratory for Foundations of Computer Science, Department of Computer Science, University of Edinburgh, UK, October 1991.

[Milner et al., 1992] Robin Milner, Joachim Parrow, David Walker. *A calculus of mobile processes (Part I and Part II).* Information and Computation, 100:1-77, 1992.

[Najm-Stefani, 1995] Elie Najm, Jean-Bernard Stefani. *A formal semantics for the ODP computational model.* Computer Networks and ISDN Systems , Vol 27, 1995.

[Nierstrasz, 1995] Oscar Nierstrasz. *Regular Types for Active Objects.* Object-Oriented Software Composition. O. Oscar Nierstrasz, D.Tsichitzis (Ed.), Prentice Hall, 1995

[Pierce et al., 1995] Benjamin C. Pierce, David Sangiorgi. *Typing and subtyping for mobile process.* Mathematical Structures in Computer Science, 1995.

[Pierce et al., 1995] Benjamin C. Pierce, David N. Turner. *PICT Language Definition.* Available electronically, 1995.

[Puntigam, 1996] Franz Puntigam. *Types for active objects based on Trace Semantics.* FMOODS'96, Chapmann and Hall.

[Takeuchi et al., 1994] Kaku Takeuchi, Kohei Honda, Makoto Kubo. *An Interaction-based Language and its Typing System.* PARLE'94, LNCS 818, Sringer-Verlag.

[Turner, 1995] David N. Turner. *The π-calculus: Types, polymorphism and implementation.* Ph.D. Thesis, LFCS, University of Edinburgh, 1995.

[Vasconcelos-Honda, 1993] Vasco T. Vasconcelos, Kohei Honda. *Principal typing schemes in a polyadic π-calculus.* CONCUR'93, July 1993.

3

A Calculus with Code Mobility

Tatsurou Sekiguchi and Akinori Yonezawa
Department of Information Science, Faculty of Science,
University of Tokyo
7-3-1 Hongo, Bunkyo-ku, Tokyo, Japan 113
e-mail: {cocoa,yonezawa}@is.s.u-tokyo.ac.jp

Abstract

Mobile agent systems have attracted a great deal of attention in recent years. Various agent systems have been proposed and implemented so far. But their systems are usually equipped with their own features that are hard to simulate by other systems even with respect to agent movement mechanisms. Therefore, a generalized framework that can describe various mechanisms in a formal manner is strongly needed. This paper proposes a simple and flexible calculus λdist, which provides a neat tool for describing movement mechanisms of code, data and execution states.

Keywords

Mobile agent; operational semantics; code movement.

1 INTRODUCTION

Distributed programming language systems in which code, data and/or execution states move around among computer systems (Cardelli 1995, Colusa 1995, White 1996, Gosling *et al.* 1995, Knabe 1995, Marzo *et al.* 1995, Straßer *et al.* 1996, Gray 1995) have been proposed with the development of telecommunication networks. Such distributed systems are often called "agent systems". There is, however, a serious terminological gap among the people who build and study agent systems. In this paper, we use *an agent system* to mean a system in which code, data and/or execution states can be transmitted *dynamically*. Many agent systems are so different to each other that it is usually the case that one agent system is hard to simulate another system and vice versa. The difference is due to the different presuppositions they assume on application, security, availability, heterogeneity, efficiency, and so on. To compare and discuss the difference among agent systems, it is necessary to have a common framework that can describe their agent movement mechanisms in a formal manner. This paper proposes a simple calculus λdist, which can describe agent movement mechanisms and a variety of mechanisms for distributed computations. We demonstrate that mechanisms of Obliq(Cardelli 1995), Telescript(White 1996), Facile(Knabe 1995) and Java(Gosling *et al.* 1995) can be described in λdist in Section 5. λdist is an extension of a call-by-value λ calculus with *agent expressions* and *data movement types*.

Existing concurrent or distributed calculi are not sufficient for modelling properties of agent movement mechanisms in distributed systems, because they give a single fixed semantics for code and data movement. Linda's tuple space(Carriero *et al.* 1989) gives a simple view of distributed computation. But since it abstracts distribution and communication mechanism, it is not suitable to describe implementational issues. Pi-calculus(Milner *et al.* 1989) and HACL(Kobayashi *et al.* 1995) can describe communication, but it lacks a notion of location or site, which is important in discussing distribution, and is also lacking a notion of identity, which is necessary for expressing autonomy of agents. The semantics of Obliq has a notion of distributed scope, which can be used as a framework to discuss data movement in a distributed system. But its expressive power is not enough to subsume all mechanisms of Telescript. Since Telescript is designed on its own metaphor, it is often inconvenient to describe familiar mechanisms such as remote references. Fournet(Fournet *et al.* 1996) proposed a process calculus with locations, in which one can describe location-dependent definitions and interaction between locations. The calculus gives static scoping rules for distributed environment.

The rest of this paper is organized as follows: Section 2 introduces our idea of representing code mobility. We illustrates the data movement mechanism in our calculus in Section 3. In Section 4, we presents the operational semantics of our calculus. We show that various mechanisms in distributed language systems can be described with our calculus in Section 5. Section 6 concludes the paper.

2   CODE MOBILITY

One of the most distinguished points of mobile agent language systems from other language systems is their facility to move code and execution states to another site *dynamically*. But semantics of code movement mechanisms in proposed mobile language systems is often given by a style in which implementational details are described. To discuss the differences between mobile language systems, we have to put them on a common formal foundation. We had investigated many proposed systems and found that their underlying mechanisms are so different that it was hard for one system to emulate another system. For instance, one needs complicated programming to implement several mechanisms of Obliq in Telescript. We design a simple calculus with code mobility, called λdist, to give a common formal framework in which many different code mobility mechanisms can be described.

2.1 Preliminary definitions

Let e_1 and e_2 be expressions, then $e_1 \equiv e_2$ denotes the fact that e_1 is syntactically equivalent to e_2. Let f be a function, then $\text{Dom}(f)$ means the domain of f. We denote by $[x_1 \mapsto y_1, ..., x_n \mapsto y_n]$ a finite map where each x_i corresponds to y_i for $i = 1, ..., n$. The concatenation of two maps s, s' denotes the following map:

$$ss'(x) = \begin{cases} s'(x) & \text{if } x \in \text{Dom}(s') \\ s(x) & \text{otherwise} \end{cases}$$

We denote by $[y_1/x_1, ..., y_n/x_n]$ the substitution that replaces each occurrence of x_i with y_i for $i = 1, ..., n$. The empty map and the empty substitution are denoted by $\emptyset$. Application of a substitution s with a finite map $[x_1 \mapsto y_1, ..., x_n \mapsto y_n]$ is defined by $[x_1 \mapsto s(y_1), ..., x_n \mapsto s(y_n)]$. Let $S_I = \{s_{i_1}, s_{i_2}, ...\}$ be an indexed set. We denote by $\text{Idx}(S_I)$ a set of indexes appearing in S_I.

2.2 Calculus overview

We use an extension of call-by-value λ-calculus, called λdist, to express code, which is defined by the following grammar:

$$e ::= p \mid \textbf{go} \mid x \mid \lambda x.e \mid ee \mid \langle e \rangle$$

where x is a variable, $\lambda x.e$ is an abstraction and ee is an application. We use p to denote a *place*, which means almost the same notion in the terminology of Telescript. It is called a *site* in the terminology of Obliq. Finally, $\langle e \rangle$ denotes an agent or a unit of code movement. In addition, we often use let-expression as a syntax sugar such that $\textbf{let } x = e_1 \textbf{ in } e_2 \equiv (\lambda x.e_2)e_1$. We denote by $\textbf{let } x_1 = e_1, ..., x_n = e_n \textbf{ in } e$ a sequence of let-expressions $\textbf{let } x_1 = e_1 \textbf{ in let } ... \textbf{ in let } x_n = e_n \textbf{ in } e$, which is so-called let* expression.

The code movement operation in our calculus is described by the following **go**-expression that is quite a similar mechanism to that of Telescript:

$$\textbf{go } p$$

where p denotes a destination. Unlike Telescript, one needs not to specify a ticket as a parameter for **go**-expression. After evaluating a **go**-expression, its continuation will be moved to destination, and then evaluated there.

$$\textbf{let } x = 3, y = \textbf{go } p \textbf{ in } (f\ x)$$

In the above code fragment, $(f\ x)$ is evaluated in place p. Multi-hopping is realized just by a repetition of **go**-expressions:

$$\text{let } x = 3, y = \textbf{go } p, z = \textbf{go } q \text{ in } (f\ x)$$

In this case, the expression moves to place p, then moves to place q, and finally $(f\ x)$ is evaluated at q.

Though an entire continuation always moves in the above examples, it is often the case that one wants to move a *part* of an expression. Such a part is called an agent or a mobile object. Thus we introduce an agent expression $\langle e \rangle$.

Definition 1 *When a **go**-expression is evaluated, the area of code moved to a destination is the innermost agent expression surrounding the **go**-expression.*

$$\text{let } x = 2, y = \langle \text{let } z = \textbf{go } q \text{ in } (g\ 3) \rangle \text{ in } (f\ x)$$

In the above example, the code to be moved, when **go** q is evaluated at a place p, is the innermost agent expression surrounding **go** q, that is $\langle \text{let } z = \textbf{go } q \text{ in } (g\ 3) \rangle$. $(g\ 3)$ is evaluated at place q and $(f\ x)$ is evaluated at place p.

Formally, the operational semantics and an agent splitting operation from an entire expression are defined as below. The semantics is defined by the following reduction rules of expressions using contexts:

$$C[(\lambda x.e)v] \longrightarrow C[[v/x]e]$$

$$C[\langle v \rangle] \longrightarrow C[v]$$

$$\text{contexts } C ::= [\ ] \mid Ce \mid (\lambda x.e)C \mid \langle C \rangle$$

where v ranges over values, defined in Figure 1. We call $[\ ]$ a hole. A context is an expression containing a single hole. Let C be a context, we denote by $C[e]$ an expression syntactically equivalent to C except that the hole in C is replaced with e. For instance, let C be context $(\lambda x.x)[\ ]$, then $C[3]$ is expression $(\lambda x.x)3$. We use a context to indicate the current evaluation point. We can specify evaluation order using contexts. Given expression e, if there is a context C such that $e \equiv C[(\lambda x.e)v]$, we regard $(\lambda x.e)v$ as the current evaluation point of the context. The second reduction rule says that an agent expression is just ignored when there is no **go**-expression in the agent expression.

When a **go**-expression is evaluated, the innermost agent expression surrounding the **go**-expression moves to the destination and the rest of the expression without the agent still remains at the current place. Suppose we evaluate $C[\textbf{go } q]$ at place p, and let $\langle A[\textbf{go } q] \rangle$ be the innermost agent expression surrounding the **go** q such that $C[\textbf{go } q] \equiv R[\langle A[\textbf{go } q] \rangle]$ for some context R. Then $A[\textbf{go } q]$ will move to the destination q and then be evaluated at q. We define that a **go**-expression is reduced to *the starting place* in view of agents. Thus, we have

$$C[\textbf{go } q] \longrightarrow A[p].$$

On the other hand, the rest of the expression R stays at the current place. We define that a go-expression is reduced to *the destination place* in view of the expression surrounding the agent. Namely, we have

$$C[\text{go } q] \longrightarrow R[q].$$

The operation of splitting an agent from an expression is precisely defined by the following function φ (where O denotes the current continuation):

$$\varphi(C) = \varphi'([],[],\langle[]\rangle,C)$$
$$\varphi'(R,A,O,[]) = (R,A)$$
$$\varphi'(R,A,O,Ce) = \varphi'(R,A[[]e],O[[]e],C)$$
$$\varphi'(R,A,O,(\lambda x.e)C) = \varphi'(R,A[(\lambda x.e)[]],O[(\lambda x.e)[]],C)$$
$$\varphi'(R,A,O,\langle C\rangle) = \varphi'(O,[],O[\langle[]\rangle],C)$$

Proposition 1 *If $\varphi(C) = (R,A)$, then $R[\langle A\rangle] \equiv \langle C\rangle$.*

This proposition says that the composition of R and A is equivalent to C, namely, φ *splits* C into R and A, correctly.

3 DATA MOBILITY

In order for code mobility to be significant, we must have a distinction between something local and something remote. If there is no difference between local evaluation and remote one, why should we move code? Thus, we introduce a local store for each place to represent local resources. Any value in a local store is accessible *only* from the code in the same place. A store is a finite map from addresses to values, written as follows:

$$[a \mapsto 7^\tau, b \mapsto \text{"abc"}^{\tau'}]$$

We denote by the above notation a store where integer 7 is stored at address a and string "abc" is stored at address b. Superscripts τ and τ' are called *data movement types*, which will be explained soon. We use the following form to express a reduction rule that expression e at place p with store s_p is reduced into expression e' and yields store s'_p:

$$s_p \vdash_p e \longrightarrow e', s'_p$$

We denote by $a@p$ a remote reference referring to address a at place p.

In addition to stores, we introduce standard operations on stores.

$$e ::= \dots \mid \text{ref } \tau e \mid !e \mid \text{set } ee$$

$$\text{contexts } C ::= \dots \mid \text{ref } \tau C \mid !C \mid \text{set } Ce \mid \text{set } vC$$

They are quite standard ones; ref allocates a new initialized memory, ! derefers an address and set updates a stored value. Formally,

$$s_p \vdash_p C[\text{ref } \tau v] \longrightarrow C[a], s_p[a \mapsto v^\tau] \qquad \text{for some } a \notin \text{Dom}(s_p)$$

$$s_p \vdash_p C[!a] \longrightarrow C[v], s_p \qquad \text{if } s_p(a) = v^\tau$$

$$s_p \vdash_p C[\text{set } av] \longrightarrow C[()], s_p[a \mapsto v^\tau] \qquad \text{where } s_p(a) = v'^\tau \text{ for some } v'$$

where concatenation of finite maps means the union of them provided that the latter map is preferred when their domains overlap.

A data movement type is associated with each address and specifies the behavior of a stored value when it is moved. When a value is moved to other place, it may be copied, become a remote reference, or be nullified. Data movement mechanisms vary widely depending on mobile language systems. For instance, Telescript and Obliq have completely different data movement mechanisms. A remote value can be referred to by a remote reference in Obliq, while one cannot have a remote reference in principle in Telescript. We introduce 6 kinds of data movement type, which are sufficient to describe movement mechanisms of almost all proposed mobile language systems. We denote a data movement type for an address by a superscript of a value stored at the address.

Copy type:

	before movement	after movement	translation
current place p	$[a \mapsto v^{\text{copy}}, ...]$	$[a \mapsto v^{\text{copy}}, ...]$	$\emptyset$
destination q	$[...]$	$[a' \mapsto v'^{\text{copy}}, ...]$	$a \mapsto a'$ (a' fresh)

This table shows how stores at the current place and the destination place change before and after data movement. The translation means how the moved address changes after the movement. The address translation at the current place is empty since one needs no address translation at the current place for this type. Because value v referred to by address a before the movement will be referred by address a' after the movement at the destination, we have to replace address a with address a' at the destination. This type of movement is so-called copy operation. A value v is copied to a new space a'. If v is also an address, the rules of data movement are applied recursively. In this case, the value referred to by address v is also copied to an newly allocated address v'.

Resident type:

	before movement	after movement	translation
current place p	$[a \mapsto v^{\text{resident}}, ...]$	$[a \mapsto v^{\text{resident}}, ...]$	$\emptyset$
destination q	$[...]$	$[...]$	$a \mapsto a@p$

This type represents data which always stay at the current place, are never moved to other place, and are only referred to as a remote reference from the outside. The address translation at the destination place maps address

a to remote reference $a@p$ since v is referred as a remote reference from the outside.

Carry type:

	before movement	after movement	translation
current place p	$[a \mapsto v^{\text{carry}}, ...]$	$[a \mapsto v^{\text{carry}}, ...]$	$a \mapsto a'@q$
destination q	$[...]$	$[a' \mapsto v'^{\text{carry}}, ...]$	$a \mapsto a'$ (a' fresh)

Conversely, this type represents migration. A value of carry type is moved to the destination and is referred to by a remote reference from the previous place. The address translation at the current place replaces every address a appeared in place p with remote reference $a'@q$ since value v is supposed to be moved to the destination place q.

Proper type:

	before movement	after movement	translation
current place p	$[a \mapsto v^{\text{proper}}, ...]$	$[a \mapsto v^{\text{proper}}, ...]$	$\emptyset$
destination q	$[...]$	$[...]$	$a \mapsto \perp$

where $\perp$ denotes the undefined value. This type represents a value that never goes out and that cannot be referred to from the outside. Every address a is replaced with the undefined value $\perp$ at the destination, thus the agent cannot refer to the value after the movement. An instance of this type is non-owner reference in Telescript.

Takeaway type:

	before movement	after movement	translation
current place p	$[a \mapsto v^{\text{takeaway}}, ...]$	$[a \mapsto v^{\text{takeaway}}, ...]$	$a \mapsto \perp$
destination q	$[...]$	$[a' \mapsto v'^{\text{takeaway}}, ...]$	$a \mapsto a'$

A value of this type is considered to be inherently possessed by an agent. Every address a in the start place p is replaced with the undefined value $\perp$ so that anyone there cannot refer to the data after the movement. It looks as if a value is gone with an agent.

Ubiq type:

	before movement	after movement	translation
current place p	$[a \mapsto v^{\text{ubiq}}, ...]$	$[a \mapsto v^{\text{ubiq}}, ...]$	$\emptyset$
destination q	$[a \mapsto v'^{\text{ubiq}}, ...]$	$[a \mapsto v'^{\text{ubiq}}, ...]$	$\emptyset$

This type is used to represent a common interface, say system calls, supposed to be facilitated by both the start and the destination place. Facile(Knabe 1995), which is an extension of Standard ML added mobility, has this type of values.

$p \in \text{Place}$

$\tau \in \text{DMType} ::= \textbf{copy} \mid \textbf{resident} \mid \textbf{carry} \mid \textbf{proper} \mid \textbf{takeaway} \mid \textbf{ubiq}$

$a \in \text{Addr}$

$a@p \in \text{RemAddr} = \text{Addr} \times \text{Place}$

$v \in \text{Val} ::= () \mid \textbf{go} \mid p \mid a \mid a@p \mid \lambda x.e$

$e \in \text{Exp} ::= v \mid x \mid ee \mid \textbf{fix } x.e \mid \langle e \rangle \mid \textbf{ref } \tau e \mid !e \mid \textbf{set } ee$

$s \in \text{Store} = \text{Addr} \xrightarrow{\text{fin}} \text{Val} \times \text{DMType}$

$C ::= [] \mid Ce \mid (\lambda x.e)C \mid \langle C \rangle \mid \textbf{set } Ce \mid \textbf{set } vC \mid !C \mid \textbf{ref } \tau C$

Figure 1 Domains

4   OPERATIONAL SEMANTICS

The semantic objects for our calculus are listed in Figure 1. The expressions are lambda expressions extended by a fixed-point operator, an agent expression and operations on stores. Let-expressions are syntactic macro such that let $x = e_1$ in $e_2 \equiv (\lambda x.e_2)e_1$. Unit is the singleton set whose element is denoted by (). A place p abstracts a location where expressions are evaluated. A store s is a finite map from addresses to values.

The operational semantics is described in the usual call-by-value fashion(Figure 2). Given a place p and a store s_p, the operational semantics associates an expression e with the reduced expression e' and the possibly updated store s'. This is denoted by $s_p \vdash_p e \longrightarrow e', s'$. Contexts are used to determine the redex of given expressions.

4.1   Formulating data movement

We formulate data movement operations by effects of data movement. An effect is a quadruple (σ, θ, s, v) where

σ : the effect on the current store,

θ : the effect on the moving agent,

s : the possibly updated destination store, and

v : the moving data.

Data movement is defined by a relation $\Rightarrow$ on effects $\times$ effects listed in Figure 3, where $\sigma, \theta, s, v \Rightarrow \sigma', \theta', s', v'$ the lefthand side expresses the effect before moving v and the righthand side expresses the effect after the movement. Suppose that we move value v from place p to place q, if $\emptyset, \emptyset, s_q, v \Rightarrow \sigma, \theta, s'_q, v'$

$$s_p \vdash_p C[(\lambda x.e)v] \longrightarrow C[[v/x]e], s_p$$
$$s_p \vdash_p C[\text{fix } x.e] \longrightarrow C[[(\text{fix } x.e)/x]e], s_p$$
$$s_p \vdash_p C[\langle v \rangle] \longrightarrow C[v], s_p$$
$$s_p \vdash_p C[\text{ref } \tau v] \longrightarrow C[a], s_p[a \mapsto v^\tau] \quad \text{for some } a \notin \text{Dom}(s_p).$$
$$s_p \vdash_p C[!a] \longrightarrow C[v], s_p \quad \text{if } s_p(a) = v^\tau.$$
$$s_p \vdash_p C[\text{set } av] \longrightarrow C[()], s_p[a \mapsto v^\tau] \quad \text{where } s_p(a) = v'^\tau \text{ for some } v'.$$

Figure 2 Operational semantics without **go**

holds, then after moving v the current and the destination stores will become σs_p and s'_q, respectively. The value v will change into θv in place q.

4.2 Formulating distributed computations

We have never mentioned so far about concurrency of distributed computations. But there are a lot of places in the world and each place may have several activities so that we must perform *several* evaluations. Concurrent computation in λdist is represented as a transition relation $\rightsquigarrow$ between global states that are a pair of a finite set S of stores, and a finite set E of expressions with their places.

$$s_p \in S \subseteq \mathcal{S} = \text{Store} \times \text{Place}$$
$$\vdash_p e \in E \subseteq \mathcal{E} = \text{Place} \times \text{Exp}$$

A global state is written $s_p, s_q, s_r, ... | \vdash_p e, \vdash_q e', ...$

The relation between local computations and distributed computations are established by the following rule:

$$\frac{s_p \vdash_p e \longrightarrow e', s'_p}{(s_p | \vdash_p e) \rightsquigarrow (s'_p | \vdash_p e')}$$

This implies that any possible local computation can be a global transition. Let T be a set of stores and F be a set of execution states. The following rule implies that we can add unrelated stores and states to a transition freely.

$$\frac{(S|E) \rightsquigarrow (S'|E') \quad \text{Idx}(S) \cap \text{Idx}(T) = \emptyset}{(S, T|E, F) \rightsquigarrow (S', T|E', F)}$$

4.3 The movement mechanism of code and execution states

The agent movement algorithm proceeds as follows:

$$\frac{s_p(a) = v \, \text{copy} \quad \sigma, \theta[a \mapsto a'], s_q, v \Rightarrow \sigma', \theta', s_q', v'}{\sigma, \theta, s_q, a \Rightarrow \sigma', \theta', s_q'[a' \mapsto v' \, \text{copy}], a'} \quad \text{(COPY)}$$

$$\frac{s_p(a) = v \, \text{resident}}{\sigma, \theta, s_q, a \Rightarrow \sigma, \theta[a \mapsto a@p], s_q, a@q} \quad \text{(RESIDENT)}$$

$$\frac{s_p(a) = v \, \text{carry} \quad \sigma[a \mapsto a'@q], \theta[a \mapsto a'], s_q, v \Rightarrow \sigma', \theta', s_q', v'}{\sigma, \theta, s_q, a \Rightarrow \sigma', \theta', s_q'[a' \mapsto v' \, \text{carry}], a'} \quad \text{(CARRY)}$$

$$\frac{s_p(a) = v \, \text{proper}}{\sigma, \theta, s_q, a \Rightarrow \sigma, \theta[a \mapsto a'], s_q[a' \mapsto \bot], a'} \quad \text{(PROPER)}$$

$$\frac{s_p(a) = v \, \text{takeaway} \quad \sigma, \theta[a \mapsto a'], s_q, v \Rightarrow \sigma', \theta', s_q', v'}{\sigma, \theta, s_q, a \Rightarrow \sigma'[a \mapsto \bot], \theta', s_q'[a' \mapsto v' \, \text{takeaway}], a'} \quad \text{(TAKEAWAY)}$$

$$\frac{s_p(a) = v \, \text{ubiq}}{\sigma, \theta, s_q, a \Rightarrow \sigma, \theta, s_q, a} \quad \text{(UBIQ)}$$

provided that $a \in \text{Addr}$, $a \notin \text{Dom}(\theta)$ and $a' \notin \text{Dom}(s_q)$ for all the above rules.

$$\frac{r = q}{\sigma, \theta, s_q, a@r \Rightarrow \sigma, \theta, s_q, a} \quad \text{(REM1)} \qquad \frac{r \neq q}{\sigma, \theta, s_q, a@r \Rightarrow \sigma, \theta, s_q, a@r} \quad \text{(REM2)}$$

$$\frac{a \in \text{Addr} \quad a \in \text{Dom}(\theta)}{\sigma, \theta, s_q, a \Rightarrow \sigma, \theta, s_q, \theta a} \quad \text{(ADDR)}$$

Figure 3 The movement relation $\Rightarrow$

1. Split the innermost agent expression by φ. Let A be the agent and R be the rest of the expression.
2. Move all the addresses referred to from A to the destination store.
3. Apply the effects of data movement to the current store, the destination store, and the agent.

As illustrated in Section 2, given a context C, if $\varphi(C) = (R, A)$, then A is the innermost agent expression and R is the rest. All the addresses referred to from A are computed by the following mapping ADDR : Context $\rightarrow$ Addr $\cup$ RemAddr:

$$\text{ADDR}([]) = \text{ADDR}(()) = \text{ADDR}(p) = \text{ADDR}(x) = \emptyset$$

$$\text{ADDR}(a) = \{a\}, \qquad \text{ADDR}(a@p) = \{a@p\}$$

$$\text{ADDR}(\lambda x.e) = \text{ADDR}(e.l) = \text{ADDR}(\text{fix } x.e) = \text{ADDR}(e)$$

$$\text{ADDR}(\langle e \rangle) = \text{ADDR}(\text{ref } \tau e) = \text{ADDR}(!e) = \text{ADDR}(e)$$

$$\text{ADDR}(e_1 e_2) = \text{ADDR}(\text{set } e_1 e_2) = \text{ADDR}(e_1) \cup \text{ADDR}(e_2)$$

Step 3 of the agent movement algorithm above is described by the following algorithm MOVE, which computes the effects on the current store, the effect on the moving agent, and the possibly updated destination store.

Algorithm MOVE : Store $\times$ Store $\times$ Context $\rightarrow$ Subst $\times$ Subst $\times$ Store

Input: (s_p, s_q, A).
Initially: $X = \text{ADDR}(A), \sigma = \emptyset, \theta = \emptyset, s = \emptyset$.

while X is not empty **do:**
Choose any a from X;
If $\exists \sigma', \theta', s', a'$ s.t. $\sigma, \theta, s, a \Rightarrow \sigma', \theta', s', a'$
then $\sigma \leftarrow \sigma'; \theta \leftarrow \theta'; s \leftarrow s'; X \leftarrow X \setminus \{a\}$;
otherwise fail

Output: (σ, θ, s).
Termination of this algorithm is obvious since X is initially finite and decreases for one step in the while loop.

Finally, the movement mechanism of code and execution states is formulated by the following rule:

$$\frac{\varphi(C) = (R, A) \quad \text{MOVE}(s_p, s_q, A) = (\sigma, \theta, s'_q)}{(s_p, s_q \mid \vdash_p C[\textbf{go } q]) \rightsquigarrow (\sigma s_p, s'_q \mid \vdash_p \sigma R[q], \vdash_q \theta A[p])}$$

5 EXPRESSIVE POWER

In this section, we demonstrate that several language constructs for distributed mobile computation can be encoded in λdist and that it can emulate several mobile languages by translating them into λdist in a simple way.

5.1 Synchronization

First of all, we define simple constructs for synchronization and communication. They are quite simple because the central issue that we want to address is not synchronization or communication mechanisms of distributed languages, either. We take a position that one can add his favorite communication constructs to λdist in a consistent manner. We shall use the following syntactic macros:

$$\text{channel} \equiv \lambda x.\textbf{ref resident } 0$$

$$\text{is-ready?} \equiv \lambda a.!a \neq 0$$

$$\text{put-value} \equiv \lambda a.\lambda v.\textbf{set } a \; v$$

$$\text{get-value} \equiv \textbf{fix } f.\lambda a.\textbf{if } !a = 0 \textbf{ then } fa \textbf{ else } !a$$

where fix is a fixed point operator. Channel macro creates a new address initialized integer zero of resident type. Is-ready? macro is reduced to true if the argument is a reference to non-zero value, and reduced to false otherwise. We interpret integer zero as an empty channel. Put-value macro stores a value to a given address. Get-value macro waits until a value stored in the address changes from integer zero.

5.2   Remote evaluation

A macro that evaluates an expression at a remote place is defined as follows:

$$\text{leval}(e) \equiv e$$

$$\text{reval}(e, p) \equiv \text{let}\ c = \text{channel}()$$
$$\_ = \langle \text{let from} = \textbf{go}\ p, \text{result} = e$$
$$\text{in}\ \langle \text{let}\ \_ = \textbf{go from in put-value}\ c\ \text{result} \rangle \rangle$$
$$\text{in get-value}\ c$$

Leval macro defines local evaluation of expression e, which is just e itself. Reval(e,p) defines evaluation of expression e at place p. To evaluate e at place p, a channel is created and the execution at the current place waits until a return value will be stored in the channel. An agent goes to place p, evaluates e there, returns to his birthplace and stores a return value in the channel. We would like to define the semantics of remote procedure call here. Let $a@p$ be a remote reference of a function defined at place p. Then invocation of remote function is defined as follows:

$$\text{rpc}(a@p, v) \stackrel{def}{=} \text{reval}(!av, p)$$

$$\text{rpc}(a, v) \stackrel{def}{=} \text{leval}(!av)$$

For convenience, we extend the semantics of RPC for local functions. If a is a local or remote reference of a function, then rpc(a, v) works with the above definition.

5.3   Remote reference

In our calculus, we can define the semantics of remote references in an intuitively natural way.

$$!a@p \stackrel{def}{=} \text{reval}(!a, p)$$

$$\text{set}\ a@p\ v \stackrel{def}{=} \text{reval}(\text{set}\ a\ v, p)$$

Namely, dereference of remote reference $a@p$ is defined as dereference of address a at place p. Similarly, assignment of value v to remote reference $a@p$ is defined as assignment of value v to address a at place p.

5.4 Encoding Obliq

Obliq(Cardelli 1995) is a simple distributed object-oriented language, which has a significant feature that one can specify the locations of values and controls in a completely implicit manner. Mobility of values is defined by mutability. One has disjoint immutable values and mutable values in Obliq. When an immutable value migrates, it is just copied. When a mutable value migrates, it is not copied and it will be referred to as a remote reference from remote places. Consistency of a mutable value is guaranteed since it is never duplicated. The difference of execution sites is due to the difference between functions and methods. In Obliq, a function is always executed locally since a function is copied across sites as a function is a value. On the other hand a method is executed remotely since an object, and of course its methods, is referred as a remote reference from the outside of its resident site. The following fragment of Obliq code is a variation of the example used in (Cardelli 1995) to illustrate difference of execution sites:

Server Site:

```
net_export("ComputeServer", Namer,
    {reval => meth(s, p)p()end, leval => proc(p)p()end});
```

Client Site:

```
let server = net_import("ComputeServer", Namer);
var x = 0;
server.reval(proc()x := x + 1end);
server.leval(proc()x := x + 1end);
```

where **Namer** means the name of the computing object. When one invokes **reval**, the function passed to **reval** as an argument that increments x will be evaluated at the server site since a function is copied when referred to from a remote site. When one invokes **leval**, it will be evaluated at the client site since **leval** is a function and it is copied to the client site when referred to.

We show that a subset of Obliq, called coreObliq, can be implemented in our calculus by constructing a simple translation map from coreObliq into our calculus. The syntax of coreObliq is as follows:

x	identifiers
$\text{proc}(x)\ s\ \text{end}$	procedures
$a(b)$	procedure invocation
$\text{meth}(x, x')\ s\ \text{end}$	methods
$\{l_1 => a_1, ..., l_n => a_n\}$	objects
$a.l$	field selection
$a.l(b)$	method invocation

$$a.l := b \qquad\qquad\qquad \text{field update/method override}$$
$$\mathbf{clone}(a) \qquad\qquad\qquad \text{object cloning}$$
$$a; b \qquad\qquad\qquad\qquad \text{sequential execution}$$

We give the semantics of coreObliq by translating it into λdist(One can find an informal explanation in (Cardelli 1995)). The translation map is defined by $\mathcal{T}$: coreObliq → λdist defined below.

$$\mathcal{T}[\![x]\!] = x, \ \mathcal{T}[\![\mathbf{proc}(x) \ s \ \mathbf{end}]\!] = \lambda x.\mathcal{T}[\![s]\!], \ \mathcal{T}[\![a(b)]\!] = \mathcal{T}[\![a]\!]\mathcal{T}[\![b]\!]$$

$$\mathcal{T}[\![\mathbf{meth}(x, x') \ s \ \mathbf{end}]\!] = \mathbf{ref} \ \mathbf{resident} \ (\lambda(x, x').\mathcal{T}[\![s]\!])$$

A procedure is translated into a lambda abstraction. Because a lambda abstraction is copied whenever moved to the outside of a place, a procedure is always executed locally. While a method is referred to by a reference of **resident** type so that it is executed in its resident place. The first parameter of the method is the object they belongs to (often called self or this variable) and the second parameter is an argument for the method.

$$\mathcal{T}[\![\{l_i => a_i\}]\!] = \mathbf{ref} \ \mathbf{resident} \ \{l_i => \mathcal{T}[\![a_i]\!],$$
$$\text{update\_}l_i => \mathbf{ref} \ \mathbf{resident} \ (\lambda(s, v).\mathbf{set} \ (s.l_i)v)\}$$

Because an object in Obliq is referred to by a remote reference from the outside of its resident place and it is not copied at migration, it is referred to by a reference of **resident** type in λdist. We add a special additional method for each field of a record to support field update operation.

$$\mathcal{T}[\![a.l]\!] = (!\mathcal{T}[\![a]\!]).l$$
$$\mathcal{T}[\![a.l(b)]\!] = \mathbf{let} \ o = \mathcal{T}[\![a]\!] \ \mathbf{in} \ \mathrm{rpc}((!o).l, (o, \mathcal{T}[\![b]\!]))$$
$$\mathcal{T}[\![a.l := b]\!] = \mathcal{T}[\![a.\text{update\_}l(b)]\!]$$

Field selection is translated straightforwardly. Method invocation is translated into a remote procedure call. Field update operation is translated to a call to a special update method. Since field update operation must be executed in its resident place, it is a reference of **resident** type (see the translation rule for objects given above).

$$\mathcal{T}[\![e_1; e_2]\!] = \mathbf{let} \ \_ = \mathcal{T}[\![e_1]\!] \ \mathbf{in} \ \mathcal{T}[\![e_2]\!]$$
$$\mathcal{T}[\![\mathbf{clone}(a)]\!] = \mathbf{ref} \ \mathbf{resident} \ (!\mathcal{T}[\![a]\!])$$

The translation rule for sequential execution is obvious. Object cloning operation creates a local clone object whether the original object is at a local or a remote place. Dereference of an object reference copies the object to a local place. But this translation does not work since methods are not copied to a local place because they have **resident** type. So, to accomplish cloning operation one has to copy each method explicitly.

Now we would like to see how the translation works by translating the

ComputeServer example above. The example is translated into the following expressions:

Server Site:

net_export("ComputeServer", Namer,

ref resident{reval => ref resident $\lambda(s,p).p()$, leval => $\lambda p.p()$,

update_reval => ..., update_leval => ...})

Client Site:

let server = net_import("ComputeServer", Namer)

x = ref resident 0

_ = let o = server in rpc$((!o).$reval, $(o, \lambda().$set x $(x + 1)))$

in $(!$server$).$leval$(\lambda().$set x $(x + 1))$

provided that update methods are omitted since they are not used in this example. Apparently, invocation of reval is compiled to a remote procedure call so that it is executed at the resident place, while the invocation of leval is compiled to just a function application so that it is executed locally.

5.5 Encoding Java class loader

We describe the class loading mechanism of Java. The unit of code transmission in Java is a class. Until a method of an unloaded class is called, the code of the class is not transmitted from a server. For simplicity, we show the case of a function that is not copied until it is called. To describe such a code movement mechanism, the following stores suffice:

server : $[a \mapsto f^{\text{resident}}]$

client : $[\text{stub} \mapsto (\lambda x.\text{let } \_ = \text{set stub } (!(a@\text{server})) \text{ in } !\text{stub } x)^{\text{copy}}]$

We have a function f at the server place and a stub of the function at the client place. When we invoke the stub with a value v like !stub v, it copies the function f to the client place and then invoke f with v. We can generate a stub of a function f by evaluating the following macro at the server place:

genstub$(f) \equiv$ let a = ref resident f, stub = ref copy 0,

$f' = \lambda x.\text{let } \_ = \text{set stub } (!a) \text{ in } !\text{stub } x,$

$\_ = \text{set stub } f'$

in stub

When evaluating genstub at the server place to generate the stub of f, the function f is referred to by an address a of resident type from the stub, so that the function itself is referred to as a remote reference from the client place. A client has a stub of the function instead.

6   CONCLUSION AND FUTURE WORK

We have proposed a simple and powerful framework λdist in which various distribute mechanisms can be modeled. It has been shown that agent movement mechanisms can be modelled formally by agent expressions and the notion of data movement types. We have also demonstrated that the important mechanisms of several distributed language systems can be described in our calculus. The motivation for our general distributed calculus arose from the difference of the underlying movement mechanisms of various agent language systems. Because they are so different, it is hard for one agent language system to simulate another agent language systems. Thus we needed a new framework that can describe a variety of language mechanisms. We are currently exploring type systems capturing ownership and distribution.

REFERENCES

Cardelli, L. (1995) A Language with Distributed Scope. *Conference Record of the 22th Symposium on POPL*, 286–297.

Carriero, N. and Gelernter, D. (1989) Linda in Context. *Communications of the ACM*, **32**(4), 444–458.

Colusa. (1995) Omniware: A Universal Substrate for Mobile Code. *Colusa Software white paper.*

Fournet, C. and Gonthier, G. (1996) A Calculus of Mobile Agents. *CONCUR'96: Concurrency Theory*, LNCS **1119**, 406–421.

Gosling, J. and McGilton, H. (1995) The Java Language Environment. White paper, Sun Microsystems. Available at `http://java.sun.com/`.

Gray, R. S. (1995) Agent Tcl: A transportable agent system. *Proceedings of the CIKM Workshop on Intelligent Information Agents.*

Knabe, F. C. (1995) Language Support for Mobile Agents. PhD thesis, Computer Science Carnegie Mellon University.

Kobayashi, N. and Yonezawa, A. (1995) Higher-Order Linear Logic Programming. *Workshop on Theory and Practice of Parallel Programming(TPPP).*

Marzo, G. D., Muhugusa, M., Tschudin, C. and Harms, J. (1995) The Messenger Paradigm and its Implications on Distributed Systems. *Workshop on Intelligent Computer Communications.*

Milner, R., Parrow, J. and Walker, J. (1989) A Calculus of Mobile Processes Part I. Technical report, University of Edinburgh. ECS-LFCS-89-85.

Straßer, M., Baumann, J. and Hohl, F. (1996) Mole – a java based mobile agent system. *ECOOP'96 Workshop on Mobile Object Systems.*

White, J. E. (1996) Mobile Agents. *Software Agents*(ed. J. Bradshaw). The MIT Press.

4

A proof-theoretic approach to the design of object-based mobility

Carlos H. C. Duarte
Department of Computing, Imperial College
180 Queen's Gate, London, SW7 2BZ, United Kingdom
e-mail: `cd7@doc.ic.ac.uk`, tel: **+44 171 594 8341**, fax: **+44 171 581 8024**

Abstract. With the advent of technologies to realise parallel computing in mobile sometimes portable platforms, it is now possible to fulfil requirements related to the very dynamic and mutable user location. Designing the required applications calls for improved formal methods to treat mobility while assuring correctness. In this paper, we argue that mobile systems can be specified and verified in an effective modular manner using a logic which allows us to deal with object creation and reconfiguration. Capitalising on our previous work on the specification and verification of actor systems using a temporal logic of objects, here we show that our approach can be used to formally design location dependent applications.

Keywords: Actors, Specification, Verification, Mobile Systems, Location Management.

1 INTRODUCTION

We are currently facing a radical change in the way users interact with software systems and in the underlying distributed software architectures. Thanks to the advent of technologies like cellular phones, personal digital assistants and active badges, users are no longer required to go to specific access points to take advantage of some locally provided functionality. Such devices have become increasingly more personal and can be carried by their owners. In turn, software systems may now be used at any time and place, and can provide location dependent functionality such as ubiquitous message delivery, transportable user sessions and others [8]. What is essentially novel in this new operational environment is the very presence of mobility. The way to support the new requirements mobility poses is the management of location information.

The need to manage both location information and mobility brings with it new problems to be addressed in the design of parallel/distributed systems. For example, it is now important to develop quantitative models to predict network performance according to user mobility [10]. We are particularly concerned here with the more fundamental question on how to develop such systems so as to correctly provide the required functionality. As complexity substantially grows with the autonomy and heterogeneity presented by mobile devices and mobile applications become more and more open — characteristics that must be accounted for in some design step — introducing errors during the development process turns out to be easier and costly. The only way to assure software correctness is the development of a formal, qualitative model of the system, to be refined in a step-by-step process until an implementation is produced. As a result, it is possible to clarify any matter of concern through proof or refutation.

Unfortunately, most of the existing formal methods are not so good in designing mobile applications. Well established methods like VDM [9] and Z [17] do not address at all the inherent concurrency of mobile systems. In some other cases, concurrency is actually treated but the design process is organized in terms of notions like processes [13] or programs [3, 20], which certainly provide important insights on how an implementation should work but poorly support understanding and representing the problem domain in an organised manner. On the other hand, the use of object-based notions like attributes, actions and encapsulation as in [7] seems to bridge this gap. Even then, expressibility concerns arise since the basic notion of mobility has to be captured.

We have developed a logic to support the design of actor systems [4]. Actors are computational objects with encapsulated state which openly interact through asynchronous point-to-point message passing [1]. As a result of processing messages, new concurrent actors can be created and actor names can be communicated, supporting in this way dynamic reconfiguration of actor communities. Our logic uses temporal theories as object descriptions and theory morphisms as specification connectors particularising [7], although it presents an additional mode of interaction between objects to capture reliable asynchrony. These characteristics make it possible to produce specifications in isolation to be subsequently combined as well as to decompose proofs of global properties in lemmas about single objects to be verified in a localised manner. We claim here that these characteristics are sufficient to guarantee an effective and modular formal design of object-based mobile systems. We are not aware of any similar work in the literature.

In order to support our claim, we have chosen to study here the design of a location management architecture for networks of mobile users and devices. For simplicity, we ignore the important issues of dependability, authenticity and security [18] and concentrate just in the management of location information. In the next section, we introduce our actor-based design approach. Subsequently, we informally describe the requirements of location management applications and devote two sections to their design, namely to their specification and verification. We conclude the paper providing a brief evaluation of our achievements and suggesting further research.

2   A PROOF-THEORETIC APPROACH TO ACTOR DESIGN

We adopt full many-sorted first-order branching time logic with equality as the underlying foundation of our work. A good survey on the subject appears in [5]. Theory signatures and presentations are used to define respectively the language and description of object behaviours and to produce modularised designs. We identify therein the basic object-based notions of attribute, action and encapsulation, and connect these entities using signature and theory morphisms as proposed in [7]. The actor-based formalism is obtained as a particularization of this generic framework. In [4] we provide the rationale leading us to propose the formalism as such and the description of its proof-theory. Here we only present the relevant notions to mobile systems design.

Before introducing technical definitions, let us present in Figure 1 the specification of region tree nodes, instances of the spatial hierarchical data structures described in [16].

Actor REGIONTREENODE
   data types $addr, direc, int, bool$ (T, F : $bool$; $0, 1, 3$: int; $+$: $int \times int \rightarrow int$)
   attributes me, to : $addr, reg[direc]$: $addr, void$: $bool, ans$: int
   actions $cnt(addr, addr)$: **local birth**;
             $node(addr, addr)$: **local + extrn birth**;
             $inc, updt(addr, addr, addr, addr)$: **local comput**;
             $ack(addr, addr, addr, addr)$: **extrn message**;
             $split(addr), in(addr, addr), rep(addr, addr, bool)$: **local + extrn message**
   axioms $k, n, p, q, r, s, t, u, x, y, z$: $addr, d$: $direc, v$: int, b : $bool$

$$node(n, p) \lor cnt(n, p) \rightarrow me = n \land to = p \land void = \text{T} \land \forall d \cdot reg[d] = n \land ans = 0 \tag{1.1}$$

$$updt(\vec{n}) \land me = p \land to = q \land ans = v \rightarrow \mathbf{X}(me = p \land to = q \land ans = v \land void = \text{F} \land \vec{reg} = \vec{n}) \tag{1.2}$$

$$inc \land me = n \land to = p \land \vec{reg} = \vec{q} \land ans = v \rightarrow \mathbf{X}(me = n \land to = p \land \vec{reg} = \vec{q} \land ans = v + 1) \tag{1.3}$$

$$split(n) \land me = p \rightarrow \mathbf{X}(\exists \vec{q} \cdot new(node, q_i, q_i, p) \land updt(\vec{q}) \land send(ack, n, \vec{q})) \tag{1.4}$$

$$(in(n, p) \land r = n \lor rep(r, s, \text{T}) \land to = p) \land me = n \rightarrow \mathbf{X}(send(rep, p, r, n, \text{T})) \tag{1.5}$$

$$in(n, p) \land me = q \land q \neq n \land void = \text{F} \land \vec{reg} = \vec{r} \rightarrow \mathbf{X}(\exists! s \cdot new(cnt, s, q, p) \land send(in, r_i, n, s)) \tag{1.6}$$

$$in(n, p) \land me = q \land q \neq n \land void = \text{T} \rightarrow \mathbf{X}(send(rep, p, n, q, \text{F})) \tag{1.7}$$

$$rep(n, p, \text{F}) \land to = r \land me = q \rightarrow \mathbf{X}(ans = 3 \land send(rep, r, n, q, \text{F}) \lor inc) \tag{1.8}$$

$$\exists \vec{n}, \vec{p} \cdot new(node, n_i, p_i, q) \lor updt(\vec{n}) \lor send(ack, r, \vec{n}) \leftarrow split(r) \land me = q \tag{1.9}$$

$$inc \leftarrow \exists n, p \cdot rep(n, p, \text{F}) \land ans \neq 3 \tag{1.10}$$

$$send(rep, n, p, q, \text{T}) \leftarrow (in(q, n) \land p = q \lor \exists s \cdot rep(p, s, \text{T}) \land to = n) \land me = q \tag{1.11}$$

$$send(rep, n, p, q, \text{F}) \leftarrow (in(p, n) \land p \neq q \land void = \text{T} \lor \exists s \cdot rep(p, s, \text{F}) \land to = n \land ans = 3) \land me = q \tag{1.12}$$

$$\exists n \cdot new(cnt, n, q, r) \lor send(in, p_i, s, n) \leftarrow in(s, r) \land me = q \land q \neq s \land void = \text{F} \land \vec{reg} = \vec{p} \tag{1.13}$$

$$node(k, n) \lor cnt(k, n) \rightarrow \mathbf{G}(\mathbf{E}(deliv(split(p))) \land \mathbf{E}(deliv(in(q, r))) \land \mathbf{E}(deliv(rep(s, t, b)))) \tag{1.14}$$

$$node(k, n) \lor cnt(k, n) \rightarrow \mathbf{XG}(inc \lor updt(p, q, r, s) \lor \mathbf{E}(split(t)) \land \mathbf{E}(in(u, x)) \land \mathbf{E}(rep(y, z, b))) \tag{1.15}$$

End

Figure 1   Specification of region trees.

These will be used in designing location management later on. At the top of each theory presentation, we can see sequences of sorts plus the associated constants and operations (data types), attribute and action symbols, denoting respectively fixed meaning data objects, the local state and the messages and computations dealt with by these actors. For instance, a region can receive a request to divide itself (*split*) in sub-regions (*reg*) organised according to the four directions of the compass points (direc). Eventually, continuations (*cnt*) are created in order to answer inclusion queries (*in*). These symbols belong the language of region tree nodes and are generically formalised as follows:

Definition 1 (Actor Signature) An actor signature Δ is a triple of disjoint and finite families $(\Sigma, \mathcal{A}, \Gamma)$ where:

- $\Sigma = (S, \Omega)$ is an universe signature, i.e., S is a set of sort symbols and Ω is an $S^* \times S$-indexed family of operation symbols. We require that the sort of mail addresses $addr \in S$;

- $\mathcal{A}$ is an $S^* \times S$-indexed family of attribute symbols;

- $\Gamma = (\Gamma_e, \Gamma_l, \Gamma_c)$, S^*-indexed sets of action symbols with $(\Gamma_e \cup \Gamma_l) \cap \Gamma_c$ empty. Γ_c is a set of local computation symbols. Γ_e and Γ_l represent respectively sets of events to be requested from the environment and provided locally. These two sets consist in collections of message and birth computation symbols, e.g. $\Gamma_l - \Gamma_{l_b}$ and Γ_{l_b} respectively.

For ϵ as the empty sequence, we write an $\epsilon \times s$-indexed family of symbols as if s were its index. Given a set X, we denote the sub-set of X symbols of sort $\langle s_1, \ldots, s_n \rangle \times s$ as $X_{\langle s_1, \ldots, s_n \rangle, s}$. We shall also operate with subscripts ($\Gamma_{e_b \cap l_b}$) to denote operations on sub-sets of Γ ($\Gamma_{e_b} \cap \Gamma_{l_b}$).

Axioms defined out of terms and formulae are also present in the previous specification. An example is that asserting both the invariance of the attributes holding the mail addresses of the node and its parent (terms in 1.2) and the creation of sub-regions in the instant following an update (*updt*). Likewise, we use a formula to say that after the birth of a node requests for *split*, *in* or *rep* may always be delivered (1.14), although they may only be consumed if the node is not busy sequentially performing local computations (1.15). To define these notions, we assume that an infinite family of rigid variables and its classification Ξ according to the sorts of a signature Δ are given:

Definition 2 (Terms) The S-indexed set of terms $T_\Delta(\Xi)$ is defined as follows, provided that $q \in \Xi_s \cup \Omega_s \cup \mathcal{A}_s$, $p \in \Omega_{\langle s_1, \ldots, s_n \rangle, s}$, $f \in \mathcal{A}_{\langle s_1, \ldots, s_n \rangle, s}$ and $t_i \in T_\Delta(\Xi)_{s_i}$:

$$t ::= q \mid p(t_1, \ldots, t_n) \mid f(t_1, \ldots, t_n)$$

Definition 3 (Formulae) The set $F_\Delta(\Xi)$ of formulae is defined by the mutual recursion below, provided that $c \in \Gamma_{\langle s_1, \ldots, s_n \rangle}$, $t_i \in T_\Delta(\Xi)_{s_i}$, $y \in \Xi_s$ and $g_i \in F_\Delta(\Xi)$:

$$g ::= \mathbf{beg} \mid c(t_1, \ldots t_n) \mid t_1 =_s t_2 \mid \mathbf{E}g' \mid g_1 \to g_2 \mid \neg g_1 \mid \exists y \cdot g_1$$
$$g' ::= g \mid \mathbf{X}g'_1 \mid g'_1 \mathbf{U}g'_2 \mid g'_1 \to g'_2 \mid \neg g'_1 \mid \exists y \cdot g'_1$$

Terms consist in variables, nulary function and attribute symbols; or function and attribute symbols applied to terms. We usually write a sequence of similar terms $t_1, \ldots, t_n$ as $\vec{t}$. Formulae stand for the initial instant (**beg**); action occurrences; term equality; the occurrence of a property in some possible behaviour (**E**), in the next instant (**X**) or until another property holds (**U**); or formulae aggregation using first-order logic connectives. These are the original CTL* constructs [5] enriched to express object-based notions.

The reader may wonder why **new**, **deliv** and **send** do not appear in our definitions above. Actually, they stand for the abbreviation of logical actions as defined in [4]. Much in the way that ASCCS is a subset of SCCS [13], our calculus — which captures synchronous object creation and reliable asynchronous message passing — can be seen as a particularization of the synchronous object calculus of [7]. The aforementioned connectives are definable therein and have the following informal meaning:

FOR	OF TYPE	FORMULA	READS
$c, n, \vec{v_c}$	$\Gamma_{e_b \cup l_b}, \mathbf{addr}, T_\Delta(\Xi)$	$\mathbf{new}(c, n, \vec{v_c})$	creation of an actor with a given name
$c, n, \vec{v_c}$	$\Gamma_{(e - e_b) \cup (l - l_b)}, \mathbf{addr}, T_\Delta(\Xi)$	$\mathbf{send}(c, n, \vec{v_c})$	dispatch of a message to a specific actor
$c, \vec{v_c}$	$\Gamma_{l - l_b}, T_\Delta(\Xi)$	$\mathbf{deliv}(c, \vec{v_c})$	delivery of a message

As for the actor primitive **become**, which allows actors to have mutable state space, there is no treatment here. If it does not receive a higher-order interpretation, this primitive is definable within the core actor theory as noticed in [1]. In designing mobile systems, such a higher-orderness is not required.

Specifications characterise communities of actors with similar behaviour. What make these actors different from each other are their distinct names (of sort **addr**, prefixed to terms when talking about global properties) and their potentially distinct interactions with the environment. Specifications are defined as theory presentations comprising a signature and a set of axioms explicitly provided by the specifier:

Definition 4 (Actor Specification) An actor specification is a pair $\Phi = (\Delta, \Psi)$ where Δ is an actor signature and Ψ is a finite set of formulae over Δ (the specification axioms).

To each specification Φ is assigned a set of additional logical axioms called Ax_Φ. These axioms are provided in schematic form in the Appendix and constrain the behaviour of the environment relative to the specified actors. Such axioms become necessary, together with the deductive system of the branching time temporal logic of objects and the additional inference rules of the Appendix, in the verification of actor properties.

For notational convenience, formulae containing derived first-order logic connectives and inequalities stand for the usual translations. Moreover, free variables in axioms are considered to be universally quantified. Other admissible connectives are defined as:

FOR	OF TYPE	FORMULA	READS	REPRESENTS
g	$F_\Delta(\Xi)$	$\mathbf{A}g$	in any behaviour	$\neg \mathbf{E}\neg g$
g	$F_\Delta(\Xi)$	$\mathbf{F}g$	eventually in the future	$(g \to g)\mathbf{U}g$
g	$F_\Delta(\Xi)$	$\mathbf{G}g$	always in the future	$\neg \mathbf{F}\neg g$
g_1, g_2	$F_\Delta(\Xi)$	$g_1 \mathbf{W} g_2$	unless	$\mathbf{G}(g_1 \wedge \neg g_2) \vee g_1 \mathbf{U} g_2$
g_1, g_2, p	$F_\Delta(\Xi)$	$g_1 \overset{i}{\leftarrow}_p g_2$	initially precedes	$p \to (\neg g_1)\mathbf{W}(g_2 \wedge \neg g_1)$
g_1, g_2, p	$F_\Delta(\Xi)$	$g_1 \leftarrow_p g_2$	precedes	$g_1 \overset{i}{\leftarrow}_p g_2 \wedge g_1 \to \mathbf{X}(g_1 \overset{i}{\leftarrow}_{(p \to p)} g_2)$

The unary connectives above are non-strict (they include the present) and usually appear in branching-time logics. Conversely, the precedence connectives are strict and forbid the simultaneous occurrence of some properties. In specifications, where p usually stands for the occurrence of some birth action (which we write as init), their index is omitted. All these temporal connectives are used, e.g., to state and reason about causality relations: that a query *in* for the inclusion of a node n in a region, when consumed by any non-empty node distinct from n, in the next instant results not only in the dispatch of many similar queries to the respective sub-regions but also in the creation of a continuation actor to process their answers (1.6), something that cannot happen otherwise (1.13).

Given some independently specified actor communities, we may want to interconnect them to define communities of heterogeneous cooperating actors. This can be done by providing language translations between their theory presentations obeying what follows:

Definition 5 (Signature Morphisms) Given two actor signatures $\Delta_1 = (\Sigma_1, \mathcal{A}_1, \Gamma_1)$ and $\Delta_2 = (\Sigma_2, \mathcal{A}_2, \Gamma_2)$, a signature morphism $\tau : \Delta_1 \to \Delta_2$ consists of:

- a morphism of algebraic structures $\tau_\upsilon : \Sigma_1 \to \Sigma_2$ such that $\tau_\upsilon(\mathsf{addr}_1) = \mathsf{addr}_2$;

- for each $f \in \mathcal{A}_{1_{(s_1, \ldots, s_n), s}}$, an attribute symbol $\tau_\alpha(f) : \tau_\upsilon(s_1) \times \ldots \times \tau_\upsilon(s_n) \to \tau_\upsilon(s)$ in $\mathcal{A}_2$;

- for each $c \in \Gamma_{1_{(s_1, \ldots, s_n)}}$, an action symbol $\tau_\gamma(c) : \tau_\upsilon(s_1) \times \ldots \times \tau_\upsilon(s_n)$ in Γ_2 such that $\tau_\gamma(\Gamma_{e_1}) \subseteq \Gamma_{e_2}$, $\tau_\gamma(\Gamma_{l_1}) \subseteq \Gamma_{l_2}$, $\tau_\gamma(\Gamma_{c_1}) \subseteq \Gamma_{c_2}$, where $\tau_\gamma(\Gamma_{e_{b_1}}) \subseteq \Gamma_{e_{b_2} \cup l_{b_2}}$ and $\tau_\gamma(\Gamma_{e_1 - e_{b_1}}) \subseteq \Gamma_{(e_2 - e_{b_2}) \cup (l_2 - l_{b_2})}$, $\tau_\gamma(\Gamma_{l_{b_1}}) \subseteq \Gamma_{l_{b_2}}$ and $\tau_\gamma(\Gamma_{l_1 - l_{b_1}}) \subseteq \Gamma_{l_2 - l_{b_2}}$ so that $\tau_\gamma(\Gamma_{e_1 \cap l_1}) \subseteq \Gamma_{e_2 \cap l_2}$.

It is straightforward to define inductively the translation of symbols, classifications, terms, formulae and sets thereof under τ.

From the first item, we can see that interconnected actor communities are named using the same sort, namely addr. In addition, the third item says that event symbols representing requests of a community to its environment can be associated with events internally provided by a distinct community (eg. $\tau_\gamma(\Gamma_{e_{b_1}}) \subseteq \Gamma_{e_{b_2} \cap l_{b_2}}$), meaning that messages can be dispatched to and actors created within one community from another.

Technically, the translation of connected theory presentations induced by signature morphisms does not capture the expected interconnection of actor behavior in a precise way. This is due to the existing logical axioms of the original theories, which are not translated by such morphisms. We are obliged to use an alternative notion:

Definition 6 (Theory or Specification Morphisms) Given two actor specifications $\Phi_1 = (\Delta_1, \Psi_1)$ and $\Phi_2 = (\Delta_2, \Psi_2)$, a specification morphism $\tau : \Phi_1 \to \Phi_2$ is a signature morphism such that $\vdash_{\Phi_2} \tau(g)$ for every $g \in \Psi_1 \cup Ax_{\Phi_1}$.

We should stress that connecting specifications using theory morphisms is analogous to providing links between identifiers in distinct actor programs [1]. As we shall see later, to verify actor properties, we also need to identify which actors are assumed to exist in the environment and which are able to receive messages from the outside, as in [2].

3   INFORMAL REQUIREMENTS OF LOCATION MANAGEMENT

A central problem in designing and implementing software systems for networks of mobile users and devices is how to manage object location. An extensive description of the problem can be found in the literature (cf. [8, 12, 18]). In this section, we provide an informal list of requirements strictly imposed by mobility. In the next section, we enumerate some design decisions based on this list and propose a formal specification for the corresponding mobile architecture.

We can classify the requirements of location management in three families, the first concerning the nature of location information and located objects, the second about the process of acquiring location information and the third on how to deal with it. In what follows, we ignore real time issues and provide a partial list of functional requirements:

1. A *location information* must be *dynamic*, in the sense that, at each time, it may be a distinct instance of a class of objects.

2. A *location information* must be *mutable*, in the sense that, at each time, it may be an instance of a distinct class of objects.

3. *Located objects* may be *users* or *devices*, at least.

4. *Location information* acquisition must be *unintrusive*, which means that the acquisition process cannot intrude user behaviour nor require user intervention.

5. *Location information* acquisition must offer support to *multiple location observations*, which means that simultaneous observations producing distinct location information for the same object may occur.

6. *Location information* management must support *indeterminacy*, which means that location information for some objects may not be available at some instant.

7. *Location information* management must offer support to *object naming*, which is the assignment of meaningless unique names to located objects.

Actor SENSOR
  data types   addr, int $(0, 1, \text{MAX} : \text{int}; + : \text{int} \times \text{int} \rightarrow \text{int})$
  attributes $srv, obj, loc, id : \text{addr}; tick : \text{int}$
  actions $sens(\text{addr, addr, addr, addr}) : \textbf{local} + \textbf{extrn birth};$
          $reloc(\text{addr}), set(\text{int}), obs : \textbf{local comput};$
          $detect(\text{addr, addr}), unreach(\text{addr, addr}) : \textbf{extrn message}$
  axioms $n, p, q, r : \text{addr}, v : \text{int}$

$$sens(n, p, q, r) \rightarrow srv = n \wedge obj = p \wedge loc = q \wedge id = r \wedge tick = 0 \tag{2.1}$$

$$reloc(n) \wedge srv = p \wedge obj = q \wedge id = r \wedge tick = v \rightarrow \mathbf{X}(srv = p \wedge obj = q \wedge id = r \wedge tick = v) \tag{2.2}$$

$$reloc(n) \rightarrow \mathbf{X}(loc = n) \tag{2.3}$$

$$set(v) \wedge srv = n \wedge obj = p \wedge loc = q \wedge id = r \rightarrow \mathbf{X}(srv = n \wedge obj = p \wedge loc = q \wedge id = r) \tag{2.4}$$

$$set(v) \rightarrow \mathbf{X}(tick = v) \tag{2.5}$$

$$obs \wedge srv = n \wedge loc = p \wedge (obj = q \vee id = q) \rightarrow \mathbf{X}(set(0) \wedge \mathbf{send}(detect, n, q, p)) \tag{2.6}$$

$$sens(n, p, q, r) \rightarrow \mathbf{G}(\neg obs \wedge tick = \text{MAX} \leftrightarrow \mathbf{send}(unreach, srv, obj, loc)) \tag{2.7}$$

$$sens(n, p, q, r) \rightarrow \mathbf{G}(\neg obs \wedge tick = \text{MAX} \rightarrow set(0)) \tag{2.8}$$

$$sens(n, p, q, r) \rightarrow \mathbf{G}(\neg set(0) \wedge \neg obs \leftrightarrow set(tick + 1) \wedge \mathbf{send}(detect, srv, id, loc)) \tag{2.9}$$

$$\mathbf{send}(detect, n, p, q) \rightarrow src = n \wedge loc = q \wedge (id = p \wedge set(v) \vee obj = p \wedge set(0)) \tag{2.10}$$

$$set(0) \leftarrow obs \vee tick + 1 = \text{MAX} \tag{2.11}$$

End

Figure 2   Specification of sensors.

The first two items should not be confused. While it is obvious that mobile object locations may change as time passes, meaning that they are dynamic, it is not so obvious that they should also be mutable. The reason for this is that a location service may provide information with distinct accuracies or that multiple services may be used [12].

4   LOCATION MANAGEMENT DESIGN IN A FORMAL SETTING

Based on the previous requirements list, we make our first design decision following [8] by using references to objects denoting spatial regions instead of dealing with location information directly. In this way, each located object acquires a new attribute (*loc*), which is annotated with the mail address of an object representing a location space region. If we use region trees as described in Section 2 for this purpose, we treat both the dynamic and mutable character of location information with this decision: as an attribute, location information can always be changed; as a reference, it does not constrain the shape and size of location observations. We make, however, the simplifying assumption that spatial regions are disjoint squares, due to the structure of such trees.

In order to treat the requirements concerning location information acquisition and management, we adopt the specification of sensors in Figure 2. Each sensor is created with knowledge of a location service mail address (*srv*) and is responsible for producing sequential observations (*obs*) of a named user (*obj*) in the specific region. Sensors are mobile as well and detect themselves in the monitored region (2.7 and 2.9). We omit their straightforward generalisation to deal with the observation of several distinct objects.

Each sensor keeps an internal clock which is reset — *set*(0) — after MAX cycles or when the user is observed. Axiom 2.11 guarantees that resets do not happen unless this

Actor MOBILEAGENT

    data types addr, bool (T, F : bool)

    attributes me, id, loc, to : addr, $fwdg$: bool

    actions $ag(\text{addr}, \text{addr}, \text{addr})$: **local + extrn birth**;

              $redir(\text{addr})$: **local comput**;

              $sub(\text{addr}, \text{addr})$: **extrn message**;

              $fwd(\text{addr}), mv(\text{addr}, \text{addr}), cp(\text{addr}, \text{addr}, \text{addr})$: **local + extrn message**

    axioms $n, p, q, r, s, t, u, x, y, z$: addr

$$ag(n, p, q) \rightarrow me = n \wedge id = p \wedge loc = q \wedge fwdg = \text{F} \wedge to = n \tag{3.1}$$

$$redir(n) \wedge me = p \wedge id = q \wedge loc = r \rightarrow \mathbf{X}(me = p \wedge id = q \wedge loc = r \wedge fwdg = \text{T} \wedge to = n) \tag{3.2}$$

$$fwd(n) \rightarrow \mathbf{X}(redir(n)) \tag{3.3}$$

$$mv(n, p) \wedge fwdg = \text{F} \wedge me = q \wedge id = r \rightarrow \mathbf{X}(redir(q) \wedge \text{send}(cp, n, p, q, r)) \tag{3.4}$$

$$mv(n, p) \wedge fwdg = \text{T} \wedge to = q \rightarrow \mathbf{X}(\text{send}(mv, q, n, p)) \tag{3.5}$$

$$cp(n, p, q) \wedge loc = r \rightarrow \mathbf{X}(\exists! s \cdot \mathbf{new}(ag, s, s, q, r) \wedge \text{send}(fwd, p, s) \wedge \text{send}(sub, n, s, r)) \tag{3.6}$$

$$redir(n) \vee \exists p, q \cdot (mv(p, q) \wedge me = n \wedge fwdg = \text{F}) \tag{3.7}$$

$$\exists n, p \cdot \mathbf{new}(ag, n, p, q, r) \vee \text{send}(fwd, s, n) \vee \text{send}(sub, t, n) \leftarrow cp(t, s, q) \wedge loc = r \tag{3.8}$$

$$\text{send}(mv, n, p, q) \leftarrow mv(p, q) \wedge to = n \wedge fwdg = \text{T} \tag{3.9}$$

$$\text{send}(cp, n, p, q, r) \leftarrow mv(n, p) \wedge me = q \wedge id = r \wedge fwdg = \text{F} \tag{3.10}$$

$$ag(n, p, q) \rightarrow \mathbf{G}(\mathbf{E}(\text{deliv}(fwd, s)) \wedge \mathbf{E}(\text{deliv}(mv, t, u)) \wedge \mathbf{E}(\text{deliv}(cp, x, y, z))) \tag{3.11}$$

$$ag(n, p, q) \rightarrow \mathbf{X}\mathbf{G}(\neg redir(r) \rightarrow \mathbf{E}(fwd(s)) \wedge \mathbf{E}(mv(t, u)) \wedge \mathbf{E}(cp(x, y, z))) \tag{3.12}$$

End

Figure 3   Simplified specification of mobile agents.

condition is fulfilled. Indeterminacy is treated by this clocking mechanism, which signs to the location service that the user is unreachable (*unreach*) whenever an observation does not happen before the deadline MAX (2.7). A *detect* message with the user location is sent to the service otherwise (2.6). Multiple observations are obtained by having many sensors dealing with the same located object. Unintrusivity is also treated as there is no causal connection between the production of observations and user behaviour.

If we realise the sensors of Figure 2 as optical devices connected to the architecture through radio frequency links, for instance, software mobility arises only when located object agents are considered. Such agents are meant to follow located objects through the architecture providing location dependent functionality such as ubiquitous message delivery and transportable user sessions [18]. Although we leave this additional functionality unspecified here, we present a specification of mobile agents in Figure 3.

We choose to capture mobility as localised agent replication. A mobile agent may receive a request to move to the location of another agent (*mv*), presumably located closer to the object the former represents. If an agent is currently moving to a new location, such requests will be delayed by self-forwarding until the agent finishes to move (3.5). In order to move, the original agent issues a request for the correctly located agent to create a local copy (*cp*) of its own (3.4), supplying in the message any required information for the copy (here in particular its name). After consuming this kind of message, an agent creates the desired replica and notifies both the original agent and the requesting service that the located object representative can be substituted (3.6).

In order to ensure coordination between sensors and agents, a location service must guarantee that the asynchronous messages they exchange are correctly addressed and

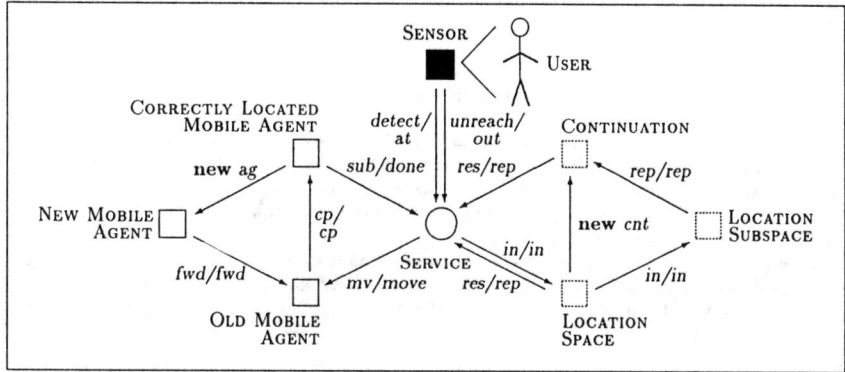

Figure 4   Internal event flow of the Location Manamegent Architecture.

ordered. The situation is better explained by the diagram in Figure 4. Once a located object is detected in a new region (*at*), the location service has to find a mobile agent in that region to request the creation of a replica of the moving agent there. The location space is recurrently queried (*in*) until such an agent is found. Then, the service requests the agent of the relocated object to move the place of the correctly located agent (*move*). At the end, the service is notified (*sub*) so that the old agent can be discarded while forwarding all possibly dispatched additional move requests to the new agent (*fwd*).

Since the location service has to associate located object names (*id*) with mobile agents, has to keep track of their location (*loc*) and has to put agents in contact to support mobility, we consider that servers providing compartmentalised bits of this functionality for each object are organised in circular lists, adopting the specification in Figure 5. Each server also records if there is no location information available for the object (*nl*). This knowledge is used to postpone answering location queries (?) until the object location becomes known (4.7 and 4.8).

Every request to the location service circulates around the linked list until the correct recipient is found. In case an observation from a sensor arrives carrying a new object location (*at*), a request for the rest of the list to find some agent placed therein is issued aiming to support moving to that location (4.9). For each located object registered in the service, the location space will be queried in a two step process: a continuation actor to process the answer of the query will be created (4.13), and this new actor will either request the original agent to move (4.14), if the current agent is located accordingly, or forward the query to the next list element (4.15).

Although illustrative, the informal description of the relationship between each pair of specifications should not substitute their formal composition, which is still missing here. The diagram in Figure 4 gives a good clue on what remains to be defined: the "physical communication channels", which are formally defined using specification morphisms. For every pair of specifications, each of them represented by a distinguished geometric figure,

Actor SERVER

   data types   addr, bool (T, F : bool)

   attributes   *me, nxt, loc, id, ag* : addr, *nl* : bool

   actions   *srv*(addr, addr, addr, addr, addr) : local + extrn birth;

               ch(addr, addr, addr, bool) : local comput;

               in(addr, addr), *move*(addr), *@*(addr, addr), *ack*(addr) : extrn message;

               ins(addr, addr, addr, addr), *done*(addr, addr), *res*(addr, addr, bool) : local + extrn message;

               at(addr, addr), *out*(addr, addr) : local + extrn message;

               mvrq(addr, addr, addr), *?*(addr, addr) : local + extrn message

   axioms n, p, q, r, s, t, u, x : addr, b : bool

$$srv(n, p, q, r, s) \rightarrow me = n \wedge nxt = p \wedge id = q \wedge loc = r \wedge ag = s \wedge nl = F \tag{4.1}$$

$$ch(n, p, q, b) \wedge me = r \wedge id = s \wedge nl = b \rightarrow X(me = r \wedge id = s \wedge nl = b) \tag{4.2}$$

$$ch(n, p, q, b) \rightarrow X(nxt = n \wedge loc = p \wedge ag = q \wedge nl = b) \tag{4.3}$$

$$ins(n, p, q, r) \wedge nxt = s \wedge loc = t \wedge ag = u \wedge nl = b \rightarrow X(\exists x \cdot new(srv, x, x, s, n, p, q) \wedge ch(x, t, u, b)) \tag{4.4}$$

$$ins(n, p, q, r) \wedge nxt = s \rightarrow X(\exists t \cdot new(srv, t, t, s, n, p, q) \wedge send(ack, r, t)) \tag{4.5}$$

$$done(n, p) \wedge nxt = q \rightarrow X(ch(q, p, n, F)) \tag{4.6}$$

$$?(n, p) \wedge id = n \wedge me = q \wedge loc = r \rightarrow X(nl = F \wedge send(@, p, n, r) \vee send(?, q, n, p)) \tag{4.7}$$

$$?(n, p) \wedge id \neq n \wedge nxt = q \rightarrow X(send(?, q, n, p)) \tag{4.8}$$

$$at(n, p) \wedge id = n \wedge me = q \wedge nxt = r \wedge ag = s \wedge (loc \neq p \vee nl = T) \rightarrow X(send(mvrq, r, s, p, q)) \tag{4.9}$$

$$at(n, p) \wedge id \neq n \wedge nxt = q \rightarrow X(send(at, q, n, p)) \tag{4.10}$$

$$out(n, p) \wedge id = n \wedge nxt = q \wedge loc = r \wedge ag = s \rightarrow X(ch(q, r, s, T)) \tag{4.11}$$

$$out(n, p) \wedge id \neq n \wedge nxt = q \rightarrow X(send(out, q, n, p)) \tag{4.12}$$

$$mvrq(n, p, q) \wedge nxt = r \wedge loc = s \wedge ag = t \rightarrow X(\exists! u \cdot new(srv, u, q, r, p, n, t) \wedge send(in, s, p, u)) \tag{4.13}$$

$$res(n, p, T) \wedge me = q \wedge id = r \wedge ag = s \rightarrow X(send(move, r, s, q)) \tag{4.14}$$

$$res(n, p, F) \wedge me = r \wedge nxt = s \wedge loc = t \wedge id = u \rightarrow X(send(mvrq, s, u, t, r)) \tag{4.15}$$

  ⋮

  and the usual axioms to guarantee absence of unsolicited responses and enableness

  ⋮

End

Figure 5   Specification of location service nodes.

that diagram shows how to relate their message symbols. For instance, the messages *mv* and *sub* of agents should be respectively associated with *move* and *done* of servers. Notice that relating external to local symbols yields the only possible direction of the message flow. Also notice in our example that we cannot produce a direct translation either from the theory of agents into that of servers or in the opposite direction. Therefore, to interconnect these entities we need to define mediating theory presentations to serve as connectors. Their nature is illustrated by the diagram in Figure 6.

To define the mobile architecture in a formal manner, we call each mediating specification in Figure (6.a) a CONNECTOR. Each of them contains two external message symbols only (without axioms as well) and hence translations including their contents after necessary renamings into the connected theories trivially exist. Taking connectors, connected theories and the morphisms betweeem them, the composite theory presentations are defined up to isomorphism by categorical constructions called pushouts (amalgamed sums), which always exist and are finite for any finite number of actor specifications [4]. Defined in this way, each COMPONENT in the figure contains all the renamed symbols and axioms of the connected theories, but the symbols identified by the connectors are

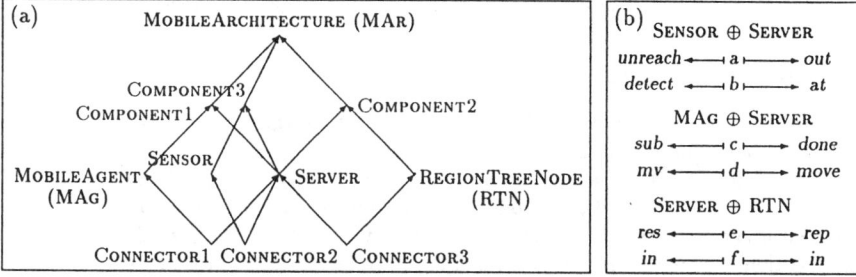

Figure 6 Composition of the architecture: Shared actors (a) and action symbols (b).

equalised. That is why, e.g., a message *move* from servers can be understood as *mv* when it is dispatched to an agent, no matter its name in the composite component. The detailed definition of connectors and their morphisms appear in Figure (6.b).

5 VERIFYING LOCATION MANAGEMENT PROPERTIES

As we have already mentioned, the existence of morphisms to connect separated actor specifications does not guarantee that interaction between actors in the respective communities can occur. Here, an analogy with telecommunication systems is useful to illustrate the situation: even if the physical cables to connect the private equipment of a customer to the network exist, it cannot receive phone calls unless assigned to an appropriate number. Therefore, "logical channels" are also needed in dealing with any kind of global property. These logical channels are captured here as an assumption on the configuration of the heterogeneous actor community.

Since the location space is a relatively separated component of our architecture, let us use it to exemplify how a property can be verified. Assume that the environment always creates region tree nodes configured correctly, providing the right name for the node and its parent (which is the node itself in the case of the root) in the creation:

$$r.\text{new}(reg, s, t, u) \rightarrow s = t \wedge (u = r \vee u = s) \qquad (1)$$

The main functionality provided by the location space is the support to queries. Hence, every node s should eventually answer queries addressed to it:

$$r.\text{send}(in, s, t, u) \rightarrow \mathbf{F}(\exists x, v \cdot x.\text{send}(rep, r, s, t, v)) \qquad (2)$$

The usual procedure for proving interaction properties of actor communities is: (a) find a local invariant of the recipient; (b) show that the invariant guarantees that this actor will eventually become enabled for delivery/consumption of the message; (c) prove that the dispatch eventually leads to the message delivery and this guarantees the consumption, which in turn produces the outcome of the interaction. The reader should notice that these steps correspond to the application of the rules *COM* followed by *RESP*, both described in the Appendix, which capture the fairness requirements of reliable message passing and finite consumption delay present in the Actors model [1].

Returning to the location space example, it is not difficult to see that the invariant of each region tree node is:

$$Inv \equiv (\forall d \cdot reg[d] = me \lor void = \text{F}) \land (0 \le ans \le 3) \tag{3}$$

To prove this, first observe that Inv is a precondition for the occurrence of birth actions of RTN, *node* and *cnt*, in Axiom 1.1. Moreover, each local computation preserves this property, meaning that occurrences of *updt* do not change the logical value of the first conjunct while keeping *ans* unmodified, according to 1.2, and *inc* is only allowed to happen if the value of *ans* remains in the interval [0,3], from Axioms 1.3 and 1.10. An application of rule $SAFE$, also described in the Appendix, shows that Inv always holds.

Now, suppose that s, the recipient, is an empty node. Let p as in rule $RESP$ be $Inv \lor$ **init**. Since Inv is invariant, its disjunction with **init** is preserved by every local computation of the actor and therefore the first premise of that rule holds. In passing, notice that *updt* and *inc* cannot continuously happen since they are causally connected to the occurrence of *split* or *rep* and the axiom scheme 10. in the Appendix says that such actions can only happen one at a time. From Axiom 1.15, it is clear that p guarantees the eventual enableness for query consumption, thus the third premise in the aforementioned rule also holds. Making q in the same rule be the reply, we can see from Axioms 1.6 and 1.8 that it will eventually happen. A similar but simpler rationale can be used to show through the application of COM that once the query is dispatched, it will eventually be delivered. Chaining these results, we conclude a step in the verification of (2).

To complete the verification of (2), consider now those cases where the cell has already been split, and so answering a query may require creating auxiliary continuation actors. Verifying such cases of chained interaction requires the following derived rule [5]:

$$[WELL] \quad 1. \quad \dfrac{\mathbf{G}(P(x) \to \mathbf{F}(\exists y \cdot (y \preceq x \land P(y)) \lor Q))}{\mathbf{G}(\exists x \cdot P(x) \to \mathbf{F}Q)} \qquad \text{where } \preceq \text{ is a (rigid or variable) well-founded relation}$$

We usually take $P(x)$ as a message dispatch by an actor with mail address x and Q as the outcome of its consumption. To prove the non-trivial case of (2), we need to apply such rule twice, one to show that the query eventually reaches the correct region or that all the location space terminal nodes have been queried without matching the parameter of the message, and the other to show that the reply will proceed through the created continuations until it reaches the original actor. For the second case, e.g., the required well-founded relation can ben taken as $R_p(x,y) \equiv y.to = x \land y \ne p$. The anti-reflexivity of R_p derives from the use of Axiom 1.6 in rule $EXIT$, which shows that a node is prevented from being a continuation of itself, and since *to* is invariant, R_p indeed defines a well-founded relation. Therefore, we can apply $WELL$ and complete the proof.

Although we have omitted this detail in the informal proof above, the assumption (1) is fundamental to ensure that the answer to the query will be addressed to the correct actor, be it dispatched by the recipient itself or by a continuation actor. It is also important to mention that, formally, queries are not expected to come from within the community of region tree nodes, but from that of sensors. Due to the morphisms connecting these communities to each other, the complex properties of the system can be decomposed and translated into lemmas to be proved in a localised manner, using the axioms of just one theory presentation.

In a more realistic context, a reasonable assumption for our architecture would consider for instance that for every region there is a named server associated to an agent and dissociated from any sensor, which could be regarded as the meta-level actors [19] required to exist to support resource management activities, in our case the access to some location. The formal verification of properties of our architecture considering these assumptions appears elsewhere. Of course, (2) could be (re)used there and other properties would be verified almost in the same manner.

6   FINAL REMARKS

In this paper, we have presented an approach to the design of object-based mobile systems using a temporal logic specially tailored to the Actors model [1]. In [4] we investigated the requirements the model poses to the definition a logic and proposed a proof-theory which constrains the synchronous value-passing calculus of [7] to deal only with synchronous object creation and asynchronous message passing for named reconfigurable objects. It turns out that this logic can be used without any modification to capture mobility. Basically, our approach consists in annotating located objects with an additional attribute, containing a reference to location objects as in [8]. The advantage of approaching mobility in this extra-logical manner is that specifications can be defined in isolation to be subsequently combined and global proofs can be decomposed in lemmas to be locally verified much in the way we design any system using the same logic.

A few related work can be gathered in the literature. In [20] additional notation is suggested to treat mobility using the programming logic of UNITY [3]. During the refinement process, specifications are augmented with logical variables to handle time and action, which are built in here, and with concrete locations. As shown in the literature, there are many benefits in using referential location information instead [8, 12]. In addition, if mobility arises in a set of requirements, that approach would not be so effective: initial specifications are required before any mobility aspect can be considered.

In the process calculi literature, mobility has received a lot of attention, motivating the evolution of the static process configurations of SCCS [13] to the dynamic ones of the π-calculus [14]. We have shown here that many applications of the π-calculus can be treated using our logic. For instance, we can simulate recursion creating continuation actors and exchanging asynchronous messages; dynamic data structures can be represented as objects and so on. More importantly, mobility receives a rather different treatment as process in the π-calculus are modelled as terms and here objects are represented by theory presentations. This is due to different design decisions: while it is relatively easier to define notions of simulation and reduction for processes, it is possible to specify and reason about objects as first-class entities using the same more expressive language, which we feel more appropriate to represent the real world. The same may only be achieved by adjoining a modal or temporal logic to process calculi. It would be interesting to compare our logic to those presented in [15] in terms of expressive power.

As a case study, we presented here the design of a location management architecture for networks of mobile users and devices. We are currently investigating the occurrence of failures in this framework in order to access if our logic is still convenient for specifying fault-tolerant behaviour. An alternative direction for further work is to refine the proposed specifications to obtain a concrete implementation. Refinement theories for logics of actions and objects already exist [6, 11] and could be adapted to our case. A challenge in such an effort would be to achieve a compositional development process, where the refinement of components and interconnections in isolation always yield an implementation for the whole system.

Acknowledgements: The author would like to thank two anonymous referees for their valuable suggestions to improve an earlier version of this paper. This work has been financially supported by CNPq, the Brazilian National Research Council.

REFERENCES

[1] G. Agha. *Actors: A Model of Concurrent Computation in Distributed Systems.* MIT Press, 1986.

[2] G. Agha, I. A. Mason, S. Smith, and C. Talcott. A foundation for actor computation. *Journal of Functional Programming*, 7(1), 1997.

[3] K. M. Chandy and J. Misra. *Parallel Program Design, A Foundation.* Addison-Wesley Publishing Company, 1988.

[4] C. H. C. Duarte. Towards a proof-theoretic foundation for actor specification and verification. In P.-Y. Schobbens, editor, *Formal Models of Agents — Proc. Esprit WG Workshop ModelAge* (to appear), Lecture Notes in Artificial Intelligence. Springer-Verlag, 1997.

[5] E. A. Emerson. Temporal and modal logic. In J. Van Leeuwen, editor, *Handbook of Theoretical Computer Science*, pages 996–1072. North Holland, 1990.

[6] J. Fiadeiro and T. Maibaum. Sometimes "tomorrow" is "sometime": Action refinement in a temporal logic of objects. In D. Gabbay and H. Ohlbach, editors, *Temporal Logic*, volume 827 of *Lecture Notes in Artificial Inteligence.* Springer Verlag, 1994.

[7] J. Fiadeiro, C. Sernadas, T. Maibaum, and G. Saake. Proof-theoretic semantics of object-oriented specification constructs. In R. Meersman and W. Kent, editors, *Object-Oriented Database Analysis, Design and Construction.* North Holland, 1992.

[8] A. Harter and A. Hopper. A distributed location system for the active office. *IEEE Network*, 8(1):62–70, January 1994.

[9] C. B. Jones. *Systematic Software Development Using VDM.* Prentice Hall International, New York, 2nd edition, 1990.

[10] D. Lam, J. Jannink, D. C. Cox, and J. Widom. Modeling location management in personal communication systems. In *Proc. of Internatinal Conference on Universal Personal Communications (ICUPC'96)*, 1996.

[11] L. Lamport. The Temporal Logic of Actions. *ACM Transactions on Programming Languages and Systems*, 16(3):872–923, 1994.

[12] U. Leonhardt and J. Magee. Towards a general location service for mobile environments. In *Proc. 3rd International Workshop on Service in Distributed and Networked Environments (SDNE'96)*, June 1996.

[13] R. Milner. Calculi for synchrony and asynchrony. *Theoretical Computer Science*, 25:267–310, 1983.

[14] R. Milner, J. Parrow, and D. Walker. A calculus of mobile processes, I and II. *Information and Computation*, 100(1):1–40 and 41–77, September 1992.

[15] R. Milner, J. Parrow, and D. Walker. Modal logics for mobile processes. *Theoretical Computer Science*, 114(1):149–171, 1993.

[16] H. Samet. The quadtree and related hierarchical data structures. *ACM Computing Surveys*, 2(2):187–260, June 1984.

[17] J. M. Spivey. *The Z Notation: A Reference Manual.* International Series in Computer Science. Prentice-Hall, 1989.

[18] M. Spreitzer and M. Theimer. Architectural considerations for scalable, secure, mobile computing with location information. In *Proc. 14th International Conference on Distributed Computing Systems*, pages 29–38. IEEE Computer Society Press, June 1994.

[19] N. Venkatasubramanian and C. Talcott. A meta architecture for distributed resource management. In *Proc. Hawaii International Conference on System Sciences*, pages 124–133. IEEE Computer Society Press, January 1993..

[20] C. D. Wilcox and G.-C. Roman. Reasoning about places, times and actions in the presence of mobility. *IEEE Transactions on Software Engineering*, 22(4):225–247, April 1996.

APPENDIX

We provide here the set of axiom schemes and inference rules which make our logic suitable for specifying and reasoning about actor communities. These are meant to particularize the axiomatisation of the underlying many sorted first order branching time temporal logic with equality adopted. As in [7] we also require that variables are rigid, attributes have a functional time-dependent interpretation and actions denote atomic events. The reader is referred to [4] for further details.

Definition 7 (Axiom Schemes) Given an actor specification $\Phi = ((\Sigma, \mathcal{A}, \Gamma), \Psi)$, the following are logical axiom schemes for Φ-actors:

1. $\bigvee_{c \in \Gamma_c} \exists \vec{v_c} \cdot n.c(\vec{v_c}) \vee \bigwedge_{f \in \mathcal{A}} \forall \vec{v_f}, k \cdot n.f(\vec{v_f}) = k \rightarrow \mathbf{X}(n.f(\vec{v_f}) = k))$

2. $\bigwedge_{c \in \Gamma_{(e-e_b) \cup (I-I_b)}} \forall \vec{v_c} \cdot \mathbf{beg} \rightarrow \mathbf{G}(\neg n_1.\mathbf{init}) \vee \bigwedge_{n \in \vec{v}_{c_{add}} \cup \{n_2\}} n_1.Wait(n, \neg \mathbf{send}(c, n_2, \vec{v_c}))$

3. $\bigwedge_{c \in \Gamma_{I-I_b}} \forall \vec{v_c} \cdot \mathbf{beg} \rightarrow (\neg n.\mathbf{deliv}(c, \vec{v_c}))\mathbf{W}(n.\mathbf{init})$

4. $\bigwedge_{c \in \Gamma_{(I-I_b) \cup c}} \forall \vec{v_c} \cdot \mathbf{beg} \rightarrow (\neg n.c(\vec{v_c}))\mathbf{W}(n.\mathbf{init})$

5. $\bigwedge_{c \in \Gamma_{e_b \cup I_b}} \forall \vec{v_c} \cdot \mathbf{beg} \rightarrow \mathbf{G}(\neg n_1.\mathbf{init}) \vee \bigwedge_{n \in \vec{v}_{c_{add}} \cup \{n_2\}} n_1.Wait(n, \neg \mathbf{new}(c, n_2, \vec{v_c}))$

6. $\bigwedge_{c \in \Gamma_{I_b}} \mathbf{beg} \rightarrow \mathbf{G}(\exists n_1, n_2, \vec{v_c} \cdot \mathbf{E}(n_1.\mathbf{new}(c, n_2, \vec{v_c})))$

7. $\bigwedge_{c \in \Gamma_{I_b}} \forall \vec{v_c} \cdot (\exists n_1 \cdot n_1.\mathbf{new}(c, n_2, \vec{v_c}) \rightarrow \mathbf{X}(n_2.c(\vec{v_c}))) \wedge (n_2.c(\vec{v_c}) \xleftarrow{i}_{\mathbf{beg}} \exists n_1 \cdot n_1.\mathbf{new}(c, n_2, \vec{v_c}))$

8. $\bigwedge_{\substack{c,d \in \Gamma_{I_b} \\ d \neq c}} \forall \vec{v_c} \cdot n_1.\mathbf{new}(c, n_2, \vec{v_c}) \rightarrow \not\exists n_3, \vec{v_c}', \vec{v_d} \cdot ((n_3 \neq n_1 \vee \vec{v_c}' \neq \vec{v_c}) \wedge n_3.\mathbf{new}(c, n_2, \vec{v_c}')) \vee n_3.\mathbf{new}(d, n_2, \vec{v_d})$

9. $\bigwedge_{\substack{c,d \in \Gamma_{I-I_b} \\ d \neq c}} \forall \vec{v_c} \cdot n.\mathbf{deliv}(c, \vec{v_c}) \rightarrow \not\exists \vec{v_c}', \vec{v_d} \cdot (\vec{v_c}' \neq \vec{v_c} \wedge n.\mathbf{deliv}(c, \vec{v_c}')) \vee n.\mathbf{deliv}(d, \vec{v_d})$

10. $\bigwedge_{\substack{c,d \in \Gamma_{(I-I_b) \cup c} \\ c \neq d}} \forall \vec{v_c} \cdot n.c(\vec{v_c}) \rightarrow \not\exists \vec{v_c}', \vec{v_d} \cdot (\vec{v_c}' \neq \vec{v_c} \wedge n.c(\vec{v_c}')) \vee n.d(\vec{v_d})$

where:

$$Wait(n, g) \equiv (g)\mathbf{W}(\mathbf{init}) \wedge (g)\mathbf{W}(Acq(n))$$

$$Acq(n) \equiv \bigvee_{d \in \Gamma_{I-I_b}} \exists \vec{v_d} \cdot (\mathbf{deliv}(d, \vec{v_d}) \wedge n \in \vec{v_d}) \vee \bigvee_{d \in \Gamma_{I_b}} \exists \vec{v_d} \cdot (d(\vec{v_d}) \wedge n \in \vec{v_d}) \vee \bigvee_{d \in \Gamma_{e_b \cup I_b}} \exists \vec{v_d} \cdot \mathbf{new}(d, n, \vec{v_d})$$

The first scheme says that each actor has encapsulated state; only its local computations can change attribute values. The next four schemes say that either an actor is not created within a community or dispatch, delivery and consumption of messages plus local computations and requests for creation do happen before its birth. Notice that the second and fifth schemes are more liberal if the actor is never created but are more restrictive otherwise requiring actor names to become known first due to delivery of a message, the birth of the source or the creation of the target before they could be used in the task. The sixth scheme says that it is always possible to create some new actors and the seventh states that requests for creation and actual births are causally connected. The last three schemes constrain concurrency, i.e. that actors cannot be simultaneously created with the same name; that messages cannot be delivered in parallel to the same actor; and that messages and local computations cannot be processed at the same time; the last axiom being supplied only to simplify specification and reasoning.

Definition 8 (Rules of Inference) Given an actor specification $\Phi = ((\Sigma, \mathcal{A}, \Gamma), \Psi)$, the following are inference rules for deriving properties of existing Φ-actors, where each p, p' and q is an arbitrary formula over a single actor and n, n' and m are terms of sort **addr**:

$$
\begin{array}{l}
[EXIST] \quad \begin{array}{ll} 1. & p' \to n'.\mathbf{new}(d, m, \vec{v_d}) \\ 2. & p \to q \vee \bigvee_{c \in \Gamma_{l_b}} \exists \vec{v_c} \cdot n.\mathbf{new}(c, m, \vec{v_c}) \end{array} \\[2ex]
d \in \Gamma_{l_b} \quad \rule{6cm}{0.4pt} \\
\qquad\qquad\qquad p' \to \mathbf{XG}(p \to q)
\end{array}
$$

$$
\begin{array}{l}
[SAFE] \quad \begin{array}{ll} 1. & \bigwedge_{c \in \Gamma_{l_b}} \forall \vec{v_c} \cdot n.c(\vec{v_c}) \to q \\ 2. & \bigwedge_{c \in \Gamma_c} \forall \vec{v_c} \cdot n.c(\vec{v_c}) \wedge q \to \mathbf{X}q \end{array} \\[2ex]
\rule{4cm}{0.4pt} \\
\qquad\qquad\qquad \mathbf{G}q
\end{array}
\qquad
\begin{array}{l}
[INV] \quad \begin{array}{ll} 1. & \bigwedge_{c \in \Gamma_c} \forall \vec{v_c} \cdot n.c(\vec{v_c}) \wedge q \to \mathbf{X}q \end{array} \\[2ex]
\rule{4cm}{0.4pt} \\
\qquad\qquad q \to \mathbf{G}q
\end{array}
$$

$$
\begin{array}{l}
[RESP] \quad \begin{array}{ll} 1. & \bigwedge_{c \in \Gamma_c} \forall \vec{v_c} \cdot n.c(\vec{v_c}) \wedge p \to \mathbf{X}(p \vee n.d(\vec{v_d})) \\ 2. & n.d(\vec{v_d}) \to \mathbf{F}q \\ 3. & p \to \mathbf{FE}(n.d(\vec{v_d})) \end{array} \\[2ex]
d \in \Gamma_{l-l_b} \quad \rule{6cm}{0.4pt} \\
\qquad\qquad n.\mathbf{deliv}(d, \vec{v_d}) \to \mathbf{X}(\mathbf{F}p \to \mathbf{F}q)
\end{array}
$$

$$
\begin{array}{l}
[COM] \quad \begin{array}{ll} 1. & \bigwedge_{c \in \Gamma_c} \forall \vec{v_c} \cdot n.c(\vec{v_c}) \wedge p \to \mathbf{X}(p \vee n.\mathbf{deliv}(d, \vec{v_d})) \\ 2. & n.\mathbf{deliv}(d, \vec{v_d}) \to \mathbf{F}q \\ 3. & p \to \mathbf{FE}(n.\mathbf{deliv}(d, \vec{v_d})) \end{array} \\[2ex]
d \in \Gamma_{l-l_b} \quad \rule{7cm}{0.4pt} \\
\qquad\qquad n'.\mathbf{send}(d, n, \vec{v_d}) \to \mathbf{X}(\mathbf{F}p \to \mathbf{F}q)
\end{array}
$$

The rule *EXIST*, based on the fact that an actor name cannot be reused once it is given to some actor, guarantees a local safety property from the configuration of the actors in the environment. *SAFE* and *INV* are the usual rules for verifying local safety and invariance properties. Rules *COM* and *RESP* capture the fairness requirements for actors and are to verify that a delivery or a message consumption eventually happen due to an interaction if the actor ever becomes enabled for a similar task in the future.

BIOGRAPHY

Carlos H. C. Duarte received the BMath degree from the University of Juiz de Fora (1992) and the MSc degree from the Pontifical Catholic University of Rio de Janeiro (1994), both in Brazil. He is currently researching formal methods as a PhD candidate at the Department of Computing, Imperial College, London. From 1989 to 1992, he worked for the Brazilian Farming and Cattle Breeding Research Company (EMBRAPA). Since 1993, he has been an employee of the Brazilian National Bank of Social and Economic Development (BNDES). He is a member of the Brazilian Computing Society (SBC) and the IEEE Computer Society.

PART TWO

Concurrent OO Specification and Programming

5

CO-OPN/2: A Concurrent Object-Oriented Formalism

O. Biberstein[1,2], D. Buchs[2], N. Guelfi[2]
[1] *CUI, University of Geneva*
CH-1211 Genève, Switzerland, Olivier.Biberstein@cui.unige.ch

[2] *EPFL-DI-LGL*
CH-1015 Lausanne, fax: (4121)6935079, Didier.Buchs@di.epfl.ch

Abstract

In this article we present the concurrent object-oriented specification language CO-OPN/2 which extends the CO-OPN (Concurrent Object Oriented Petri Nets) formalism, destined to support the specification of large distributed systems. The CO-OPN/2 approach proposes a specification language, based on the object-oriented paradigm, which includes a fine description of true concurrent behavior.

This hybrid approach (model and property-oriented) allows for a description of the concurrent aspects through the use of high-level Petri nets which includes data structures expressed as algebraic abstract data types and a synchronization mechanism for building abstraction hierarchies. This latter notion is the concept which is used in application structuring.

Some nice properties of CO-OPN/2, such as the progressive refinement of specifications, allow for an incremental building of systems. In this article, we introduce CO-OPN/2 informally, by means of a typical example of distributed system, the *transit node*, in order to introduce each useful and innovative mechanism of the language.

Keywords

Formal specification language, object orientation, strong subtyping, concurrency, Petri nets, algebraic specification, refinement.

1 INTRODUCTION

For important applications, distributed processing provides the most general, flexible and evolutionary approach for computer processing.

However, distributed systems introduce several new considerations that must necessarily be taken into account. Interactions between concurrent components give rise to issues such as non-determinism, contention, and synchronization. This implies that particular attention must be paid to languages used to specify distributed systems requirements. In a formal and rigorous specification development, it is necessary to have a sound mathematically-based formalism allowing for the expression of all the particularities of distributed systems. Moreover, when we are faced with large problems it is also necessary to have complete structuring facilities. The objective of the specification phase is to clearly state which is the set of functionalities offered by the software. The ever-increasing complexity of software systems thus imposes a progressive procedure of adaptation based on abstraction, refinement and enrichment. It is preferable to have access to structuring primitives so as to effectively control this procedure.

We demonstrate in the following sections that constructing a formal specification using an incremental approach is a very efficient way of working. Indeed, the initial perception of the system to be built may be very vague. As the analysis and the simultaneous validation progress, the definition of the architecture, of the algorithms and of the associated data structures gradually improve and the final implementation may finally take form.

In our proposition, we have chosen object orientation as the structuring paradigm. We have defined a general language which may express both abstract and concrete aspects of systems, with emphasis on the description of structure, concurrency and data. This approach, called CO-OPN/2, is presented in this paper in an intuitive way and the description is oriented around expressiveness aspects and methodological considerations rather than around theoretical matters which are tackled in [5] and [3].

This article describes the context for the development of distributed system specifications, using refinement and the CO-OPN/2 language. This is illustrated by a comprehensive example: *the Transit Node*. Beginning with a very abstract view of the system we progressively introduce several of its concrete dimensions in order to show how to use the CO-OPN/2 specification language in the development of real distributed systems.

2 CO-OPN/2 PRINCIPLES

The two underlying formalisms of CO-OPN/2 are the algebraic specifications and the Petri nets which are combined in a way that is similar to algebraic nets [11]. The former is used to describe the data structures and the functional aspects of a system, while the latter serves to model its concurrent features.

However, both these formalisms are not suitable to specify "in the large". To compensate for the lack of structuring capabilities in the Petri nets, the object paradigm has been adopted. Thus, a system is considered as being a collection of independent entities which interact and collaborate together in order to accomplish the various tasks of the system.

In order to overcome some limitations of the first version of CO-OPN, CO-OPN/2 introduces some notions peculiar to object-orientation such as the notions of class, inheritance, and subtyping. For the sake of homogeneity regarding the notion of subtyping, order-sorted algebraic specifications [7] have been adopted for the description of the data structures.

Object and Class An object is considered as an independent entity composed of an internal state and which provides some services to the exterior. The only way to interact with an object is to ask for its services; the internal state is then protected against uncontrolled accesses. Our point of view is that encapsulation is an essential feature of object-orientation and there should be no way of violating it.

CO-OPN/2 defines an object as being an encapsulated algebraic net in which the places compose the internal state and the transitions model the concurrent events of the object. A place consists of a multi-set of algebraic values. The transitions are divided into two groups: the parameterized transitions, also called the methods, and the internal transitions. The former correspond to the services provided to the outside, while the latter compose the internal behaviors of an object. Contrary to the methods, the internal transitions are invisible to the exterior world and may be considered as being spontaneous events.

An important characteristic of the systems we want to consider is their potential dynamic evolution in terms of the number of objects they may include. Thus, the dynamic creation of objects is a major objective. A class describes all the components of a set of objects and is considered as an object template. Thus, all the objects of one class have the same structure.

Object Interaction In our approach, the interaction with an object is synchronous, although asynchronous communications may be simulated. Thus, when an object requires a service it asks to be synchronized with the method (parameterized transition) of the object provider. The synchronization policy is expressed by means of a synchronization expression, which may involve many partners joined by three synchronization operators (one for simultaneity, one for sequence, and one for alternative or non-determinism). For example, an object may simultaneously request two different services of two different partners, followed a service request to a third object.

Concurrency Intuitively, each object possesses its own behavior and concurrently evolves with the others. The Petri net model naturally introduces both inter-object and intra-object concurrency into CO-OPN/2 because the objects are not restricted to sequential processes.

The step semantics of CO-OPN/2 allows for the expression of true concurrency which is not the case of interleaving semantics. A set of method calls can be concurrently performed on the same object.

Object Identity Within CO-OPN/2 framework, each class instance has an identity, which is also called an object identifier, that may be used as a reference. Moreover, a type is explicitly associated with each class. Thus, each object identifier belongs to at least one type. An order-sorted algebra of object identifiers is constructed in order to reflect the subtyping relation which is established between the classes types, i.e. two carrier sets of object identifiers are related by inclusion if, and only if, the two corresponding types are related by subtyping. Since object identifiers are algebraic values, it is possible to define data structures which are built upon object identifiers, e.g. a stack or a queue of object identifiers. Obviously, the places of algebraic nets may contain object identifiers.

Inheritance and subtyping We believe that inheritance and subtyping are two different notions which are used for two different purposes. Inheritance is considered as being a syntactic mechanism which frees the specifier from the necessity of developing classes from scratch and is mainly employed to reuse parts of existing specifications. A class may inherit all the features of another and may also add some services or change the description of some services already defined.

Our subtyping relationship is based upon the strong version of the substitutability principle [8]. This principle implies that, in any context, any class instance of a type may be substituted for a class instance of its supertype while the behavior of the whole system remains unchanged. In other words, the instances of the sub-type have a strong semantic conformance relationship with the super-type definition. This conformance relationship is based, in CO-OPN/2, upon the bisimulation between the semantics of the super-type and the semantics of the sub-type restricted to the behavior of the super-type.

Both inheritance and subtyping relationships must be explicitly given but the respective hierarchies generated by these relationships do not necessarily coincide. Identifying both inheritance and subtyping hierarchies leads to several limitations as stated by Snyder [12] and America [1].

Syntactic aspects of CO-OPN/2 A CO-OPN/2 specification consists of two kinds of modules: the algebraic abstract data type modules and the class modules. Both kinds of modules are composed of three parts: a header, which includes the information about inheritance and genericity; an interface, which describes what is accessible when another module uses it; and a body, which primarily conceals the properties of the operation, the behavior and the state of the objects.

When a non-generic class is developed from scratch, its header comes down

to the keyword **Class**** followed by the name of the class. When a class is used only for classification, or when it will not be completely implemented and the creation of some of its instances makes no sense, we preface the keyword **Class** by the keyword **Abstract**. In the **Interface** section, the field **Use** declares all the modules used by the current class interface definition. The **Type** field declares the name of the instances type which is used whenever an object identity has to be defined. This field has been introduced in order to avoid any confusion between the name of a module or a class and the name of its type, especially in cases where inheritance and subtyping are required. Both names are often very similar but address two different concepts. Usually classes are used to dynamically create new instances but it is also possible to declare static instances by means of the **Objects** field. All the services provided by the class instances are declared within the field **Methods**. Note that the mix-fix and the applicative notation has been adopted for the profile of the methods. The final field **Creation** included in the interface section concerns the dynamic creation of the class instances. Within this field are listed the particular creation methods which create and initialize the objects; these methods may be used only once for a given object. A pre-defined creation method **create** is provided when the **Creation** field is empty or absent. The **Body** section includes a **Use** section and some internal or spontaneous transitions declared under the **Transitions** field as well as the attributes of the instances within the **Places** field. The **Initial** field describes the initial marking or the static initialization of each instance while the properties of the methods and the internal transitions are described by means of behavioral axioms within the **Axioms** field. It is necessary to recall that a transition (method or internal transition) may ask to be synchronized with other partners by means of a synchronization expression. The synchronization expressions are declared after the **with** keyword. The usual dot notation has been adopted and three synchronization operators have been provided: '//' for simultaneity '..' for sequence, '+' for alternative.
A behavioral axiom is established as follows

$$[\,Cond \Rightarrow\,]\,Event\,[\,\textbf{with}\,\,Sync\,]:Pre \rightarrow Post$$

where *Cond* is an optional condition imposed on the algebraic values involved in the axiom, *Event* is either an internal transition name or a method with parameters, and *Sync* is an optional synchronization expression. *Pre* and *Post*, respectively, correspond to what is consumed and what is produced in the different places composing the net. Finally, all the variables used within the body section are grouped together in the **Where** field.

---

*Specification in Figure 3 may help the reader to understand the meaning of the various keywords introduced in this subsection.

3   THE TRANSIT NODE CASE STUDY

This section introduces the CO-OPN/2 language by means of a well-known case study, the Transit Node (TNode for short). A transit node is a node in a communication system which receives messages on various input ports and routes them onwards through various output ports, according to some designated route. This case study was defined in the RACE project 2039, and one may find assorted specifications of the transit node in [10, 9]. Slight changes have been made to the TNodes definition in the RACE project and an informal description of the transit node is given in section 3.1.

Our aim is to build a heterogeneous distributed communication system (i.e. which presents some interconnected TNodes with wires including different kinds of TNodes and wires) and to progressively introduce all the syntactic and semantics aspects of the CO-OPN/2 language. We begin by describing a basic and abstract version of the TNodes and the wires. Therefore, this allows us to present the main ideas of the language and to progressively enrich and refine this version.

3.1   Informal Description of the Transit Node

The RACE project has defined a Transit Node as being a node in a communication system which receives messages on its input ports and then routes them onwards on its output ports according to some designated route. A TNode (Figure 1(b)) consists of N data input ports, M data output ports, one control input port and one control output port. Each port is serialized and represents a specific entity which is concurrent to all others. The node is "fair", i.e. all messages are likely to be treated equally when a selection must be made. Furthermore, all messages will eventually leave the node, or be placed within a collection of faulty messages. The control ports can be used to configure the transit node or to get statistical informations such as the number of faulty messages, the average transit time . . . .

A system of many interconnected TNodes is called a distributed communication system. The interconnection is realized by means of wires, each of which links the input port of one TNode to an output port of another TNode. Figure 1(a) shows a distributed communication system composed of three interconnected TNodes.

3.2   Presentation Overview

Since a TNode based distributed communication system is viewed as composed of a communication layer (the wires) and a transmission layer (the

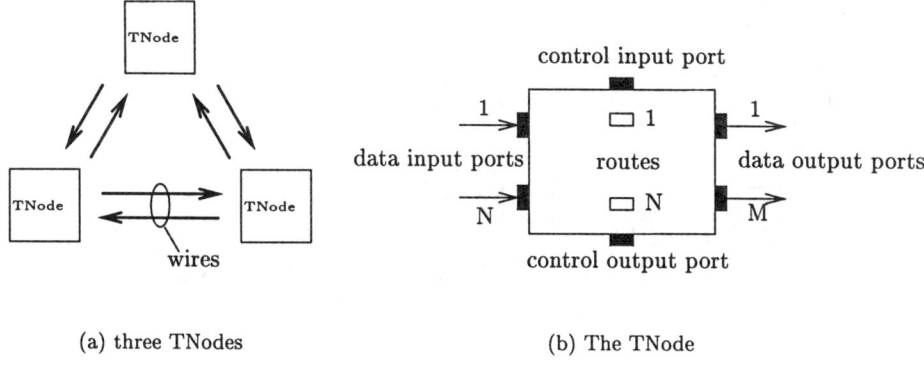

(a) three TNodes   (b) The TNode

Figure 1 TNode and Systems of TNodes

nodes), we have two parts in the specification corresponding to these two layers.

This informal introduction of CO-OPN/2 is divided into two stages, which are then split into different steps: In the first stage we begin to specify, on one hand, the data types to be used in the example by means of algebraic specifications, and on the other hand a basic version of the TNodes in which all TNodes can transmit messages to each other. The second step is concerned with the enrichment of the basic system with respect to the data input/output ports, while in the third step the routes are added. The fourth step introduces the input/output control ports and a communication system involving two different kinds of TNodes is finally built; At the second stage, a more realistic example of a heterogeneous distributed communication system is added. It takes into account the timing informations, the error detection algorithms and the heterogeneous distributed systems gateway. Due to the limited space available, this part is not described here, interrested readers should consult the full paper.

4 ADT, CLASS AND OBJECTS: BASIC TRANSIT NODE

This section explains the first step of the progressive presentation of the language. This first step, which introduces the basic TNodes and the basic communication layer, requires the definition of many fundamental concepts such as the concept of abstract data type, the concept of class and the notion of cooperation between class instances using synchronization expressions. All these concept definitions as well as a graphic representation of the classes are given in the following subsections.

A basic TNode consists in a simple, unstructured buffer in which messages arrive at an input port and remain there until they are routed onwards through

```
Adt Nat;
Interface
  Use Bool;
  Sorts nat;
  Generators
  0 : nat;
  succ _ : nat → nat;
  Operations
   _+_ : nat nat → nat;
   ... other operations ...
  Body
   Axioms
   0+n = n;
   succ(n)+m = succ(n+m);
   ... and their axioms ...
   Where n, m : nat;
End Nat;
```

```
Adt Bool;
Interface
  Sorts bool;
  Generators
  true, false: → bool;
  Operations
  not _    : bool → bool;
  _ and _ : bool bool → bool;
  Body
   Axioms
   not(true)   = false;
   not(false)  = true;
   true  and x = x;
   false and x = false;
   Where  x : bool;
End Bool;
```

```
Adt Message;
Interface
  Sorts message
End Message;
```

(a) Naturals   (b) Booleans   (c)   Mes-
                                                          sages

Figure 2 Abstract data types

the output port. At this step, a communication is viewed as the passing of
a message from one TNode to another by synchronizing the events "message
output" and "message input".

CO-OPN/2 specification: Adt modules are devoted to the data structures of the specification. In Figure 2, one can see the well-known booleans'
specification with its operations and their arity under the field **Operations**
and **Generators**. The underscore character indicates the position of respective arguments position.

The abstract data type of the natural numbers is given in Figure 2. Note
that the **Use** field is followed by the list of all the modules used in the module
itself, here the module Bool.

Throughout this paper our specifications will be as simple as possible so as
not to hide the main ideas which we wish to express. Hence, the messages of
the basic version will have no data and, shown in Figure 2, the abstract data
type module of messages only use the sort message. The first entities of our
example which have an internal state are the basic TNodes. Figure 3 shows
the specification of this kind of object. One may notice that the **Type** field is
followed by the type of the class. The operations supported on the reference
type t, its supertype sp and its subtype sb are _ = _ : t, t -> boolean
, new: t -> t, super t -> sp, sub t -> sb. Such type information
will be used to define reference variables with respect to the objects and to
the subtyping hierarchy.

The **Methods** field lists all the parameterized events which are visible
from the outside. Methods are used in a rendezvous like manner. The last
three fields remain in the **Body** section, ensuring encapsulation. In the field
Places , the attribute msg is a multi-set containing values of sort message.
The axioms in the field **Axioms** are quite simple and do not entail the formulation of conditions nor of synchronization expressions. They express the fact

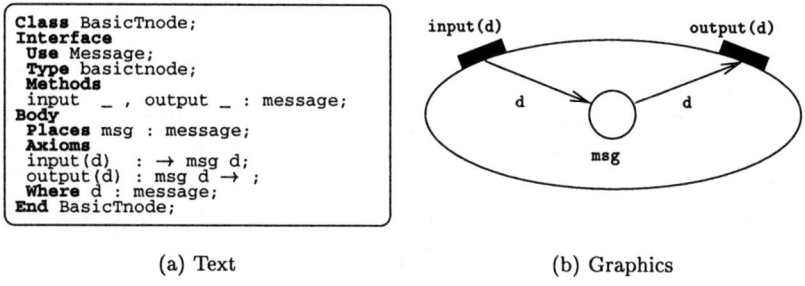

(a) Text   (b) Graphics

Figure 3 The basic TNodes

that the formal method argument d has been added to the place msg (input axiom) or has been removed (output axiom).

A natural but partial graphic representation of the classes is provided. The following conventions are used:

- the inside of the ellipses represents what is encapsulated,
- the black rectangles represent the methods,
- the white rectangles correspond to the internal transitions,
- the grey rectangles correspond to a special method taking care of the dynamic creation and the initialization of objects,
- the circles identify the places or the object attributes,
- the solid arrows indicate the data flow.

Figure 3(b) gives a graphical representation of the class BasicTnode. The communication layer consists of an instance of the class BasicWire given in Figure 4. This class must reflect that any TNode may exchange messages at any time with any other TNode. This is accomplished by means of the internal transition msg-passing, which asks for a simultaneous synchronization between two TNodes. The behavioral axiom should read "the event msg-passing behaves the same as both of the simultaneous external events, input and output, of both of the partners that can be identified by different tn1 and tn2". We use the usual dot notation for method call. Figure 4(b) gives the graphical illustration of the class BasicWire. Both of the dash arrows show the clientship between the communication layer and both the TNodes. However, the arrow direction does not express the data flow but the dependency relationship between the modules.

Discussions: We have defined a very simple version of the transit node which is reduced to a buffer with several undistinguished communication ports. Unbounded concurrency between the communication ports is modeled by the self-concurrency of the methods. The communication layer is defined at a very abstract level. All the wires are modeled using only one instance of

```
Class BasicWire;
Interface
  Use BasicTnode;
  Type basicwire;
Body
  Transitions msg-passing;
  Axioms
  (tn1=tn2)=false ⇒
    msg-passing with
      tn1.input(d) // tn2.output(d);
  Where
  d : message;  tn1, tn2 : basictnode;
End BasicWire;
```

(a) Text

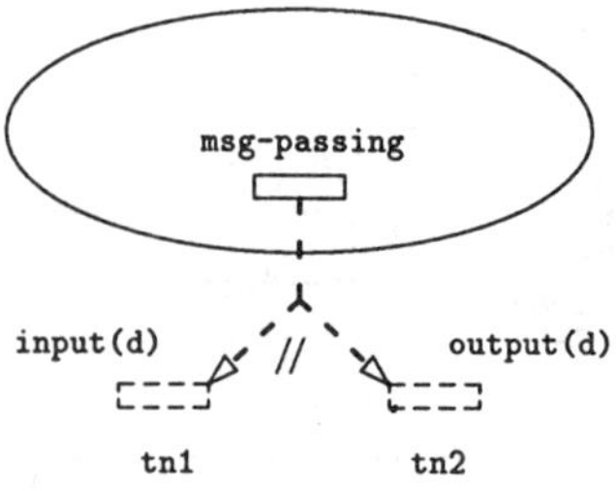

msg-passing

input(d)   //   output(d)

tn1   tn2

(b) Graphics

Figure 4 Basic communication layer class

```
Adt Link;
Interface
  Use Port, PortTnode;
  Sorts link;
  Generators <_ _ _ _> : porttnode port  porttnode port → link;
End Link;
```

Figure 5 An abstract data type of a link

the class BasicWire. All the communications are the result of the concurrent occurrences of the same event. The questions that must be considered are: how could we refine the TNode in order to have explicitly different communication ports instead of a global and abstract mechanism; how could we specify the definition of a distributed system topology.

5   ENRICHMENT AND OBJECT IDENTIFIERS: TNODES AND COMMUNICATION PORTS

This step covers of the enrichment of the basic TNodes and communication. The notion of port is added, i.e. a message arrives at one of the N input ports of a TNode and leaves the TNode onwards from one of the M output ports. Moreover, attribute of a wire contains the references of the two TNodes it connects.

A communication is viewed as the passing of a message through a wire which links one of the N input ports and one of the M output ports of two TNodes. A TNode can communicate to itself and several messages can be received or transmitted through the same port. However, the input and output ports are different even if they use the same port number.

CO-OPN/2 Specifications: Both classes BasicTnode and PortTnode are similar. The introduction of the notion of port induces, of course, a new profile for the methods input and output, as well as a new behavior. These changes are reflected in Figure 6. The specification of sort port (not detailed here) is built from the natural numbers and the constants M and N.

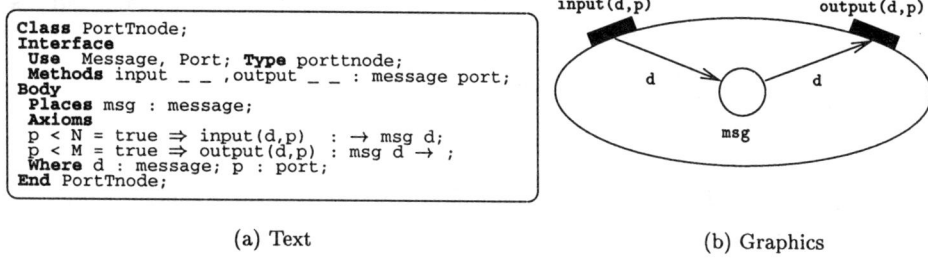

(a) Text

(b) Graphics

Figure 6 The TNodes with communication ports

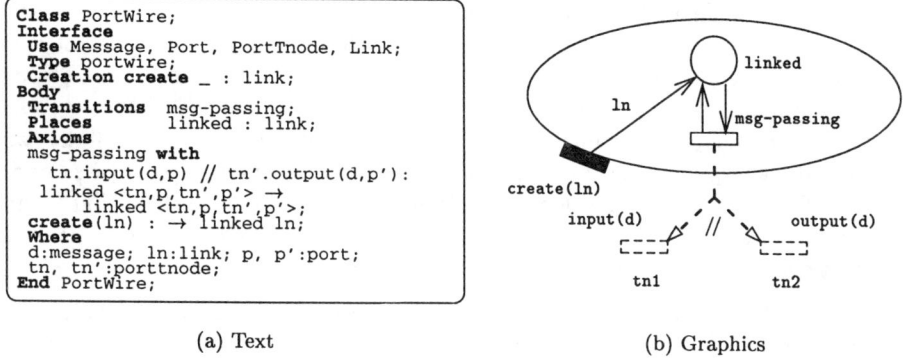

(a) Text

(b) Graphics

Figure 7 Wires between TNodes with ports

For the class `PortWire` we need the specification of a link between two TNodes. This is realized by the module `Link` in Figure 5. The sort `link` represents a cartesian product of four components, the two Tnodes references and their respective input/output ports. Figure 7 gives the specification of the class `PortWire`. One observes, on the one hand, the event `msg-passing` taking account of the attribute `linked` and, on the other hand, the special method **create**, which takes charge of the dynamic creation and initialization of the objects, and its parameter `ln` of sort `link` which is used to initialize the attribute. This initialization ensures that only one quadruplet is present in the place `linked`.

Discussions: We have reached a more realistic TNode definition in which each TNode has now several distinguished communication ports. The communications are made through wires which are instances of the class `PortWire` and which link explicitly input ports to output ports. Thus it is possible to define the network topology since wire parameters are given at the instance creation. A wire does not support simultaneous message passing and no routing technique is used inside a TNode. A port can be self-concurrently accessed inside a TNode. The questions that must be considered are how we could easily serialize the access to the TNode ports and add routing features, and what a specification of unreliable distributed systems which could lose or corrupt messages could be.

6 INHERITANCE: TRANSIT NODES AND ROUTING

This enrichment step has the objective of introducing the routing part of the TNode. Inheritance is used in order to derive the new classes corresponding to the TNode and to the wire template. Both preceding classes are reused, and some services as well as new attributes are added or redefined.

Each TNode includes information for the routing of its messages to the output ports. This information associates each input port with a set of permissible output ports. It is essential that the routing information of a TNode be modifiable if necessary.

Communication remains almost identical except that a wire now links two TNodes equipped with the new type of routing. The following sort `portset`, which is necessary to develop the specifications associated to the TNodes and the wires, are not detailed; it is obtained by instantiation of a generic algebraic sets specification with the module `Port` as actual parameter.

CO-OPN/2 specifications: New attribute and two new external events are inserted. These are the route definition method `routedef` and the internal events. Their respective behavioral axioms are added in the class `RouteTnode`. The class `RouteTnode` in Figure 8 is not developed from scratch, it uses the inheritance mechanism which allows to reuse or redefine some components. It must be noted that this example is rather poor in term of reuse, because only state definition is reused. In practice this form of reuse will be prohibited. The keyword **Redefine** expresses that the inherited method and its behavioral axioms are ignored and redefined by means of the new ones given in the derived class. This is the case of methods `input` and `output`. Moreover, the principle may also be applied to inherited attributes. Now routes are used to determine the message port destination. The new internal event `loss-msg` indicates that some messages can be lost in the TNode according to a given criterion. At this step the criterion is very abstract and represented by means of a function called `loss-crit`. We assume that this function is defined in the module `LossCriteria` which is not detailed here. The route wire is not changed, we only define a new class which inherits all the preceding properties.

Discussions: The routing and communication interferences have been easily obtained from a previous development step using CO-OPN/2 refinement techniques based on inheritance. The serialization of the input ports is performed through the routing mechanism, which is decided at message arrival, while no output order is imposed. The route initialization is implemented us-

```
Class RouteTnode;
Inherit PortTnode;
Rename porttnode → routetnode;
Interface
  Use Message, PortSet;
  Methods
  Redefine input  _ _ , output _ _ : message port;
    routedef _ _ : port portset;
  Redefine create _ : nat;
Body
  Use LossCriteria;
  Places routes : port portset;
  Redefine msg : message port;
  Transitions lossmsg;
  Axioms
  p < N = true ⇒ input(d,p) : routes p,ps → msg d,select(ps), routes p,ps;
  p < M = true ⇒ output(d,p) :  msg d,p → ;
  routedef(p,ps) :  routes p,ps' → routes p,ps;
  losscrit(d) = true ⇒ lossmsg :  msg d,p → ;
  create(0) :  → route p,ps;
  p < N=true ⇒ create(succ(p))with self.create(p): → route p,ps;
  Where n : nat; d : message; p : port; ps, ps' : portset;
End RouteTnode;
```

Figure 8 Transit nodes with routes

ing a recursive method call. The questions that must be considered now are how to refine the TNodes in order to implement the communication interferences using an error detection code, and how we can collect messages to be recovered by the system manager. Moreover, we would like to achieve this specification without modifying the definition of the communication layer.

7 INHERITANCE VERSUS SUBTYPING: DESIGN OF THE LOSS OF MESSAGES

The previous versions of the running example led to a simple distributed system composed of transit nodes of class RouteTnode linked using basic wires. The RouteTnode class inherits from PortTnode without any subtype relation. Thus, it is not possible to define a distributed system made of RouteTnode and PortTnode objects linked by means of the wire class because they are not substitutable. In this version, we define a new class of transit node CtrlTnode in order to illustrate the notion of subtype, which is used to specialize the previous TNode into a particular error detection mechanism.

These new transit nodes filter erroneous messages as before, from the point of view of the output port. But these messages are redirected to a special port which can be used to collect faulty messages. This new class of TNode allows the customers to recover corrupt messages thanks to an error detection algorithm such as the Hamming method.

We define a new class CtrlTnode which is a subtype of RouteTnode. As previously explained, the subtyping relation is a semantic constraint. In our case, it means that each CtrlTnode behaves at least as a RouteTnode, in order to satisfy the substitutability principle. This notion of subtyping is strongly related to the notion of observational equivalence of objects and de-

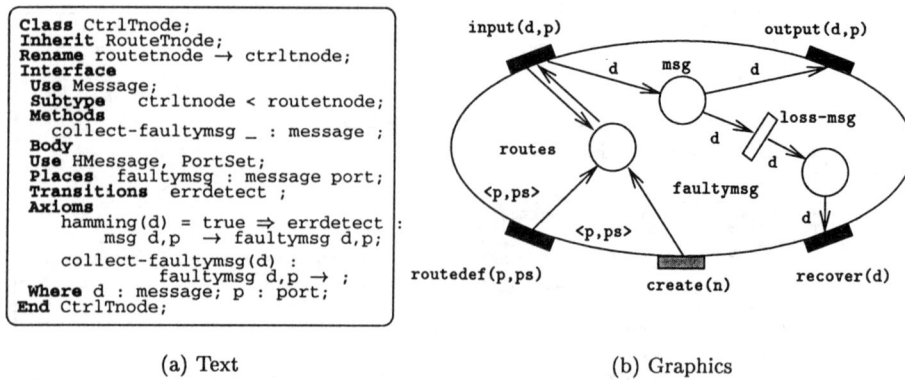

(a) Text   (b) Graphics

Figure 9 TNode with error message and recovering

pends on the observers, which are generally the methods defined in the object interface. In our case, it is easy to show that the general class RouteTnode has a greater number of behaviors, due to the weak constraint imposed to the loss function, than the particular hamming function.

In our case the objective is also to allow the use of polymorphic wires which can link RouteTnode as well as CtrlTnode because the message passing does not depend on the filtering semantics. Thus the communication layer is exactly the same as in the previous part.

The state and event part is enriched from the previous version in order to collect all the faulty messages. The figure 9 shows the CO-OPN/2 specification of this class.

A class CtrlTnode inherits from the class RouteTnode and is extended with the following components: The specification of the sort message is enriched by an operation hamming which tells for a data message whether an error has been detected or not; A place faultymsg contains values of the sort messageport which are the faulty messages; An internal transition errdetect selecting the erroneous messages; A method collect-faulty-msg with a parameter of sort message which can be called in order to collect the faulty messages.

Discussions: Refinement using subtyping is exploited in order to easily create new kinds of TNodes which are compatible with respect to their supertypes. This allows to modify the TNode without changing the communication layer.

8   CONCLUSION

In this paper we have presented a formal approach called CO-OPN/2 which can be used for the development of large distributed systems. This approach is flexible enough to be adapted to modeling of many different kinds of applications including the development of distributed algorithms , the development of parallel algorithms [4], etc. The transit node has been used to describe completely and informally particular aspects of the CO-OPN/2 language.

CO-OPN/2 supports modularity and the expression of various kinds of concurrency. In this paper we have shown how CO-OPN/2 can handle concurrency issues and classification requirements. The following major points of our approach have been presented: We express intra-object and inter-object true concurrency; We differentiate inheritance as a syntactic mechanism from subtyping which is a semantic principle; We give a way to express hierarchies of abstraction; We allow modular construction of specifications; We give a formal model of specification refinement.

Nevertheless, important research topics around CO-OPN/2 have not been presented here such as verification issues that can be found in [2] and validation issues that are presented in [6]. Complete CO-OPN/2 semantics is presented in [3].

With the progressive development of the Transit node, we have shown the possible ways of refining specifications, starting from an abstract description and progressively introducing concrete aspects. The validation of each step of the refinement process was possible through the use of prototyping or proving tools available in the SANDS environment [4]. Future studies will be conducted along three lines: Tools to support the new features of the CO-OPN/2 language for the already developed environment SANDS will be developed; A distribution model of CO-OPN/2 objects will be defined for heterogeneous systems with its operational semantics; Experiment will be performed with CO-OPN/2 on different practical case studiesespecially in CSCW.

Acknowledgments

We wish to thank S. Barbey, J. Hulaas and P. Racloz for their numerous and valuable comments during the evolution of this study. We also thanks J. Vachon for help with our English and all DeVa reviewers for careful reading of an earlier version of this paper. This work has been sponsored partially by the Esprit Long Term Research Project 20072 "Design for Validation" (DeVa) with the financial support of the OFES (Office Fédéral de l'Éducation et de la Science), and by the Swiss National Science Foundation project 2000.40583.94 "Formal Methods for Concurrent Systems".

REFERENCES

[1] Pierre America. Inheritance and subtyping in a parallel object-oriented language. In J. Bézivin, J.-M. Hullot, P. Cointe, and H. Lieberman, editors, *ECOOP'87: European conference on object-oriented program-*

ming: proceedings, volume 276 of *LNCS*, pages 234–242, Paris, France, June 1987. Springer-Verlag.

[2] Stéphane Barbey, Didier Buchs, and Cécile Péraire. A theory of specification-based testing for object-oriented software. In *Proceedings of EDCC2 (European Dependable Computing Conference)*, LNCS (Lecture Notes in Computer Science), Taormina, Italy, October 1996. Springer Verlag. Also available as Tech. Report (EPFL-DI-LGL No 96/163).

[3] Olivier Biberstein. *CO-OPN/2: An Object-Oriented Formalism for the Specification of Concurrent Systems*. PhD thesis, University of Geneva, April 1997. To appear.

[4] D. Buchs, P. Racloz, M. Buffo, J. Flumet, and E. Urland. Deriving parallel programs using sands tools. *Transputer Communications*, 3(1):23–32, January 1996.

[5] Didier Buchs and Nicolas Guelfi. CO-OPN: A concurrent object-oriented Petri nets approach for system specification. In M. Silva, editor, *12th International Conference on Application and Theory of Petri Nets*, pages 432–454, Aahrus, Denmark, June 1991.

[6] Didier Buchs and Jarle Hulaas. Evolutive prototyping of heterogeneous distributed systems using hierarchical algebraic petri nets. In *Proceedings of the International Conference on Systems, Man and Cybernetics*, Beijing, China, October 1996. IEEE.

[7] Joseph A. Goguen and José Meseguer. Order-sorted algebra I: Equational deduction for multiple inheritance, overloading, exceptions, and partial operations. *TCS: Theoretical Computer Science*, 105(2):217–273, 1992. (Also in technical report SRI-CSL-89-10 (1989), SRI International, Computer Science Lab).

[8] Barbara Liskov and Jeanette M. Wing. A behavioral notion of subtyping. *ACM Transaction on Programming Languages and Systems*, 16(6):1811–1841, November 1994.

[9] A. Mauboussin, H. Perdrix, M.Bidoit, M.C. Gaudel, and J. Hagelstein. From an erae requirements specification to a pluss algebraic specification: a case study. In J.A Bergstra and L.M.G. Feijs, editors, *Algebraic methods II: Theory, Tools and Applications*, volume 490 of *LNCS*. vrije University, Springer Verlag, 1991.

[10] S. Mauw and F. Wiedijk. Specification of the transit node in psfd. In J.A Bergstra and L.M.G. Feijs, editors, *Algebraic methods II: Theory, Tools and Applications*, volume 490 of *LNCS*. vrije University, Springer Verlag, 1991.

[11] Wolfgang Reisig. Petri nets and algebraic specifications. In *Theoretical Computer Science*, volume 80, pages 1–34. Elsevier, 1991.

[12] Alan Snyder. Encapsulation and inheritance in object-oriented programming languages. In *Proceedings OOPSLA '86*, volume 21, 11 of *ACM SIGPLAN Notices*, pages 38–45, November 1986.

6

Analysis for Concurrent Objects

Paolo Di Blasio
Dipartimento di Informatica e Automazione, Università di Roma Tre
Via della Vasca Navale 84, I-00146 Roma, Italy, ph. +39-6-55177052,
fax: +39-6-5573030, email: `diblasio@inf.uniroma3.it`

Kathleen Fisher
AT&T Labs Research
600 Mountain Ave, Murray Hill, New Jersey, 07974, USA, ph. +1-908-582-7584, fax: +1-908-582-4113, email: `kfisher@research.att.com`

Carolyn Talcott
Computer Science Department, Stanford University
Stanford, California, 94305, USA, email: `clt@sail.stanford.edu`

Abstract

We present a set-based control flow analysis for an imperative, concurrent object calculus extending the Fisher-Honsell-Mitchell functional object-oriented calculus described in (Fisher *et al.* 1993). The analysis is shown to be sound with respect to a transition-system semantics.

Keywords: concurrent object-oriented, control-flow analysis, soundness, prototype-based

1 INTRODUCTION

A well-designed set of computation abstractions for concurrent object-oriented programming, equipped with a well-developed formal semantics, is important because it (*i*) helps write simpler, easier to manage programs; (*ii*) facilitates correct and efficient implementations; (*iii*) provides a basis for tools to help programmers; and (*iv*) aides program specification and verification. As we move to the world of mobile code and agents, a clear formal semantics becomes even more crucial (cf.(Magic 1996, McGraw *et al.* 1996)).

A well-developed formal semantics includes: operational semantics with clear computational meaning; abstract compositional semantics upon which to base specifications and transformations; and sound formal analyses such as type systems, effect systems, and data and control flow analyses. Program analyses are particularly important and difficult to perform in languages which provide first-class objects and possibly first-class functions in addition to concurrency and dynamic process creation. For example, in the presence of first-class objects, the target of a method invocation is generally not known until run-time. Thus the code to be executed is dynamically determined. Similarly,

in the case of first-class functions, the function part of an application is determined at run-time. Hence analyses and program manipulations in this situation may require a control flow analysis. In the case of higher-order programs
such an analysis determines (an upper bound on) the set of functions (closures)
that may be returned as values at each program point. In the case of first-class
objects, a control flow analysis means determining the objects (and their possible method tables) that may be returned as values from each program point.
Extensive work has been done to develop techniques to perform control flow
analysis for sequential higher-order (*e.g.*, (Shivers 1991, Flanagan *et al.* 1995))
and object-oriented (*e.g.*, (Palsberg *et al.* 1991, Holzle *et al.* 1995)) languages.
Formal analysis methods for concurrent programs typically do not treat dynamic creation of processes or higher-order entities. Two exceptions are work
on CML (Concurrent ML) (Nielson *et al.* 1994, Colby 1995). Although not
explicitly object-oriented, CML exhibits many of the challenges of concurrent
object-oriented languages. Regarding formal analysis work that treats the
combination of concurrency and object-oriented programming, (Kobayashi *et
al.* 1995) presents a static analysis of communication which may be used to
reduce the cost of implementing message passing, and (Plevyak *et al.* 1995)
describes optimization techniques (*e.g.* inlining, state caching) based on type
inference information.

Our objective is to carry-out an in-depth semantic study of a representative concurrent object-oriented language that is simple enough to manage
but complete enough to exhibit the problems and challenges. The language
we have chosen is an imperative and concurrent extension of the Fisher-
Honsell-Mitchell functional object-oriented calculus described in (Fisher *et
al.* 1993). This calculus belongs to the family of so-called prototype-based
object-oriented languages, in which new objects are created from existing
ones via the inheritance primitives of object extension and method override.
Objects interact by message passing, which results in method invocation. Concurrency is achieved through a combination of asynchronous method invocation and the identification of objects and processes. Without asynchronous
communication the amount of concurrency would be static, and it would be
necessary to add some form of spawning primitive. We have also chosen to
directly support synchronous (rpc) method invocation because it is a commonly used pattern of interaction and it is not conveniently encodable via
asynchronous message passing, although translations of synchronous communication into asynchronous have been demonstrated (Amadio 1994, Mason
et al. 1997). Synchronization constraints, which describe restrictions on the
availability of methods, are specified through guards. We adopted guards because they provide one of the most natural ways to define synchronization
code and require minimal additional syntax.

In (Di Blasio *et al.* 1996) we define an operational semantics for our language
and develop a static analysis for it that includes a type inference system to
detect *message-not-understood* errors and an effect system to guarantee that

synchronization code is side-effect free. We prove the soundness of the type and effect system with respect to the operational semantics. In this paper, we describe a control-flow analysis technique for our language. In developing our analysis, we extend the approach outlined in (Flanagan *et al.* 1995) by Flanagan and Felleisen for a fragment of the higher-order language Scheme. We consider this work as a first step towards the application of formal techniques for program optimization to concurrent object-oriented languages. Along these lines, we show some examples of applying these techniques. Much of this material is based on the Ph.D. thesis of one of the authors (Di Blasio 1997), where proofs are presented in detail.

2 EXAMPLES

Eager invocation. To provide some intuition for programming in our calculus, we describe how to encode *eager invocation*, a communication mechanism that is a compromise between synchronous and asynchronous communication. Eager invocation increases concurrent activity with respect to synchronous communication because like asynchronous communication, sender objects do not block when they send messages. Unlike the asynchronous case, however, eager invocation allows the result of an invocation to be returned to the sender, by means of a *future variable*. When the sender needs a result, it accesses the relevant future variable. If the result has not yet been stored in the future variable, the sender blocks and waits for it. Future variables are encoded as objects as follows:

$$
\begin{array}{rll}
\textit{future-var} \;\; = \;\; \langle \;\; & get & = \;\; when(\lambda self.\, false) \;\; \lambda self.\, \lambda arg.\, c, \\
& set & = \;\; \lambda self.\, \lambda arg. \langle self \leftarrow get = \lambda s.\, \lambda a.\, arg \, \rangle; \\
& & \qquad \langle self \leftarrow set = when(\lambda s.\, false) \;\; \lambda s.\, \lambda a.\, nil \, \rangle; \\
& & \quad nil \;\; \rangle
\end{array}
$$

The above code describes a concurrent object with two methods: *get* and *set*. Formally a method definition has two parts: a guard of the form $when(\lambda s.\, e_g)$ and a body $\lambda s.\, e_b$, with e_b of function type. In both cases the variable s is called the *self* variable since it is bound to the object executing the method. We omit the guard if it is the constant *true*. When a method is invoked, first the guard is evaluated, then, if it returns true, the body is executed. The *get* method represents the storage of the future variable. Its guard $(when(\lambda self.\, false))$ insures that when the object is created the value contained in the future variable cannot be read. Constant c is an initial value of the proper type. The *set* method is immediately available when the future variable is created. In fact its guard (which is omitted for convenience) is constantly *true*. When invoked it stores the received value *arg* by modifying the *get* method and modifies itself so that it can no longer be invoked. The new *get* method is now available because its (omitted) guard is true, and simply returns the stored value *arg*. Expressions such as $\langle self \leftarrow get = \lambda s.\, \lambda a.\, arg \, \rangle$ are method override expressions. An override executed by an object on itself (called self-inflicted) is a method update and is the mechanism for state change.

As an example of the use of eager invocation, let us consider a program fragment in which we read values from a buffer via a future variable.

$$\text{let} \quad f = \textit{future-var}, \; b = \textit{buffer}$$
$$\text{in} \quad b \Leftarrow_a \textit{read}(f); \ldots \textit{code} \ldots \text{let} \; x = f \Leftarrow_s \textit{get in} \; \ldots$$
where
$$\textit{buffer} = \quad \langle \quad \textit{read} \quad = \quad \textit{when}(\lambda \textit{self.} \ldots) \, \lambda \textit{self.} \, \lambda a. \ldots . a \Leftarrow_a \textit{set}(v) \ldots,$$
$$\ldots \qquad \rangle$$

The process which executes this program fragment first invokes the buffer's *read* method with the future variable f as an argument. This invocation is asynchronous ($\Leftarrow_a$), so the *code* part of the program can be executed concurrently with the *read* operation in the buffer object. When the program needs the value from the buffer, it accesses the future variable f by synchronously invoking ($\Leftarrow_s$) f's *get* method. If the buffer has not yet stored the result (v) in the future variable, then (and only then) the program blocks to wait for it.

Control flow analysis. To provide some intuition about the program analysis we are interested in, let us consider the following fragment of code. For brevity, we omit guards.

$$\text{let} \quad o_1 = \quad \langle m_1 = \lambda s_1. \, e_1, \; m_2 = \lambda s_2. \, \lambda a_2. \, s_2 \Leftarrow_a m_1(\_) \rangle,$$
$$o_2 = \quad \langle o_1 \leftarrow m_2 = \lambda s_3. \, e_3 \rangle,$$
$$f = \quad \lambda x. \, x \Leftarrow_a m_2(\_)$$
$$\text{in} \quad f \, o_1;$$

Our goal is to approximate the values that variables may assume at runtime. With this aim, set-based analysis produces two finite tables: a *set environment* $\mathcal{H}$ and a *set object environment* $\mathcal{A}$.

A *set environment* maps each program variable x to a set of abstract values, $\mathcal{H}(x)$, which approximates the values that that program variable may assume. In the example, variable o_1 assumes as values the addresses of the objects created at site o_1. We statically approximate these object addresses by the variable o_1 itself, returning $\mathcal{H}(o_1) = \{o_1\}$. At the site denoted by o_2, we have a method override operation. The semantics of method override depends on whether the operation is *self-inflicted* or not. We say that an operation is self-inflicted if the target object is the same as the object that invokes the operation. If $\langle o_1 \leftarrow m_2 = \lambda s_3. \, e_3 \rangle$ is self-inflicted, we replace o_1's m_2-method with the new one and return the modified object. If it is not self-inflicted, we return a new object obtained by overriding o_1's m_2-method. In general, we do not know statically if the operation is self-inflicted or not; consequently, the values that o_2 may assume are the addresses of objects created at sites o_1 and o_2. Thus our analysis returns $\mathcal{H}(o_2) = \{o_1, o_2\}$. Finally, variable x gets the values of f's actual parameter o_1; hence, $\mathcal{H}(x) = \mathcal{H}(o_1) = \{o_1\}$.

A *set object environment* maps each program variable that denotes a possible object creation site (*e.g.* o_1, o_2) to a set of abstract method tables (denoted by $\mathcal{A}(o_1)$, $\mathcal{A}(o_2)$). Each such set approximates the method tables that objects created at that site can have at run-time. In the example, objects created at site o_1 may have abstract method tables $\{m_1 = \lambda s_1. e_1, m_2 = \lambda s_2. \lambda a_2. s_2 \Leftarrow_a m_1(\_)\}$ and $\{m_1 = \lambda s_1. e_1, m_2 = \lambda s_3. e_3\}$ (the second one if the method override in line 2 is self-inflicted). Objects created at site o_2 may have abstract method table $\{m_1 = \lambda s_1. e_1, m_2 = \lambda s_3. e_3\}$ (if the method override is non-self-inflicted).

This analysis allows us to produce a control flow graph of the program. In fact we know the code possibly invoked at any function or method call site. For example, we know that the method tables of the objects that x may refer to are the ones associated with the objects created at site o_1. Thus the bodies possibly invoked by $x \Leftarrow_a m_2(\_)$ are $\lambda s_2. \lambda a_2. s_2 \Leftarrow_a m_1(\_)$ and $\lambda s_3. e_3$.

3 THE LANGUAGE AND ITS A-NORMAL FORM

3.1 Language Syntax

The syntax of our language is given by the following grammar.

Expressions:

$$e \in Exp ::= \quad x \mid c \mid \lambda x. e \mid e_1 e_2$$
$$\mid \langle\rangle \mid \langle e_1 \longleftrightarrow m = when(e_2)e_3 \rangle \mid \langle e_1 \leftarrow m = when(e_2)e_3 \rangle$$
$$\mid e_1 \Leftarrow_a m(e_2) \mid e_1 \Leftarrow_s m(e_2)$$

In the definition of *Exp*, x is a variable, c is a constant symbol, $\lambda x. e$ is a lambda abstraction, and $e_1 e_2$ is function application. The remaining syntactic forms are the object primitives. Expression $\langle\rangle$ creates an empty object while $\langle e_1 \longleftrightarrow m = when(e_2)e_3 \rangle$ returns a new object obtained by extending e_1 with a new method m having guard e_2 and body e_3. If $\langle e_1 \leftarrow m = when(e_2)e_3 \rangle$ is self-inflicted, we replace e_1's m-method with the new one. If it is not self-inflicted, we return a new object obtained by overriding e_1's m-method. Note that non-self-inflicted extension and override operations allow us to support width and depth inheritance respectively. Expression $e_1 \Leftarrow_a m(e_2)$ sends message m asynchronously with argument e_2 to e_1. The corresponding synchronous invocation is $e_1 \Leftarrow_s m(e_2)$. λ is a binding construct and *let* abbreviates lambda application as usual. An operational semantics for our language is given in (Di Blasio *et al.* 1996).

In the rest of the paper we use the following meta-notations: $\Leftarrow$ stands for both $\Leftarrow_a$ and $\Leftarrow_s$, and $\leftarrow\!\circ$ stands for both $\longleftrightarrow$ and $\leftarrow$. We also write $\langle m_1 = when(e_1)e_1', \ldots, m_k = when(e_k)e_k' \rangle$ for $\langle \ldots \langle\langle\rangle \longleftrightarrow m_1 = when(e_1)e_1' \rangle \ldots \longleftrightarrow m_k = when(e_k)e_k' \rangle$.

3.2 A-Normal Form

Performing program analysis on source code can be a complex task, particularly in a calculus like ours in which functions and objects are first-class data and expressions can be arbitrarily nested. Consider, for example, the following code fragment.

$$let\ f = h\,y\ in \quad let\ x = ((f\,g) \Leftarrow_s m(v)) \Leftarrow_s n(w)\ in\ e \qquad (*)$$

To gather information about the values that variable x may assume, we first have to locate the subexpressions in the expression defining x and give each of them a label: l_1 for $f\,g$ and l_2 for $l_1 \Leftarrow_s m(v)$. Then we associate the labels with intermediate values. In other words, we associate l_1 with the values resulting from the function application $(f\,g)$ and l_2 with the values resulting from the invocation of m. More generally, we need a mechanism for referring to program points to talk about the corresponding intermediate values. One approach is to modify the syntax by labelling each subexpression (cf. (Nielson *et al.* 1997)). The approach we take is to transform programs to equivalent programs in A-normal form (Flanagan *et al.* 1995, Flanagan *et al.* 1993). In this transformation, a variable is introduced for each subexpression while preserving the execution order. For example, an A-normal form of the code fragment $(*)$ is

$$let\ f = h\,y\ in$$
$$\quad let\ l_1 = f\,g\ in$$
$$\quad\quad let\ l_2 = l_1 \Leftarrow_s m(v)\ in$$
$$\quad\quad\quad let\ x = l_2 \Leftarrow_s n(w)\ in\ \ M \qquad \text{where } M \text{ is an A-normal-form of } e$$

The syntax of the A-normal-form (ANF) expressions for our language is given by the following grammar.

ANF Expressions:

$$
\begin{aligned}
M, N \in \Lambda_a \ &::= \ s \mid let\ x = r\ in\ M \mid let\ x = N\ in\ M \\
r \ &::= \ c \mid \lambda y.\,N \mid s\,s' \mid s \Leftarrow m(s') \mid \langle\rangle \mid \langle s \leftarrow\!\bullet\ m = when(s')s'' \rangle \\
s \in Sval \ &::= \ x \mid l
\end{aligned}
$$

Intuitively, an ANF expression can be either a simple value s or a *let* expression. Simple values are either variables x or locations l. We introduce locations into the syntax so that we may express computation syntactically. There exists a simple procedure, that we omit here for space considerations, that translates an expression $e \in Exp$ to its corresponding ANF (Di Blasio 1997). Intuitively, an expression e, which can be uniquely decomposed as $C[e_{rdx}]$ for some context C and redex e_{rdx}, is equivalent to the expression $let\ x = e_{rdx}\ in\ M$, where M is the ANF of $C[x]$.

In the ANF syntax we introduce the more general form $let\ x = N\ in\ M$ to guarantee that terms resulting from the reduction of function applications ($let\ x = ss'\ in\ M$) will be in ANF (see Section 4.1). $FV[M]$ denotes the set of *free variables* in M. A term M is *closed* if it contains no free variables and *static* if it contains no locations. Finally, *programs*, which we will denote by meta-variable P, are closed static expressions.

Throughout the rest of the paper, we assume that all variables in an ANF term bound by either λ or *let* are distinct. We will use meta-variables L, M, N, and P to denote ANF terms arising via translation from programs typeable in the source language using the type inference system of (Di Blasio *et al.* 1996).

4 OPERATIONAL SEMANTICS

We formalize the operational semantics of ANF terms as a transition relation on configurations in the same style as for the original language (Di Blasio *et al.* 1996). A *configuration* $\langle\!\langle\ H\ |\ \alpha\ |\ \mu\ \rangle\!\rangle$ (meta-variable g ranges over configurations) consists of a *global heap* H in which the values associated with variables are allocated, an *object soup* α, containing all created objects, and a collection of *pending asynchronous messages* μ.

A heap H is a finite map from locations l to heap values h. Heap values include constants, c, functions, $\lambda x.\,M$, and object addresses, a. The set of locations *Loc* is partitioned into infinite subsets of subscripted locations Loc_x, one partition for each variable x. Every value assigned to x is allocated at a new location l_x taken from the set Loc_x.

An *object* is represented at run-time as a triple $(a, \eta, [S])$, where a is the object's *address*, η is its *method table*, and S its *state*. Similarly to the set of locations, we partition the set of object addresses *ObjAddr* into infinite subsets $ObjAddr_x$, one for each variable x. We will see the reason for this partitioning in Section 5. Method tables are finite functions from method names to pairs of locations. These locations store the guard and method body functions, respectively. The state S can be either idle $([I])$ or busy $([M])$, in which case the expression M represents the remaining computation. An object passes from the idle to a busy state in response to either a synchronous or an asynchronous method invocation. At the end of the resulting computation, it returns to the idle state.

Definition 1 (Initial Configurations) *The initial configuration corresponding to a program, P, is $g_P = \langle\!\langle\ H_0\ |\ \alpha_0\ |\ \mu_0\ \rangle\!\rangle$ where H_0 is empty, $\alpha_0 = (main, \eta_\emptyset, [P])$, where $\eta_\emptyset$ is the empty method table, and μ_0 is an empty collection of messages.*

The following grammar defines ANF redexes (M_{rdx}) and reduction contexts (E). We have the unique decomposition property – each ANF expression is either a simple value or decomposes uniquely in the form $E[M_{rdx}]$.

ANF Redexes (M_{rdx}) and Reduction contexts (E):

$$
\begin{aligned}
M_{rdx} &\quad ::= \quad let\ x = s\ in\ M \mid let\ x = r\ in\ M \\
E &\quad ::= \quad [\,] \mid let\ x = E\ in\ M
\end{aligned}
$$

4.1 Reduction Rules

We describe in detail the transition rules for some expressions to illustrate the key concepts of the transition system. The complete set of rules can be found in (Di Blasio 1997).

Function application (*apply*). The heap value at location l_y is a function. The transition stores the value of the actual parameter $H(l_z)$ at a new location

l_w in the heap and replaces all occurrences of w within N by l_w. The function new_w takes a heap and returns a new location from Loc_w.

$$\left\langle\!\left\langle\; H \mid \alpha, (a, \eta, [E[let\ x = l_y l_z\ in\ M]]) \mid \mu \;\right\rangle\!\right\rangle \longmapsto$$

$$\left\langle\!\left\langle\; H \cup \{l_w = H(l_z)\} \mid \alpha, (a, \eta, [E[let\ x = N[w \leftarrow l_w]\ in\ M]]) \mid \mu \;\right\rangle\!\right\rangle$$

where $H(l_y) = \lambda w.\, N$, and $l_w = new_w(H)$.

Empty object creation ($\langle\rangle$). A new object with empty method table and idle state is created. Its address is subscripted by the name of the variable x that marks its creation site. The function $new\text{-}a_x$ takes an object soup and return a fresh object address from $ObjAddr_x$. The value a_x is stored at a new location l_x in the heap.

$$\left\langle\!\left\langle\; H \mid \alpha, (a, \eta, [E[let\ x = \langle\rangle\ in\ M]]) \mid \mu \;\right\rangle\!\right\rangle \longmapsto$$

$$\left\langle\!\left\langle\; H \cup \{l_x = a_x\} \mid \alpha, (a, \eta, [E[M[x \leftarrow l_x]]]), (a_x, \eta_\emptyset, [I]) \mid \mu \;\right\rangle\!\right\rangle$$

where $l_x = new_x(H)$, $a_x = new\text{-}a_x(\alpha, (a, \eta, \ldots))$.

Self-inflicted, synchronous method invocation ($\Leftarrow_s \text{-}self$). Since method invocation involves a double application, we might naively reduce the expression $let\ x = l_y \Leftarrow_s m(l_z)\ in\ M$ to $let\ x = l_b\, l_y\, l_z\ in\ M$. Unfortunately, this expression is not in A-normal form. Thus we need to normalize it, a process which produces expression M_1. The normalization is achieved by introducing two fresh variables n_2 and n_3.

$$\left\langle\!\left\langle\; H \mid \alpha, (a, \eta, [E[let\ x = l_y \Leftarrow_s m(l_z)\ in\ M]]) \mid \mu \;\right\rangle\!\right\rangle \longmapsto$$

$$\left\langle\!\left\langle\; H \mid \alpha, (a, \eta, [E[M_1]]) \mid \mu \;\right\rangle\!\right\rangle$$

where $H(l_y) = a$, $\eta(m) = (l_g, l_b)$ and
$M_1 \equiv let\ n_2 = l_b l_y\ in \quad let\ n_3 = n_2\, l_z\ in \quad let\ x = n_3\ in\ M$

5   SET-BASED ANALYSIS

Set-based analysis (Heintze 1992, Flanagan *et al.* 1995) is a program analysis technique which provides static information about the values that a program variable may assume at run-time. Since calculating this information exactly is undecidable, set-based analysis computes for each program variable x a finite set of 'abstract values', which collectively represent an upper bound on the set of values that x may assume at run-time.

In this section, we adapt and extend the set-based analysis of (Flanagan *et al.* 1995) for Scheme to our calculus. The fact that our calculus permits shared mutable objects requires an elaboration of their analysis. In particular, although Flanagan and Felleisen allow assignment to variables, they do not treat reference cells as first-class values. Hence, they do not address imperative features such as memory, object addresses, and the alias problem. To remedy this deficiency, our analysis produces two finite approximations instead of just

Run-time

$$\cdots \qquad \cdots$$

$$\text{H:} \quad \frac{(l_x, h)}{(l'_x, h')} \qquad \alpha: \quad \frac{(a_y, \eta, S)}{(a'_y, \eta', S')} \qquad \eta: \quad \frac{\cdots}{(m, l_g, l_b)}$$

$$\cdots \qquad \cdots \qquad \cdots$$

$$h \in Hval ::= c \mid \lambda x. M \mid a_y \qquad \text{(Heap values)}$$

Abstraction

$$\cdots \qquad \cdots \qquad \cdots$$

$$\mathcal{H}: \quad \frac{(x, \{\ldots, \hat{h}, \hat{h}', \ldots\})}{} \qquad \mathcal{A}: \quad \frac{(y, \{\ldots, \hat{\eta}, \hat{\eta}' \ldots\})}{} \qquad \hat{\eta}: \quad \frac{(m, g, b)}{}$$

$$\cdots \qquad \cdots \qquad \cdots$$

$$
\begin{array}{llll}
\hat{v} & \in & \widehat{Val} & ::= \quad \hat{h} \mid \hat{\eta} & \text{(Ab. values)} \\
\hat{h} & \in & \widehat{Hval} & ::= \quad c \mid \lambda x. M \mid y & \text{(Ab. heap values)} \\
\hat{\eta} & \in & \widehat{MeTable} & ::= \quad \{m_1 = (g_1, b_1), \ldots, m_n = (g_n, b_n)\} & \text{(Ab. method tables)}
\end{array}
$$

Figure 1 Static approximations of run-time values

one: a *set environment* $\mathcal{H}$ and a *set object environment* $\mathcal{A}$ (see Figure 1). *Set environment* $\mathcal{H}$ maps each program variable to a set of abstract heap values. This set approximates the values that program variables may assume during an execution. $\mathcal{H}$ is an abstraction of the global heap H: locations subscripted by x (*e.g.* l_x, l'_x) are statically represented by variable x, while heap values (*e.g.* h, h') are represented by abstract heap values (*e.g.* $\hat{h}$, $\hat{h}'$), defined in Figure 1. Environment $\mathcal{H}$ is a valid approximation of heap H if every value h that H associates with a location l_x may be obtained from an abstract value $(\hat{h})$ in $\mathcal{H}(x)$ by substituting appropriate locations for free variables in $\hat{h}$.

Set object environment $\mathcal{A}$ maps each program variable that potentially denotes an object creation site (*i.e.*, variables that label empty object, method override, and extension expressions) to a set of abstract method tables. Intuitively, the set of abstract method tables for a given object creation site approximates the method tables that objects created at that site may have at run-time. An object environment $\mathcal{A}$ is an abstraction of an object soup α: addresses of objects created at site y (*e.g.* a_y, a'_y) are represented by variable y, while method tables (*e.g.* η, η') are represented by abstract method tables (*e.g.* $\hat{\eta}$, $\hat{\eta}'$), which are defined in Figure 1. $\mathcal{A}$ is a valid approximation of α if every method table η that objects created at site y assume at run-time is obtained from an abstract method table in $\mathcal{A}(y)$ by substituting appropriate locations for free variables in the abstract method table. Method tables are approximated by abstract method tables in which the locations containing the guard and method body functions are represented by the corresponding abstract locations, *e.g.*, g for l_g and b for l_b .

To formalize our analysis, we need to introduce a bit more notation. Given a program P, let Var_P be the set of variables occurring in P. Let Var_O be the set of variables labelling possible object creation points occurring in P. Let $\widehat{Hval}_P$ be the set of abstract heap values of P. Let $\mathcal{P}_{fin}$ be the finite power-set constructor, and let $\alpha_{mt}(a)$ denote the method table of object a in α.

We now define the environments $\mathcal{H}$, $\mathcal{A}$ for P and the relation $P \models \mathcal{H}, \mathcal{A}$, which establishes the conditions under which $\mathcal{H}, \mathcal{A}$ are valid for P. In particular, the relation $P \models \mathcal{H}, \mathcal{A}$ holds if $\mathcal{H}, \mathcal{A}$ are valid approximations of every heap, object soup combination H, α produced during any execution of P.

Definition 2 *Let P be a program.*

- *A set environment for P is a mapping $\mathcal{H}:$ $Var_P \to \mathcal{P}_{fin}(\widehat{Hval}_P)$.*
- *A set object environment for P is a mapping $\mathcal{A}:$ $Var_O \to \mathcal{P}_{fin}(\widehat{MeTable})$.*
- *$P \models \mathcal{H}, \mathcal{A}$ holds if $g_P \longmapsto^* \left\langle\!\left\langle\, H \mid \alpha \mid \mu \,\right\rangle\!\right\rangle$ implies $H, \alpha \models \mathcal{H}, \mathcal{A}$.*
- *$H, \alpha \models \mathcal{H}, \mathcal{A}$ holds if for all bindings $(l_x = h) \in H$*

$$\begin{cases} z \in \mathcal{H}(x) \text{ and } \alpha_{mt}(a_z) \in Cl(\mathcal{A}(z)) & \text{if } h = a_z \\ h \in Cl(\mathcal{H}(x)) & \text{otherwise} \end{cases}$$

where

$$Cl(\hat{v}) = \{\hat{v}[x_1 \leftarrow l_{x_1}, \ldots, x_n \leftarrow l_{x_n}] \mid FV[\hat{v}] = \{x_1, \ldots, x_n\}, \text{ and } l_{x_i} \in Loc_{x_i}\}$$

5.1   Set Constraints

Having defined what we mean by valid environments for a program P, we need to be able to compute such environments. To that end, we generate from P a collection of set constraints such that any $\mathcal{H}$ and $\mathcal{A}$ satisfying these constraints are valid for P. Properties of the constraints for P guarantee they have a least solution. A proof of the existence of a least solution and an algorithm that computes it can be found in (Di Blasio 1997).

Given a program P, a set constraint is composed of a premise A_P which concerns the environments $\mathcal{H}, \mathcal{A}$ and the program P, and a conclusion B which concerns only the environments $\mathcal{H}, \mathcal{A}$. A pair of environments $\mathcal{H}, \mathcal{A}$ *satisfies* such a set constraint relative to P if whenever A_P holds for $\mathcal{H}, \mathcal{A}$ and P, then B also holds for $\mathcal{H}, \mathcal{A}$.

Before proceeding with the constraints, we first introduce some notation. We write $M \in P$, to indicate that M is an ANF term that occurs in P. Given an abstract method table $\hat{\eta}$, $\hat{\eta}' = \hat{\eta}[m = (w, z)]$ is equal to $\hat{\eta}$ except for the entry m; in this case, $\hat{\eta}'(m) = (w, z)$ (see C_{ov}^P, C_{ext}^P constraints in Figure 2). The empty abstract method table is denoted by $\hat{\eta}_{\emptyset}$. Function *FinalVar*, defined below, computes for any ANF term M the variable whose value will be the result of reducing M.

$$C_{const}^P \quad \frac{(let\ x = c\ in\ M) \in P}{c \in \mathcal{H}(x)} \qquad C_{lamb}^P \quad \frac{(let\ x = \lambda y.\ N\ in\ M) \in P}{\lambda y.\ N \in \mathcal{H}(x)}$$

$$C_{emp-obj}^P \quad \frac{(let\ x = \langle\rangle\ in\ M) \in P}{x \in \mathcal{H}(x) \quad \hat{\eta}_\emptyset \in \mathcal{A}(x)} \qquad C_{var}^P \quad \frac{(let\ x = N\ in\ M) \in P}{\mathcal{H}(FinalVar[N]) \subseteq \mathcal{H}(x)}$$

$$C_{apply}^P \quad \frac{(let\ x = y\ z\ in\ M) \in P \qquad \lambda w.\ N \in \mathcal{H}(y)}{\mathcal{H}(z) \subseteq \mathcal{H}(w) \qquad \mathcal{H}(FinalVar[N]) \subseteq \mathcal{H}(x)}$$

$$C_{ext}^P \quad \frac{(let\ x = \langle y \longleftarrow m = when(w)z\rangle\ in\ M) \in P \qquad y' \in \mathcal{H}(y) \qquad \hat{\eta} \in \mathcal{A}(y')}{x \in \mathcal{H}(x) \qquad \hat{\eta}' \in \mathcal{A}(x)}$$

$$C_{ov}^P \quad \frac{(let\ x = \langle y \leftarrow m = when(w)z\rangle\ in\ M) \in P \qquad y' \in \mathcal{H}(y) \qquad \hat{\eta} \in \mathcal{A}(y')}{x \in \mathcal{H}(x) \qquad \hat{\eta}' \in \mathcal{A}(x) \qquad y' \in \mathcal{H}(x) \qquad \hat{\eta}' \in \mathcal{A}(y')}$$

$$C_{s-inv}^P \quad \frac{\begin{array}{c} (let\ x = y \Leftarrow_s m(z)\ in\ M) \in P \\ y' \in \mathcal{H}(y) \quad \hat{\eta} \in \mathcal{A}(y') \quad \hat{\eta}(m) = (g, b) \\ \lambda s_1.\ N_1 \in \mathcal{H}(g) \quad \lambda s_2.\ N_2 \in \mathcal{H}(b) \quad \lambda w.\ N_3 \in \mathcal{H}(FinalVar[N_2]) \end{array}}{\mathcal{H}(y) \subseteq \mathcal{H}(s_1) \quad \mathcal{H}(y) \subseteq \mathcal{H}(s_2) \quad \mathcal{H}(z) \subseteq \mathcal{H}(w) \quad \mathcal{H}(FinalVar[N_3]) \subseteq \mathcal{H}(x)}$$

$$C_{as-inv}^P \quad \frac{\begin{array}{c} (let\ x = y \Leftarrow_a m(z)\ in\ M) \in P \\ y' \in \mathcal{H}(y) \quad \hat{\eta} \in \mathcal{A}(y') \quad \hat{\eta}(m) = (g, b) \\ \lambda s_1.\ N_1 \in \mathcal{H}(g) \quad \lambda s_2.\ N_2 \in \mathcal{H}(b) \quad \lambda w.\ N_3 \in \mathcal{H}(FinalVar[N_2]) \end{array}}{\mathcal{H}(y) \subseteq \mathcal{H}(s_1) \quad \mathcal{H}(y) \subseteq \mathcal{H}(s_2) \quad \mathcal{H}(z) \subseteq \mathcal{H}(w) \quad nil \in \mathcal{H}(x)}$$

Figure 2 Set constraints for A-normal-form expressions

Definition 3 *The function FinalVar on ANF terms is defined as follows:*

$$FinalVar[x] = FinalVar[l_x] \quad = \quad x$$
$$FinalVar[let\ x = \ldots\ in\ M] \quad = \quad FinalVar[M]$$

Figure 2 shows the set constraints for our language. In the following, we give the intuition behind the soundness of some of the constraints.

- ($C_{emp-obj}^P$) When we execute *let* $x = \langle\rangle$ *in* M, the value associated with x is a new object address a_x, and the method table of the created object is empty. Thus we add x to $\mathcal{H}(x)$ and $\hat{\eta}_\emptyset$ to $\mathcal{A}(x)$.

- (C^P_{var}) When we execute *let* $x = N$ *in* M, the value associated with x is the value of the return variable of N. Thus we add the constraint $\mathcal{H}(FinalVar[N]) \subseteq \mathcal{H}(x)$.

- (C^P_{apply}) The two constraints on the value sets in the conclusion represent respectively the flow of values from the actual to the formal parameter of the function and from the variable containing the result of the function, $FinalVar[N]$, to the variable x.

- $(C^P_{ext,ov})$ When we execute *let* $x = \langle l_y \leftarrow\!\circ m = when(l_w)l_z \rangle$ *in* M, where $\leftarrow\!\circ$ is either $\leftarrow\!+$ or the non-self-inflicted version of $\leftarrow$, the value associated with x is a new address a_x; thus we add x to $\mathcal{H}(x)$. The method table of a_x is obtained by adding $m = (l_w, l_z)$ to the method table η of object $a_{y'}$, where $a_{y'} = H(l_y)$. Thus we add $\hat{\eta}' = \hat{\eta}[m = (w, z)]$ to $\mathcal{A}(x)$. In C^P_{ov}, we must further consider the case that the operation is self-inflicted. Here, the value associated with x is $H(l_y) = a_{y'}$. Thus we add y' to $\mathcal{H}(x)$. Since $a_{y'}$ has a new method table obtained by overriding method m in η with the new pair (l_w, l_z), we add $\hat{\eta}' = \hat{\eta}[m = (w, z)]$ to $\mathcal{A}(y')$.

- $(C^P_{s,as-inv})$ The execution of a synchronous method invocation may result in the application of a guard and a body. The constraint $\mathcal{H}(y) \subseteq \mathcal{H}(s_1)$ represents the flow of values from the actual to the formal parameter in the guard function. The method body application is a generalization of the function application case. It consists of a double application. Thus the second and third constraint in the conclusion represent the flow of values during input, while the fourth constraint represents the return flow. For the asynchronous method invocation case the value *nil* is returned immediately. Thus we add *nil* to $\mathcal{H}(x)$.

5.2   Examples

Inlining. To show how set-based analysis works and its possible applications, we consider the following ANF of the program described in Section 2.

$$
\begin{array}{ll}
let\ o_3 = \langle\rangle\ in & let\ l_1 = \lambda s_1.\,N_1\ in \\
\quad let\ o_4 = \langle o_3 \leftarrow\!+ m_1 = l_1 \rangle\ in & \quad let\ l_2 = \lambda s_2.\,N_2\ in \\
\quad\quad let\ o_1 = \langle o_4 \leftarrow\!+ m_2 = l_2 \rangle\ in & \quad\quad let\ l_3 = \lambda s_3.\,N_3\ in \\
\quad\quad\quad let\ o_2 = \langle o_1 \leftarrow m_2 = l_3 \rangle\ in & \quad\quad\quad let\ f = \lambda x.\,N_4\ in \\
\quad\quad\quad\quad let\ x_1 = f\ o_1\ in\ x_1
\end{array}
$$

where $N_2 \equiv let\ f_2 = \lambda a_2.\,N_2'\ in\ f_2$, $N_2' \equiv let\ i_1 = s_2 \Leftarrow_a m_1(\_)\ in\ i_1$, and $N_4 \equiv let\ i_2 = x \Leftarrow_a m_2(\_)\ in\ i_2$ and N_1, N_3 are ANFs of e_1, e_3 respectively.

We apply the set constraints in Section 5.1 to the program in ANF. Some of the resulting constraints, one for each type of expression in the example, are showed in Figure 3. We annotate each constraint with the name of the set constraint that produced it and with the variable denoting the expression to which the constraint was applied. We assume that the unspecified code

Set Constraints

$C_{emp-obj}^{P}$ on o_3:	$o_3 \in \mathcal{H}(o_3)$, $\eta_\emptyset \in \mathcal{A}(o_3)$	
C_{lamb}^{P} on l_1:	$\lambda s_1. N_1 \in \mathcal{H}(l_1)$	
C_{ext}^{P} on o_4:	$o_4 \in \mathcal{H}(o_4)$, and for all $y' \in \mathcal{H}(o_3)$, $\hat{\eta} \in \mathcal{A}(y')$, $\hat{\eta}[m_1 = (\_, l_1)] \in \mathcal{A}(o_4)$	
C_{as-inv}^{P} on i_1:	for all $y' \in \mathcal{H}(s_2)$ and $\hat{\eta} \in \mathcal{A}(y')$, with $\hat{\eta}(m_1) = (\_, b)$ and	
	$\lambda s. N \in \mathcal{H}(b)$, we have $\mathcal{H}(s_2) \subseteq \mathcal{H}(s)$ and $nil \in \mathcal{H}(i_1)$	
C_{ov}^{P} on o_2:	$o_2 \in \mathcal{H}(o_2)$, and for all $y' \in \mathcal{H}(o_1)$ and $\hat{\eta} \in \mathcal{A}(y')$, we have $y' \in \mathcal{H}(o_2)$,	
	$\hat{\eta}[m_2 = (\_, l_3)] \in \mathcal{A}(o_2)$, and $\hat{\eta}[m_2 = (\_, l_3)] \in \mathcal{A}(y')$	
C_{apply}^{P} on x_1:	for all $\lambda w. N \in \mathcal{H}(f)$, $\mathcal{H}(o_1) \subseteq \mathcal{H}(w)$, $\mathcal{H}(FinalVar[N]) \subseteq \mathcal{H}(x_1)$	

Set environment

$\mathcal{H}(o_1) = \{o_1\}$	$\mathcal{H}(o_2) = \{o_1, o_2\}$	$\mathcal{H}(o_3) = \{o_3\}$	$\mathcal{H}(o_4) = \{o_4\}$	$\mathcal{H}(x) = \{o_1\}$
$\mathcal{H}(l_1) = \{\lambda s_1. N_1\}$	$\mathcal{H}(f) = \{\lambda x. N_4\}$	$\mathcal{H}(s_1) = \{o_1\}$	$\mathcal{H}(s_2) = \{o_1\}$	$\mathcal{H}(s_3) = \{o_1\}$
$\mathcal{H}(l_2) = \{\lambda s_2. N_2\}$	$\mathcal{H}(f_2) = \{\lambda a_2. N_2'\}$	$\mathcal{H}(x_1) = \{nil\}$	$\mathcal{H}(i_1) = \{nil\}$	$\mathcal{H}(i_2) = \{nil\}$
$\mathcal{H}(l_3) = \{\lambda s_3. N_3\}$	$\mathcal{H}(a_2) = \{\_\}$			

Set object environment

$\mathcal{A}(o_1) = \{\hat{\eta}_2, \hat{\eta}_3\}$	$\mathcal{A}(o_2) = \{\hat{\eta}_3\}$	$\hat{\eta}_1(m_1) = (\_, l_1)$	$\hat{\eta}_2(m_1) = (\_, l_1)$	$\hat{\eta}_3(m_1) = (\_, l_1)$
$\mathcal{A}(o_3) = \{\hat{\eta}_\emptyset\}$	$\mathcal{A}(o_4) = \{\hat{\eta}_1\}$		$\hat{\eta}_2(m_2) = (\_, l_2)$	$\hat{\eta}_3(m_2) = (\_, l_3)$

Figure 3 Inlining Example

N_1, N_3 does not affect the analysis. Figure 3 contains also the least solution $\mathcal{H}, \mathcal{A}$ of the constraints.

This analysis reveals the code that may be invoked at any function or method call site. We might take advantage of this control flow information to inline code. Inlining is an optimization technique which replaces a function or a method call with the called body (Jagannathan *et al.* 1996). The main benefit of inlining is that it eliminates the cost caused by call overhead. Inlining is particularly important for languages which provide objects, as run-time method lookup is a significant source of program inefficiency.

Inlining code at a call site is sound if there is a unique function or method applicable at that site. For example, we cannot inline $x \Leftarrow_a m_2(\_)$ because there is no unique applicable body (see Section 2). Note however, that even if the body were unique, we could inline it only if the language were sequential; in a concurrent setting we could not preserve the semantics of the original code because we would then be inlining a non-self-inflicted method invocation, thus changing the process that executes the code. In contrast, inlining $s_2 \Leftarrow_a m_1(\_)$ in the body of method m_2 is safe because the body is unique and the operation is self-inflicted. This example shows that the benefit from set-based analysis can be substantially enhanced by an in-depth study of self and non-self-inflicted operations. (See (Di Blasio 1997) for a first step.)

Colocation of objects in a distributed setting. The following fragment of code describes an object o_1 which executes a particular task using a helper object o_2. The task starts invoking method *do* on o_1. Method *do* creates the helper object o_2, sets the value of the *helper* method, and starts a computation in o_2 in parallel with the one in o_1. When o_2 completes the execution of its *continue* method, it acknowledges o_1. Depending on some conditions (omitted in the code) method *ack* of o_1 can either invoke another round of the *continue* method on itself and o_2 or alternatively return a result to some object y.

$$
\begin{aligned}
let \; o_1 \; = \langle \quad do \qquad &= \quad \lambda s_1.\, \lambda a.\, let \; o_2 = helper\_ob \\
&\quad in \; \langle s_1 \leftarrow helper = \lambda s.\, \lambda a.\, o_2 \rangle; \\
&\qquad o_2 \Leftarrow_a start(s_1); \; s_1 \Leftarrow_s continue, \\
helper \quad &= \quad \lambda s.\, \lambda a.\, c, \\
ack \quad &= \quad \lambda s_2.\, \lambda a.\ldots let \; x = (s_2 \Leftarrow_s helper) \; in \; x \Leftarrow_a continue \ldots \\
&\qquad \ldots s_2 \Leftarrow_s continue \ldots y \Leftarrow_a return(\_), \\
continue \quad &= \quad \lambda s.\, \lambda a.\ldots do \; the \; task \ldots \; \rangle \\
in \; o_1 \Leftarrow_a do(\_) &
\end{aligned}
$$

where

$$
\begin{aligned}
helper\_ob \; = \langle \quad start \qquad &= \quad \lambda s_3.\, \lambda a.\langle s_3 \leftarrow helped\_ob = \lambda s.\, \lambda x.\, a \rangle; \\
&\qquad s_3 \Leftarrow_s continue, \\
helped\_ob \quad &= \quad \lambda s.\, \lambda a.\, c', \\
continue \quad &= \quad \lambda s_4.\, \lambda a.\ldots do \; the \; task \ldots \\
&\qquad let \; x' = (s_4 \Leftarrow_s helped\_ob) \; in \; x' \Leftarrow_a ack(\_) \; \rangle
\end{aligned}
$$

with c, c' constants of the proper type.

If we analyze the code above using the technique described in the previous example, we get for the variables on which methods are invoked the following set environment:

$$
\begin{aligned}
&\mathcal{H}(o_2) = \{o_2\} \quad \mathcal{H}(s_1) = \{o_1\} \quad \mathcal{H}(s_2) = \{o_1\} \qquad \mathcal{H}(x) = \{c, o_2\} \\
&\mathcal{H}(s_3) = \{o_2\} \quad \mathcal{H}(s_4) = \{c, o_2\} \quad \mathcal{H}(x') = \{c', o_1\}
\end{aligned}
$$

The first line shows that all method invocations by objects created at the program site denoted by o_1, except $y \Leftarrow_a return(\_)$, refer either to themselves or to objects created at site o_2. We have the analogous situation for the methods invoked by o_2. In a distributed setting, we can take advantage of this information by locating objects created at program sites o_1 and o_2 in the same network site.

Introducing parallelism. Consider the following object:

$$ o \; = \; \langle \ldots m = \lambda s.\, \lambda a.\ldots let \; x = o' \Leftarrow_s n(\_) \; in \; \ldots f(x) \ldots \rangle. $$

We can take advantage of the parallelism of our language by transforming the synchronous method invocation into an eager invocation following the example in Section 2. However, the transformation makes sense only if the invocation is always non-self-inflicted, *i.e.* the value of o' is always different from the value of the self parameter s. This condition is guaranteed if $\mathcal{H}(s) \cap \mathcal{H}(o') = \emptyset$.

5.3   Soundness of the Set Constraints

Given a program P and environments $\mathcal{H}, \mathcal{A}$ which satisfy the set constraints relative to P, proving the soundness of the set constraints means showing that for any configuration $g = \left\langle\!\!\left\langle \; H \mid \alpha \mid \mu \; \right\rangle\!\!\right\rangle$ to which the initial state for

P (g_P) may reduce, $\mathcal{H}, \mathcal{A}$ are sound approximations of H and α, in the sense of Definition 2. Following the outline of the soundness proof in (Flanagan *et al.* 1995), the proof is done by induction on the length of the reduction. To prove the inductive step we must first define an invariant, $g \models_P \mathcal{H}, \mathcal{A}$, for our configurations which contains the relation $H, \alpha \models \mathcal{H}, \mathcal{A}$ and is preserved by the reduction.

Definition 4 *Given a configuration* $g = \big\langle\!\big\langle\, H \mid \alpha \mid \mu \,\big\rangle\!\big\rangle$, $g \models_P \mathcal{H}, \mathcal{A}$ *holds if*

$$H, \alpha \models \mathcal{H}, \mathcal{A} \quad and \quad \alpha \models_P^H \mathcal{H} \quad and \quad \mu \models_P \mathcal{H}.$$

The above definition says that g is valid with respect to $\mathcal{H}, \mathcal{A}$, and P if each of its component is valid as well. An object soup α is valid for P and $\mathcal{H}$ ($\alpha \models_P^H \mathcal{H}$) if for every object in α with state $[M]$, M obeys the constraints on the set environments. A pending queue μ is valid for P ($\mu \models_P \mathcal{H}$) if every message in μ comes from an asynchronous method invocation in P. The formal specifications of these relations can be found in (Di Blasio 1997).

The following is the main Lemma we use to prove the induction step in the soundness theorem.

Lemma 1 *If* $g \models_P \mathcal{H}, \mathcal{A}$, $g \longmapsto g'$, *and* $\mathcal{H}, \mathcal{A}$ *satisfy the constraints relative to* P, *then* $g' \models_P \mathcal{H}, \mathcal{A}$.

Theorem 1 (Soundness of Constraints) *If* $\mathcal{H}, \mathcal{A}$ *satisfies the set constraints relative to* P, *then* $P \models \mathcal{H}, \mathcal{A}$.

6 CONCLUSIONS

In this paper, we have presented a set-based analysis technique for an imperative, concurrent object calculus and shown it to be sound. This analysis provides static information about the values that variables may assume; this information may be used to define a program control flow graph, which is the starting point for most other analysis. We showed some examples that demonstrate how this analysis may be applied to problems that arise in a concurrent setting.

This work is intended as a first step towards the development of formal methods for program optimization for concurrent object-oriented languages. However, more work needs to be done to establish the applicability and limitations of this technique. Our analysis corresponds to what is called 0-CFA (Shivers 1991), the least precise and least complex of a family of control flow analyses. One direction for future research is to improve the precision and efficiency of this analysis, for example, along the lines developed in (Nielson *et al.* 1997). Another potential focus for research is the problem of statically detecting (non-)self-inflicted operations. Such an analysis would attempt to determine when the target object of an operation is the same as the object executing the request. As the inlining example illustrates, such information may be used to avoid the cost of complex remote invocation protocols.

REFERENCES

Amadio, R. M. (1994) Translating core facile. *Tech. Rep.* ECRC-1994-3, European Computer-Industry Research Centre.

Colby, C. (1995) Analyzing the communication topology of concurrent programs. In Proc. of ACM *Symposium on Partial Evaluation and Semantics-Based Program Manipulation*, pp. 202–213.

Di Blasio, P. (1997) A calculus for concurrent objects: design and control flow analysis. *PhD thesis*, Università di Roma "La Sapienza".

Di Blasio, P. and Fisher, K. (1996) A calculus for concurrent objects. In Proc. of *CONCUR'96: Concurrency Theory*, vol. LNCS 1119, Springer, pp. 655–670.

Fisher, K., Honsell, F. and Mitchell, J. C. (1993) A lambda calculus of objects and method specialization. *Nordic J. Computing (*formerly *BIT)*.

Flanagan, C., Sabry, A., Dubra, B. and Felleisen, M. (1993) The essence of compiling with continuation. In Proc. of *PLDI*, pp. 237–247.

Flanagan, C. and Felleisen, M. (1995) Set-based analysis for full scheme and its use in soft-typing. *Tech. Rep.* TR95-253, Department of Computer Science, Rice University.

Heintze, N. (1992) Set based program analysis. *PhD thesis*, Carnegie-Mellon University.

Holzle, U. and Ungar, D. (1995) Reconciling responsiveness with performance in pure object-oriented languages. *ACM Transaction on Programming Languages and Systems*, 18, 4, 355–400.

Jagannathan, S. and Wright, A. (1996) Flow-directed inlining. In Proc. of *PLDI'96*, pp. 193–205.

Kobayashi, N., Nakade, M. and Yonezawa, A. (1995) Static analysis of communication for asynchronous concurrent programming languages. In Proc. of *SAS'95*.

Magic, G. (1996) Telescript technology: Mobile agents. *White paper*.

Mason, I. A. and Talcott, C. L. (1997) A semantically sound actor translation. To appear in *ICALP'97*.

McGraw, G. and Felten, E. W. (1996) *Java Security*. Wiley Computer Publishing.

Nielson, H. and Nielson, F. (1994) Higher-order concurrent programs with finite communication topology. In Proc. of *POPL'94*.

Nielson, H. and Nielson, F. (1997) Infinitary control flow analysis: a collecting semantics for closure analysis. In Proc. of *POPL'97*.

Palsberg, J. and Swartzbach, M. (1991) Object-oriented type inference. In Proc. of *OOPSLA'91*, pp. 146–161.

Plevyak, J., Zhang, X. and Chien, A. (1995) Obtaining sequential efficiency for concurrent object-oriented languages. In Proc. of *POPL'95*.

Shivers, O. (1991) Control-flow analysis of higher-order languages or taming lambda. *PhD thesis*, Carnegie-Mellon University.

7

Specifying Distributed Information Systems: Fundamentals of an Object-Oriented Approach Using Distributed Temporal Logic

G. Denker and H.-D. Ehrich
Abteilung Datenbanken, Technische Universität Braunschweig
Postfach 3329, D-38023 Braunschweig, Germany
Voice: +49 +531 391{3103 |3271}, Fax: +49 +531 3913298
{G.Denker|HD.Ehrich}@tu-bs.de

Abstract

We present fundamentals of an approach to object-oriented specification of distributed information systems. We do not assume global time for concurrent object systems. For specifying those systems we propose DTL, a distributed temporal logic. The main contribution is that DTL is capable of specifying complex constraints about the behavior of distributed systems and communication between concurrent objects. For instance, we distinguish different kinds of synchronous communication such as immediate calling vs. deferred calling. The ideas are illustrated by examples given in TROLL, a formal object-oriented specification language. We introduce notations for formally specifying object-based distributed systems. Finally, we briefly explain how DTL is semantically explained in terms of a true concurrent model, i.e., labelled event structures, and which concepts for in-the-large specification are covered by our approach.

Keywords

Object orientation, specification language, distributed system, semantical model, distributed temporal logic, formal description technique, information system, concurrency

1   INTRODUCTION

Most of the work currently done in the field of distributed systems is about networks of processors. Questions concerning hardware, process communication, distributed operating systems, etc. are investigated. Concurrency comes along with distribution. We assume that units of distribution are understood as granules of concurrent behavior. There is a bunch of mathematical models for concurrent processes such as petri nets, event structures, and transition systems together with process logics such as temporal logics or Hennessy-Milner-Logik. Process languages (e.g. CSP or CCS) as well as programming languages (e.g. OCCAM or UNITY) arose from these efforts.

We adopt an object-oriented approach to system specification. Object-oriented concepts are already widely accepted for software programming. Recently object-oriented concepts take the step into early design phases of software systems including analysis and specification. A sound theoretical underpinning is necessary to deal with design issues such as verification, refinement, model checking, etc. Event structures and temporal logics are used as semantical framework in this paper.

We present fundamentals of an approach to specifying distributed information systems. A system is modeled as an interacting collection of sequential objects that operate concurrently. The approach combines ideas from algebraic data type specification, conceptual modeling, database design, behavior modeling, specification of reactive systems, and concurrency theory.

A specification language named TROLL is being developed that is based on these ideas. TROLL is a language especially designed for being used in early phases of information system design. The description of static and dynamic features is integrated in object descriptions. Theses object descriptions are the basic building blocks of system specifications. TROLL supports different abstractions, such as classes, inheritance, static and dynamic specialization, composition, etc. The latest published version is (Jungclaus *et al.* 1995). A new version, still under development, emphasizes distribution and concurrency (Hartel 1996).

In this paper, we elaborate on distribution ideas that are being discussed for inclusion in the next TROLL version. The emphasis here is on theoretical foundations rather than language features. We illustrate our ideas by example using ad-hoc notations.

For specifying concurrent object behavior, we envisage to use DTL, a distributed temporal logic that is an extension of linear temporal logic. DTL is based on n-agent logic (Lodaya *et al.* 1992). Interpretation structures are locally sequential labelled event structures. Objects correspond to sequential substructures. Synchronous and symmetric "handshake" communication is modelled by letting the sequential substructures overlap at shared events. So we have local linear times within objects and synchronization points shared by objects. There is no global time.

The distinguishing feature of this model is that it expresses non-interleaving concurrency while the logic does not explicitly talk about concurrency: we are still close to linear temporal logic with the added possibility of "talking about other objects".

This is achieved by specifying from the local viewpoints of objects in the system, not that of an external observer with a bird's view of the system.

There are some similarities to the approaches of TLA (Abadi *et al.* 1995) a temporal logic of actions to specify concurrent systems, and CTR (Bonner *et al.* 1996) a concurrent transaction logic in which communication is specifiable. In contrast to TLA and to CTR our models are true concurrent models whereas TLA is based on a sequential model and CTR also uses an interleaving model. We use event structures as models. Using them means that we do not assume interleaving semantics. This model implies that we do not have global time nor global space, i.e., shared data. Thus, the way of describing the properties of an open distributed system, e.g. global constraints such as behaviour relations between concurrent objects, is done by communication.

Abadi and Lamport use so called assumption/guarantee specifications which assert that the system provides a guarantee if its environment satisfies an assumption. We only use the communication principle to specify the behavior of concurrent systems. Another difference to their approach is that they are concerned with proving implementation relations between specifications. One of their main issue is to reason about composition of assumption/guarantee specifications. Since our approach is purely based on communication, we will face different problems. It is subject to future work to see how we can deal with implementation problems mentioned in (Abadi *et al.* 1995).

The approach of Bonner and Kifer differs from our in the sense that they specify several concurrent processes which synchronize on a common database. Our approach uses the concept of objects to encapsulate data. Thus, we do not have a central data space on which all processes have to synchronize. Their approach is similar to ours in the respect that they also specify behavior locally to processes, but the programming paradigm of CTR differs from the one that is induced by using distributed temporal logics as the basis for object-oriented languages.

Among other approaches to developing object specification languages with a sound theoretical basis, MAUDE (Meseguer 1993, Clavel *et al.* 1996) also deals with concurrency. It is based on rewriting logic (Meseguer 1992) that is a uniform model of concurrency. GNOME (Sernadas *et al.* 1994) is based on temporal logic, it is most closely related to TROLL although the semantics is not capable of expressing non-interleaving concurrency.

The paper is organized as follows: In the following section we examine the issues that arise in the specification of distributed systems. In particular, we discuss different kinds of synchronous communication in distributed systems by means of examples. Our examples are given in TROLL, a formal object-oriented specification language. In Section 3 we propose a linear time temporal logic (DTL) which can deal with distribution. We show how DTL can be used to give semantics to TROLL specifications. This is done by translating language features to temporal formulas. Thus, TROLL only provides a convenient way of temporal specification. In Section 4 we briefly describe event structure semantics, and in Section 5 we mention the expressive power of our approach since it incorporates different in-the-large concepts for specification. We conclude with a summary and future work.

2   BASIC LANGUAGE CONCEPTS

We demonstrate basic specification concepts by means of a toy example, using an ad-hoc notation in the spirit of TROLL. Some language features for specifying distributed systems have been proposed in Hartel (1996).

Example 1 (BankWorld) We have object classes `Account` and `Bank`. *money* is a predefined data type, *data type* `acct# = [100..999]` is a user-defined data type.

An account has an attribute `balance` and actions `open`, `close`, `credit`(*nat*) and `debit`(*nat*). * and + express that the actions may only occur at the beginning or at the end of a life cycle, respectively. The behavior of an object is specified separately from the signature. The effects of actions are specified in *do-od* clauses. Preconditions for actions are specified in *onlyIf* clauses.

```
object class   Account
  attributes
    balance: nat initialized 0;
  actions
    * open;
      credit(money);
      debit(money);
    + close;
  behavior
    credit(m) do balance:=balance+m od;
    debit(m)
       onlyIf balance>=m
       do balance:= balance-m od;
end
```

A bank has a `founder` as attribute of type *string* which will be set when it is created. The founder is a *constant* attribute. Thus, it cannot be changed during the life of a bank object. A bank is a complex object with accounts as components that are identified by account numbers. `transfer(m,from,to,at)` is an action for transferring an amount m of money from a local account to an account at another bank with bank code number at. Bank code number is a user-defined data type: *data type* `bankCode# = [1..9]`.

`receive(m,from,at,to)` is an action for receiving an amount m from an account at bank to a local account. The receiver bank acknowledges the receipt of money (`acknowledge(m,from,at,to)`) and the sender bank receives the acknowledgement (`receiveAckn(m,from,to,at)`).

The two banks involved in a transfer operate concurrently. We will come back to distribution issues in the context of `BankWorld`.

```
object class   Bank
  attributes founder: string constant;
    components Accs(acct#): Account;
    actions
        * establish(founder: string);
        transfer(m: money,from: acct#,to: acct#,at: bankCode#);
        receive(m: money,from: acct#,at: bankCode#,to: acct#);
        acknowledge(m: money,from: acct#,at: bankCode#,to: acct#);
        receiveAckn(m: money,from: acct#,to: acct#,at: bankCode#);
        + liquidate;
    behavior
        establish(f) do founder:=f od;
        transfer(m,from,to,at) do Accs(from).debit(m) od;
        receive(m,from,at,to)
            do eventually Accs(to).credit(m),
                next acknowledge(m,from,at,to) od;
    constraints
        transfer(m,from,to,at) and transfer(m,from,to2,at2)
        implies (to=to2 and at=at2);
        transfer(m,from,to,at) implies ( not transfer(m,from,to,at)
            until receiveAckn(m,from,to,at));
        receive(m,from,at,to) implies (Accs(to).credit(m)
            before Accs(x).debit(n));
end
```

The behavior specification says that if a transfer happens then the money will be debited in the same step. Thus, the bank communicates with its component. If money is received then it will eventually be put on the appropriate account and the receipt will be acknowledged in the next step.

Furthermore, there are three constraints ensuring correct transaction management: (1) there must not be two different transfers of the same amount from the same account at the same time, and (2) as long as receipt of a transfer is not acknowledged, the same amount may not be transferred from the same account to the same account, and (3) as long as an account is not credited no other account may be debited. The first constraint ensures that no debit is lost, and the second constraint ensures that every transfer is acknowledged. The last constraint ensures that money which has been received will be credited before any other account will be debited. All constraints are implicitly universally quantified and all temporal operators are implicitly bound to bank self. In more detail, the second constraint says that the same transfer in self is forbidden *until* the acknowledgement has been received so that self is sure that bank at has received the money. The meaning of the temporal *until* operator ensures that the acknowledgement must eventually happen, so at must eventually have received the money.

In the example, object classes are specified as generic structure and behavior templates of objects of the same type. Instances of object classes are called objects.

Objects are concurrently put together to form a distributed system. TROLL offers a way to specify concurrent objects in an *object system*. After declaring all data types and object classes, the possible *objects* of the described world are defined. The *objects* construct provides a set *Id* of object identities. BankWorld consists of several concurrently existing Banks. These are distinguished by bank code number. Each bank object is a complex object composed of several accounts.

object system BankWorld
    data type acct# ...; bankCode# ...;
    object class Account ... *end*;
    object class Bank ...   *end*;
    objects Banks(nr:bankCode#): Bank;
    behavior ...
end

Dynamic creation and abortion of objects is treated by * and + actions. So we have a fixed set of "possible" objects while there is a time-varying subset of "actual" objects.

We distinguish between sequential and concurrent objects. Objects are understood as sequential units, even if they are composed. E.g. a bank consisting of several accounts is semantically modelled as a sequential process. Though in each step different accounts may perform actions, the overall behavior of a bank is a sequential process. Furthermore, there is communication between the components of the composed object as well as between the compound object and its components. For example, a bank communicates with an account while transferring money by debiting the account. In our model communication is synchronous. The mentioned example will be reflected in the model by a sequence of actions where in one step both actions (transfer and debit) happen. Due to the underlying sequential model this kind of communication can be described using linear temporal logics. In a distributed system we need new description techniques for communication.

Different objects are concurrently put together to form a distributed system. For example the banks of BankWorld behave concurrently. Going from sequential objects to concurrent systems opens new possibilities. For instance, assume there is a system requirement that can only be fulfilled if different actions from various concurrent objects are executed. Since in our model there is no global observer we can only assure such a requirement by communicating the execution of the corresponding actions among the objects which are incorporated. But such an approach leaves freedom for two parameters: the communication point between the objects and the execution time of the actions they are asked for. To formalize such requirements we need a formalism which does not demands global view. For this purpose we will introduce DTL in the following section. Before doing so we will illustrate communication in distributed systems by means of examples.

For instance, given two banks Banks(1) and Banks(2) and a transfer action Banks(1).transfer(m,x,y,Banks(2)). We want to specify that the transferred money will be received, i.e., Banks(2).receive(m,x,Banks(1),y) happens. As-

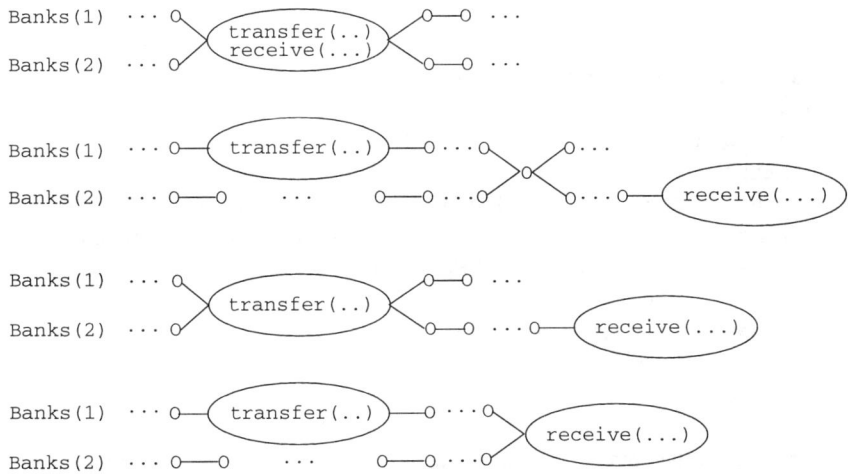

Figure 1 Possible synchronous communication schemas between concurrent objects

suming synchronous communication there are four possibilities to fulfill the requirement that Banks(1) is sure that Banks(2) received the money:

1. The banks meet and perform the corresponding actions together. This corresponds to synchronous communication in sequential systems. In other words, the transfer action of Banks(1) **immediately calls** Banks(2) and the latter **immediately executes** the receive action.
2. The meeting point between Banks(1) and Banks(2) is somewhere in-between the states where the actions transfer(...) and receive(...), respectively, are executed. This corresponds to a **deferred calling** from Banks(1) to Banks(2) and a **deferred execution** of both actions.
3. When Banks(1) transfers the money it **immediately calls** Banks(2) to get the information that the latter will receive the money sometime in the future (**deferred execution**). This is a special case of 2 where the communication point coincides with the execution of Banks(1).transfer(...). The called action Banks(2).receive(...) will be executed later.
4. Another special case of 2 is that the communication point coincides with the execution of the called action, i.e., **deferred calling** and **immediate execution**. Sometime after transferring the money Banks(1) contacts Banks(2) which is currently receiving the money.

These possibilities are illustrated in Figure 1. Object states are depicted by circles put in sequence by connecting lines, and communication points are depicted by shared states.

Therefore, we have four possible communication structures between concurrent objects. Communication in a sequential object is always immediate. Appropriate language features are still under development, but a possible description is given below.

object system `BankWorld`
    `. . .`
   objects `Banks(nr:bankCode#): Bank;`
   communication
    `Banks(snd)->Banks(rcv):`
     `transfer(m,from,to,rcv) >> receive(m,from,snd,to)`
     `transfer(m,from,to,rcv) >F>` *eventually* `receive(m,from,snd,to)`
     `transfer(m,from,to,rcv) >>` *eventually* `receive(m,from,snd,to)`
     `transfer(m,from,to,rcv) >F> receive(m,from,snd,to)`
    `Banks(rcv)->Banks(snd):`
     `acknowledge(m,from,snd,to) >F> receiveAckn(m,from,to,rcv)`
end

The first part of the communication refers to the objects involved, i.e., directed synchronous communication between Banks(snd) and Banks(rcv).

The second part concerns the actions involved and the kind of communication: immediate communication (>>) or deferred communication. For deferred communication there is the choice between >X>, i.e., communication in the next state, or >F>, i.e., communication in some following state. Moreover, either the called action is immediately executed when the communication takes place or the action is executed in the next (*next*) or some following state (*eventually*).

The last communication axiom says that the acknowledgement given by the receiving bank will sometime be received by the sender bank. The communication will be at the time of receipt.   ∎

The focus of our work is on mathematical foundations of distributed specifications. Thus, in the following section we will propose a distributed temporal logic (DTL) in which situations like the one given above can be formulated. An appropriate logic must be able to express formulae from the viewpoint of different objects. We show how the given bank example is translated into logical terms using DTL. The general idea is that TROLL only provides a convenient way of specification. Semantically, TROLL specifications are treated by translation to DTL. Thus, one can use TROLL as a language tool for temporal system specification. In the following section we will present how different TROLL features are translated to DTL. Because of lack of space we only illustrate the translation in terms of our example. The full semantics of TROLL is presented in Hartel (1996). The expresiveness of the current language TROLL is much more restrictive than of the underlying logics. This is due to the fact that TROLL is also used in a real-size case study (Krone *et al.* 1996) where intuitive understanding and executability of specification is of great importance.

In our example we focussed on features for specifying distributed, interacting object systems. TROLL offers also in-the-large features like inheritance, generalization, etc. We briefly mention these issues in section 5.

3   DTL SPECIFICATION

Let $\Sigma = (S, \Omega)$ be a data signature where S is a set of sorts and Ω is an $S^* \times S$-indexed set family of operation symbols. A TROLL specification determines structural as well as behavioural aspects of the system. An in-the-small TROLL system specification is formalized as a pair $SysSpec = (\Sigma_I, \Phi)$. $\Sigma_I = (Id, At, Ac)$ is a signature covering all structural aspects, consisting of a set Id of identities, an $Id \times S$-indexed set family $At = \{At_{i,s}\}_{i \in Id, s \in S}$ of attribute symbols, and an Id-indexed set family $Ac = \{Ac_i\}_{i \in Id}$ of action symbols. Without loss of generality we assume that actions do not have parameter. $\Phi = \{\Phi_i\}_{i \in Id}$ is an Id-indexed set family of DTL formulae. A detailed construction of Σ_I from a given TROLL specification is given in Ehrich *et al.* (1995).

Example 2 (BankWorld signature) Let Banks(1), Banks(2), ... be banks. Let 100, 101, ... be account numbers. The instance signature of the bank specification is $\Sigma_I = (Id, At, Ac)$, where $Id = \{$Banks(1), Banks(2), ...$\}$, and for j,k = 1,2, ..., we have

- $At_{\text{Banks(j)},string} = \{$Banks(j).founder$\}$,
- $At_{\text{Banks(j)},money} = \{$Banks(j).Accs(100).balance,
                          Banks(j).Accs(101).balance, ... $\}$,
- $Ac_{\text{Banks(j)}} = \{$Banks(j).establish,
           Banks(j).transfer(50,100,999,Banks(k)), ... ,
           Banks(j).receiveAckn(50,100,999,Banks(k)), ... ,
           Banks(j).Accs(100).debit(50), ... $\}$.

The attributes and actions of the components of the bank, i.e., its accounts, are imported in the signature of Banks(j).   ∎

The DTL formulae $\Phi = \{\Phi_i\}_{i \in Id}$ are obtained by translating corresponding language fragments: initialization declarations, action effects and preconditions, synchronization axioms, constraints, etc.

Before we introduce DTL we start with a simpler logic, a basic distributed temporal logic DTL_B, in which immediate calling and execution can be treated easily. We illustrate by example how to translate the sequential part of our example as well as immediate communication to DTL_B formulae. Afterwards we define DTL in which immediate as well as deferred calling and execution, respectively, can be formalized appropriately. But DTL is not more expressive than DTL_B in the sense that every DTL formula can be translated to a DTL_B formula introducing further communication actions (Ehrich *et al.* 1997). DTL is only more convenient for specifying distributed systems. Wrt communication DTL_B corresponds to the actual TROLL version. There communication is expressed by directly defining the actions on which objects synchronise. In DTL it is possible to define deferred communication and execution, among others. DTL_B is defined as follows:

$$DTL_B ::= \{DTL_B{}^i\}_{i \in Id},$$

$$DTL_B{}^i ::= i.H_b^i \mid i.C_b^i,$$

$$H_b^i ::= \text{ATOM} \mid (H_b^i \Rightarrow H_b^i) \mid (H_b^i \, \mathcal{U} \, H_b^i) \mid (H_b^i \, \mathcal{S} \, H_b^i),$$

$$\text{ATOM} ::= false \mid T_\Sigma \, \theta \, T_\Sigma \mid AT_\Sigma \, \theta \, T_\Sigma \mid \triangleright AC \mid \odot AC,$$

$$C_b^i ::= \ldots \mid (C_i \Rightarrow j.C_j) \mid \ldots \qquad \text{for } i, j \in Id, \ j \neq i.$$

Basic distributed temporal logic is split into the local *home*-logic $i.H_b^i$ of object i and the communication logic $i.C_b^i$. The local home logic is a propositional linear time temporal logic. An atom is a boolean constant, a pair of data terms, an attribut-data term pair, or a predicate on an action expressing that an action is enabled ($\triangleright a$) or has occurrred ($\odot a$). T_Σ, AT_Σ and AC denote data, attribute and action terms, respectively; θ denotes a comparison operator like $=, \leq, \ldots$. $\mathcal{U}$ and $\mathcal{S}$ are the temporal operators *until* and *since*. The other temporal operators can be derived from these, for instance $X \, \varphi = false \, \mathcal{U} \, \varphi$ for *next*, $F \, \varphi = \varphi \vee true \, \mathcal{U} \, \varphi$ for *sometime in the future*, $\varphi \, \mathcal{P}^+ \, \psi = \neg((\neg\varphi) \, \mathcal{U}^\circ \, \psi)$ for *before*, $G \, \varphi = \neg(F(\neg\varphi))$ for *always*, $P \, \varphi = \varphi \vee true \, \mathcal{S} \, \varphi$ for *sometime in the past*, etc. (cf. (Ehrich 1996) for details). Let C_i be communication predicates of object i $\in$ Id, e.g. $\odot a, \triangleright b \wedge \odot a$. The communication logic $i.C_b^i$ can express immediate communication. For instance, $i.(\odot a \Rightarrow j. \odot b)$ is a communication formula that means that whenever a happens in object i it immediately synchronizes with object j which will perform b.

The following example shows how some of the relevant specification fragments of the object classes Bank and Account in example 1 are expressed in DTL$_B$.

Example 3 (BankWorld axioms)   Referring to the signature Σ_I given in example 2, let Banks(1) be an instance of class Bank. Local formulae of object Banks(1) have the form Banks(1).(φ), where φ is a home- or a communication formula. We omit the local identity and give in the following only φ to increase readability. From the class specifications in example 1, we obtain the following formulae in $\Phi_{\text{Banks}(1)}$.

initialization:
The first axiom translates the *initialized* statement in *object class* Account. The second axiom corresponds to the effect of the birth action establish:
$\odot$Banks(1).Accs(acct#).open $\Rightarrow$ Banks(1).Accs(acc#).balance $= 0$,
$\odot$Banks(1).establish(f) $\Rightarrow$ Banks(1).founder $=$ f.

constant attributes:
After a founder has been set he/she will remain unchanged for the rest of the bank's life:
Banks(1).founder $=$ f$\Rightarrow$(G Banks(1).founder $=$ f $\mathcal{U}$ $\odot$Banks(1).liquidate).

effects of actions:
Specifying effects of actions requires to refer to attribute values in the state before the action was executed:
($\odot$Banks(1).Accs(from).debit(m) $\wedge$ Y Banks(1).Accs(from).balance $=$ b)
$\quad \Rightarrow$ Banks(1).Accs(from).balance $=$ b $-$ m.

preconditions of actions:
$\odot$Banks(1).Accs(from).debit(m) $\Rightarrow$ Banks(1).Accs(from).balance $>=$ m.

constraints:

Constraints are mainly translated one to one from the specification, e.g.

$\odot$Banks(1).receive(m, from, to, at)$\Rightarrow$

 $\odot$ Banks(1).Accs(from).credit(m) $\mathcal{P}^+$ $\odot$Banks(1).Accs(x).debit(n).

immediate communication:

We have to distinguish between immediate and deferred communication. Immediate communication between concurrent objects can easily be formalized with DTL_B. Deferred communcation will be illustrated after introducing DTL. The following formula corresponds to the first possibility given in Figure 1.

$\odot$Banks(1).transfer(m, x, y, z) $\Rightarrow$ $\odot$Banks(2).receive(m, x, Banks(1), y). $\blacksquare$

We introduce DTL to express deferred kinds of communication. DTL is defined as follows. We concentrate on the propositional fragment:

$$\text{DTL} ::= \{\text{DTL}^i\}_{i \in Id},$$

$$\text{DTL}^i ::= i.H^i,$$

$$H^i ::= \text{ATOM} \mid (H^i \Rightarrow H^i) \mid (H^i \, \mathcal{U} \, H^i) \mid (H^i \, \mathcal{S} \, H^i) \mid C^i,$$

$$\text{ATOM} ::= false \mid T_\Sigma \, \theta \, T_\Sigma \mid AT_\Sigma \, \theta \, T_\Sigma \mid \triangleright AC \mid \odot AC,$$

$$C^i ::= \ldots \mid \text{DTL}^j \mid \ldots \qquad \text{for } i, j \in Id, \; j \neq i.$$

The main difference to DTL_B is the possibility of arbitrarily changing the viewpoint between concurrent objects in a formula. Change of viewpoint implies that communication takes place. For instance, $i.(\odot a \Rightarrow F \, j.(\odot b))$ means that whenever object i executes a it will sometime in the future meet object j (deferred calling) and object j will execute b at the meeting point (immediate execution). More examples are given below.

DTL as well as DTL_B are propositional, linear time temporal logics. As already mentioned DTL is not more expressible than DTL_B, but it is more appropriate for different kinds of communication specification in distributed systems including immediate as well as deferred calling and execution. So far there is no proof system for our logic. This is subject to further research. But nevertheless, since we have a well-defined semantics we can proof semantically properties of the distributed system. Since in a distributed system communication is the technique to exchange information, the interesting properties to be dealt with are those which involve several concurrent objects. Requirements which basically depend on the communication between such objects are of major interest. Our logic provides a way to specify such requirements. Moreover, it is still a linear temporal logic. Properties expected to be analyzable are those which express specific protocols between objects, behavior patterns in distributed systems, behavior that emerges when concurrent objects are composed to systems, etc.

Example 4 (BankWorld axioms (contd)) With the help of DTL the communication axioms given in BankWorld are as follows:
immediate and deferred communication:

Banks(snd).[⊙Banks(snd).transfer(m, from, to, rcv)
        ⇒ ⊙Banks(rcv).receive(m, from, snd, to)],

Banks(snd).[⊙Banks(snd).transfer(m, from, to, rcv)
        ⇒ F Banks(rcv).[F ⊙Banks(rcv).receive(m, from, snd, to)]],

Banks(snd).[⊙Banks(snd).transfer(m, from, to, rcv)
        ⇒ Banks(rcv).[F ⊙Banks(rcv).receive(m, from, snd, to)]],

Banks(snd).[⊙Banks(snd).transfer(m, from, to, rcv)
        ⇒ F ⊙Banks(rcv).receive(m, from, snd, to)],

Banks(rcv).[⊙Banks(rcv).acknowledge(m, from, snd, to)
        ⇒ F ⊙ Banks(snd).receiveAckn(m, from, to, rcv)].   ∎

DTL achieves its expressiveness by changing the local view in a temporal formula. Though the logics does not explicitly talk about distribution, distributed systems are formalizable. DTL is by far more expressive than what we exploit to give semantics to the four different communication styles in the TROLL example. To give an idea of its capability we present some formulae involving two distributed objects. We introduce $@u \,\hat{=}\, u.X^? \, true$ (at u, *locality*) as an abbreviation. This means that there is currently communication with u. Let $i, u \in Id$ *("I, you")* be object identities:

$i. \mathsf{G}(@u \Rightarrow \mathsf{X}\,\varphi)$ whenever I talk to you, you have φ the next day
$i.u.\mathsf{P}\,\varphi$ I hear that φ was once valid for you
$i.(\varphi \Rightarrow \mathsf{X}\,@u)$ if φ is true, then I will talk to you tomorrow
$i.u.\mathsf{X}\,@i$ you tell me that you will contact me tomorrow

DTL is semantically defined with the help of a non-interleaving model. In the following section we briefly present event structure semantics and satisfaction of DTL.

4   EVENT STRUCTURE SEMANTICS

Interpretation structures for DTL are locally sequential labelled event structures (cf. (Winskel *et al.* 1995) for event structures). In (Sassone *et al.* 1993) a classification of concurrency models is given and event structures are considered to be a fair choice when a non-interleaving denotational model is needed. Single objects are modelled by sequential labelled event structures. Models for single objects are sets of sequential life cycles where in each step a set of (component) actions may be performed. This captures the intuition that objects have a state and perform actions changing state.

Formally single objects are modelled by labelled sequential event structures $\mathbf{C} = (E, \lambda, P)$. A sequential event structure $E = (Ev, \rightarrow^*, \#)$ consists of a set of events e, a partial ordering $\rightarrow^*$ (causality) and a symmetric and irreflexive binary relation

(conflict). There is no concurrency in sequential event structures. The labelling $\lambda : Ev \to P$ maps events to the state logic P, i.e., assertions about attribute values, enabled and occurring actions, etc. Interpretation of DTL will be given in life cycles. A life cycle L of a sequential object is a maximal totally ordered sub-event structure of E, i.e., L is a sequence of events of a sequential object.

For modelling distributed systems, sequential event structures are composed concurrently to form a locally sequential event structure, i.e., an union of sequential event structures overlapping at shared interaction events. Thus, a possible system run corresponds to a distributed life cycle, a concurrent composition of sequential life cycles of system objects.

One of the main differences between sequential objects and concurrent systems is that of concurrency. The concept of concurrency is derived in event structures. Let $E = (Ev, \to^*, \#)$ be an event structure. Two events $e, e' \in Ev$ are concurrent, $e \; co \; e'$, iff $\neg(e \to^* e' \vee e' \to^* e) \wedge \neg e \# e'$. A system $\mathbf{S}$ of objects $i \in Id$ is a triple $\mathbf{S} = (E, \lambda, P)$, where $\mathbf{C}_i = (E_i, \lambda_i, P_i) \subseteq \mathbf{S}$ is a local class of object i in $\mathbf{S}$.

As usual, models of a system specification $SysSpec = (\Sigma_I, \Phi)$ are interpretation structures satisfying the axioms.

Satisfaction of DTL is locally defined. For a given system model $\mathbf{S}$ we define satisfaction of a formulae for a distributed life cycle L at a specific event $e \in L$.

Let $\mathbf{S} = (E, \lambda, P)$ be a system Let L_i be a life cycle of object i and $L = L_1 \cup \cdots \cup L_n$ be a distributed life cycle in $\mathbf{S}$. Let $e \in L$ be an event of the system. Satisfaction $\vDash_i$ is locally defined for object i. Let $\varphi, \psi \in H^i$, $\gamma \in C^j = D^j$.

$\mathbf{S}, L, e \vDash i.\varphi$	holds iff $e \in L_i$ and $\mathbf{S}, L, e \vDash_i \varphi$.
Let $e \in L_i$:	
$\mathbf{S}, L, e \vDash_i p$	holds iff $\lambda_i(e) \vDash_{P_i} p$,
$\mathbf{S}, L, e \vDash_i false$	does not hold,
$\mathbf{S}, L, e \vDash_i (\varphi \Rightarrow \psi)$	holds iff $\mathbf{S}, L, e \vDash_i \varphi$ implies $\mathbf{S}, L, e \vDash_i \psi$,
$\mathbf{S}, L, e \vDash_i (\varphi \; \mathcal{U} \; \psi)$	holds iff there is a future event $e' \in L_i$, $e \to^+ e'$, where $\mathbf{S}, L, e' \vDash_i \psi$ holds, and $\mathbf{S}, L, e'' \vDash_i \varphi$ holds for all $e'' \in L_i, e \to^+ e'' \to^+ e'$,
$\mathbf{S}, L, e \vDash_i (\psi \; \mathcal{S} \; \varphi)$	holds iff there is a past event $e' \in L_i$, $e' \to^+ e$, where $\mathbf{S}, L, e' \vDash_i \varphi$ holds, and $\mathbf{S}, L, e'' \vDash_i \psi$ holds for all $e'' \in L_i, e' \to^+ e'' \to^+ e$,
$\mathbf{S}, L, e \vDash_i j.\gamma$	holds iff $e \in L_j$ and $\mathbf{S}, L, e \vDash_j \gamma$.

At an event e a communication formulae holds iff both objects are sharing e. Thus, whenever the view is changed in a formulae it requires that the involved objects communicate.

With an appropriate morphism concept for locally sequential event structures, the model category of a specification has a final element that may serve as a canonical semantics (Caleiro 1996).

5   IN-THE-LARGE SPECIFICATION

The power of our approach comes to fruition when dealing with in-the-large concepts like inheritance, hiding and composition such as generalization and sequential and concurrent aggregation (Ehrich 1996).

Relevant relationships between models are given by event structure morphisms (Winskel *et al.* 1995). For example, semantic inheritance is modeled by event structure inclusion morphisms expressing that the inheriting object has obtained some of the attributes and actions from the original one. Overriding is covered because event structure morphisms are partial. Hiding, i.e., the relationships between objects and their interfaces, generalized objects and their constituents, complex objects and their components, etc. are also modeled by event structure morphisms.

Event structures and their morphisms form a category **ev**. Sequential event structures and their (total) morphisms form a category **evs**. The main results are the following:

- **ev** is complete and has coproducts.
- **evs** is complete and cocomplete.
- Coproducts in **evs** coincide with those in **ev**.
- These coproducts express generalization, i.e. alternative choice.
- Products in **ev** express concurrent aggregation.
- Products in **evs** express sequential aggregation, i.e. component composition.

The coproduct construction is given in (Winskel *et al.* 1995). For products in **ev**, an elegant construction is given in (Vaandrager 1989). We are not aware that the completeness and cocompleteness results have been published elsewhere. The constructions in **evs** are simple, though. Equalizers in **ev** also have a simple construction. Pullbacks may be utilized for modeling aggregation with sharing, e.g. event sharing for handshake communication. Pushouts in **evs** may be utilized for modeling generalized objects with shared constituents. Coequalizers in **ev** do not exist in general but there are interesting cases where they exist and may be utilized.

6   CONCLUDING REMARKS

The theory outlined here is taking shape but is not complete, it has to be elaborated and refined in several respects. For instance, in-the-large composition aspects have to be worked out. The goal is to establish a high-level algebraic treatment and optimization of concurrent systems construction.

Further research will focus on interaction and modularization concepts. A richer spectrum of interaction concepts is needed, including asynchronous directed interaction. Investigations towards extending the theory to asynchronous communication have been done in (Denker *et al.* 1996). A modularization concept is needed providing generic building blocks with external and internal interfaces related by reification. As for reification, cf. (Denker 1995).

Acknowledgements The authors are grateful to their colleagues for many discussions. Special thanks are due to Amílcar Sernadas and Carlos Caleiro for many suggestions and hints concerning the distributed temporal logics. Juliana Küster Filipe did a great job in working out proof details. The work of Peter Hartel has been the starting point to introduce concurrency into TROLL. The remarks of the anonymous referees helped to improve the paper.

Work reported here was partially supported by the EU under ESPRIT BRA ASPIRE 22704 and DFG under Eh75/11-1.

REFERENCES

Abadi, M. and Lamport, L. (1995) Conjoining Specifications. *ACM Transactions on Programming Languages and Systems*, 17(3):507–533, May.

Bonner, A.J. and Kifer, M. (1996) Concurrency and Communication in Transaction Logic, in *Proc. Joint Int. Conf. and Symp. on Logic Programming (JICSLP96), Bonn, Germany* (ed. Maher, M.). The MIT Press.

Caleiro, C. (1996) *Personal communication.*

Clavel, M. and Eker, S. and Lincoln, P. and Meseguer, J. (1996) Principles of Maude, in *Rewriting Logic and its Applications, First International Workshop, Asilomar Conference Center, Pacific Grove, Ca, September 3-6, 1996* (ed. Meseguer, J.), pages 65–89.

Denker, G. (1995) Transactions in Object-Oriented Specifications, in *Recent Trends in Data Types Specification, Proc. 10th Workshop on Specification of Abstract Data Types joint with the 5th COMPASS Workshop, S.Margherita, Italy, May/June 1994, Selected papers* (eds. Astesiano, E. and Reggio, G. and Tarlecki,A.), pages 203–218, Springer, Berlin, LNCS 906.

Denker, G. and Küster Filipe, J. (1996) Towards a Model for Asynchronously Communicating Object, in *Proc. 2nd Int. Baltic Workshop on Databases and Information Systems, Tallinn, June 12-14, 1996* (eds. Haav, H.-M. and Thalheim, B.), pages 182–193. Institute of Cybernetics.

Ehrich, H.-D. (1996) *Object Specification.* Technical Report, Technische Universität Braunschweig, URL://www.cs.tu-bs.de/idb/welcome_e.html.

Ehrich, H.-D. and Sernadas, A. (1995) Local Specification of Distributed Families of Sequential Objects, in *Recent Trends in Data Types Specification, Proc. 10th Workshop on Specification of Abstract Data Types joint with the 5th COMPASS Workshop, S.Margherita, Italy, May/June 1994, Selected papers* (eds Astesiano, E. and Reggio, G. and Tarlecki,A.), pages 219–235. Springer, Berlin, LNCS 906.

Ehrich, H.-D. and Caleiro, C. and Sernadas, A. and Denker, G. (1997) Logics for Specifying Concurrent Information Systems, *forthcoming.*

Hartel, P. (1996) *Konzeptionelle Modellierung von Informationssystemen als verteilte Objektsysteme.* Reihe DISDBIS, infix-Verlag, Sankt Augustin.

Jungclaus, R. and Saake, G. and Hartmann, T. and Sernadas, C. (1995) TROLL – A Language for Object-Oriented Specification of Information Systems. *ACM Transactions on Information Systems*, 14(2):175–211, April.

Krone, M. and Kowsari, M. and Hartel, P. and Denker, G. and Ehrich, H.-D. (1996) Developing an Information System Using TROLL: an Application Field Study, in *Proc. 8th Int. Conf. on Advanced Information Systems Engineering (CAiSE'96)* (eds. Constantopoulos, P. and Mylopoulos, J. and Vassiliou, Y.), pages 136–159, Springer, Berlin, LNCS 1080.

Lodaya, K. and Ramanujam, R. and Thiagarajan, P.S. (1992) Temporal Logics for Communicating Sequential Agents. *Int. Journal of Foundations of Computer Science*, 3(2):117–159.

Meseguer, J. (1992) Conditional Rewriting Logic an a Unified Model of Concurrency. *Theoretical Computer Science*, 96(1):73–155.

Meseguer, J. (1993) A Logical Theory of Concurrent Objects and its Realization in the Maude Language, in *Research Directions in Concurrent Object-Oriented Programming* (eds. Agha, G. and Wegner, P. and Yonezawa, A.), pages 314–390. The MIT Press.

Sassone, V. and Nielsen, M. and Winskel, G. (1993) A Classification of Models for Concurrency, in *CONCUR'93, Proc. 4th International Conference on Concurrency Theory, Hildesheim, Germany, August 1993* (ed. Best, E.), pages 325–392. Springer, LNCS 715.

Sernadas, A. and Ramos, J. (1994) The GNOME Language: Syntax, Semantics and Calculus. Technical Report, Instituto Superior Téchnico (IST), Dept. Mathemática, Av. Roviso Pais, 1096 Lisboa Codex, Portugal.

Vaandrager, F.W. (1989) A simple definition for parallel composition of prime event structures. Technical Report CS-R8903, Centre for Mathematics and Computer Science, P.O. Box 4079, 1009 AB Amsterdam, The Netherlands.

Winskel, G. and Nielsen, M. (1995) Models for Concurrency, in *Handbook of Logic in Computer Science, Vol. 4, Semantic Modelling* (eds. Abramsky, S. and Gabbay, D.M. and Maibaum, T.S.E), pages 1–148. Oxford Science Publications.

Biography

Grit Denker received a degree in mathematics in 1991 and received her PhD in 1995 at Technical University of Braunschweig, Germany. Since 1996 she is assistant professor at TU Braunschweig. Her interests are in object-oriented specification of distributed systems, formal methods and models for system specification, concurrency, reification.

Hans-Dieter Ehrich studied mathematics at the University of Kiel, Germany. He received his PhD in 1970 at the University of Hannover, Germany. Since 1982 he is professor at the Technical University of Braunschweig. He leads the database group. His interests are information systems design from theory to applications.

PART THREE

Actors

8

A Set-Constraint-based analysis of Actors

J-L. Colaço, M. Pantel and P. Sallé
LIMA/ENSEEIHT/IRIT
2 rue Camichel 31071 TOULOUSE, FRANCE.
{colaco, pantel, salle}@enseeiht.fr
http://www.enseeiht.fr/Recherche/Info/Logiciel/vestale/vestale.html

Abstract

This paper presents a type inference system for a primitive actor calculus (CAP) based on set-constraints resolution. In contrast with concurrent objects, actors can change dynamically their interface (the set of messages they can handle). Therefore, the CAP calculus reduction rules can lead to orphan messages which will never be handled. The aim of the inference system is to detect statically many orphan messages and to produce information leading to the dynamic detection of the others. In this purpose, we define a flattening operation which abstracts the various behaviors of an actor. This static analysis is based on Aiken and Wimmers set-constraints resolution. It gives slightly better results than Vasconcelos or Yonezawa kinded types based analysis for concurrent objects.

Keywords

Actors, concurrent objects, type inference, set constraints.

1 INTRODUCTION

The actor model proposed by Hewitt and Agha (Agha 1986, Hewit, Bishop and Steiger 1973) generally leads to dynamically typed languages in the LISP tradition (Marcoux, Maurel and Sallé 1988, Yonezawa 1990). These languages involve many dynamic type checks. "Soft typing" has been introduced to reduce the number of dynamic type check (see (Cartwright and Fagan 1991, Aiken, Wimmers and Lakshman 1994)). Types for functional and object-oriented languages have been the subjects of active investigations, but many recent studies focus on typing concurrence (Kobayashi and Yonezawa 1994, Vasconcelos and Tokoro 1993, Pierce and Sangiorgi 1995, Nielson and Nielson 1993, Puntigam 1996). Our study addresses the problem of type inference for Actor based languages.

Actors are self-contained computational agents interacting via asynchronous message passing. An actor is composed of a *mail address* (that identifies the actor) and a *behavior*. When handling a message, an actor can *create* new actors; *send* messages to its acquaintances; and *modify* its behavior. These acquaintances can be stored in the local state of the actor or given as arguments of a message. These possibilities lead to a dynamic topology of communication. In the Actor model, messages are guaranteed to be received but their order of arrival is unknown.

In contrast with concurrent objects, "Actor" means that the behavior of the entities can change during the computation and so, its interface (the set of messages that an actor can handle) can change dynamically. In this paper, we use set constraints to type actor programs. This static analysis cannot detect all the orphan messages (*messages that will never be handled*). But information to detect dynamically remaining orphans can be derived from inferred types; then this system can be classified as a *"soft typing system"*.

This paper is organized as follows. Section 1 describes the formalism used to study actors, i.e. CAP * (Colaço, Pantel, Sallé and Senteni 1996), a kind of "actor-oriented process calculus". Section 2 presents the definition of types and the type system. The use of types information for run-time is outlined in section 3. Then in section 4 we compare our type system to Vasconcelos' one (Vasconcelos and Tokoro 1993). Finally, some insights in our future works are proposed.

2   A PRIMITIVE ACTOR CALCULUS (CAP)

The kernel of actor languages is based on asynchronous communication between actors. Existing actor languages also contain predefined data structures and sequential control structures. Communication is yet sufficient to express all possible computations and actors can represent data structures (see (Colaço, Pantel, Sallé and Senteni 1996)). We advocate that our calculus is primitive, because it only contains communication to express computation.

As in Milner's π-calculus (Milner 1991) and in Honda's v-calculus (Honda and Tokoro 1991) the mail address of an actor is represented by a name. The actor behavior is represented by an interface containing methods and private fields. Methods can be accessed by communication and private fields are only reachable by the actor itself. Private fields can be seen as a private record associated to a behavior. This record contains data local to an actor.

CAP does not respect all the principles of Agha's actors, but it contains the notion of behavior and the notion of address as primitives that allow to express very easily actor programs. It also contains behavior communication (which is not in the original Actor model, but allows a simple but useful form of reflection (Colaço, Pantel and Sallé 1996, Colaço, Pantel, Sallé and Senteni 1996)) and the sharing of the same address by several different actors. A programming discipline in the use of CAP leads to classic actor programs; this discipline can be enforced by a linearity analysis described in (Colaço, Pantel and Sallé 1997). In (Colaço, Pantel, Sallé and Senteni 1996), we have shown how to translate a "classical" actor language in CAP.

Here is an example of a CAP expression :

---

*in french:*Calcul d'Acteurs Primitifs*

$$\nu a,b,d \ \big(a \triangleright [read(c) = \zeta(e,s)(c \triangleleft rep(s.val) \parallel e \triangleright s)$$
$$write(v) = \zeta(e,s)e \triangleright s.val \Leftarrow v$$
$$, val = b] \parallel a \triangleleft write(d) \parallel \cdots) \parallel \cdots$$

The ν operator creates three names (a,b,d) whose scope is represented by the outside parenthesis. The main part of the expression describes an actor whose mail address is a and whose behavior both accepts two different messages (or methods) : $read(c)$ and $write(v)$ and has a private field called val. This is the behavior of a buffer cell. A message $write(d)$ is sent to a.

The $\zeta(e,s)$ (zeta) captures the *current address* (called *ego*) and the *current behavior* (called *self*) when the actor accepts a message. This operator is inspired by the ς (sigma) defined by Abadi and Cardelli to formalize self-substitution in objects. In our context, the capture of *self* and *ego* is used to formalize behavior changes.

2.1 Syntax

Programs are build from names and behaviors using the following constructors: messages, actors, concurrent composition, name creation (or scope restriction), inaction, field modification and field selection. Here are the main constructions:

- Message sending: $T_1 \triangleleft m(\tilde{T}_2)$, messages are addressed to a single actor identified by his name (or mail address) (T_1), they carry a message label (m) and a tuple $(\tilde{T}_2)$ representing the content of the message.
- Actors: $T_1 \triangleright T_2$, actors have a single identifier (the mail address) (T_1) associated to a behavior (T_2).
- Behaviors: $[m_1(\tilde{x}_1) = \zeta(e_1,s_1)C_1 \cdots, p_1 = T_1 \cdots]$, behaviors encapsulate a finite set of message labels with the associated formal arguments, and a set of private fields without formal arguments (no substitution is possible during the field selection).

The complete syntax is given in the following definition.

Definition 21 (Syntax) *Let N be an infinite set of names, V an infinite set of variables and L a finite set of labels. N^* (resp. V^*) represents the set of finite sequences over N (resp. over V).*

Convention: in the following, $a,b,c,\ldots \in N$ - $v,x,e_i,s_i,\ldots \in V$ - $m_i,p_j \in L$ - $\tilde{x},\tilde{x}_i,\ldots \in V^*$ - $C,D,C_i,\ldots$ are configurations (actors and messages composed with parallel

composition) - $T, T_i, \ldots$ are terms.
CAP *syntactic rules are:*

$$
\begin{array}{lll}
Config & ::= & \phi & ; \textit{Empty (Inaction)} \\
       & | & \textit{va Config} & ; \textit{restriction} \\
       & | & \textit{Actor} \parallel \textit{Config} & ; \textit{parallel composition} \\
       & | & \textit{Message} \parallel \textit{Config} & ; \textit{parallel composition} \\
       & | & (\textit{Config}) & ; \textit{parenthesis} \\
Message & ::= & \textit{Term} \lhd m(\widetilde{\textit{Term}}) & ; \textit{message construction} \\
Actor & ::= & \textit{Term} \rhd \textit{Term} & ; \textit{actor construction}
\end{array}
$$

$$
\begin{array}{lll}
Term & ::= & [m_i(\tilde{x}_i) = \zeta(e_i, s_i)C_i^{i \in I}, p_j = T_j^{j \in J}] & ; \textit{behavior} \\
     & | & Term.p_j \Leftarrow Term & ; \textit{modification} \\
     & | & Term.p_j & ; \textit{selection} \\
     & | & a & ; \textit{name} \\
     & | & x & ; \textit{variable} \\
     & | & (Term) & ; \textit{parenthesis}
\end{array}
$$

2.2   Semantics

As we enable the communication of behaviors, we need to differentiate names and variables. Variables represent identifiers that can be substituted by *names* or *behaviors*; no substitution can be done on names.

Definition 22 (Free variables and free names) *This notion is defined in an usual way. "v" is the only binder for names, "$\zeta(e, s)$" and "$m(\tilde{x}_i)$" are the two binding forms for variables. The set of free names (resp. free variables) of an expression is given by the function $\mathcal{FN}()$ (resp. $\mathcal{FV}()$).*

Definition 23 (Substitution) *Let $\sigma = \{T/X\}$ (where T is a term and X a variable) be a substitution. Let A be the expression on which σ is to be applied; it is supposed that every name and variable of A have been renamed in order to avoid capture of free names or free variables of T.*

$$
\begin{array}{rcl}
(\textit{va } C)_\sigma & = & \textit{va } C_\sigma \\
(C \parallel D)_\sigma & = & C_\sigma \parallel D_\sigma \\
(T \lhd m(\tilde{U}))_\sigma & = & T_\sigma \lhd m(\tilde{U}_\sigma) \\
(T \rhd U)_\sigma & = & T_\sigma \rhd U_\sigma \\
(T.p_j \Leftarrow U)_\sigma & = & T_\sigma.p_j \Leftarrow U_\sigma \\
(T.p_j)_\sigma & = & T_\sigma.p_j \\
[m_i(\tilde{x}_i) = \zeta(e_i, s_i)C_i^{i \in I}, p_j = T_j^{j \in J}]_\sigma & = & [m_i(\tilde{x}_i) = \zeta(e_i, s_i)C_{i\sigma}^{i \in I}, p_j = T_{j\sigma}^{j \in J}]
\end{array}
$$

$$a_\sigma = a$$

$$x_\sigma = \begin{cases} T & if\ x = X \\ x & otherwise \end{cases}$$

To define the reduction rules in a more concise way, we first introduce a congruence relation between the expressions of the calculus.

Definition 24 (Congruence) *"≡" is the smallest congruence over* CAP *expressions defined by the following rules:*

1. $C \equiv D$ *if C is* α*-convertible to D (*α*-conversion of names and variables).*
2. $C \parallel \phi \equiv C$
3. $C \parallel D \equiv D \parallel C$
4. $(C \parallel D) \parallel E \equiv C \parallel (D \parallel E)$
5. $T \triangleright T_1 \equiv T \triangleright T_2$ *if* $T_1 \equiv T_2$
6. $[m(\tilde{x}) = \zeta(e,s)C_m\ n(\tilde{y}) = \zeta(e,s)C_n,\ p = T_p\ q = T_q]$
   $\equiv [n(\tilde{y}) = \zeta(e,s)C_n\ m(\tilde{x}) = \zeta(e,s)C_m,\ p = T_p\ q = T_q]$
   $\equiv [m(\tilde{x}) = \zeta(e,s)C_m\ n(\tilde{y}) = \zeta(e,s)C_n,\ q = T_q\ p = T_p]$
7. $va\ C \parallel D \equiv va(C \parallel D)$ *if* $a \notin \mathcal{FN}(D)$

Definition 25 (Reduction rules) *The reduction denoted by "*⟶*" is the smallest relation generated by the rules:*

$$STRUCT: \frac{D \equiv C \quad C \longrightarrow C' \quad C' \equiv D'}{D \longrightarrow D'}$$

$$SELECT\text{-}CONTEXT: \frac{T \longrightarrow T'}{T.p_k \longrightarrow T'.pk} \qquad AFFECT\text{-}CONTEXT: \frac{T_1 \longrightarrow T'_1}{T_1.p_k \Leftarrow T_2 \longrightarrow T'_1.pk \Leftarrow T_2}$$

$$MESS\text{-}CONTEXT: \frac{T \longrightarrow T'}{a \triangleleft m(\ldots T \ldots) \longrightarrow a \triangleleft m(\ldots T' \ldots)}$$

$$PAR: \frac{C \longrightarrow C'}{C \parallel D \longrightarrow C' \parallel D} \qquad RES: \frac{C \longrightarrow C'}{vxC \longrightarrow vxC'}$$

$$MESS: \frac{T \longrightarrow T'}{T \triangleleft m(\tilde{t}) \longrightarrow T' \triangleleft m(\tilde{t})} \qquad ACT: \frac{T_1 \longrightarrow T'_1 \quad T_2 \longrightarrow T'_2}{T_1 \triangleright T_2 \longrightarrow T'_1 \triangleright T'_2}$$

$$COMM: \begin{cases} if\ k \in I\ and\ len(\tilde{v}) = len(\tilde{x}_k), \\ a \triangleleft m_k(\tilde{v}) \parallel a \triangleright [m_i(\tilde{x}_i) = \zeta(e_i,s_i)C_i{}^{i\in I}, p_j = T_j{}^{j\in J}] \\ \longrightarrow C_k\{a/e_k\}\{[m_i = \cdots, p_j = \cdots]/s_k\}\{\tilde{v}/\tilde{x}_k\} \end{cases}$$

SELECT: $if\ k \in J,\ [m_i(\tilde{x}_i) = \zeta(e_i,s_i)C_i^{i\in I}, p_j = T_j^{j\in J}].p_k \longrightarrow T_k$

REDEF: $\begin{cases} if\ k \in J, & [m_i(\tilde{x}_i) = \zeta(e_i,s_i)C_i^{i\in I}, p_j = T_j^{j\in J}].p_k \Leftarrow T \\ & \longrightarrow [m_i(\tilde{x}_i) = \zeta(e_i,s_i)C_i^{i\in I}, p_j = T_j^{j\in J\backslash\{k\}} \ \ p_k = T] \end{cases}$

These reduction rules allow to send a message that cannot be understood by the target actor. In this case, the message remains in the expression. A future behavior of the target actor will perhaps be able to handle it, otherwise this message is called *orphan*.

This notion of *orphan messages* breaks the fairness principle of Agha's model. However, we can define a weaker form of fairness adapted to our semantics. We require that: *if a message "$a \lhd m(\cdots)$" is present in the medium, then "a" can only handle a finite number of messages labeled "m" before handling this message.*

CAP syntax and reduction rules are not very restrictive and they allow some undesirable expressions to be written or to appear during computation. These *ill-formed expressions* are of the following forms:

- ill-formed Actors : $[m_i(\tilde{x}_i) = \cdots, p_j = \cdots] \rhd T, T \rhd a$
- ill-formed Message : $[m_i(\tilde{x}_i) = \cdots, p_j = \cdots] \lhd m(\tilde{T})$
- ill-formed Terms : $\begin{cases} [\cdots, p_j = T_j^{j\in J}].p_k \text{ with } k \notin J, \\ [\cdots, p_j = T_j^{j\in J}].p_k \Leftarrow T \text{ with } k \notin J. \end{cases}$

Expressions leading to ill-forms will be statically eliminated by the type system.

2.3   Examples

To illustrate the calculus, two short examples are given.

Example 21 (the one-place buffer) *This example shows a two-states actor with an* empty buffer *behavior and a* full buffer *behavior.*

$$OPbuff \stackrel{def}{=} [\ put(v) = \zeta(e_{empty}, s_{empty})$$
$$(e_{empty} \rhd [get(c) = \zeta(e_{full}, s_{full})(c \lhd rep(v) \parallel e_{full} \rhd s_{empty})]) \ ]$$

When installing this behavior on a mail address, the actor behaves as an empty buffer that can only accept the message "put". After receiving such a message, the actor only accepts a "get" message and then behaves as an empty buffer. The variable s_{empty} *is associated to the empty-buffer behavior and the variable* s_{full} *is associated to the full-buffer behavior containing the stored value.*

Example 22 (2-D point) *This second example presents the use of private fields:*

$$point(x,y) \overset{\text{def}}{=} [\quad getx(c) = \zeta(e,s)(c \lhd rep(s.xc) \parallel e \rhd s)$$
$$gety(c) = \zeta(e,s)(c \lhd rep(s.yc) \parallel e \rhd s)$$
$$move(xx,yy) = \zeta(e,s)(e \rhd (s.xc \Leftarrow xx).yc \Leftarrow yy)$$
$$, \quad xc = x$$
$$yc = y]$$

Coordinates of the point are stored in private fields. When a point receives a "move" message, the actor updates its data.

More examples are given in (Colaço, Pantel, Sallé and Senteni 1996, Colaço, Pantel and Sallé 1996).

3 TYPES

The work presented in this paper is the first step in our studies on type inference for actors; this first proposition focuses on indeterminism and behavior changes. Therefore we only consider names communications. The type system will reject behaviors communications as not typable in this system.

We define three kinds of types representing *name types, behavior types* and *configuration type*. "$\wp$" is the type of all well-typed configurations. A name type represents all the messages a name (or a variable that will be substituted by a name) can potentially receive. "Potentially" means that during the computation an actor can change his behavior and therefore lose the ability to handle some messages that he could handle before. A behavior type contains two components: a name type that will be associated to the name on which it will be installed and a record type representing its private fields.

Definition 31 (Types syntax) *Let $\mathcal{V}_\tau$ be an infinite set of type variables, the set $\mathcal{T}$ of types is defined by the following grammar.*

$$\tau \quad ::= \quad \alpha \mid \beta \mid \wp$$
$$\alpha \quad ::= \quad t \mid \langle m_i(\tilde{\alpha}_i)^{i \in I} \rangle \qquad \text{(where I and J are finite sets)}$$
$$\beta \quad ::= \quad \alpha \rhd [p_j : \tau_j^{j \in J}]$$

$\mathcal{T}$ is the union of name types $\mathcal{T}_{name}$, behavior types $\mathcal{T}_{beh}$ and process type $\wp$: $\mathcal{T} = \mathcal{T}_{name} \cup \mathcal{T}_{beh} \cup \{\wp\}$. In this paper, we use the following conventions: $\tau, \tau', \tau_i, \ldots$ range over $\mathcal{T}$; $\alpha, \alpha', \alpha_i, \ldots$ over $\mathcal{T}_{name}$; $\beta, \beta', \beta_i, \ldots$ over $\mathcal{T}_{beh}$ and $t, \ldots$ range over $\mathcal{V}_\tau$. Sequences of types are denoted with a tilde ($\tilde{\tau}$).

Note that a type variable can only denote a name type.

A **name type** $\langle m_i(\tilde{\alpha}_i)^{i \in I} \rangle$ is interpreted as a set of m_i messages (with constraints $\tilde{\alpha}_i$ on arguments) that can be sent to an actor identified by this name. Our set operator definitions are based on this interpretation.

Name types can also be seen as sets of names identifying actors that can accept the same set of messages $\{m_i\}$ with constraints on arguments represented by $\tilde{\alpha}_i$.

3.1   Flattenings and actor types

The structure we use for name types is frequently used in the context of distributed objects (objects correspond to entities that do not change their interface like actors do). In these cases, types in methods are contravariant to ensure that no *"message not understood"* error occurs due to name passing. In our context, to prevent all orphan messages, one simple approach consists in ensuring that any message sent can be handled by every behavior the target actor can adopt (the intersection of all behaviors). This policy allows actors to change their behavior, but the types lack the parts of the interface which are not common to all the future interfaces. This solution is safe but too restrictive; for example, the type of the *"one-place buffer"* will be $\langle\rangle$, no message is common to all behavior, therefore the type system forbids every emission to any *"one-place buffer"* actor. Our purpose is to type actors with changing interface, so intersection is not an adequate solution. The union is neither the solution; as contravariance leads to an unsafe relaxation of constraints on parameters (by doing intersection instead of union). We define a new operation called *flattening* that behaves like a union for the set of labels and like an intersection at the parameters level. This choice relaxes the constraints on the interface of the actors, but not on the parameters.

In the context of actors and concurrent objects, there is a natural notion of *subsumption* meaning that an actor can always be replaced by an other one offering more services. Our aim is to express such a condition with an inclusion relation (which will be defined below).

3.2   Operations and relations on Types

Notation: operations with a tilde represent the sequence obtained by applying each component of one sequence to each component of the other; relations with a tilde represent the logical "and" between each component. Both require sequences of the same length.

Definition 32 (Name type inclusion "$\subseteq$") *The contravariant inclusion is defined by:*

$$\langle m_i(\tilde{\alpha}_i)^{i\in I}\rangle \subseteq \langle m_j(\tilde{\alpha}'_j)^{j\in J}\rangle \iff I \subseteq J \wedge (\forall k \in I)\tilde{\alpha}_k \tilde{\supseteq} \tilde{\alpha}'_k$$

Definition 33 (Type equality) *With the previous inclusion, we can define an equality between two types:*

$$\alpha = \alpha' \iff \alpha \subseteq \alpha' \wedge \alpha \supseteq \alpha'$$
$$\alpha \triangleright [p_j : \beta_j{}^j \in J] = \alpha' \triangleright [p_j : \beta'_j{}^j \in J] \iff \alpha = \alpha' \wedge (\forall j \in J, \beta_j = \beta'_j)$$

Definition 34 (Union and intersection) *Because of the contravariance on message arguments, union and intersection have the following form:*

$$\langle m_i(\tilde{\alpha}_i)^{i\in I}\rangle \cup \langle m_j(\tilde{\alpha}'_j)^{j\in J}\rangle = \langle m_k(\tilde{\alpha}_k\tilde{\cap}\tilde{\alpha}'_k)^{k\in I\cap J}\, m_i(\tilde{\alpha}_i)^{i\in I\setminus J}\, m_j(\tilde{\alpha}'_j)^{j\in J\setminus I}\rangle$$
$$\langle m_i(\tilde{\alpha}_i)^{i\in I}\rangle \cap \langle m_j(\tilde{\alpha}'_j)^{j\in J}\rangle = \langle m_k(\tilde{\alpha}_k\tilde{\cup}\tilde{\alpha}'_k)^{k\in I\cap J}\rangle$$

Definition 35 (Flattening and flat inclusion) *The flattening operation is denoted by* " $\uplus$ ":

$$\langle m_i(\tilde{\alpha}_i)^{i\in I}\rangle \uplus \langle m_j(\tilde{\alpha}'_j)^{j\in J}\rangle = \langle m_k(\tilde{\alpha}_k\tilde{\cup}\tilde{\alpha}'_k)^{k\in I\cap J}\, m_i(\tilde{\alpha}_i)^{i\in I\setminus J}\, m_j(\tilde{\alpha}'_j)^{j\in J\setminus I}\rangle$$

we also define an order on flattenings denoted by " $\underline{\underline{\Subset}}$ "

$$\langle m_i(\tilde{\alpha}_i)^{i\in I}\rangle \underline{\underline{\Subset}} \langle m_j(\tilde{\alpha}'_j)^{j\in J}\rangle \Longleftrightarrow I\subseteq J \wedge (\forall k\in I, \tilde{\alpha}_k\tilde{\subseteq}\tilde{\alpha}'_k)$$

The union "$\cup$" (resp. flattening " $\uplus$ ") is a least upper bound operator for the inclusion "$\subseteq$" (rep. flat order " $\underline{\underline{\Subset}}$ "); this result is immediate from previous definitions.

3.3 The type system

In this type system, judgments are of the form $E \vdash A : \tau$ where E is a type environment for names and variables, A is a CAP expression and τ a type. Some rules are labeled with a constraint between name types at the right side; if these constraints are satisfied, no reduction can lead to *ill-formed expressions*. Moreover, no arity problem may occur (in the COMM reduction rule it is no more needed to check if "$len(\tilde{v}) = len(\tilde{x}_k)$"). In our approach, these constraints are collected and used to compute most general types which inform on the allowed use of actors.

$$\text{(Beh)}\ \frac{\begin{array}{c}(\text{let } \beta_e = \alpha_e \triangleright [p_j : \tau_j]) \\ E \vdash T_j : \tau_j\ (\forall j \in J) \\ E, \tilde{x}_i : \tilde{\alpha}_i, s_i : \beta_e, e_i : \alpha_{e_i} \vdash C_i : \wp\ (\forall i \in I)\end{array}}{E \vdash [m_i(\tilde{x}_i) = \zeta(e_i, s_i)C_i,\ p_j = T_j] : \beta_e}\ \left(\alpha_e \supseteq\!\supseteq \langle m_i(\tilde{\alpha}_i)^{i\in I}\rangle \uplus \Big(\underset{i\in I}{\uplus}\ \alpha_{e_i}\Big)\right)$$

$$\text{(Actor)}\ \frac{E \vdash T_1 : \alpha_1 \qquad E \vdash T_2 : \beta_2 \qquad (\text{where } \beta_2 = \alpha_2 \triangleright [\cdots])}{E \vdash T_1 \triangleright T_2 : \wp}\ \big(\alpha_1 = \alpha_2\big)$$

$$\text{(Message)}\ \frac{E \vdash T_1 : \alpha_1 \qquad E \vdash \tilde{T}_2 : \tilde{\alpha}_2}{E \vdash T_1 \triangleleft m(\tilde{T}_2) : \wp}\ \big(\alpha_1 \supseteq \langle m(\tilde{\alpha}_2)\rangle\big)$$

$$\text{(Empty)}\ \frac{}{E \vdash \phi : \wp} \qquad\qquad \text{(Restriction)}\ \frac{E, a : \alpha_a \vdash C : \wp}{E \vdash va\, C : \wp}$$

$$(\text{Parallel}) \frac{E \vdash C : \wp \quad E \vdash D : \wp}{E \vdash C \parallel D : \wp} \qquad (\text{Select}) \frac{E \vdash T : \alpha \triangleright [p_j : \tau_j{}^{j \in J}] \quad k \in J}{E \vdash T.p_k : \tau_k}$$

$$(\text{Affect}) \frac{E \vdash T : \alpha \triangleright [p_j : \tau_j{}^{j \in J}] \quad E \vdash T' : \tau \quad k \in J}{E \vdash T.p_k \Leftarrow T' : \alpha \triangleright [p_j : \tau_j{}^{j \in J}]} \quad \left(\tau_k = \tau \right)$$

The main rule of this system is the Beh rule; because it shows how to build an object type for an actor by flattening present and future behaviors. Most of the time, the constraint on α_e given in this first rule is recursive; this comes from the fact that actors usually adopt several times the same behavior. This recursive equation is of the form: $\alpha \supseteq \alpha_1 \uplus \cdots \uplus \alpha_n \uplus \alpha$; there is a least solution in the sense of flat inclusion given by the least upper bound of $\alpha_1 \cdots \alpha_n$ i.e. $\alpha = \alpha_1 \uplus \cdots \uplus \alpha_n$. The types computed by solving the set of constraints are the least solution of such equations; they are the most precise. If this system was used to type-check expressions, the user could give other solutions of the recursive problem and the analysis would be less precise; but we do not consider type-checking, our system has been designed to infer types.

The other important rule is the Message rule expressing constraints on the target of the message and on effective arguments ($\bar{\alpha}_2$) whose types must be subsets of types required by the target actor.

In practice, the inference is done in two steps: *collecting constraints* and *solving constraints*.

3.4   Some properties of the type system

If an expression is well-typed within this system, then it will never reduce to an ill-formed expression.

Lemma 31 (Structural preserving of type assignment) *If $A \equiv B$ and $E \vdash A : \tau$ then $E \vdash B : \tau$.*

PROOF : By induction on the size of the derivation.

Theorem 31 (Subject reduction) *If $E \vdash A : \tau$ and $A \longrightarrow A'$ then $\exists E'$ such that $E' \vdash A' : \tau'$.*

PROOF : By induction on the definition of the reduction, using the previous lemma.

Lemma 32 *Ill-formed expressions and communication of behaviors are not typable.*

Theorem 32 (Soundness of the type system) *If A is typable, the reduction of an expression will never produce Ill-formed expressions.*

3.5 Constraints resolution

The inference process introduces type variables for all type names and generates three kinds of constraints on these variables: *equalities, flat inclusions* and *inclusions*. The first ones are used for type unification; the second ones express the construction of behavior types; the third ones are used to check if a message has some chance to be treated. We use a set constraint solver to compute the solution of the system. This solver is derived from the works of Aiken and Wimmers (Aiken and Wimmers 1993). It uses the operators definition and properties to decompose complex constraints and then it combines constraints on variables using transitivity. This scheme is applied until no more new simplified constraints may be introduced. Such an approach has also been used in (Pantel 1994) to type a functional object-oriented language. In order to compute the least solution of recursive problems, constraints of the form $t \supseteq \tau_1 \uplus t \uplus \tau_2$ are replaced by $t \supseteq \tau_1 \uplus \tau_2$. When the closure of the system is computed, as we are interested in the least solution, flat inclusions on type variables are replaced by equalities. At the end of the resolution, the solved system is composed of:

- equalities for the type variables of an actor name or an *ego* variable: $t = \langle m_i(\tilde{t}_i)^{i \in I} \rangle$
- intervals for the type variables attached to formal parameters: $\tau_1 \subseteq t \subseteq \tau_2$

Example 31 (The one-place buffer actor typing) *Using the behavior "OPbuff" previously defined, the typing of the actor "a ▷ OPbuff" leads to the following system:*

$$\{ \ t_a = t_{e_{empty}} = t_{e_{full}} = \langle get(\langle rep(t_v) \rangle) \ put(t_v) \rangle \ \}$$

Example 32 (linear-cell) *This example presents a cell that can only be assigned once, and then only accepts to give its value, but not to modify it.*
$$a \triangleright [init(v) = \zeta(e,s)(e \triangleright [get(c) = \zeta(e',s')(c \triangleleft rep(v) \ \| \ e' \triangleright s')])]$$
The final system for this actor is:

$$\left\{ \begin{array}{l} t_a = \langle init(t_v) \ get(t_c) \rangle \\ t_e = t_{e'} = \langle get(t_c) \rangle \\ t_c \supseteq \langle rep(t_v) \rangle \end{array} \right\}$$

The type t_a contains the whole initial potential of the actor; the type t_e represents the actor type after handling a message "init", this type is smaller in the sense of flat inclusion: $t_e \subsetneq t_a$. The remaining inclusion requires that the argument of get must be a name which is attached to an actor understanding the message "rep". The stored value v is not really used in this actor, it is only transmitted to c; in this context any value is correct.

3.6   Interpretation of the system after resolution

When the system is applied on a configuration, its answer is either "∅" or "*Type Error*", which is not very informative. The approach proposed by Vasconcelos in (Vasconcelos and Tokoro 1993) is to give as result the environment containing the type of each free identifier in the configuration. This answer is interesting in the context of open systems; free names and variables correspond to ports reachable from the outside, these types inform us on how to use these ports.

We can do the same by giving the set of constraints on the types of free identifiers to the user. Although constraints give very rich information, they are still difficult to read. Inclusion constraints express an idea of polymorphism because there is a solution for every set of types satisfying the set of constraints. Moreover, the notion of subsumption can help us for the simplification of constraints set. Fuh and Mishra (Fuh and Mishra 1989) have proposed some solutions to the problem of types simplification in a functional context with subtyping. It is sometimes possible to saturate a constraint and then replace an inclusion by an equality and give an equivalent system. This possibility comes from the notion of subsumption: a name of type τ can be used every time a smaller type is required. Intuitively, in order to represent all the solutions of the system, we have to minimize the constraints on formal arguments in the behaviors. But due to the contravariance of the inclusion, minimizing a type can lead to maximize an other one. These two phenomena can appear simultaneously on the same type variable. In these cases, no saturation can be done on this variable and the constraint cannot be simplified.

For example, let us apply the saturation to the linear-cell. The types t_v and t_c have to be minimized, the minimization of t_c implies the maximization of t_v then t_c can be saturated and replaced by $\langle rep(t_v) \rangle$ and t_v remains in the system.

Example 33 (the linear-cell in a context) *In order to present the typing of an actor in a context, this example considers the previous linear-cell actor in parallel with a message and another actor:*

$$a \triangleright [init(v) = \zeta(e,s)(e \triangleright [get(c) = \zeta(e',s')(c \triangleleft rep(v) \parallel e' \triangleright s')])]$$
$$\parallel \ a \triangleleft get(b) \parallel b \triangleright [rep(w) = \zeta(e'',s'')(w \triangleleft mess() \parallel e'' \triangleright s'')]$$

The resulting system after saturation is:

$$\left\{ \begin{array}{l} t_a = \langle init(t_v) \ get(\langle rep(t_v) \rangle) \ \rangle \\ t_e = t_{e'} = \langle get(\langle rep(t_v) \rangle) \ \rangle \\ t_b = t_{e''} = \langle rep(\langle mess() \rangle) \ \rangle \\ t_v \supseteq \langle mess() \rangle \end{array} \right\}$$

In this system, we can see more intuitively why t_v cannot be saturated. Its lower bound depends on its use by the context. Without any context: t_v is not constrained *(or has the constraint* $"t_v \supseteq \langle\rangle"$*). In this specific context* t_v *is constrained by* $"t_v \supseteq \langle mess()\rangle"$*. The lower bound of* t_v *is therefore context-dependent.*

4 PRACTICAL USE OF TYPES DURING RUN-TIME

As our type system is not able to detect statically all the orphan messages, we propose to add some annotations (derived from inferred types) on the terms of the calculus, to allow dynamic detection of remaining orphan messages.

4.1 Annotated calculus

Definition 41 (Interface) *The interface is the set of message labels an actor can potentially accept. It can be calculated from name types using the function* $I()$*:*
$$I(\langle m_i(\tilde{\alpha}_i)^{i\in I}\rangle) = \{m_i^{i\in I}\}$$
In the following we use $\sigma, \sigma', \sigma_i$ *to denote interfaces.*

These decorations only appear in the actor constructor, they are computed from the type of the term at the actor name position. An actor is now denotated as $t \overset{\sigma}{\rhd} t'$ where $\sigma = I(\tau)$ and τ is the type computed for the term t by the inference system.

Now that we have annotated terms, we can extend the semantics. There are, at least, two possibilities to deal with orphans when they are detected: *Run-time error* or *Memory deallocation*.

(a) Extension with run-time error
In this case the emergence of an orphan message is considered as something undesirable during the computation. In this semantics, we need to add a constant *Error* and to choose if errors are propagated or not in the rest of the expression. The following rules detects errors:

ORPH-ERROR : if $m \notin \sigma$ then $a \lhd m(\tilde{v}) \parallel a \overset{\sigma}{\rhd} [m_i(\tilde{x}_i)\cdots] \longrightarrow Error$

(b) Extension with garbage collection
If we don't consider an orphan message as something dangerous or undesirable, the decoration can be used to free the memory space where such messages are stored.

ORPH-COLLECT : if $m \notin \sigma$ then $\begin{cases} a \lhd m(\tilde{v}) \parallel a \overset{\sigma}{\rhd} [m_i(\tilde{x}_i)\cdots] \\ \longrightarrow a \overset{\sigma}{\rhd} [m_i(\tilde{x}_i)\cdots] \end{cases}$

Example 41 (Annotated linear-cell) *Using the types computed for this behavior, the linear-cell actor can be annotated :*

$$a \overset{\{init,get\}}{\rhd} [init(v) = \zeta(e,s)(e \overset{\{get\}}{\rhd} [get(c) = \zeta(e',s')(c \lhd rep(v) \parallel e' \overset{\{get\}}{\rhd} s')])]$$

5   APPLYING THE SYSTEM TO AN OBJECT CALCULUS

This type inference system can be used in the more restrictive context of objects. In this section, we show how to encode Vasconcelos' calculus of objects (Vasconcelos and Tokoro 1993) and we compare the types obtained by both systems. We could also compare our approach with the works of Yonezawa (Kobayashi and Yonezawa 1994). But this work is quite similar with Vasconcelos' one, therefore our comparison also holds for this work.

5.1   Vasconcelos' calculus of concurrent objects

The syntax of this calculus is given by the following rule:

$$P ::= a \triangleleft l(\tilde{v}) \mid a \triangleright [l_1(\tilde{x}_1).P_1 \& \cdots \& l_n(\tilde{x}_n).P_n] \mid P_1, P_2$$
$$\mid \mathsf{v}xP \mid !a \triangleright [l_1(\tilde{x}_1).P_1 \& \cdots \& l_n(\tilde{x}_n).P_n] \mid 0$$

The syntax and semantics are very close to CAP, except that no orphan message is allowed and an object can't change its interface. This object calculus can be encoded in CAP with the function $\mathcal{T}[\![]\!]$ defined by:

$$\mathcal{T}[\![a \triangleleft l(\tilde{v})]\!] = a \triangleleft l(\tilde{v})$$
$$\mathcal{T}[\![a \triangleright [l_1(\tilde{x}_1).P_1 \& \cdots]\!] = a \triangleright [l_1(\tilde{x}_1) = \zeta(\_,\_)\mathcal{T}[\![P_1]\!] \cdots]$$
$$\mathcal{T}[\![!a \triangleright [l_1(\tilde{x}_1).P_1 \& \cdots]\!] = a \triangleright [l_1(\tilde{x}_1) = \zeta(e_1,s_1)(e_1 \triangleright s_1 \parallel \mathcal{T}[\![P_1]\!]) \cdots]$$
$$\mathcal{T}[\![P_1, P_2]\!] = \mathcal{T}[\![P_1]\!] \parallel \mathcal{T}[\![P_2]\!]$$
$$\mathcal{T}[\![\mathsf{v}xP]\!] = \mathsf{v}x(\mathcal{T}[\![P]\!])$$
$$\mathcal{T}[\![0]\!] = \phi$$

5.2   Comparison of the two systems

As the interface of objects does not change, the flattening operation is equivalent to an intersection. In this case, the type constraints resolution detects all the messages that cannot be understood. Our system is more powerful than Vasconcelos' one because it uses subsumption instead of kinded types. Kinds express constraints, but they are limited to the level of interface; for the arguments of messages type equality is required. The following example cannot be typed with Vasconcelos' type system:

$$!b \triangleright [quest(x).(x \triangleleft rep())]$$
$$, \quad !a \triangleright [rep().0]$$
$$, \quad b \triangleleft quest(a)$$
$$, \quad !c \triangleright [rep().0 \ \& \ mess().0]$$
$$, \quad b \triangleleft quest(c)$$

The typing process fails because the two messages $b \lhd quest(a)$ and $b \lhd quest(c)$ contain arguments of different types. But this program can be executed without error because the object b only requires an argument that can, at least, accept the message $rep()$. The translation in CAP of this example can be typed because the system only requires that a and c accept $rep()$ and this requirement is satisfied.

What we want to show with this simple example is that subsumption seems better adapted for concurrent programming than kinded types.

CONCLUSION

In this paper we have presented a static analysis based on type reconstructions using constraints for our primitive actor calculus (CAP). The main problem was the mix of indeterminism and behavior changes; for this purpose, we have define a flat union that allows to deal with all the possible futures of a behavior. This system can statically detect some orphan messages, but some of them are still hidden for the types. We have shown how to use the computed types (an abstract version) to dynamically detect the remaining orphans. The advantages of this approach are: *soft and expressive types*, *no type declaration* and a *relatively low cost analysis*. A prototype of the analysis has been implemented in CaML-Light.

We have also proposed in (Colaço et al. 1997) an analysis of linearity inspired by (Kobayashi, Pierce and Turner 1996). This analysis checks that one name identifies at last one actor; which corresponds to the usual actor discipline in writing CAP expressions.

The complete detection of orphan message needs a more precise abstraction than the presented types. We are working on an extension of structure of types presented here that detect every orphan, but which introduces some limitations on the use of behavior changes.In order to solve this problem, others static analysis techniques like *abstract interpretation* or *effect systems* have to be explored; works on their use in the context of concurrent computation have already been developed (Andreoli, Pareschi and T.Castagnetti 1993, Nielson and Nielson 1993).

REFERENCES

Agha, G.: 1986, *Actors: A model of concurrent computation in distributed systems*, MIT Press, Cambridge, Mass.

Aiken, A. and Wimmers, E.: 1993, Type inclusion constraints and type inference, *Proc. of the ACM Symp. on FPCA.*

Aiken, A., Wimmers, E. and Lakshman, T.: 1994, Soft typing with conditional types, *Proc. of the 21st. ACM Symp. on PoPL.*

Andreoli, J.-M., Pareschi, R. and T.Castagnetti: 1993, Abstract interpretation of linear logic programming, *Proc. of the International Logic Programming Symposium*, pp. 315–334.

Cartwright, R. and Fagan, M.: 1991, Soft typing, *Proc. of the ACM Symp. on PLDI.*

Colaço, J.-L., Pantel, M. and Sallé, P.: 1996, CAP: An actor dedicated process calculus, *ECOOP'96 Workshop on Proof Theory of Concurrent Object-Oriented Programming*.

Colaço, J.-L., Pantel, M. and Sallé, P.: 1997, Analyse de linéarité par typage dans un calcul d'acteurs, *Actes des Journées Francophones des Langages Applicatifs*.

Colaço, J.-L., Pantel, M., Sallé, P. and Senteni, A.: 1996, Un calcul d'acteurs primitifs (CAP), *Actes des Journées Francophones des Langages Applicatifs*, pp. 25–43.

Fuh, Y.-C. and Mishra, P.: 1989, Polymorphic subtype inference, closing the theory-practice gap, *Proc. of the Symp. on TAPSOFT*.

Hewit, C., Bishop, P. and Steiger, R.: 1973, An universal modular actor formalism for artificial intelligence, *Proc. of the IJCAI'73*.

Honda, K. and Tokoro, M.: 1991, An object calculus for asynchronous communication, *in* P. America (ed.), *Proceedings ECOOP '91*, LNCS 512, Springer-Verlag, Geneva, Switzerland, pp. 133–147.

Kobayashi, N., Pierce, B. C. and Turner, D. N.: 1996, Linearity and the pi-calculus, *Proceedings of the ACM Symposium on Principles of Programming Languages*.

Kobayashi, N. and Yonezawa, A.: 1994, Type-theoretic foundations for concurrent object-oriented programming, *Proceedings of ACM SIGPLAN Conference on Object-Oriented Programming Systems, Languages, and Applications (OOPSLA'94)*, pp. 31–45.

Marcoux, A., Maurel, C. and Sallé, P.: 1988, A language for distributed applications, *IEEE Workshop on Future Trends of Distributed Systems in the 90's*.

Milner, R.: 1991, The polyadic π-calculus: a tutorial, *Technical Report ECS-LFCS-91-180*, Laboratory for Foundations of Computer Science, Department of Computer Science, University of Edinburgh, UK. Also in *Logic and Algebra of Specification*, ed. F. L. Bauer, W. Brauer and H. Schwichtenberg, Springer-Verlag, 1993.

Nielson, F. and Nielson, H.: 1993, From cml to process algebras, *CONCUR'93*, LNCS 715, pp. 493–508.

Pantel, M.: 1994, *Représentation et Transformation : Un modèle de la réutilisabilité dans les langages fonctionnels à objets*, PhD thesis, Institut National Polytechnique de Toulouse.

Pierce, B. and Sangiorgi, D.: 1995, Typing and subtyping for mobile processes, *Mathematical Structures in Computer Science* .

Puntigam, F.: 1996, Type for active objects based on trace semantics, *Proc. of Formal Methods for Open Object-based Distributed Systems (FMOODS'96)*, pp. 5–20.

Vasconcelos, V. T. and Tokoro, M.: 1993, A typing system for a calculus of objects, *Proceedings of the International Symposium on Object Technologies for Advanced Software*, LNCS 742, Springer-Verlag, pp. 460–474.

Yonezawa, A.: 1990, *ABCL: An Object-Oriented Concurrent System*, MIT Press.

9

Actors and Virtual Time: an Experience using Time Warp, Timed Petri Nets and Cellular Networks

R. Beraldi, L. Nigro, F. Pupo
Dipartimento di Elettronica, Informatica e Sistemistica
Università della Calabria, I-87036 Rende(CS) - Italy
Voice: +39-984-494748, Fax: +39-984-494713
email: {r.beraldi,l.nigro,f.pupo}@unical.it

Abstract

This paper describes an actor based framework which allows the development of distributed time dependent applications. The approach is centred on light-weight actors and on reflection techniques which provide a time-sensitive message scheduling structure. As a significant application, the paper reports an experience using an implementation of a Time Warp mechanism upon which an effective simulation model of generalised timed Petri nets (TPN) is built. The TPN structure is then used to analyse a formal model for large cellular networks. Some performance measures are finally provided.

Keywords

Actors, reflection, modularity, virtual time, Time Warp, timed Petri nets, cellular networks

1   INTRODUCTION

The Actor model (Agha, 1986) is a well established computational framework suitable for general, time independent, distributed applications. The model is based on the concept of actors, i.e., autonomous agents which communicate to one another by asynchronous and buffered messages. An actor hosts an internal thread on behalf of which a message at a time, received into the actor mail queue, is fetched and processed. Message execution is atomic. At the end of a message execution the actor is ready to process the next message and so forth.

In the last years, Agha et al. (Ren, 1995) (Ren, 1996) (Saito, 1995) have proposed extensions to the actor model in order for it to be applicable to real-time systems. The extensions rely on capturing message *interaction patterns* among actors through a *RTsynchroniser* construct (Ren, 1995). RTsynchronisers are declarative in character. They involve a group of actors and specify constraints on the execution of relevant messages. Constraints reflect upon timing issues of messages (i.e., its invocation time) and the visible state of observed actors. RTsynchronisers provide an elegant and formal tool to control message exchanges. Their concrete application depends on the possibility of ensuring a global time notion in a distributed system. Moreover a selected scheduling structure is required in order to guarantee the timing constraints of messages are ultimately met. A major benefit of the use of RTsynchronisers is modularity, which stems from a separation of functional from timing requirements in a development. Actors are firstly defined according to functional issues only. Then timing aspects are separately specified through RTsynchronisers which affect scheduling.

The work described in this paper aims at experimenting with the concept of actor programming for distributed time sensitive applications. An architecture - DART (Nigro, 1995) (Nigro, 1996a) (Kirk, 1997) Distributed Architecture for Real Time- has been designed to be exploitable in popular object-oriented languages (e.g., C++, Java, ...). Key factors of DART are a light-weight notion of actors which are orchestrated by a customisable reflective control machine which regulates scheduling on a per processor basis. The control machine can reason upon virtual or real time. The control machines of a distributed application can co-operate according to an interaction policy in order to fulfil system-wide timing constraints. A control machine depends on a scheduler actor whose responsibility is to filter messages and to reason upon timing constraints of groups of related application actors. A scheduler can implement a set of RTsynchronisers in the sense of (Ren, 1995).

This paper focuses on virtual time. Section 2 gives an overview to DART concepts using the sequential simulation paradigm. In section 3 the basic principles of distributed simulation are briefly reviewed and the features of an implemented actor-based Time Warp mechanism described. Section 4 presents a timed Petri nets formalism together with an effective execution model which fits naturally in the context of the developed Time Warp kernel. As a significant application, section 5 shows the use of TPNs in the modelling and analysis of the time behaviour of large cellular systems. The speedup of the distributed simulation of a modelled cellular system on a heterogeneous computing environment is shown. Finally, the conclusions and some directions of future work are presented.

2 AN OVERVIEW OF DART

DART (Nigro, 1995) (Nigro, 1996a) (Kirk, 1997) is a specialisation of the Actor model designed for the development of distributed and time dependent systems. At the programming in-the-large level a system consists of a collection of subsystems, one per processor, interconnected through a (possibly deterministic) communications system (e.g., CAN (Kirk, 1995)). Each subsystem hosts a control machine, a scheduler actor and a set of application actors. Discipline of programming in-the-large can suggest having only one *administrator* actor being referenced at the system level by a unique identifier. It receives message requests and possibly delegates its processing to inner subsystem actors. The control machine provides the necessary support for scheduling and message select and dispatch.

Actors are the basic building blocks in-the-small. An actor is modelled as a finite state machine which evolves through a lifecycle (Shlaer, 1992), i.e., a succession of states (behaviours). It is a reactive object which responds to incoming messages on the basis of current state and message contents. Message reception is implicit. An actor is at rest until a message arrives. Message processing triggers a state transition and the execution of an action. Action execution is atomic and cannot be pre-empted nor suspended. Messages can be unexpected in the current state. Their processing can be postponed by remembering them on local data or in states of the lifecycle. Basic operations on actors are: (a) *new*, for creating a new actor, (b) *become*, for changing the actor state, (c) (non blocking) *send* for transmitting messages to acquaintances actors (including itself). Differences from the Actor model are:

- an absence of an internal thread in actors. DART actors are instance of a (passive) class which extends the Actor base class. Life cycle is provided by a *message handler* (Reiser, 1992) which is invoked by the control machine at message dispatch time
- an absence of the mail queue per actor. Rather, a single message queue is introduced by the control machine upon which sent messages are scheduled, selected and dispatched according to a proper control structure

- source of concurrency among actors in a same subsystem is *action interleaving* ensured by the control machine dispatching scheme. Concurrency among actors allocated upon different processors is true parallelism.

Besides actors, DART allows for the existence of passive objects too which are lacking of a life cycle. Their methods can be invoked by the usual procedure-call synchronous semantics. Since the atomic character of action execution in actors, passive objects behave naturally as monitors and there is no need to introduce conventional mutual exclusion primitives.

2.1   Reflection and control machine

The dynamic behaviour of actors is influenced by a control discipline (Nigro, 1994) adopted at the control machine level. Here the control structure can be pure event-driven as in the Actor model, useful for concurrent applications, pure time-driven suited for hard real time systems (Nigro, 1996b), or a combination of the two. The control machine relies on the concept of reflection (Maes, 1987). It can reason on a time dimension and is causally connected with application actors. Basic components of a control machine (see also Figure 1) are:

- a *clock*
- a *message queue* (i.e., a calendar, plan or event list)
- a *controller*
- a (programmer defined) *scheduler* actor.

The control machine is fed of a particular scheduler at its initialisation. The time notion can be virtual or "real" time. In a simple case, messages are coupled with a *timestamp* indicating the occurrence time of the event captured by the message. More in general, for real-time applications, a message can be accompanied by a validity time window $[t_{min}, t_{max}]$ which specifies possible delivery times (Nigro, 1996b) (Kirk, 1997). The controller method of a simulation control engine repeats a basic message loop (sequential simulation engine). At each iteration

- a message is selected from the message queue and its timestamp used to adjust the simulated clock
- the message is delivered to its target actor by invoking the relevant message handler which in turn causes a state switch and triggers an action into execution
- at the action termination, the control loop is resumed and the scheduler is given a chance to schedule the "just sent messages".

The loop is repeated until the virtual time exceeds the simulation period or the simulation deadlocks.

The controllers of a set of distributed control machines can co-operate in order to fulfil system-wide synchronisation constraints. To this purpose, they can rely on a limited set of *control messages* which are hidden to application actors.

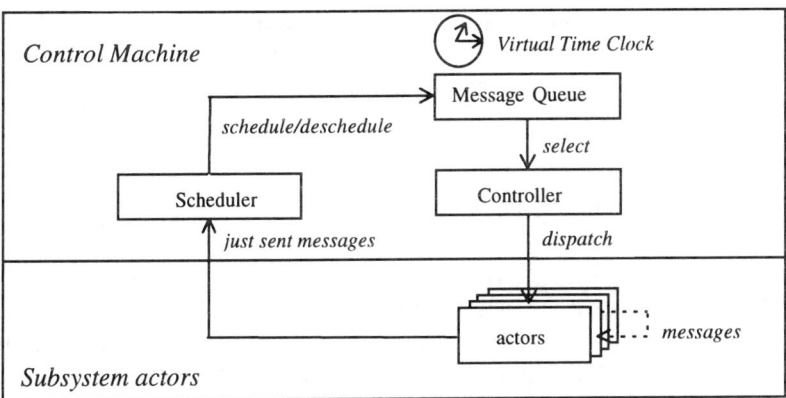

Figure 1 The organisation of a sequential simulation Control Machine

The time behaviour of a subsystem actors is specified separately from actor functional behaviour in a scheduler actor which knows application time requirements. The following interface is offered for programming a scheduler (Ren 1995) (Kirk, 1997):

- msg.*cause*() which returns the identity of the message whose processing generated msg
- msg.*iTime*() which returns the dispatch time of message msg
- *schedule*(msg, timestamp) which schedules message msg by attaching timestamp to it
- *deschedule*(msg, timestamp) which removes a previously scheduled message msg
- *now*() which returns the current time.

A scheduler hosts a set of *time clauses*. To exemplify, a cyclic actor of a class Source can have a message class Generate an instance of which is continually sent to itself for sustaining a generation process. The timestamp of the message can be defined by an exponential random variate. The following is a possible time clause for scheduling such a message (using a Java-like syntax):

```
if( msg instanceof Generate && msg.cause()==msg )
    schedule( msg, now()+Random.exponential( lambda ) );
```

where lambda is the generation mean of the physical process modelled by Source class.

A key factor of DART is a smooth transition from the event-driven to the time-driven paradigm, with actors which are not aware of the specific control discipline enforced by the control machine. This also favours modularity since actors can be reused according to different application requirements.

The control machine and scheduler actor can effectively be programmed using basic mechanisms (dynamic binding, polymorphism and runtime type identification) of an object-oriented language like Oberon-2, C++, Java ... The resultant actor programming style is safe, clean and sound (Nigro, 1995) (Kirk, 1997).

3   DISTRIBUTED SIMULATION AND TIME WARP

The simulation of complex, asynchronous and discrete event systems, e.g. formal models of large personal communications systems, can be very expensive in terms of the required computational resources (memory space and cpu time) on a single processor machine. Such simulations can potentially be accelerated by using parallel or distributed architectures. Different parts -*logical processes* LPs- of the modelled system are allocated on different processors and simulated locally by a sequential simulation engine. However, concurrent execution of messages related to different points in the simulation, on different LPs/processors, while is the source of possible speedup it also introduces a need of *synchronisation protocols* which are at the heart of the parallel or distributed simulation problem, and which can shorten the actual achieved speedup. Such protocols must ensure the ordering of events with respect to virtual time as in the sequential simulation, thus preserving the causality of events. Synchronisation mechanisms broadly fall into two categories: *conservative* and *optimistic*. Conservative protocols (Misra, 1986) (Fujimoto, 1990) prevents causality errors ever occurring (by blocking an LP would there be a chance to process an "unsafe" event, i.e., one for which causal dependencies are still pending). Optimistic protocols (Jefferson, 1987) (Fujimoto, 1990), using a *detection and recovery* approach, provide for an LP to redo the simulation of an event should it detect that premature processing of local events is inconsistent with causality conditions generated by other LPs.

Time Warp (TW) (Jefferson, 1987) (Fujimoto, 1990) is an optimistic strategy supporting distributed discrete event simulation. Every LP has a local clock (Local Virtual Time, LVT) and a separate message queue which drives a local simulation algorithm. LPs are allowed to proceed asynchronously, thus experimenting different values of LVTs at a same real time. However, if an LP receives an external message (*straggler*) whose timestamp is lower than LP's LVT, the basic synchronisation problem of TW arises. To avoid *causality errors* (the future can't influence its past!), the state of the LP must be *rolled back* at a virtual time less than or equal to straggler's timestamp, by undoing the effects of incorrectly performed forward computation. The undo process can propagate to partner LPs.

A rolled back LP proceeds again (*coasting forward phase*) into its future towards the straggler's timestamp. During this phase external messages are not re-sent. The straggler is then processed. After that forward computation is restarted according to one of two basic techniques: *aggressive cancellation* (AC) or *lazy cancellation* (LC) of previously sent (possibly erroneous) messages. *Anti-messages* are maintained for any externally sent message. AC sends immediately all the anti-messages as a prompt undo request. An anti-message gives rise to a roll back in an LP if the corresponding positive message was already processed. LC refrains from doing so. Only in the case the original LP, during its new forward computation, doesn't re-generate an external message, the corresponding anti-message is sent. AC and LC have shown their relative performance into different application domains (Fujimoto, 1990).

TW is characterised by its necessary frequent saving of LP's state (*checkpointing*). The Global Virtual Time (GVT) is the so far committed simulation time of the whole simulation model. No LP can be rolled back at a time prior to GVT. As a consequence, each LP can free the memory space of no more useful state versions (*fossil collection*).

A critical issue of TW is how frequent GVT should be updated, since a high frequency conserves memory but wastes real time, a low frequency can result in an out-of-memory exception.

3.1 A DART based Time Warp Mechanism

Time Warp can be viewed as a particular interaction policy among the control machines of a DART simulation model. The following summarises key points of an achieved TW mechanism based on C++ and PVM (Geist, 1994). An LP includes a control machine and a collection of application actors (subsystem). A single LP is allocated onto one physical processor. A message is characterised by the tuple (*sender-LP, timestamp*). The timestamp consists of the pair (ts, tr) where ts is the "send time", i.e., the LVT of the sender LP at the moment of transmission; $tr \geq ts$ is the "receive time", i.e., the virtual time at which the receiver LP should process the message.

A distinguishing factor of the implemented TW is the adoption of a *modified aggressive cancellation* (MAC) (Beraldi, 1996) technique which operates as follows. An LP affected by roll back at time τ must undo all the erroneous computation prematurely carried out at times $\tau' > \tau$. This first requires, depending on the availability of state versions, a saved state to be withdrawn with a time $\tau'' < \tau$ and re-installed. After the coasting phase from τ'' to τ, the incorrect external computation is cancelled. In order to minimise the communication overhead, a single *undo* message is transmitted to each partner LP to which messages have been sent with $ts \geq \tau$. It requires the cancellation of all the messages sent by the rolled back LP, whose $tr \geq undo.tr$. The cancellation process annihilates positive messages still pending in the receiver LP or it raises a further roll back. MAC is capable of annihilating previously sent messages very quickly, a required feature in *self-driving* simulations (Carothers, 1994).

Performance of TW is critically affected by the values of State Save Rate (SSR) and Maximum Number of State Versions (MNSV) parameters (Beraldi, 1996). A value s of SSR indicates that the LP status will be copied just before the s-th LVT change from the last saved version. A value m of MNSV specifies that the GVT updating procedure will be invoked after m state versions are stored on SVL and there is the need to checkpoint the LP again. A preliminary step for an effective use of TW consists of *parameter tuning*, which in turn is tied to load balancing conditions (see later in this paper).

4   DISTRIBUTED SIMULATION OF TIMED PETRI NETS

Petri nets (Murata, 1989) are widely used as a modelling tool for studying asynchronous concurrent systems. Quantitative properties of modelled systems can be evaluated by exploiting the notion of time explicitly added to the classical definition of Petri nets. In the following the class of timed Petri nets (TPNs) where timing information is associated to transitions (Ghezzi, 1991) (Chiola, 1993a) is considered.

A TPN is a tuple (P, T, F, W, τ, M) where

- $P = \{p_1, p_2, ..., p_{nP}\}$ is a finite set of P -elements (places),
- $T = \{t_1, t_2, ..., t_{nT}\}$ is a finite set of T-elements (transitions) with $P \cap T = \varnothing$ and a non empty set of nodes ($P \cup T \neq \varnothing$).
- $F \subseteq (P \times T) \cup (T \times P)$ is a finite set of arcs between P-elements and T-elements.
- $W: F \rightarrow N$ assigns weights $w(f)$ to elements of $f \in F$ denoting the multiplicity of unary arcs between the connected nodes.
- $\Pi: T \rightarrow N$ assigns priorities π_i to T-elements $t_i \in T$.
- $\tau: T \rightarrow \Re$ assigns firing delays τ_i to T-elements $t_i \in T$.
- $M: P \rightarrow N^{+0}$ is the marking $\mu_i^{(0)} = \mu^{(0)}(p_i)$ of P-elements $p_i \in P$.

There are three important cases of firing delays, namely the case when $\tau_i = 0$ (*immediate transitions*), the case when $\tau_i \in \Re$ is a deterministic time value (*deterministic timed transitions*) (Ghezzi, 1991), and the case when τ_i is an instance of a random variable (*stochastic timed transitions*). If T contains only stochastic timed transitions where the firing delay random variable is exponentially distributed, T belongs to the class of *Stochastic Petri Nets* (SPNs). *Generalized* SPNs *(GSPNs)* (Chiola, 1993a) allow a combination of non timed (immediate) and stochastically timed transitions.

4.1 Timed enabling and firing semantics

- Let $I(t)$ and $O(t)$ denote the set of input and respectively output places of $t \in T$. A transition $t \in T$ is *enabled* in some marking μ at time τ, iff $\forall p \in I(t), \mu(p) \geq w(p(t))$ in μ.

- If $E_\tau(\mu)$, i.e. the set of all transitions enabled in μ at time τ, contains immediate and timed transitions, then immediate transitions are always selected prior to timed transitions for firing.
- If $E_\tau(\mu)$ contains only timed transitions, and RET(t_i), $t_i \in E_\tau(\mu)$ is the remaining enabling time of t_i, then the transition t that *must* fire next is the one with $t \mid min_i RET(t_i)$.
- If $\forall p \in I(t), \mu(p) \geq cw(p(t))$, and $c > 1$, then t is said to be multiply enabled at degree c. If t can only fire one enabling at a time, it adopts the *Single Server* (SS) semantics. *Infinite Server* (IS) semantics occurs when *any* amount of enablings can be fired at a same time, expressing a notion of parallelism among them.

There is a variety of different firing rules defined upon TPNs. In SPN, for example, transition firing is atomic. A random time elapses between the enabling and the firing of a transition t_i, during which the enabling token(s) reside(s) in the input place(s). Transition t_i must be continuously enabled during the time τ_i, and fires at that time (*race or preemptive policy*). Another execution rule for TPNs occurs when an enabled transition fires in three phases: in a "start firing" phase tokens are removed from the input places, remaining invisible for a "firing in progress" phase until they are released into output places in the "end firing" phase (*preselection or non preemptive policy*).

4.2   A TPN execution model based on DART/Time Warp

TPNs easily allow modelling and analysis of dynamic discrete event systems. Basically the causality of events is directly expressed by the net structure, whereas the dynamic behaviour is modelled by associating firing delays (i.e., events) to transitions.

The execution of a TPN model can be instrumented on a single machine (sequential simulation). However, for the animation of complex systems parallel or distributed simulation should be used (Chiola, 1993b) (Ferscha, 1994). The basic idea is to separate topological TPN parts to be simulated by logical processes (LPs). For performance reasons (Chiola, 1993c) the TPN model can be partitioned into a number of *regions* in such a way that conflicting transitions together with all their input places reside in the same LP. In the following the attention will be on TPNs where, for simplicity, $w(f)=1 \ \forall f \in F$, and the enabling and firing semantics are centred on preselection and infinite server policy. Moreover, the enablings can use inhibitor arcs.

An obvious mapping of TPNs on DART/TW would be that of associating distinct actors to transitions and places of a TPN, with an initial configuration which creates all the actors of a region and initialises them according to the region topology in terms of acquaintances relations. Transition and place actors would interact by suitable messages during conflict resolution of enabled transitions and firings. However, a different architecture was actually chosen with the aims of

- allowing for the configuration of coarse grained regions adequate for a distributed system of workstations
- minimising the number of exchanged messages, which is a critical factor for the simulation of complex systems
- facilitating the saving/restoring operations of the region (LP) status required by Time Warp.

A single actor of a basic Region class is introduced per LP which is initialised by a (passive) data structure describing the region topology. The Region class implements the functional TPN execution model. The internal status of a region actor includes a vector *Marking* having an entry for each place in the region, and a vector *Firings* which registers the history of transition firings for statistical data collection. A region understands two messages: *transition_fire* and *token_arrival*. A message *transition_fire* is sent by a Region actor to itself for firing a given transition *t*. The corresponding action carries the following execution steps

1. vector *Firings* is modified by annotating the firing of transition *t*
2. vector *Marking* is changed by depositing a token into each place $p \in O(t)$
3. a *list of enabled transitions*, determined by step 2, is built and for each enabled transition a new *transition_fire* message is generated and sent. The parameter of *transition_fire* consists of a transition identifier or a list of conflicting transitions. Resolution of conflicts is out of the scope of a Region actor.

A token generated into a region *r'* but destined to a place located into a different region *r* is carried by a *token_arrival* (external) message. The corresponding action accomplishes the steps 2 and 3 of a *transition_fire* action.

A MetaRegion actor is introduced which is used as a scheduler and arbiter and reflects on the timing behaviour of a corresponding Region actor. A meta region is fed of all the timing attributes of the corresponding region net. In particular, for each timed transition its exponential random variate mean λ, and possibly priority and firing probability of immediate transitions, are given. A meta region captures *transition_fire* messages and resolves conflicting immediate transitions by applying associated priorities and then by taking into account the probability attributes. For a list of conflicting timed transitions, the meta region finds a winner t_w by evaluating a minimum timestamp among the transitions, and actually schedules the *transition_fire* with *only* t_w as a parameter. The proposed timestamp of a conflicting timed transition t_c is defined by $now()+t_c$'next sample random variate. It is worthy of note that all the *transition_fire* messages corresponding to the list of enabled transitions determined at step 2, are *instantaneous*. In other words, they are "just sent messages" which will be evaluated by the meta region starting from the same time horizon given by current value of LVT. This is important to ensure that the infinite server firing semantics is correctly applied.

The separation of concerns between the actors region and meta region favours region reusability in the presence of different timed enabling and firing semantics. The meta region could adopt, transparently, infinite or single server semantics (by dropping multiple enabled conflicting transitions carried in a same set of "just sent *transition_fire* messages"), non preemptive or preemptive policy (by descheduling a previously scheduled *transition_fire* message related to a conflicting transition whose timestamp is greater than that of a newly scheduled *transition_fire* message).

5   A MODEL FOR LARGE CELLULAR NETWORKS

A TPN model is presented for a *personal communication services* network (PCS) (Carothers, 1994), i.e., a cellular communication system providing voice services to *Mobile Subscribers* (MS). The service area has a wrap-around Manhattan-like topology partitioned into regular sub-areas called *cells*. The service within a cell is supplied by a *Base Station* (BS) which is identified by its coordinates (x,y).

The behaviour of a mobile user, modelled as a stateless token, is formalised in Figure 2 where a TPN model of a generic cell is portrayed.

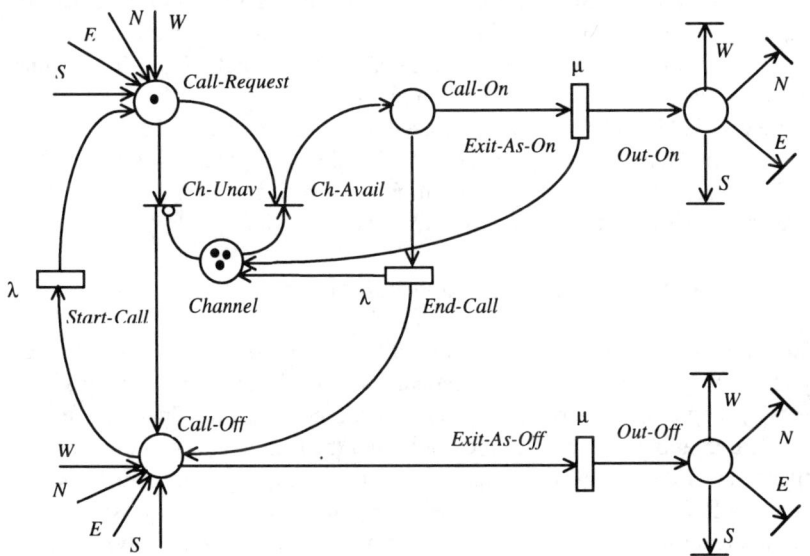

Figure 2 A TPN model of a cell.

Each cell is assigned a fixed number of channels. Movement and call issues are orthogonal one to another. A user can enter a cell with or without a call in progress. If a call is on, the old channel is released to the exiting cell and a new channel requested to the entering cell (*handover* procedure). If no channel is available, the call is blocked. Otherwise the call is continued with a newly assigned channel. Quality of service can require designing the system (e.g., number of channels per cell) so as to keep below a given value the *blocking probability* (e.g., 10^{-4}).

In Figure 2 timed transitions are portrayed by squares. A user with a call in progress is received into the Call-Request place where the handover procedure is handled. A user without a call in progress is accepted into the Call-Off place. From Call-Off a new call may initiate (transition Start-Call). Would a channel be not available (transition Ch-Unav), a user-token moves from Call-Request into the Call-Off place (a call is rejected or a handover fails). In the Call-On state the user can abandon this cell (transition Exit-As-On) or he/she can terminate the call (transition End-Call) moving to the Call-Off place. In both cases, the utilised channel is first released to the Channel pool. Finally, a user can exit the cell without a call in progress (transition Exit-As-Off). Out-On and Out-Off are exiting places respectively of users with or without a call in progress. From an output place a user-token can non deterministically move to one of the four neighbouring cells, respectively located at its North, East, South and West. The exit transitions from Out-On (resp. Out-Off) are linked to incoming arcs to Call-Request (resp. Call-Off) places of near cells.

Statistical timed behaviour is achieved by timed transitions and related exponential random variate means. Transitions Exit-As-On and Exit-As-Off have mean $1/\mu$ (*dwell time*), transitions Start-Call and End-Call have mean $1/\lambda$ (*call interarrival time*). Non determinism among input/output W, N, E, S immediate transitions is achieved by attaching to them the probability 0.25.

The cell TPN model can easily be used to collect statistical data of the physical system. To exemplify, one can estimate the blocking probability by counting the total number of firings of the transition Ch-Unav and dividing it by the total number of firings of transitions Ch-Unav and Ch-Avail.

5.1   System partitioning and load balancing

The cell model of Figure 2 has been used to build large (e.g., 30x30 cells) cellular systems by spatially replicating identical cells, and to analyse them by simulation on a heterogeneous distributed system composed of a Sun Sparc1, Sun Sparc4, Sun Sparc5 and a HP9000 Apollo connected through a standard and shared with other users Ethernet at 10 Mb/s. A system is partitioned into four rectangular regions $<R_0,R_1,R_2,R_3>$. It is intended that the cells in a region Ri can directly communicate with cells in the neighbouring region. A configuration is specified by a tuple (x_0,x_1,x_2). A region is captured in a different LP/processor with a particular pair (Region, MetaRegion) of actors. The meta region is initialised with the timing attributes of a single cell: the means of timed transitions λ and μ and a single random generator on the basis of which the exponential variates are achieved. The saving/restoring of the meta region saves/restores the state of the random generator. *transition_fire* messages identify enabled transitions through their name and unique identifier derived from the belonging cell coordinates.

Before simulating a cellular system, a configuration (x_0, x_1, x_2) capable of ensuring good performance through parameter tuning (see section 3.1) of the Time Warp engine was determined. As a measure of good static load balancing was assumed a homogeneous number of roll backs on the various processors. More precisely, a certain number of preliminary runs have been conducted to establish suitable values of checkpointing parameters SSR and MNSV. After that, some experiments have been carried out for performance evaluation of the TPN based PCS distributed simulation versus a corresponding sequential one. Some results are summarised in the next section.

5.2 Experimental results

A system with 30x30 cells was investigated under 10 channels per cell, a varying number of users and an assigned dwell time $1/\mu$ and interarrival time $1/\lambda$. A good partition was found with the configuration (11,18,25) to which corresponds the regions $R_0=[0,10]x[0,29]$ allocated on Sparc4, $R_1=[11,17]x[0,29]$ allocated on HP9000 Apollo, $R_2=[18,24]x[0,29]$ allocated on Sparc5 and $R_3=[25,29]x[0,29]$ allocated on Sparc1. The values SSR=0 and MNSV=3 improve Time Warp performance for the chosen system.

Speedup is shown in Figure 3. It was calculated as a ratio between the completion time of the sequential simulation, carried on the fastest machine (Sun Sparc4) and the completion time of the distributed simulation. Each point was determined by taking the mean of 3 runs. Good speedup is obtained for a number of users above 2700. Under such load conditions, processors do not experiment an excessive difference in their LVTs which in turn diminishes the number of observed roll backs.

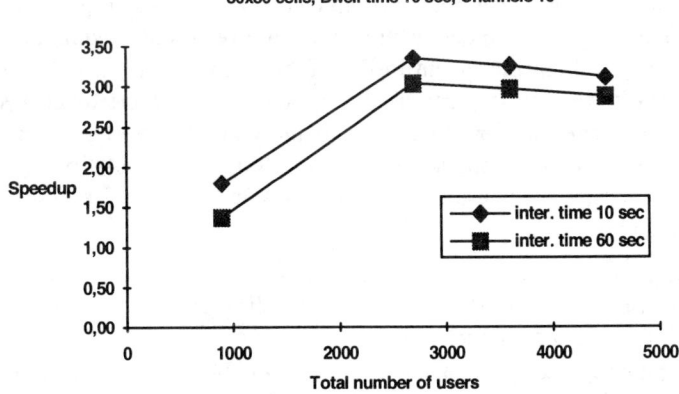

Figure 3 Speedup (4 processors)

As one can see from Figure 3, when the number of users exceeds 3000 the speedup starts decreasing. This behaviour relates to the augmented number of messages (and size of Message Queue) to be handled by each TW control machine. Although the roll back number remains equilibrated among the different processors, the state saving (checkpointing) and roll back operations (cancellation of no longer useful state versions, coasting forward, undo message handling) become more costly. Additional computational resources (processors) should be used. Table 1 depicts some measured values of the blocking probability.

Table 1 Measures of blocking probability

number of users	blocking probability
3600	1.63×10^{-4}
4500	3.75×10^{-4}

6   CONCLUSIONS

This paper describes an actor based architecture, DART, which is suited to the development of distributed and time dependent applications. DART is centred on the concepts of light-weight actors and control machine which hides a particular control engine orchestrating the evolution of a cluster of tightly coupled actors (subsystem). The control machine is capable of reasoning on local timing constraints. A programmer-defined scheduler object can host a set of timing clauses which directly affect message scheduling. Timing issues are confined and dealt with in the scheduler and control machine. This in turn improves modularity and reusability of actors which are not aware of timing.

Control machines of a distributed system can co-operate according to a system-wide interaction policy in order to fulfil system-level timing requirements. The paper summarises the achievement in DART of a Time Warp mechanism which is simple yet effective. It has been used to concurrently execute timed Petri nets, i.e., a powerful formalism for modelling and analysing discrete and asynchronous systems. As a significant application, the modelling and simulation of complex cellular networks has been described, along with some experimental results. The on-going research aims at

- extending the DART based execution model of timed Petri nets by implementing different policies of timed enabling and firing semantics, so as to widen their field of applicability
- applying the separation of concerns approach of DART to distributed real-time systems modelled, visualised and analysed both from the point of view of functional and temporal behaviour through timed Petri nets (Nigro, 1996a)

- porting DART in Java in order to improve object design, memory management and platform neutral realisations.

Acknowledgements

The authors are grateful to Francesco Tisato and Brian Kirk for the helpful discussions during the preparation of the paper.

REFERENCES

Agha G. (1986) Actors: A model for concurrent computation in distributed systems. MIT Press.

Beraldi R., Marano S., Nigro L. (1996) Distributed simulation of PCS networks using a Time Warp mechanism, in *Eurosim'96, HPCN*, Dekker L., Smit W. and Zuidervaart (eds), North-Holland, 307-314.

Carothers C.D., Fujimoto R.M., Lin Y.-B., England P. (1994) Distributed simulation of large-scale PCS networks. *Mascots'94*.

Chiola G., Ajmone Marsan M., Balbo G., Conte G. (1993a) Generalized stochastic Petri nets: a definition at net level and its implications. *IEEE Transactions on Software Engineering* **19**(2), 89-107.

Chiola G., Ferscha A. (1993b) Distributed simulation of Petri nets. *IEEE Parallel and Distributed Technology* **1**(3), 33-50.

Chiola G., Ferscha A. (1993c): Distributed simulation of timed Petri nets: exploiting the net structure to obtain efficiency. *Proc. of the 14th International Conference on Application and Theory of Petri Nets*, Chicago, Illinois. Springer Verlag.

Ferscha A. (1994) Concurrent execution of timed Petri nets. *Proc. of the 1994 Winter Simulation Conference*, Lake Buena Vista, Florida, USA, 229-236.

Fujimoto R.M. (1990) Parallel discrete event simulation. *Communications of the ACM* **33**(10), 30-53.

Geist A. et al. (1994) PVM: Parallel Virtual Machine - A users guide and tutorial for networked parallel computing. The MIT Press.

Ghezzi C., Mandrioli D., Morasca S., Pezzè M. (1991) A unified high-level Petri net formalism for time-critical systems. *IEEE Transactions On Software Engineering*, **17**(2), 160-172.

Jefferson D., et al. (1987) Distributed simulation and the Time Warp Operating System. *ACM Symposium on Operating Systems Principles*, 77-93.

Kirk B. (1995) Real time protocol design for control area networks. *Proc. of Real Time'95 Conf.*, Ostrava (Cz Rep.), 251-268.

Kirk B., Nigro L., Pupo F. (1997) Using real time constraints for modularisation. *Proc. of Joint Modular Languages Conference'97*, 19-21 March, Linz (Austria), Springer-Verlag, LNCS 1204, 236-251.

Kumar D., Harous S. (1994) Distributed simulation of timed Petri nets: basic problems and their resolution. *IEEE Transaction on Systems, Man, and Cybernetics* **24**(10), 1498-1510.

Maes P. (1987) Concepts and experiments in computational reflection. *Proc. of OOPSLA'87, ACM SIGPLAN Notices*, **22**(12), 147-144.

Misra J. (1986) Distributed discrete event simulation. *ACM Computing Surveys* **18**(1), 39-65.

Murata T. (1989) Petri nets: properties, analysis and applications. *Proc. of IEEE*, **77**(4), 541-580.

Nigro L. (1994) Control extensions in C++. *J. of Object Oriented Programming*, **6**(9):37-47.

Nigro L. (1995) A real time architecture based on Shlaer-Mellor object lifecycles. *J. of Object Oriented Programming*, **8**(1):20-31.

Nigro L., Pupo F. (1996a) Modelling and analysing DART systems through high level Petri nets. Springer-Verlag, LNCS 1091, 420-439.

Nigro L., Tisato F. (1996b) Timing as a programming in-the-large issue. *Microsystems and Microprocessors J.*, **20**(4), 211-223.

Reiser M., Wirth N. (1992) Programming in Oberon - steps beyond Pascal and Modula. Addison-Wesley.

Ren S., Agha G. (1995) RTsynchroniser: language support for real-time specification in distributed systems. *ACM SIGPLAN Notices*, **30**(11), 50-59.

Ren S., Agha G., Saito M. (1996) A modular approach for programming distributed real-time systems. *J. of Parallel and Distributed Computing*, Special issue on Object-Oriented Real-Time Systems.

Saito M., Agha G. (1995) A modular approach to real-time synchronisation, in Object-Oriented Real-Time Systems Workshop, 13-22, *OOPS Messenger, ACM SIGPLAN*.

Shlaer S., Mellor S. (1992) Object lifecycles - Modelling the world in states. Yourdon Press.

BIOGRAPHY

Roberto Beraldi has received in 1996 a PhD degree in Computer Science from University of Calabria. His research interests include: distributed systems, parallel simulation, Time Warp, performance evaluation of wireless systems.

Libero Nigro is an Associate Professor of "Fondamenti di Informatica" at the University of Calabria. He is also responsible of "Sistemi di Elaborazione". His research interests include: object oriented software engineering, actor systems, scheduling control, distributed real time systems, parallel simulation, timed Petri nets.

Francesco Pupo is a PhD student at University of Calabria. He is interested in Petri nets, actor systems, real time, discrete event simulation.

10

Formalizing Multimedia QoS Constraints Using Actors

Shangping Ren, Nalini Venkatasubramanian and Gul Agha
Open Systems Laboratory
Department of Computer Science
1304 W. Springfield Avenue
University of Illinois at Urbana-Champaign
Urbana, IL 61801, USA
Email: ren/nalini/agha@cs.uiuc.edu
Web: http://www-osl.cs.uiuc.edu

Abstract

The vision of future information systems is that different forms of information are potentially accessible at anytime through the Internet. We describe challenges in the modeling and specification of timing related multimedia (MM) services in open distributed systems. Management of multimedia services in an open system is complicated by the heterogeneity of application requirements, multimedia information, and system components. Services and systems in this environment evolve dynamically and their components interact with an environment that is not under their control. We propose a formal specification of timing related Quality-of-Service (QoS) attributes in an actor-based distributed system and describe some techniques for informally reasoning about quantitative QoS properties.

Keywords

QoS, Real-time, Multimedia, Actors, Concurrency semantics, Sessions

1 MOTIVATION

Services and systems in a global information environment like the WWW evolve dynamically and their components interact with an environment that is not under their control. Many multimedia (MM) applications can tolerate relatively minor and infrequent violations of their performance requirements; the extent to which such violations are permissible is specified as a quality-of-service (QoS) parameter. In this paper, we discuss preliminary work on the specification and semantics of timing based Quality-of-Service (QoS) parameters in multimedia applications. Our goal is to provide formal specifications of QoS in a distributed MM system and thereby reason about the formal validation of quantitative properties. QoS statements may specify constraints on timing, availability, security, and resource utilization. However, we will only address the specification of timing based QoS constraints.

Multimedia applications are characterized by the presence of QoS requirements for resolution, tolerable jitter, delays etc. With multimedia applications, correctness and semantics of the application depends heavily on the timeliness of the interactions among different system components. We focus on timing based QoS requirements and their specification in the actor model. Consider the following timing-based QoS properties:

- **End-to-end Delay (EED):** represents the sum total of the delays experienced between the service-provider and end-user . Simplistically, it can be stated as the difference in time between the sending of a message at the server sv (t^{sv}_{send}) and arrival of a message at a client cl (t^{cl}_{arr}) , assuming that the sender and receiver clocks are synchronized.

$$EED = t^{cl}_{arr} - t^{sv}_{send} \tag{1}$$

- **Jitter:** quantifies the perceived deviation from expected bit-rate, and measures the expected inter-arrival times of the frames in a stream. It is defined as the variation in delays experienced by two consecutive packets of a session transmitted across a network. Violation of jitter bounds portrays itself as flickering video, distorted audio, blanks and frozen images. The jitter j^i is defined as

$$j^i = (\delta^i - \delta^{i-1}) \tag{2}$$

where δ^i is the difference between arrival times of packets i and $i+1$; δ^{i-1} is the difference between arrival times of packet $i-1$ and i.
- **Synchronization Skew:** quantifies the skew i.e. the difference in arrival rates between 2 multimedia streams, for example audio and video. Let $t^{i,a}$ and $t^{i,v}$ be time points of two packets from different media which should be synchronized in the same time interval. The synchronization skew σ is defined as:

$$\sigma = t^{i,a} - t^{i,v} \tag{3}$$

For acceptable QoS in a distributed multimedia system, these parameters are bounded.

1.1 Actors and Multimedia

A drawback with the traditional method of QoS specification is that the specification of QoS requirements is intermixed with the service specification [6, 7]. While the correctness of program execution, e.g. data delivery, relies on meeting the QoS requirements, merging the specifications results in loss of modularity and complicates the correctness validation process. In general, models and methodologies that have been developed for sequential programming are inadequate for creating correct distributed applications: to simplify distributed programming, a model of concurrent computation that provides facilities for the modular specification of distributed interactive applications is needed.

The *Actor* model of computation [1] has a built-in notion of encapsulation and interaction, and thus it is a natural model to use as a basis for interactive applications. Actors can be viewed as a model of coordination between autonomous interacting components. Actors can be dynamically created. They communicate via message passing which is asynchronous and fair. The communication topology of an actor system is called the *acquaintance relation* and can change dynamically. Semantics of actor interactions are relatively well understood and reasoning about systems of actors has be formalized [2, 11].

Specifying QoS in the Actor based model is essentially a problem of specifying coordination constraints between distributed objects. Synchronizers [5] allow us to express these coordination constraints, i.e. local synchronization constraints or multi-actor coordination constraints [4]. Synchronizers allow us to qualitatively control the semantics of message delivery. Earlier work separates real-time constraints from the computational aspects of an application; real-time constraints are described by synchronization code between the interfaces of objects [9]. Objects in this system are defined using a real-time variant of the Actor model. A high-level programming language construct called *RTsynchronizer*, specifies a collection of temporal constraints between actors. This approach helps separate what an object does from when it does it and facilitates the ability to dynamically modify real-time constraints.

1.2 QoS Synchronizers

The structure of a unified system is depicted in Figure 1. We can visualize MM applications as a collection of autonomous, concurrent information processing actors called *media-actors* involved in a multimedia session.

A session consists of a set of media-actors that interact with each other to achieve some common goal, e.g. a video-on-demand session or a multimedia conferencing session. To incorporate the notion of a session in our model, we propose a special entity called the *QoS synchronizer* that manages QoS constraints and specifications, and verifies invariants between a group of

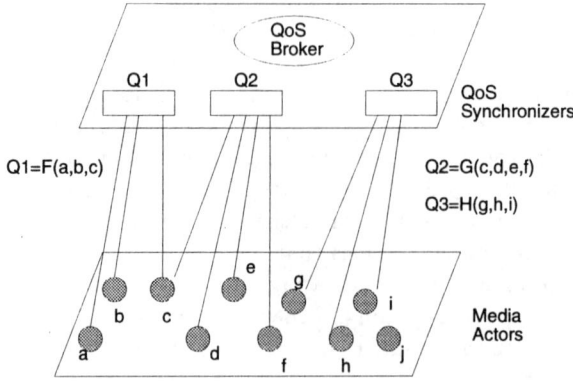

Figure 1 Separation of concerns - QoS requirements specified by system level QoS actors and application functionality encapsulated in media-actors.

media-actors. *QoS synchronizers* are special actors with QoS attributes, e.g. synchronization or timing-related information. Using the concept of QoS Synchronizers permits the modularization of QoS requirements and encourages separation of concerns. By specifying program behavior and QoS constraints separately, we reason about program and system correctness independent of the constraints. We can also separately reason about the validity and satisfiability of the QoS constraints, then embed the QoS constraints in the system in a modular way and reason about the composite correctness. Independently specifying the QoS constraints and the service gives us the ability to modify one without impacting the other. A QoS synchronizer can observe transitions on the set of media-actors it controls. It cannot send messages to or receive messages from media-actors but can affect their execution.

In general, a multimedia session consists of

- a collection of actors (media-actors) that are involved in that session (components at the server and client ends),
- a session synchronizer that holds and enforces QoS constraints between media-actors in a session.
- a notion of time that is uniform among media-actors of that session.

For the sake of simplicity, we assume that there is only one ongoing session in the system at a time. In a system with multiple sessions, we need to address the issue of satisfying QoS constraints for each session. For this, we propose an agent called the *QoS broker* [8] that acts as a coordinator for all the ongoing sessions and performs admission control for new incoming sessions. Design criteria for an actor-based QoS broker, its components and interactions in a distributed multimedia environment is beyond the scope of this paper and is a future area of research [12].

We define a QoS Synchronizer QosSync encoded in RTsynchronizer syntax to express timing-based QoS requirements by an example illustrated in Figure 2. An RTsynchronizer specification consists of

- a declaration of state variables which may be changed by observing certain events on the controlled media-actors;
- a set of constraints that need to be enforced; and
- rules for state change.

In this example, we consider a multimedia session with 2 streams of information. The streams have the following QoS requirements: (1) Jitter less than maxJitter (2) End-to-end delay EED less than maxEED; (3) Interstream synchronization skew less than maxSkew. These constraint values may be passed in as parameters to the RTsynchronizer, for example, the desired synchronization skew may be 80 milliseconds.

The group of media-actors which QosSync controls are specified in the formals. As shown in Figure 2, the **Declare** section defines and initializes the local state variables, such as the array indices, ind1, ind2, arrival times arrtime1, arrtime2, etc.. The **Constrain** section specifies a set of QoS constraints, which is of the form condition : action. The statements in the **Constrain** section ensure that messages for which EED >= maxEED or jitter >= maxJitter or skew >= maxSkew are never delivered. This infinite delay is specified by the timing condition (inf, inf) and is logically equivalent to dropping the packet. We will further explain this notation and its formal semantics in Section 2.2.

The **Update** section defines how the QosSync changes its state upon observing certain events which are defined by message patterns. Here, stream1ptrn is the pattern that identifies the stream corresponding to mediaActor1 (e.g. a videostream) and stream2ptrn is the pattern that identifies the stream corresponding to mediaActor2 (e.g. the associated audiostream).

When a packet arrives at one of the media-actors, local state variables are updated, e.g., arrtime1 is updated to be the current time. The last statement indicates that under any condition, skew is defined as arrtime1[i] - arrtime2[i], the difference in arrival time between two packets corresponding to the two different streams, specified as true:arrtime1[i] - arrtime2[i]. Structures describing the RTsynchronizer, the distribution of state information between an RTsynchronizer and its domain of control and strategies for maintaining interobject consistency are described in [10].

2 FORMALIZING MM SESSIONS

The basic Actor model captures the fundamental properties of general purpose distributed computing in which only logical time is concerned. Individ-

```
RTsynchronizer QoSSync( actor: Mediaactor1, Mediaactor2;
                real: maxEED, maxJitter, maxSkew) {

Declare
  int count;
  int ind = ind1 = ind2 = 0; /* array indices */
  real arrtime1= arrtime2 = 0;
  ...
  /* declaration of other local variables */

Constrain
  true:((*,*),(msg),(EED >= maxEED))(inf,inf);
        /* DropPacket if EED requirement is not satisfied */
  true:((*,*),(msg),(jitter >= maxJitter))(inf,inf);
        /* DropPacket if Jitter requirement is not satisfied */
  true:((*,*),(msg),(skew >= maxSkew))(inf,inf);
        /* DropPacket if skew requirement is not satisfied */


Update
stream1ptrn: /* stream corresponding to Mediaactor1 */
                arrtime1[ind1++] = currentTime;
                sendtime1[ind1++] = sendTime;
                /** defining End-to-end delay */
                EED1[ind1]       = sendtime1[ind1] - arrtime1[ind1];
                /** defining jitter */
                if (time11 == NULL) time11 = currentTime;
                 else if (time12 == NULL) time12 = currentTime;
                    else if (time13 == NULL)
                        {
                           time13  = currentTime;
                           delta11 = time12 - time11;
                           delta12 = time13 - time12;
                           jitter1 = mod(delta12-delta11);
                           time11 = time12; time12 = time13; time13 = NULL;
                        }

stream2ptrn: /* stream corresponding to Mediaactor2 */
         /* code similar to stream1 */


true: /* defining sync skew between stream1 and stream2 */
     skew = mod(arrtime1[ind]-arrtime2[ind]);
```

Figure 2 Declaring a QoS Service using RTsynchronizer

ual objects are constrained only by the computational causal order. However, in distributed multimedia systems, in addition to computational causal order that ensures the computational correctness, quantitative precedence orderings among different (possibly independent) computing units have to be enforced in order to achieve the QoS requirements. Extensions to the actor model, e.g. RTsynchronizers, allow us to specify time related QoS constraints modularly and verify the correctness of constraint satisfaction. In this section, we first state the semantics of RTsynchronizer. We then apply these semantics to a session-based multimedia system and provide a mapping for the semantic constructs. Finally, we informally reason about the correctness of this mapping.

2.1 RTsynchronizer Semantics

RTsynchronizer semantics is based on the ART model, which is an extension of the Actor model. The ART model takes the quantitative real-time aspect into consideration. In particular, statically specified constraints (which are encapsulated by RTsynchronizers), and dynamically created *expectation instances* on message invocations become the new basic integral components of ART systems configuration.

RTsynchronizers are special actors which may be in one of the following two different states:

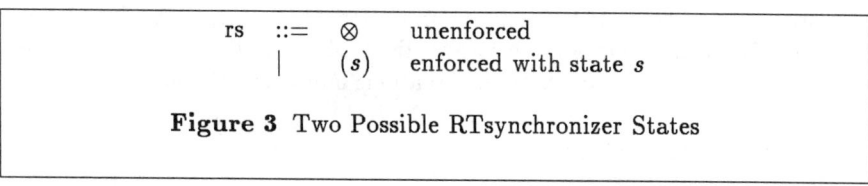

Figure 3 Two Possible RTsynchronizer States

Unlike ordinary computational actors, RTsynchronizers may not send or receive messages. The functionality of RTsynchronizers is to impose timing constraints on message invocations. An RTsynchronizer's control may be turned off by an actor performing unenforce operation. Thereafter, the RTsynchronizer reaches unenforced state $\otimes$. Before a message is invoked at its target actor, the controlling RTsynchronizer must evaluate the validity of invocation of the message. The state in which RTsynchronizers evaluate the constraints c is called evaluating state and denoted as (s). Another component in an ART system configuration is the *expectation instance set* (ς), which is a collection of expectation instances. An expectation instance is defined as follows:

Definition 1 (Expectation Instance)
The structured unit $s : p \triangleright \delta$ is called an expectation instance, where p is a message pattern defined in the enclosing RTsynchronizer s; δ specifies the time interval during which a message of pattern p is expected to be invoked.

δ is of the form $[d1, d2]$, where $d2 \geq d1 \geq 0$, denoting the starting and ending time points.

Formally, an ART configuration is a structured entity representing the system state at a given time instance. Such time instance is global with respect to the configuration. An ART system configuration is defined as:

Definition 2 (ART Configuration)

$$\left\langle\!\!\left\langle \ \alpha, \ (\tau)_{\text{Timer}} \ \mid \ \mu \ \mid \ \langle\,\sigma \,\|\, \varsigma\,\rangle \ \right\rangle\!\!\right\rangle$$

where

- α is an actor map, which maps a finite set of actor names to their behavior;
- Timer is a special actor whose state ($\tau \in \Re^+$) indicates the current global time with respect to the configuration when the snapshot is taken.
- μ is a finite multi-set of messages, yet to be processed;
- σ is an RTsynchronizer map, which maps a finite set of RTsynchronizer names to their states;
- ς is a multi-set of expectation instances on message invocations needed to be satisfied.

Before we present operational semantics for ART systems, we first define a group of notations and a group of functions based on these notations. These definitions will simplify further explanations and future formulae.

Definition 3 (Notations)

- $\mathcal{AN}$: the set of all possible actor names
- $\mathcal{AS}$: the set of all possible actor states, which contains the following information:

  - execution state (*unknown, idle* or *busy*)
  - the values of its state variables
  - the current expression it is reducing if in *busy* state

- $\mathcal{RN}$: the set of all possible RTsynchronizer names
- $\mathcal{RS}$: the set of all possible RTsynchronizer states, which contains the following information:

  - the enforcing state (unenforced, idle or validating)
  - the values of its state variables (including *ctime*())
  - the static constraint expressions

- the dynamic expectation instances
- the current constraint expression it is validating (i.e. in validating state).

- Π: the set of all possible message patterns
- $\mathcal{M}$: the set of all possible messages
- $\mathcal{D}$: the set of all possible time intervals
- $\mathcal{E}$: the set of all possible expectation instances

Definition 4 (Functions)

- α, a map which maps actor names to actor states.
  $\alpha: \mathcal{AN} \to \mathcal{AS}$
- σ, a map which maps RTsynchronizer names to RTsynchronizer states.
  $\sigma: \mathcal{RN} \to \mathcal{RS}$
- Sat_σ, a boolean function which returns T if message m satisfies pattern p under the enclosing RTsynchronizer s.
  $Sat_\sigma: \mathcal{M} \times \Pi \times \mathcal{RN} \to \{T, F\}$

$$Sat_\sigma(m, p, s) = \begin{cases} T & \text{if the evaluation of pattern matching returns T} \\ F & \text{if the evaluation of pattern matching returns F} \end{cases}$$

- $Safe_\sigma$, a boolean function which returns T if message m is safe for invoking at time τ with respect to an expectation instance $(s : p \triangleright [t, t1])$.
  $Safe_\sigma: \mathcal{E} \times \mathcal{M} \times \Re_{\geq 0} \to \{T, F\}$

$$Safe_\sigma((s : p \triangleright [t, t1]), m, \tau) = \begin{cases} F & \text{if } (\sigma(s) \neq \otimes) \wedge (Sat_\sigma(m, p, s) = T) \wedge \\ & ((\tau < t) \vee (\tau > t1)) \\ T & \text{otherwise} \end{cases}$$

- Dlv_σ, a function which returns T if message is deliverable with respect to the current expectation instance set.
  $Dlv_\sigma : \mathcal{M} \times 2^{\mathcal{E}} \times \Re_{\geq 0} \xrightarrow{\cdot} \{T, F\}$

$$Dlv_\sigma(m, \varsigma, \tau) = \begin{cases} T & \text{if } (\forall (s : p \triangleright [t1, t2]) \in \varsigma \ (Sat_\sigma(m, p, s) = F)) \vee \\ & (\exists (s : p \triangleright [t1, t2]) \in \varsigma \ ((Sat_\sigma(m, p, s) = T) \wedge \\ & (Safe_\sigma((s : p \triangleright [t1, t2]), m, \tau) = T))) \\ F & \text{if } (\exists (s : p \triangleright [t1, t2]) \in \varsigma \ (Sat_\sigma(m, p, s) = T) \wedge \\ & \forall (s : p \triangleright [t1, t2]) \in \varsigma \ (((Sat_\sigma(m, p, s) = F) \vee \\ & (Safe_\sigma((s : p \triangleright [t1, t2]), m, \tau) = F))) \end{cases}$$

- $NewEI_\sigma$, a function which returns a set of new expectation instances created by invoking a message. *

$NewEI_\sigma: \mathcal{M} \times \Re_{\geq 0} \to 2^\mathcal{E}$

$NewEI_\sigma(m, \tau) = \cup_{s \in \mathrm{Dom}(\sigma)} \{(s : p' \ \rhd \ [t1 + \tau, t2 + \tau]) \ | \\ \qquad\qquad (Sat_\sigma(m, p, s) = \mathrm{T}) \wedge ((p : p' \ [t1, t2]) \text{ is in } \sigma(s)) \}$

- *RemEI*, a function which returns a set of expectation instances satisfiable by invoking a message.

$RemEI: \mathcal{M} \times 2^\mathcal{E} \times \Re_{\geq 0} \to 2^\mathcal{E}$

$$RemEI_\sigma(m, \varsigma, \tau) = \{e \ | \ (e \in \varsigma) \ \wedge \ (Safe_\sigma(e, m, \tau) = \mathrm{T})\}$$

- *Update*, a function which changes the RTsynchronizer to new state.

$Update: \mathcal{M} \times \Re_{\geq 0} \times \mathcal{RN} \to \mathcal{RS}$

$$Update_\sigma(m, \tau, s) = \begin{cases} newState & \text{if } \exists p Sat_\sigma(m, p, s), \text{and} \\ & p \text{ is an update pattern in } s \\ \sigma(s) & \text{otherwise} \end{cases}$$

The standard operational semantics of ART systems is build on top of the basic actor semantics [3] with two extensions: (1) there is a special actor Timer whose state τ represents the current time in the system, (2) constraints are imposed on the invocation of messages.

In the standard operational semantics for actors there are two kinds of transitions: execution steps local to an individual actor and those dealing with message receipt. Execution transitions not affect other objects states in the system, — thereby the RTsynchronizer map (σ) and the expectation set (ς) remain the same as before and after the transitions. We skip the discussion of these transition rules in this paper *.

However, different from its corresponding part in the standard actor semantics, the meaning of transition rule <rcv : *m*> may be 3-fold:

- The prerequisite for taking the transition at the current configuration is that the message *m* is deliverable, namely the function $Dlv_\sigma(m, \varsigma, \tau)$ must evaluate to true.
- The same as the standard rcv rule, the receiving actor binds the actual parameters carried in the message to its formal parameters and applies its behavior to the message contents, i.e., carry out the underlying computation.
- If the message *m* is captured by a pattern in an RTsynchronizer, the invocation of such message may cause:

---

*No need to do the safe testing because the function NewEI is called only when the message is safe to be delivered at the destination. The same reason applies for RemEI and Update functions

*The details of these transition rules can be found in [10]

- RTsynchronizers state change, formulated as $Update_\sigma(m, \tau, \sigma)$;
- If m is an expected message, invoking it will satisfy an expectation instance in the system; hence, a satisfied expectation instance will be removed from the expectation instance set $(RemEI_\sigma(m, \varsigma, \tau))$;
- the message m may also be a "causal" message, thereby may create new expectation instances. Hence new expectation instances $(NewEI_\sigma(m, \tau))$ are added into the expectation set.

Because constraints are added into the ART systems, there exists the case that the system fails to satisfy the required constraints and causes system failure. In addition, time as an independent entity progresses concurrently with the underlying system activities. Two new transitions are added into ART systems to express system failure and time progression. The short version of ART system operational semantics using transition rules are given in Figure 4. The detailed semantics can be found in citeren-phd.

<rcv : a, cv>

$$\left\langle\!\!\left\langle \alpha, [R[\text{ready}(v)]]_a, (\tau)_{\text{Timer}} \mid m, \mu \mid \langle \sigma \| \varsigma \rangle \right\rangle\!\!\right\rangle \mapsto$$

$$\left\langle\!\!\left\langle \alpha, [\text{app}(v, cv)]_a, (\tau)_{\text{Timer}} \mid \mu \mid \right.\right.$$

$$\left.\left. \langle (s \to Update_\sigma(m, \tau, s))_{s\in\text{Dom}(\sigma)} \| \varsigma \cup \{ei\} \uplus NewEI_\sigma(m, \tau) \rangle \right\rangle\!\!\right\rangle$$

$$\text{if } m = <tag : a \Leftarrow cv>, Dlv_\sigma(m, \varsigma, \tau) = \text{T, and}$$

$$ei \in RemEI_\sigma(m, \varsigma, \tau)$$

<progress : d>

$$\left\langle\!\!\left\langle \alpha, (\tau)_{\text{Timer}} \mid \mu \mid \langle \sigma \| \varsigma \rangle \right\rangle\!\!\right\rangle_\chi \mapsto \left\langle\!\!\left\langle \alpha, (\tau + t_0)_{\text{Timer}} \mid \mu \mid \langle \sigma \| \varsigma \rangle \right\rangle\!\!\right\rangle$$

$$t_0 = \min\{t1 \mid \exists(s : p \triangleright [t1, t2]) \in \varsigma, \tau \le t1\}$$

<failure : >

$$\left\langle\!\!\left\langle \alpha, (\tau)_{\text{Timer}} \mid \mu, m \mid \langle \sigma, (sb)_s \| \varsigma, s : p \triangleright [t1, t2] \rangle \right\rangle\!\!\right\rangle \mapsto \perp$$

$$\text{if } Sat_\sigma(m, p, s) \wedge \tau \ge t2, \text{ where } \perp \text{ is the failure configuration.}$$

Figure 4 ART Systems Transition Rules

We formally define a session configuration as a structured entity representing the system state at a given time instance similar to an ART configuration. The time actor in this case gives us a virtual time with respect to the ongoing session in the system. Formally, a session configuration is mapped from an ART configuration as:

Definition 5 (SessionConfiguration)

$$\left\langle\!\!\left\langle\ \alpha,\ (\tau)_{\text{Session}}\ \mid\ \mu\ \mid\ \langle\ \sigma_s \parallel \varsigma\ \rangle\ \right\rangle\!\!\right\rangle$$

where

- α is an actor map that represents the media-actors in the session
- $\tau_{Session}$ is a session based virtual time, which is a special case of global time τ and
- σ_s is a mapping from a single QoS synchronizer to its state $(S)_{QoS}$. The QoS synchronizer encodes the required QoS constraints.

We apply the same mapping to the rest of the RTsynchronizer semantics to obtain the session based QoS Synchronizer semantics.

2.2   The Correctness of Session Based Semantics

In this section, we provide informal reasoning that shows that the RTsynchronizer semantics for the QoS specification provides the desired behavior, i.e satisfies the desired time related QoS requirements. We need to show that the transition rules described in the previous section implement the specified QoS requirements.

Consider the example discussed in Figure 2. There are two media streams both of which obey: (a) EED constraints (b) Jitter constraints and (c) synchronization skew constraints. Constraints in MM systems can be specified using two different formulations:

- A : B[t1,t2]: In this representation, delivery of a message satisfying pattern A at t0 creates an expectation instance B[t0+t1, t0+t2] indicating that a message of pattern B must be delivered in the interval [t0+t1, t0+t2].
- true: B[t1,t2]: In this representation, there is always an expectation instance B[t1,t2] i.e. a message of pattern B that must be delivered in the time interval [t1, t2]. Specifically if t1 and t2 are inf, i.e B[inf, inf], the message delivery is postponed indefinitely.

We use the second format to encode the jitter, EED and synchronization skew constraints by stating that if, for any message to be processed, delivering this message causes the EED, jitter and skew values to be greater than maxEED, maxJitter or maxSkew respectively, the message is indefinitely postponed (i.e. logically dropped).

In general, the arrival of a message causes the receive transition <rcv : m> to be triggered. By the transition semantics discussed in Section 2.1, the packet is considered deliverable only when the delivery clause Dlv_{σ_s} evaluates to true. σ_s is an encoding of QoS constraints within a media actor, hence jitter and end-to-end delay constraints must be satisfied before the message is considered deliverable. i.e. EED and jitter are calculated for the incoming packet. If the value of EED is greater than the maxEED value specified in the constraint, or the jitter value is greater than maxJitter, that message packet is indefinitely postponed(i.e dropped).

σ_s also encodes synchronization skew constraints between the two media-actors controlled by the QoS synchronizer. Since both actors share the same notion of time ($\tau_{session}$), synchronization skew is obtained by measuring the interarrival times of messages to the two actors in the configuration. Hence the delivery of a message at one actor can occur only if the corresponding message to the other media-actor has also been received within a time period specified by the synchronization skew constraint maxSkew. Since this is also a clause for the satisfaction of the Dlv_{σ_s} constraint, inter-actor constraints must also be met before the messages are considered deliverable.

Compositionality of the session configurations described can be achieved by synchronizing the session virtual times $\tau_{session}$ of the composing configurations. Let

$$K_1 = \Big\langle\!\!\Big\langle\ \alpha_1,\ (\tau)_{\text{Session1}}\ \big|\ \mu_1\ \big|\ \langle\ \sigma_{s1}\ \|\ \varsigma_1\ \rangle\ \Big\rangle\!\!\Big\rangle$$

and

$$K_2 = \Big\langle\!\!\Big\langle\ \alpha_2,\ (\tau)_{\text{Session2}}\ \big|\ \mu_2\ \big|\ \langle\ \sigma_{s2}\ \|\ \varsigma_2\ \rangle\ \Big\rangle\!\!\Big\rangle$$

be two independent session configurations. the composition of these 2 sessions $K_1\|K_2$ is defined as:

$$K_1\|K_2\ \textit{iff}\ \tau_{\text{Session1}} = \tau_{\text{Session2}}$$

3 CONCLUSIONS

We have proposed a formal specification of timing related QoS attributes in an actor-based distributed system and described some techniques for informally reasoning about quantitative QoS properties. When two QoS synchronizers operate on the same media actor, there may be inconsistencies that must be resolved. We are developing multisession semantics to reason about the behavior of an actor based realtime multimedia system. In practice, multiple system and application activities will need to occur concurrently, for example, scheduling, monitoring, inter and intra stream synchronization all of which may need to satisfy QoS constraints. One implementation of QoS synchronizers is through a meta-architecture framework [13]. Meta-architectures can form the basis for adaptive environments that provide customizable services needed to ensure end-to-end guarantees to multimedia traffic. We are currently exploring the specification of QoS in the presence of multiple sessions and addressing the composability of QoS constraints in this meta-architectural framework.

REFERENCES

[1] G. Agha. *Actors: A Model of Concurrent Computation in Distributed Systems.* MIT Press, Cambridge, Mass., 1986.

[2] G. Agha, I. A. Mason, S. F. Smith, and C. L. Talcott. Towards a theory of actor computation. In *The Third International Conference on Concurrency Theory (CONCUR '92)*, volume 630 of *Lecture Notes in Computer Science*, pages 565–579. Springer Verlag, August 1992.

[3] G. Agha, I. A. Mason, S. F. Smith, and C. L. Talcott. A foundation for actor computation. *Journal of Functional Programming*, 7:1–72, 1997.

[4] S. Frølund and G. Agha. A language framework for multi-object coordination. In *Proceedings of ECOOP 1993*, volume 707 of *Lecture Notes in Computer Science*. Springer Verlag, 1993.

[5] Svend Frølund. *Coordinating Distributed Objects: An Actor-Based Approach to Synchronization.* MIT Press, 1996.

[6] Peter Leydekkers and Valerie Gay. Odp view on qos for open distributed mm environments. In Jan de Meer and Andreas Vogel, editors, *4th International IFIP Workshop on Quality of Service, IwQos96 Paris, France*, pages 45–55, March 1996.

[7] Flavio Henrique de Souza Lima and Edmundo Roberto Mauro Madeira. Odp based qos specification for the multiware platform. In Jan de Meer and Andreas Vogel, editors, *4th International IFIP Workshop on Quality of Service, IwQos96 Paris, France*, pages 45–55, March 1996.

[8] Klara Nahrstedt and Jonathan M. Smith. The qos broker. *IEEE Multimedia*, 2:53–67, 1995.

[9] S. Ren, G. Agha, and M. Saito. A modular approach for programming distributed real-time systems. *Journal of Parallel and Distributed Computing*, 36(1), July 1996.

[10] Shangping Ren. *Modularization of Time Constraint Specification in Realtime Distributed Computing (to be published).* PhD thesis, Department of Computer Science, University of Illinois at Urbana-Champaign, 1997.

[11] C. L. Talcott. Interaction semantics for components of distributed systems. In *1st IFIP Workshop on Formal Methods for Open Object-based Distributed Systems, FMOODS'96*, 1996.

[12] N. Venkatasubramanian. *Composing Distributed Resource Management Activities (to be published).* PhD thesis, University of Illinois, Urbana-Champaign, 1997.

[13] N. Venkatasubramanian and C. L. Talcott. Reasoning about Meta Level Activities in Open Distributed Systems. In *Principles of Distributed Computation*, 1995.

BIOGRAPHICAL SKETCH

Shangping Ren is a doctoral candidate and research assistant in the Open Systems Laboratory at the University of Illinois at Urbana-Champaign. Her research interests include programming languages and parallel distributed real-time computing. She received her BS and MS from Department of Computer Engineering at Hefei Polytechnic University, China.

Nalini Venkatasubramanian is currently a doctoral candidate at the Open Systems Laboratory at the University of Illinois at Urbana-Champaign. She is also a member of technical staff working on interactive multimedia at Hewlett-Packard Laboratories in Palo Alto, California. Her research interests include multimedia and real-time systems, parallel and distributed technology, resource management, distributed operating systems and concurrent object programming languages. She has been working on software architectures for multimedia servers, multimedia networking protocols, real-time issues and load-balancing on distributed multimedia servers. Prior to this she has worked on various database management systems and on programming languages/compilers for high performance machines. She received an MS in Computer Science from the University of Illinois, Urbana-Champaign in 1992.

Gul Agha is Director of the Open Systems Laboratory at the University of Illinois at Urbana-Champaign. His research interests include models, languages, and tools for computing in open distributed systems. His book, *Actors: A Model of Concurrent Computing in Distributed Systems* stimulated considerable research in concurrent objects. Professor Agha is an ACM International Lecturer, Editor-in-Chief of *IEEE Concurrency*, and Associate Editor of *Theory and Practice of Object Systems* and *ACM Computing Surveys*. He is a recipient of the Incentives for Excellence Award from Digital Equipment Corporation, Naval Young Investigator Award from the US Office of Naval Research, and of a Fellowship at the University of Illinois Center for Advanced Study. Agha has served as Consulting Scientist to Microelectronics and Computer Technology Consortium and as a Visiting Professor at the University of Grenoble, France.

PART FOUR

Distributed systems: ODP and CORBA (I)

11

Invited Paper – Computational models for open distributed systems

Elie Najm[a], Jean-Bernard Stefani[b]

(a) ENST - École Nationale Supérieure des Télécommunications.
46, Rue Barrault - 75014 Paris, France,
E-mail: najm@res.enst.fr

(b) France Telecom - CNET
38-40, rue du Général Leclerc - 92131 Issy-Les-Moulineaux, France
E-mail: stefani@issy.cnet.fr

Abstract

The notion of computational model is central to the Reference Model for Open Distributed Processing. It defines an abstract model for distributed computations and characterizes at the same time the functionality of a supporting distributed virtual machine. Most current distributed systems, including e.g. the CORBA platforms, have converged on a subset of this computational model. In this paper, we review the model, providing a formal characterization for it, and we discuss various directions for extensions, considering in particular different trends in distributed systems construction.

Keywords

ODP, objects, computational model, formal techniques, quality of service, reflection

1 INTRODUCTION

The notion of computational model* is central to the Reference Model for Open Distributed Processing (RM-ODP) [20]. It provides a model of distributed computations in an ODP environment and it describes, in a programming language-independent fashion, the interface of the distributed virtual machine whose structure is described by the engineering model of the RM-ODP.

The computational model made explicit in the RM-ODP underlies most of the distributed computational models which have been proposed, explicitly or implicitly, in recent distributed language or distributed system developments. For instance, the computational model adopted in the OMG CORBA specifications [29] corresponds* to the so-called operational subset of the RM-ODP computational model. The same is true for e.g. the Network Object system [7], the Java RMI system [33] which supports distributed Java programs.

Despite the differences between these systems, the convergence of all these proposals on the operational subset of the RM-ODP computational model as the basic model for distributed computations is remarkable. It suggests that some "universal" for open distributed system programming has been identified. In this paper, we begin by reviewing and formally characterizing the operational subset of the RM-ODP computational model. That it can be showed at least as expressive as the (asynchronous) π-calculus brings further evidence to its universality.

Apart from its well-understood operational subset, the ODP computational model contains additional features, such as binding objects and environment contracts which are probably less known but which provide important additions to the basic model. The introduction of binding objects extends the basic computational model with notions of explicit communication objects. The introduction of environment contracts allows the specification of quality of service constraints associated with objects. In this paper, we characterize formally the notion of binding object and we discuss the possibility of formally capturing the notion of environment contract using failure observers.

A computational model provides an abstraction of a distributed system programming interface. It is thus interesting to look at the current developments in distributed systems research to understand what a computational model for open distributed systems should contain, and whether extensions to the ODP computational model — taken as the reference computational model — are required. The current research on distributed systems covers a lot of ground. Among the issues covered, we can highlight issues of mobility, issues associated with real-time and fault-tolerant processing, support for different forms of communication and distributed control, support for different soft-

---

*The ODP Reference Model uses "language" instead of "model". We prefer to use the latter to emphasize the language- independent nature of the notion.

*As a first approximation, leaving typing issues — which we do not consider in this paper — aside.

ware architecture styles and programming with distributed components, etc. While some of these issues can be understood within the framework of the ODP computational model, the extent of these different developments calls for additional extensions to the ODP computational model, including, notably, notions of composition and reflection. We discuss in this paper these various aspects and comment on their possible formalization. In the process, we hope to bring some insights to further the quest for a flexible, universal basis for open distributed programming.

The paper is organized as follows:

- section 2 reviews the operational subset of the computational model described in the RM-ODP, and discusses some of its features;
- section 3 reviews binding object features of the RM-ODP computational model and characterizes them formally;
- section 4 reviews environment contracts and discusses a possible approach to their formalization;
- section 5 hints at various extensions of the model and at a reflective computational model that could provide the basis for a new computational model for adaptable open distributed systems.
- section 6 concludes the paper.

2   THE BASIC COMPUTATIONAL MODEL

Following [26], the operational subset of the ODP computational model can be described formally using rewriting logic [24]. More precisely, the model can be formalized by a rewrite rule schema, a rewriting theory and a predicate that characterize valid computations. We review them below, leaving aside typing aspects. Those are covered in the original paper [26]. We call CM_{basic} the basic ODP computational model.

Since the ODP computational model abstracts away from the construction of individual objects, we consider objects as a given sort in the rewriting theory, characterizing them merely through their "visible" properties. We adopt an infix notation for operators and we use the same name for an operator and its extension to subsets of its domain. We use $\wp_f(S)$ to denote the set of finite subsets of a set S. We use S^* to denote the set of finite sequences of elements of a set S.

In what follows, we use the following sorts (to simplify notations we identify sorts and their carriers):

- Id denotes the set of interface identifiers; we use u, v, u_i, v_i to denote interface identifiers.
- Object denotes the set of objects; we use ω, ω', ω_i, ϖ, ϖ_i to denote objects.

- **Signal** denotes the set of signals; we use s, s_i to denote signals. Signals are units of interaction between objects and their environment.
- **LocConf** denotes the set of local configurations; we use c, c_i to denote local configurations. Local configurations are used to describe the behavior of objects.
- **DisConf** denotes the set of so-called "distributed configurations"; we use d, d_i to denote distributed configurations. Distributed configurations represent parallel compositions of objects that conform to the ODP computational model. The set of distributed configurations has two special elements, noted $\emptyset$ and $\perp$, that represent, respectively, an empty configuration and an invalid configuration.
- **Message** denotes the set of messages; we use m, m_i to denote messages. Messages represent asynchronous messages that are used to convey operation invocations between objects in the ODP computational model.
- **Name** denotes the set of signal names; we use n, n_i to denote signal names.

We use the following operators:

- $\|$: DisConf $\times$ DisConf $\to$ DisConf. $\|$ is the asynchronous parallel operator implied by the ODP computational model. A distributed configuration in the basic computational model consists in a set of objects and messages in parallel.
- Lhs : Object $\times$ $\wp_f$(Signal) $\to$ LocConf. Lhs is used to specify the behavior of objects.
- Rhs : Object $\times$ $\wp_f$(Signal) $\times$ $\wp_f$(Object) $\to$ LocConf. Rhs is used to specify the behavior of objects.
- L : Object $\to$ $\wp_f$(Id). $\omega.L$ yields the set of interface identifiers of object ω.
- K : Object $\to$ $\wp_f$(Id). $\omega.K$ yields the set of interface identifiers known by object ω.
- tgt : Signal $\to$ Id. $s.tgt$ is the identifier of the interface to which s is sent.
- arg : Signal $\to$ Id*. $s.arg$ is the sequence of interface identifiers that appear as arguments of s.
- nm : Signal $\to$ Name. $s.nm$ is the name of s.

We also have "injection" operators that turn a signal into a message, and that turn an object or a message into a distributed configuration. To simplify the notations, we do not make these obvious operators explicit and we just use objects and signals in these different contexts.

The following laws are assumed to hold. In the rest of this section, terms appearing in rewrite rules must be understood modulo the equations (E) below, i.e. as denoting their E-equivalence class.

- Abelian monoid laws for $\|$, with $\emptyset$ as a neutral element (i.e. $d \| \emptyset = d$) and $\perp$ as an absorbing element (i.e. $d \| \perp = \perp$).

- Reduction to $\perp$ for distributed configurations: $\omega_1.L \cap \omega_2.L \neq \emptyset \Rightarrow \omega_1 \parallel \omega_2 = \perp$.

The last law above specifies the conditions leading to an invalid distributed configuration. In other terms, no two objects in a valid distributed configuration may have an interface with the same identifier.

2.1 Characterizing objects

An object in CM_{basic} (as well as in other computational models discussed later) is characterized by a set of rewrite rules that obey the rewrite rule schema $(\Diamond)$, with Ω a (possibly empty) finite set of objects, and $\aleph$, $\aleph'$ (possibly empty) finite sets of signals:

$$(\Diamond) \quad Lhs(\omega, \aleph) \to Rhs(\omega', \aleph', \Omega)$$

where:

In the rest of the paper we use the notation $\Diamond(\omega, \aleph, \omega', \aleph', \Omega)$ to stand for $Lhs(\omega, \aleph) \to Rhs(\omega', \aleph', \Omega)$. In addition, the following conditions must hold, i.e. $\Diamond(\omega, \aleph, \omega', \aleph', \Omega) \Rightarrow C_0$, where C_0 is the conjunction of the conditions below:

- $\aleph.tgt \subseteq \omega.L$
  Signals on the left-hand side of the rule schema are targeted at interfaces of object ω.
- $\omega.L \subseteq \omega'.L$
  The set of interfaces of an object may increase over time. Notice that an object may have several interfaces (or access point). This condition indicates that ω *evolves* into ω' during the rewrite step, since the identity of an object is determined by its interfaces.
- $\neg(\omega' \parallel \Omega \to \perp)$
  The second side of the rule schema should not degenerate into an invalid distributed configuration. This is a local condition on the generation of new interface identifiers. It is complemented by the condition (†) introduced below, which prevents the evolution of distributed configurations into invalid ones.
- Let $A = \omega.K \cup \aleph.arg \cup \omega'.L \cup \Omega.L$. Then:

$$\omega'.K \subseteq A \wedge \aleph'.tgt \subseteq A \wedge \aleph'.arg \subseteq A \wedge \Omega.K \subseteq A$$

This constraint characterizes encapsulation, expressing that an object knowledge of its environment can grow only through interactions.
- Let ρ be a bijection on Id. We note $\omega.\rho$ the object resulting from substituting $\omega.L.\rho$ and $\omega.K.\rho$ to $\omega.L$ and $\omega.K$, respectively. In the same way, $s.\rho$ denotes the signal obtained from signal s by replacing $s.tgt$ by $s.tgt.\rho$ and $s.arg$ by

$s.arg.\rho$. Then:

$$\Diamond(\omega, \aleph, \omega', \aleph', \Omega) \;\Leftrightarrow\; \Diamond(\omega.\rho, \aleph.\rho, \omega'.\rho, \aleph'.\rho, \Omega.\rho)$$

This condition expresses the independence of object behavior and object definition from the actual choice of interface identifiers.

Intuitively, each rule conforming to ($\Diamond$) describes a possible state transition of an object ω. An equivalent reading is that an object is characterized by a transition system whose states are local configurations.

2.2   Characterizing distributed computations

Distributed configurations conforming to CM_{basic} can be characterized by a rewrite rule ($\clubsuit$) and an axiom ($\dagger$). Rule ($\clubsuit$) is the conditional rewrite rule given below:

$$(\clubsuit) \qquad \frac{\Diamond(\omega, \aleph, \omega', \aleph', \Omega)}{\omega \parallel \aleph \to \omega' \parallel \aleph' \parallel \Omega}$$

Rule ($\clubsuit$) defines the allowed transitions between distributed configurations. It needs to be complemented with another constraint to ensure the uniqueness of interface identifiers in distributed configurations. This is the role of predicate WF, given by ($\dagger$), that defines valid distributed configurations:

$$(\dagger) \quad \mathsf{WF}(d) \equiv \neg(d \to \bot)$$

A distributed configuration d that conforms to CM_{basic} is thus a valid distributed configuration (i.e. a distributed configuration d such that $\mathsf{WF}(d)$) that is obtained through the application of rule ($\clubsuit$).

2.3   Discussion

CM_{basic} described above corresponds to the so-called operational subset of the ODP computational model. Actually, the ODP computational model comprises two kinds of operation invocations: one way asynchronous invocations and two-way asynchronous invocations. Only the first have been presented in the formal model above but the second can easily be recovered as a specialization of the above rewrite rule schema[*].

CM_{basic} exhibits "mobility" in much the same way as in the π-calculus. This statement can be made formal by comparing the expressive power of the (asynchronous) π-calculus (as defined e.g. in [3]) and of CM_{basic}. Let us say, as in [17], that a computational model M_2 equipped with equivalence $\approx_2$ is

---

[*]See [26] for details.

more expressive than a computational model M_1 equipped with equivalence $\approx_1$ when there is a fully abstract encoding $\mathcal{T}$ from M_1 to M_2, i.e. when for all P and Q in M_1 we have $P \approx_1 Q \Leftrightarrow \mathcal{T}(P) \approx_2 \mathcal{T}(Q)$

Theorem 1 CM_{basic} *is more expressive than the asynchronous π-calculus up to their weak barbed congruences.*

Weak barbed congruence is defined for the asynchronous π-calculus in [3]. It can be readily defined for CM_{basic} as follows. Observability in distributed configurations is first defined through predicates $\downarrow_s$ on distributed configurations, where $s \in \mathsf{Signal}$: $s \downarrow_s$, and $d_1 \parallel d_2 \downarrow_s$ if $d_1 \downarrow_s$ or $d_2 \downarrow_s$.

We then define weak barbed bisimulation for distributed configurations as follows:

Definition 1 *A symmetric relation R_ρ on distributed configurations, indexed by a bijection ρ on Id, is a weak barbed bisimulation if, for all $(d_1, d_2) \in R_\rho$, we have:*

1. *if $d_1 \to d_1'$, then $d_2 \to d_2'$ such that $(d_1', d_2') \in R_\rho$;*
2. *if $d_1 \downarrow_s$ then $d_2 \downarrow_{s.\rho}$.*

Two distributed configurations d_1 and d_2 are weak barbed bisimilar if there is a weak barbed bisimulation R_ρ such that $(d_1, d_2) \in R_\rho$.

Finally we define weak barbed congruence for distributed configurations:

Definition 2 *Two distributed configurations d_1 and d_2 are weak barbed congruent if, for all distributed configurations d such that $d.K \subseteq d.L \cup d_1.L$ and $d \parallel d_1 \neq \perp$, there is a weak barbed bisimulation R_ρ such that $(d \parallel d_1, d.\rho \parallel d_2) \in R_\rho$.*

Loosely speaking, we can paraphrase the above in saying that two distributed configurations are weak barbed congruent if they are bisimilar under all closed*, valid contexts.

The proof of Theorem1 is too long to be reproduced here in detail but it is relatively standard. It involves the encoding of each π-calculus process as an object, the interpretation of a term $P_1 \mid P_2$ of the π-calculus as the creation of a an object associated with process P_2, and the modeling of each π-calculus

---

*Intuitively, a closed context for a distributed configuration d_1 is a distributed configuration d that does not have any references to other objects than those appearing in d itself or in d_1. Notice that it is always possible to close a distributed configuration with appropriate dummy objects (i.e. objects that cannot change state) bearing the required interface identifiers. The constraint that contexts be closed is required to avoid the illicit "guess", by the context, of future interface identifiers generated by transitions from d_1.

channel by an object that provides a non-deterministic dispatching of signals towards registered objects. The latter is required to capture the ability of a π-calculus channel to be used for reception by several distinct processes. The appendix to this paper provides more details on this encoding.

Mobility in CM_{basic} is achieved through the communication of interface identifiers, but it may be implemented in different ways. For instance, the model may, classically, be implemented using some form of RPC protocol. But it may also be implemented by moving (client) objects to the site where their correspondents (servers) reside (e.g. as an instance of the so-called client-agent-server architecture formalized in [18]). As such, CM_{basic} does not provide explicit control over the physical locations or resources required to support objects and distributed configurations. Instead, the model can be understood as taking a view of "maximum distribution" of objects: an object performs local computations (as represented by rules conforming to the schema ($\Diamond$)) and communicate asynchronously by exchanging signals with a priori distant other objects.

The formal model presented in the previous sections has a lot in common with the abstract actor model described in [34]. In fact, except for the specification of the fairness conditions for actor computations, the basic computational model provides a direct modeling of actors as objects with a single interface.

3   BINDING OBJECTS

CM_{basic} is limited in its ability to directly capture features of distributed platforms and in its modeling capabilities. In particular, even though it makes an assumption of maximum distribution, it remains limited in its capacity to describe the different forms of communications and of failures that may occur in an actual distributed setting. Failures that can be taken into account are essentially local failures (through the definition of appropriate faulty object behavior). The model does not account for the diversity of network situations that may arise in a real environment, with different assumptions and guarantees being provided by underlying networks. The basic form of communication supported by CM_{basic} does not account for other forms of communications required and supported in actual distributed environments, that exhibit a large variety of communication semantics and topologies (e.g. point-to-point or multipoint continuous media communication, blackboard communication schemes, group communication schemes, etc.). Even within the basic RPC-like model of asynchronous operation invocation, a large variety of semantics can be provided and made available to applications (see e.g. systems such as Spring [19], SOR [30], or Spin [6]. A distributed system that has adopted the notion of binding as a means to provide a highly flexible distributed platform is described in [11].

For all these reasons, the ODP computational model has been defined to comprise an explicit notion of communication object, called a binding object. A binding object mediates communication between other objects. The ODP computational model does not prescribe a particular communication semantics for binding objects. Instead, a binding object, just as any other object, may have an arbitrary behavior. If necessary, a binding object can encapsulate behavior corresponding to faulty network behavior. A formal presentation of the full ODP computational model with binding objects can be given using the elements introduced in the previous section. The resulting model is called $CM_{bindings}$.

Distributed configurations are now restricted to be finite sets of objects in parallel, i.e. terms of the form $\omega_1 \parallel \ldots \parallel \omega_n$. Distributed configurations d conforming to $CM_{bindings}$ are characterized by the conditional rewrite rule ($\spadesuit$) together with the constraint that they be valid (i.e. that $WF(d) \equiv true$, where WF is given by ($\dagger$)):

$$(\spadesuit) \quad \frac{\Diamond(\omega, \aleph, \omega', \aleph', \Omega) \wedge \forall j \in J \Diamond(\omega_j, \aleph_j, \omega'_j, \aleph'_j, \Omega_j) \wedge C_1}{\omega \parallel_{j \in J} \omega_j \to \omega' \parallel \Omega \parallel_{j \in J} (\omega'_j \parallel \Omega_j)}$$

where J is a finite index set and C_1 is the condition defined as follows:

$$C_1 \equiv (\aleph = \bigcup_{j \in J} \aleph'_j) \wedge (\aleph' = \bigcup_{j \in J} \aleph_j)$$

The resulting model is thus a synchronous model, where, intuitively, synchronous interaction are local interactions, and where communications between distant objects are systematically mediated by bindings. Notice that the above formalization does not make any difference between objects and binding objects. In the general case, it is merely the intention associated with the conveyance of information from one object to another that characterizes a binding object.

Note that the rule ($\spadesuit$) constitutes a particular modeling choice, since the RM-ODP only specifies, informally, that a "signal occurrence is an atomic action". We could have made other choices, compatible with the informal assertions of the RM-ODP. For instance, we could have an alternative rule where the interactions remain synchronous but exhibit a clear causal relation, with some object being the initiator of the transition.

Because of its synchronous nature, rule ($\spadesuit$) may be difficult or even impossible to implement in presence of arbitrary failures if ω, appearing in the rule, is to be implemented as a distributed object: in this case, the rule would require a form of distributed rendez-vous. The condition Dis defined below imposes an additional condition on distributed objects that solves the problem, confining synchronous interaction as a primitive for local communication only.

The rule ($\spadesuit$) does not capture a distinction between local and distant objects. In CM_{basic}, the distinction was made explicit through the use of asynchronous messages. We can capture the basic asynchrony present in CM_{basic} through the predicate Dis, defined as follows. Let ω be an object and let

$J = \omega.L$. Let t_j $(j \in J)$ be triples of the form $(\omega_j, \aleph'_j, \Omega_j)$. Let $\aleph_0$, $\aleph_j$ $(j \in J)$ be finite sets of signals such that $\forall j \in J$, $\aleph_j.tgt = j$ and $\aleph_0 = \bigcup_{j \in J} \aleph_j$. Finally let $\alpha_1, \dots, \alpha_n, \beta, \gamma_1, \dots, \gamma_n$ be the following assertions:

- $\alpha_j \equiv Lhs(\omega, \aleph_j) \rightarrow Rhs(t_j)$
- $\beta \equiv Lhs(\omega, \aleph_0) \rightarrow Rhs(t_0)$
- $\gamma_j \equiv Lhs(\omega, \aleph_0 \setminus \aleph_j) \rightarrow Rhs(t_0)$

We can then define:

$$\mathsf{Dis}(\omega) \equiv \wedge \bigwedge_{j \in J} \alpha_j \Rightarrow \exists t_0, \ \beta \wedge (\bigwedge_{j \in J} \gamma_j)$$

$$\wedge \ \beta \Rightarrow \exists t_1, \dots, t_n, \ (\bigwedge_{j \in J} \alpha_j) \wedge (\bigwedge_{j \in J} \gamma_j)$$

Predicate Dis corresponds to a kind of confluence property and constitutes a condition for objects to be implementable in a general distributed setting, in presence of arbitrary forms of failures. Interfaces of object ω in the above definition can be construed as distant interfaces, located on different sites*. Interactions on one interface do not depend on interactions on other distant interfaces. Such interactions can be implemented without the recourse to a distributed consensus protocol which would be expensive or even impossible [16, 21] to implement in the presence of arbitrary failures. In other terms, in presence of arbitrary failures, a distributed atomic interaction — as would be mandated by the general case of rule (♠) — cannot be taken as primitive and must be explicitly modeled as a set of more primitive actions. Predicate Dis, as a characterization of distributable objects, confines synchronous interactions to the local case, thus making rule (♠) a suitable, primitive model for distributed computations.

CM$_{\text{bindings}}$ constitutes a conservative extension of CM$_{\text{basic}}$ in that all features of CM$_{\text{basic}}$ can be fully recovered* in CM$_{\text{bindings}}$. This merely involves modeling the sending of an asynchronous message in CM$_{\text{basic}}$ as the creation of an elementary binding object whose sole responsibility is to invoke a signal on a given target interface.

4   QUALITY OF SERVICE

Apart from binding objects, the ODP computational model also identifies the notion of *environment contracts*. Intuitively, an environment contract specifies non-functional aspects of an object behavior, notably its quality of ser-

---

*ω is thus a distributed object, with each interface on a different site. This is again taking a view of maximum distribution.
*I.e. by a fully abstract encoding.

vice (QoS) constraints. The notion of quality of service should be understood here in a broad sense, covering both real-time constraints such as delay or throughput constraints, and dependability constraints such as availability or fault-tolerance constraints. The ODP computational model [20] thus indicates that a computational object specification may contain additional behavioral constraints in the form of an environment contract specification. The introduction of environment contracts is motivated by the desire to provide a complete characterization of a distributed system or application from a computational point of view, abstracting away from the detailed mechanisms used to meet quality of service constraints. Such a characterization is required for a wide variety of applications, from real-time process control, and mission-critical applications to interactive multimedia applications, whose correct behavior depend on the provision of a suitable quality of service. The introduction of environment contracts also follows an obvious trend in distributed systems construction to explicitly support quality of service constraints. A detailed presentation of system issues involved and some proposals can be found in the recent book [8].

Whereas the notion of binding is well specified in [20], the notion of environment contract associated with an object remains under-specified. We discuss in this section some possibilities for formalizing the notion.

The term *contract* emphasizes a deontic view of such constraints (see e.g. [32] for a discussion): an environment contract specifies at the same time expectations from an object on its environment and guarantees or obligations that the object will fulfill in return. Informally, the behavior of an object conforms to a given contract if it does not violate the obligations of the contracts at least as long as the contract expectations are fulfilled by the object environment.

Formalizing this notion of contract requires, as a minimum, the capture of appropriate notions of failures, and the introduction of an explicit notion of time in the computational model. For instance, capturing dependability constraints requires the definition of certain failure modes for objects and groups of objects. First attempts in this direction, that introduce simple primitives for failures* in the context of mobile process calculi include [2] and [18]. Another attempt, dealing especially with real-time constraints, is [12], which extends the ODP computational model with time and interaction failures, and formalizes environment contracts as specific (failure) observers.

Drawing on these ideas, it would be interesting to define an extension of $CM_{bindings}$ dealing explicitly with the possibility of failures. Let us call $CM_{failures}$ the envisaged model, and let us consider some possible features of the model. A first feature of $CM_{failures}$ would be the explicit presence of time in the model. This can be done using timed rewriting logic models discussed in [25] or directly on our object model. For instance, adopting the basic ideas behind timed automata (see e.g. [1]), we could consider pairs (ω, u) as the

---

*Essentially, primitives for fail-silent objects.

basic units in distributed computations, where ω is an object whose state depends on u, a clock variable that takes its values in a time domain Time. Object behaviors would now be described using two kinds of rewrite rules:

- a modified ($\Diamond$) rule schema of the form:

$$(\Diamond_t) \quad Lhs((\dot{\omega}, u), \aleph) \to Rhs((\omega', u), \aleph', \Omega)$$

  where Ω is now a finite set of pairs of the form (ϖ, v).
- new labeled rules for allowing time to pass of the form:

$$t : (\omega, u) \to (\omega, u + t)$$

where $t \in$ Time.

Notice that the explicit presence of rules with time labels is required in the definition of an object to allow time to pass. This allows an object to not let time pass, which in turn allows to express urgency for transitions. The model could be completed with time-related axioms such as *time determinism*, *time additivity*, *deadlock-freeness*, *action persistence*, etc. (see e.g. [1, 27] for a discussion of properties of timed models).

We would then have the following modified ($\spadesuit$) rule for distributed configurations:

$$(\spadesuit_t) \quad \frac{\Diamond_t(\omega, u, \aleph, \omega', \aleph', \Omega) \wedge \forall j \in J \ \Diamond_t(\omega_j, u_j, \aleph_j, \omega'_j, \aleph'_j, \Omega_j) \wedge C_1}{(\omega, u) \ \|_{j \in J} (\omega_j, u_j) \to (\omega', u) \ \| \ \Omega \ \|_{j \in J} ((\omega'_j, u_j) \ \| \ \Omega_j)}$$

And a new labeled rule for timed distributed configurations:

$$\frac{t : d_1 \to d'_1 \ \wedge \ t : d_2 \to d'_2}{t : d_1 \ \| \ d_2 \to d'_1 \ \| \ d'_2}$$

A second feature of $CM_{failures}$ would be to provide the ability to capture and detect various forms of failures. For instance, we could capture the notion of a fail-silent object through the following rules:

1. $\Downarrow (\omega) : (\omega, u) \to (\Downarrow \omega, u)$
2. the behavior of $\Downarrow \omega$ is given by the set of rules: for all $t \in$ Time, $t :$ ($\Downarrow \omega, u) \to (\Downarrow \omega, u + t)$

Added to the behavior of an object ω, these rules allow ω to fail silently (i.e. to evolve into the failed object $\Downarrow \omega$). Note that the only behavior of a failed object is to let time pass. More complex behaviors would be possible to reflect different failure modes, or to model the possibility of recovery. The label $\Downarrow (\omega)$ can be interpreted as a predicate asserting that object ω has failed.

In the same vein, we could capture the notion of interaction failure, or missed rendez-vous, through the definition of a predicate $\Downarrow (s)$, where s is a signal, asserting that s could not be exchanged successfully between two objects. Let $\mathsf{locked}_t(\omega, u)$ assert that (ω, u) can only progress by letting time pass:

$$\mathsf{locked}_t(\omega, u) \equiv \neg \exists \aleph, \omega', \aleph', \Omega, \ \Diamond_t(\omega, u, \aleph, \omega', \aleph', \Omega)$$

Let us define also a predicate urgent as:

$$\mathsf{urgent}(\omega, u, \omega', s, \Omega) \quad \equiv \quad \wedge \ \Diamond_t(\omega, u, \emptyset, \omega', \{s\}, \Omega) \tag{1}$$
$$\wedge \ \neg(\exists t, \ t : \ (\omega, u) \to (\omega, u + t)) \tag{2}$$

We can now define the following rule:

$$\frac{\mathsf{urgent}(\omega_1, u_1, \omega_1', s, \Omega_1) \ \wedge \ s.tgt \in \omega_2.L \ \wedge \ \mathsf{locked}_t(\omega_2, u_2)}{\Downarrow (s) : \ \omega_1 \parallel \omega_2 \to \omega_1' \parallel \omega_2 \parallel \Omega_1}$$

The rule expresses the fact that an interaction failure occurs, with signal s failing to be transmitted from ω_1 to ω_2, when ω_1 must urgently transmit s while ω_2 cannot receive it.

The $\Downarrow$ predicates just introduced, can be used to define failure detectors, much in the same sense as failure detectors introduced in [9]. $\Downarrow (s)$, in particular, provides just the basic observation of interaction failure that is required by the calculus of object contracts introduced in [12]. A contract, as defined in this work, is merely a process which observes and arbitrates the collective behavior of configurations of objects. A contract observes objects and depending on the outcome of their interactions (success or failure) may identify and incriminate faulty objects. [12] also defines other contract-related notions such as assumption, obligation, realization an refinement, and it provides a composition theorem which gives sufficient conditions for collections of objects, each satisfying a given contract, to be composed. A similar, more general endeavor may be attempted for CM$_\mathsf{failures}$, thus resulting in a formalization of environment contracts as specialized failure observers. In particular, one could envisage introducing observers corresponding to the different forms of failure detectors discussed in [9], providing a suitable basis for specifying fault-tolerant distributed systems.

5 TOWARDS A REFLECTIVE COMPUTATIONAL MODEL

We have reviewed in previous sections the present contents of the ODP computational model. With its combination of binding objects and environment contracts, the model obtained is already quite rich, but is it sufficient ? Considering the ODP computational model as the abstract programming interface of a (distributed) virtual machine incites us to review different issues currently under investigation in the distributed systems and distributed programming language communities.

5.1   Components and software architecture styles

The ODP computational model as defined by CM$_{bindings}$ remains fairly low-level. Several authors argue in favor of a higher-level, *component-based* approach to the construction of complex software systems, and distributed systems in particular (see e.g. [23,'28, 5, 31]). They also point out the necessity to support different architecture styles such as those discussed in [31], e.g. dataflow, object-based, event-based, data-centered, rule-based, to account for the diversity of design possibilities. Most of these proposals suggest the introduction of different forms of connectors to construct arbitrary configurations of components. Leaving aside language issues, which do not concern us here, we can remark that components in these proposals, whether primitives or composite, can invariably be modeled as computational objects, i.e. whose behavior can be defined as a set of transitions conforming to the ($\Diamond$) schema. A more important observation is that the notion of binding object, as defined in section3, subsumes that of connector, e.g. as envisaged in [23] or [31]. The ODP computational model does not prescribe the way objects are constructed, but nothing prevents the specification of an object as a parallel composition of a set of computational objects interconnected by binding objects. Since binding objects can embody any communication semantics, any form of connector can be construed as a binding object. For instance, binding objects can be defined that provide event multicasting facilities (e.g. similar to event channels defined in [29]), or a form of generative communication à la Linda [14]. The latter example could be realized, e.g. following the formal specification of generative communication provided in [10]. In that respect, CM$_{bindings}$ provides an excellent support for component-based approaches to distributed system programming.

5.2   Object migration, resource control and reflection

The ODP computational model adopts a standpoint which we have qualified as a *maximum distribution* standpoint, where each object is potentially located on a different site. In fact, since each object may indeed be a distributed object, as we saw in section 3, each interface can be considered as potentially located on a different site. This view is appropriate if no finer control over object location is required or can be left for a later phase in system construction. An explicit modeling of locations is necessary, however, if such a control is required — as would be the case e.g. for purposes of explicit load-balancing. [2] and [18] provide two examples of how locations can be introduced in mobile process-calculi. In our context, the notion of location could be introduced in a similar way to that of the join-calculus [18]. To provide even more flexibility and control, we could go beyond the mere modeling of locations as names and consider locations as objects in their own right, with a specific operator for associating (standard) computational objects and locations.

As a simple illustration, we consider the following extension to CM_{basic}. Distributed configurations are now parallel ($\parallel$) compositions of messages and of localized configurations of the form $l : [d]$, where l is a location and d is a standard CM_{basic} distributed configuration*. We extend the notion of message to cover also messages of the form $l.n\langle a_1, \ldots, a_n \rangle$, where l is a location, $n \in$ Name is an operation name, and each of the a_i is either an interface identifier or an object. Object transitions now obey the following rewrite rule schema:

$$(\Diamond_l) \quad Lhs(\omega, \aleph) \rightarrow Rhs(\omega', \aleph', \Omega)$$

where Ω is a set of localized configurations of the form $l : [\varpi_1 \parallel \ldots \parallel \varpi_n]$. Assuming that all locations respond at least to the message $go\langle . \rangle$, our new model can be defined by the following rewrite rules:

- Let Ω be such that $\Omega = \Omega_l \cup \Omega_{\bar{l}}$ with $\Omega_l = l : [\omega_1 \parallel \ldots \parallel \omega_n]$ and $\Omega_{\bar{l}} = \{l_1 : [d_1], \ldots, l_p : [d_p]\}$ with $l_j \neq l$ for all $j \in \{1, \ldots, p\}$. Let $\aleph'$ be such that $\aleph' = \aleph'_l \cup \aleph'_{\bar{l}}$ with $\aleph'_l.tgt \subseteq (\omega' \parallel \Omega_l).L$, and $\aleph'_{\bar{l}}.tgt \cap (\omega' \parallel \Omega_l).L = \emptyset$. Then:

$$(\clubsuit_l) \quad \frac{\Diamond_l(\omega, \aleph, \omega', \aleph', \Omega)}{l : [\omega \parallel \aleph] \rightarrow l : [\omega \parallel \aleph'_l \parallel_{i=1\ldots n} \omega_i] \parallel \aleph'_{\bar{l}} \parallel \Omega_{\bar{l}}}$$

Rule schema $(\Diamond_l)$ specifies the general form object transitions may take. Notice that new objects may be created in different locations (which must be known to the object ω that create them, i.e. one must have $\{l_1, \ldots, l_p\} \subseteq \omega.K^*$). Object behaviors may additionally comprise transitions of the form:

$$\omega \rightarrow l.go\langle \omega \rangle$$

With the above rewriting rule an object may send itself to a different location.

-

$$\frac{\aleph.tgt \subseteq d.L}{l : [d] \parallel \aleph \rightarrow l : [d \parallel \aleph]}$$

This rule just specifies that messages flow asynchronously from locations to locations.

-

$$l : [d] \parallel l.go\langle \omega \rangle \rightarrow l : [d \parallel \omega]$$

This rule specifies the semantics of the go message, which is to transmit an object to a target location.

---

*Notice that we could just as simply have allowed extended distributed configurations inside a $l : [.]$ context, thereby allowing trees of locations, much as in [18].

*To simplify notations, we identify here locations and identifiers of their pre-determined interface, i.e. that which supports the go operation.

Specifying locations on which objects may reside provides applications explicit control over the placement of objects, presumably for purpose of load-balancing, security, fault-tolerance or availability. The ability to migrate objects from location to location as illustrated above with the $go\langle\rangle$ message, offers dynamic control over object placement, allowing communication trade-offs. Some comments are in order. Note first that the last two rules are in fact instances of ($\clubsuit_l$) if one note that $l : [d]$ corresponds to a composite object as per the above section, with d as an internal state, and which "exports" interfaces from d. Note then that the resulting model is a conservative extension of CM_{basic}: the latter can be recovered simply by having one object per location, formalizing the implicit "maximum distribution" principle discussed in section 2.

Controlling the placement of objects is not sufficient for applications that require quality of service guarantees, notably real-time guarantees. Such applications typically require the ability to access and control the characteristics of the resources they use, e.g. processors, memory, I/O. Even in non real-time contexts, applications may require e.g. fine-grained control over the degree of parallelism with which they run, thus requiring access to thread and process-like entities within a given node or location. More generally, providing applications with fine-grained control over their implementation can be useful to easily adapt applications to different environments, and to different non-functional requirements. This has lead researchers to consider using reflection in distributed systems, e.g. [4, 13, 15, 22, 35]. Having reflection as a feature of our reference computational model for open distributed systems is attractive for it promises to unify in one setting the different extensions we have discussed in this paper. Let us just mention what that would entail. First, introducing reflection in the model will require reifying the different mechanisms implicit in distributed computations, corresponding roughly to the seven *aspects* of [22]: sending, accepting, queuing, and receiving messages; describing object state and object transitions; executing object transitions. Second, we can note that reifying execution aspects, would mean reifying most of the ODP engineering model [20], with its notions of nodes, capsules, clusters, and threads. Locations, as described above, already provide a first approximation of RM-ODP nodes. Capsules, clusters and threads can be introduced similarly. In particular, note that all can be understood as specialized forms of locations, with exclusive access in the case of threads. Lastly, let us note that the above rules for location, already capture some features of message acceptance and queuing.

6   CONCLUSION

We have reviewed the ODP computational model and its formal semantics, developed using a conditional rewriting logic approach. This lead us to a series of computational models, from the most basic one, CM_{basic}, that corresponds to the operational subset of the ODP computational model, and $CM_{bindings}$ that captures the full model with binding objects. Capturing the ODP notion of environment contracts and their associated quality of service constraints has been discussed as part of $CM_{failures}$. Prominent features of $CM_{failures}$ were discussed, suggesting ways to model failures in the model and suggesting the use of specific failure observers to implement or characterize environment contracts. Finally, we have discussed several issues in open distributed systems construction which, taken together, call for a more radical approach to the flexibility and adaptability — in other terms, to the openness — of distributed systems. This leads us to consider elements of a reflective computational model, $CM_{reflect}$, which may provide an elegant way to integrate numerous features under one highly flexible distributed computational model. Going beyond "basic" and "bindings", towards "failures" and "reflect", looks like a promising research agenda.

REFERENCES

[1] R. Alur, D. Hill: "A theory of timed automata" – Theoretical Computer Science 126, pp 183-235, 1994.

[2] R. Amadio, S. Prasad: "Localities and Failures" – in Proceedings 14th Conference on Foundations of Software Technology and Theoretical Computer Science, LNCS 880, Springer Verlag, 1994.

[3] R. Amadio, I. Castellani, D. Sangiorgi: "On bisimulations for the asynchronous π-calculus" – in Proceedings CONCUR '96, LNCS 1119, Springer Verlag, 1996.

[4] Barga, R., C. Pu, "Reflection on a Legacy Transaction Processing Monitor" – Proceedings of Reflection 96, G. Kiczales (ed), pp 63-78, San Francisco, USA, April 1996.

[5] L. Bellissard, S. Ben Atallah, A. Kerbrat, M. Riveill: "Component-based programming and application management with Olan" – Object-based parallel and distributed computation, LNCS 1107, Springer Verlag, 1995.

[6] Bershad, B.N., S. Savage, P.Przemyslaw, E.G. Sirer, M.E. Fiuczynski, D. Becker, C. Chambers, S. Eggers, S.: "Extensibility, Safety and Performance in the SPIN Operating System" – in Proceedings of the 15th ACM Symposium on Operating Systems Principles (SOSP '95), pp 267-284, Copper Mountain CO, U.S.A., December 1995.

[7] A. Birrell, Greg Nelson, Susan Owicki, Edward Wobber: "Network Objects", SRC Research Report 115, Digital Systems Research Center, December 1995.

[8] G. Blair, J.B. Stefani: "Open Distributed Processing and Multimedia" – Addison-Wesley 1997.

[9] T.D. Chandra, S. Toueg: "Unreliable failure detectors for reliable distributed systems" – Journal of the ACM, Vol. 43, No. 2, pp. 225-267,

March 1996.

[10] P. Ciancarini, R. Gorrieri, G. Zavattaro: "Towards a calculus for generative communication" – Proceedings 1st IFIP Workshop on Formal Methods for Open Object-based Distributed Systems (FMOODS '96), E. Najm and J.B. Stefani eds, Chapman & Hall 1996.

[11] F. Dang Tran, V. Perebaskine, J.B. Stefani, B. Crawford, A. Kramer, D. Otway: "Binding and Streams: the ReTINA approach", in Proceedings TINA '96 International Conference, Heidelberg, Germany, September 1996.

[12] A. Fevrier, E. Najm, J.B. Stefani: "Contracts for ODP" – in Proceedings 4th AMAST Workshop on Real-Time Systems, Concurrent, and Distributed Software, Ciudad de Mallorca, Mallorca, Spain, May 1997.

[13] Forman, I. R., S. Danford, H. Madduri, "Composition of Before/ After Metaclasses in SOM" – Proceedings of OOPSLAÕ94, pp427-439, ACM, 1994.

[14] D. Gelernter: "Generative communication in Linda" – ACM Transactions on Programming Languages and Systems 7(1), pp. 80-112, 1985.

[15] Gowing, B., V. Cahill, "Meta-Object Protocols for C++: The Iguana Approach" – Proceedings of Reflection 96, G. Kiczales (ed), pp 137-152, San Francisco, USA, April 1996.

[16] M. Fischer, N. Lynch, M. Paterson: "Impossibility of distributed consensus with one faulty process" – Journal of the ACM 32(2), pp. 374-382, April 1985.

[17] C. Fournet, G. Gonthier: "The reflexive chemical abstract machine and he join-calculus" – 23rd ACM Symposium on Principles of Programming Languages (POPL), January 1996.

[18] C. Fournet, G. Gonthier, J.J. Levy, L. Maranget, D. Remy: "A calculus of mobile agents" – in Proceedings CONCUR '96, LNCS 1119, Springer Verlag, 1996.

[19] G. Hamilton, M. Powell, J. Mitchell: "Subcontract: a flexible base for distributed programming", Proceedings of the 14th Symposium on Operating Systems Principles, Asheville NC, December 1993.

[20] ITU-T Recommendation X.903 | ISO/IEC International Standard 10746-3: "ODP Reference Model: Prescriptive Model" – 1995.

[21] N. Lynch: "Distributed Algorithms" – Morgan Kaufmann, 1996.

[22] McAffer, J., "Meta-Level Architecture Support for Distributed Objects" – Proceedings of Reflection 96, G. Kiczales (ed), pp 39-62, San Francisco, USA, April 1996.

[23] J. Magee, N. Dulay, J. Kramer: "Specifying distributed software architectures" – Proceedings European Software Engineering Conference, LNCS, Springer Verlag, 1995.

[24] J. Meseguer: "Conditional rewriting logic as a unified model of concurrency" – Theoretical Computer Science 96, pp. 73-155, 1992.

[25] J. Meseguer: "Rewriting logic as a semantic framework for concurrency: a

progress report" – in Proceedings CONCUR '96, LNCS 1119, Springer Verlag, 1996.

[26] E. Najm, J.B. Stefani: "A formal semantics for the ODP computational model" – Computer Networks and ISDN Systems 27, pp.1305-1329, 1995.

[27] X. Nicollin, J. Sifakis: "An overview and synthesis on timed process algebras" – Proceedings 3rd Workshop on Computer-Aided Verification, Alborg, Denmark, July 1991.

[28] O. Nierstrasz, J.G. Schneider, M. Lumpe: "Formalizing composable software systems - A research agenda" – Proceedings 1st IFIP Workshop on Formal Methods for Open Object-based Distributed Systems (FMOODS '96), E. Najm and J.B. Stefani eds, Chapman & Hall 1996.

[29] Object Management Group: "The Common Object request Broker: Architecture and Specification", CORBA V2.0, July 1995.

[30] M. Shapiro: "A binding protocol for distributed shared objects", 14th International Conference on Distributed Computer Systems (ICDCS), Poznan, Poland, June 1994.

[31] M. Shaw, D. Garlan: "Software architecture: perspectives on an emerging discipline" – Prentice-Hall 1996.

[32] J.B. Stefani: "Computational Aspects of QoS in an object-based, distributed systems architecture" – Proceedings 3rd International Workshop on Responsive Computer Systems, Lincoln, NH, USA, September 1993.

[33] Sun Microsystems: "Java Remote Method Invocation Specification", Technical Report, Sun Microsystems, Moutain View CA, USA, May 1996.

[34] C. Talcott: "Interaction semantics for components of distributed systems" – Proceedings 1st IFIP Workshop on Formal Methods for Open Object-based Distributed Systems (FMOODS '96), E. Najm and J.B. Stefani eds, Chapman & Hall 1996.

[35] Yokote, Y., "The Apertos Reflective Operating System: The Concept and Its Implementation" – Proceedings of OOPSLAÕ92, vol. 28 of ACM SIGPLAN Notices, pp 414-434, ACM Press, 1992.

APPENDIX 1   A FULLY ABSTRACT ENCODING OF THE π-CALCULUS IN CM_{BASIC}

In this section, we sketch an encoding of the asynchronous π-calculus in CM_{basic}. We consider the version of the calculus described in [3], with a guarded choice operator. The exact syntax used is given by the following grammar:

$$P \quad ::= \quad \bar{a}b \mid P \mid P \mid \nu a P \mid !G \mid G \tag{3}$$
$$G \quad ::= \quad 0 \mid a(b).P \mid \tau.P \mid G + G \tag{4}$$

The encoding, T, uses an auxiliary environment ρ, to record the association of π-calculus names with certain interface identifiers. An environment ρ thus takes the form: $\rho = \{a_1 \mapsto u_1, \ldots, a_n \mapsto u_n\}$, where $a_i \in$ Name (Name the set of π-calculus names) and where $u_i \in$ Id. We note $\rho[a \mapsto u]$ the environment which is like ρ except on a: if $a \mapsto v \in \rho$, then $\rho[a \mapsto u] = \rho\backslash\{a \mapsto v\}\cup\{a \mapsto u\}$; if $a \notin domain(\rho)$, then $\rho[a \mapsto u] = \rho \cup \{a \mapsto u\}$. We note $\bar{\rho}$ an environment that has the same domain than ρ and a range that comprises entirely fresh identifiers.

In the definition of the encoding, $\mathcal{T}_\rho(P)$ denotes an object ω such that $\omega.L = \psi(range(\rho))$, and $\omega.K = domain(\rho)$. We assume the existence of two injections ϕ and ψ, from Name into Id, such that $range(\phi) \cap range(\psi) = \emptyset$. We also denote messages thus: $t.u$, where t is the target of the message and u is the (only) argument. Notice that we do not make use of signal names: we just assume a default signal name used for each message. Notice also that, as a convention, when we make use of identifier w in the encoding, it refers to a fresh identifier (i.e. one which is different from the interface identifiers of the current object, and whose uniqueness is ensured through (†)). Instead of using the cumbersome $\Diamond$ notation, we specify directly the behavior of auxiliary objects in the encoding using rewrite rules between distributed configurations, an abuse of notation that is fully justified by rule (♣). Finally, we assume that in a π-calculus process P, all variables have been given suitably different names to avoid name capture.

The encoding is defined inductively as follows ($I = \{1, \ldots, n\}$):

- $\mathcal{T}_\rho(\bar{a}b) = \psi(a).\psi(b)$ if $a, b \in domain(\rho)$.
- $\mathcal{T}_\rho(P) = D_a^\emptyset \parallel \mathcal{T}_{\rho[a \mapsto w]}(P)$ if $a \notin domain(\rho)$ and $a \in FV(P)$.
- $\mathcal{T}_\rho(P_1 \mid P_2) = \mathcal{T}_\rho(P_1) \parallel \mathcal{T}_{\bar{\rho}}(P_2)$ if $\forall a \in FV(P_1 \mid P_2),\ a \in domain(\rho)$
- $\mathcal{T}_\rho(\nu a P) = \text{nu}(a, P, \rho)$
- $\mathcal{T}_\rho(!G) = \mathcal{T}_\rho(!G) \parallel \mathcal{T}_{\bar{\rho}}(G)$
- $\mathcal{T}_\rho(\sum_{i\in I} \alpha_i.P_i) = \text{choice}(\sum_{i\in I} \alpha_i.P_i, \rho) \parallel_{i\in I} m_i(\alpha)$ with $m(\tau) = \emptyset$ and $m(a(b)) = \phi(a).\rho(a)$

with auxiliary objects defined thus (to simplify notations, we identify objects which can no longer evolve with $\emptyset$):

- $\text{nu}(a, P, \rho) \to D_a^\emptyset \parallel \mathcal{T}_{\rho[a \mapsto w]}(P)$
- $\text{choice}(\sum_{i\in I} \alpha_i.P_i, \rho) \to \mathcal{T}_\rho(P_i)$ if $\alpha_i = \tau$
- $\text{choice}(\sum_{i\in I} \alpha_i.P_i, \rho) \parallel \rho(a).\psi(c) \to \mathcal{T}_{\rho[c \mapsto w]}(P_i\{c/b\})$ if $\alpha_i = a(b)$
- $D_a^A \parallel \phi(a).u \to D_a^{A \cup \{u\}}$
- $D_a^A \parallel \psi(a).u \to D_a^A \parallel v.u$ with $v \in A$

Notice in the encoding above that there are no rules pertaining to $\tau.P$ and $a(b).P$ as these are covered by the rule for the choice operator $\sum_{i\in I} \alpha_i.P_i$.

12

Using SDL to develop CORBA object implementations

Björkander, M.
Telelogic AB
Box 4128, S-203 12 Malmö, Sweden
Phone: + 46 40 17 47 31
E-mail: morgan.bjorkander@telelogic.se

Abstract

This paper describes a methodology for using SDL in conjunction with CORBA; more specifically the CORBA-oriented approach is explored, where SDL is used to implement a predefined IDL description. This approach also includes the definition of mapping rules between IDL and SDL.

An in-depth description of how the implementation of the CORBA-oriented approach is carried out is also provided, focusing particularly on how an SDL design tool is made to interwork with an existing ORB implementation.

Keywords

CORBA, IDL, SDL, OOA, OOD

1   INTRODUCTION

With the advent of the Common Object Request Broker Architecture (CORBA) to manage large, heterogeneous object systems, the need to be able to specify, verify, and test such systems is becoming increasingly important. It is not sufficient to be able to specify object behaviour using for example C++, because the behaviour of such systems are not easily verifiable. Instead, by using a formal description language such as the Specification and Description Language, SDL (ITU-T Recommendation Z.100, 1995), to define the object behaviour all the commonly used techniques for verification and validation become available.

The work presented in this paper is based on results obtained when implementing CORBA-support in the SDL Design Tool (SDT) developed by Telelogic, and is considered both from a methodology and an implementation point of view.

In Section 2, a CORBA-oriented approach of using SDL together with CORBA is presented. This approach is also incorporated in the SDL-based Object Modelling Technique (SOMT), which provides a framework for how object-oriented analysis and SDL-based design might be used in a coherent way. The necessary mapping rules between the Interface Definition Language (IDL) and SDL are also summarised. This section is concluded with some examples of how object models might be mapped to IDL.

Section 3 goes more into detail regarding how an SDL application designed according to the principles outlined in Section 2 can be implemented together with an existing ORB implementation.

The last section describes the conclusions obtained from this work, and also some directions for future work.

2   DEFINING OBJECT BEHAVIOUR USING SDL

2.1   Using CORBA and SDL

As is described in (Olsen, 1995), there are basically two ways to use SDL with CORBA, depending on whether SDL is viewed as a specification or design language:

- The CORBA-oriented approach, where SDL is supported as implementation language for the definition of behaviour and treated in the same way as the already supported implementation language C++.
- The SDL-oriented approach, where a CORBA platform is used as execution system for SDL processes.

When to use the different approaches

The above approaches may seem very similar at first, but the methodologies used are quite different.

The SDL-oriented approach is primarily used when SDL is viewed as a specification language. Once the specification (of the entire distributed system) has been performed in SDL, it should be partitioned into arbitrarily small subsystems that are then (optionally) implemented on different machines and communicate with each other using CORBA. The smallest unit of partitioning should be at the process level. In this case, IDL descriptions are automatically generated from the SDL system; it might, however, be necessary to limit the available SDL concepts when using this approach.

The CORBA-oriented approach, on the other hand, is primarily used when SDL is viewed as an implementation language, and a given IDL description should be designed using SDL. That IDL description is used to generate a stub (skeleton) in SDL, to which behaviour is then added.

This paper focuses on the CORBA-oriented approach, whereas a more thorough description of the SDL-oriented approach can be found in (Olsen, 1995).

The CORBA-oriented approach

The most important parts of the development process when using the CORBA-oriented approach are shown in Figure 1.

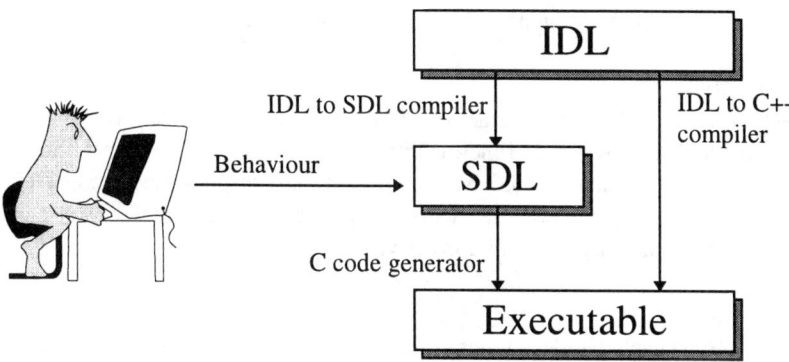

Figure 1 The activities of the CORBA-oriented approach.

The activities that can be clearly distinguished when using this approach are:
1. Convert an IDL description to an SDL stub.
2. Define the behaviour of the SDL system using the generated SDL stub as the starting point.
3. Generate the C/C++ code that is used to implement the SDL application.
4. Generate the C++ code that is needed by the ORB.

In reality, several additional activities occur, but these will be further outlined in Section 3.1. The only activities that should concern the developer are the first two; the user must be aware that the SDL system is a CORBA server (i.e., an object implementation that can also act as a client by requesting services from other object implementations).

By using the mapping rules described in Section 2.3 it is possible create a tool that automatically converts an IDL description to an SDL stub. The latter two activities described above are also performed automatically, and it is necessary to adhere to the IDL to C++ mapping rules that are defined in (OMG, 1996).

2.2   The SOMT method

The SDL-based Object Modelling Technique (SOMT) has been developed to manage the entire development process with the key focus of integrating object-oriented analysis with SDL design (Telelogic, 1996). The main activities of the SOMT method are summarised below.

Requirements analysis
In the requirements analysis activity, the requirements of the system that is going to be built are captured. The system itself is viewed as a black box, where only the interaction with objects and concepts defined outside its boundaries are of interest. The foremost modelling concepts of the requirements analysis are use cases (using MSCs) and an object model, but additional concepts also exist.

System analysis
The system analysis describes the architecture of the system as well as the objects that are needed to provide its functionality. It is only the objects that must be present in the system that should be modelled at this stage; objects outside the system were modelled in the requirements analysis. Use cases and an object model are the primary models used within this activity.

System design
The purpose of the system design activity is to define how the system architecture is going to be implemented. It is necessary to define both the architecture of the resulting application and the interfaces between the different components. The main modelling notation for system design is SDL, where the architecture is defined using system and block diagrams, and the interfaces are reflected using signals and remote procedures.

Object design
During the object design the functionality of the system is defined in detail, and the behaviour of the active objects in the system is described using SDL process graphs.

Implementation

The goal of the implementation is to produce the final application. This is done by automatically producing code directly from the SDL design.

Implementation links and Paste As

One of the features of the SOMT method is implementation links (*implinks*); these are used to maintain a relationship between objects in different models to indicate that one object can be seen as the implementation of another.

The main purpose of implinks is to provide traceability of the development through the different activities mentioned above, thereby facilitating for example consistency checks etc.

One way to create implinks is to make use of the *Paste As* functionality that is defined by SOMT. This is a tool supported concept that allows a developer to copy an object in one model, and then paste it into another model as something else, while at the same time creating an implink between the two objects. By selecting one of these objects it is then possible to follow the link to the other object.

As an example, an object model class can be copied from the analysis model and then pasted as an SDL process type in the system design model. This means that the paste as functionality can be seen as the implementation of a set of transformation rules.

According to the SOMT method, it is possible to copy/paste as arbitrarily large collections of objects and relationships, but current tool support restricts this by only allowing single object entities to be copied/pasted as.

Introducing CORBA in SOMT

When the CORBA-oriented approach is introduced in SOMT, IDL is mainly used to describe the interfaces between the different parts in the system design. In fact, the definition of IDL interfaces replaces the ordinary system design activity that was mentioned above. Each of the components that are arrived at can then be implemented using the preferred language, and the subsystems that should be implemented using SDL are part of a larger system that is connected through CORBA.

Ordinarily, when going from analysis to design, the SOMT method advocates that objects model classes are pasted from the system analysis, and then pasted as SDL entities in the system design. In the CORBA-oriented approach, however, the object model concepts that are part of the component interfaces should first be pasted as IDL. Some specific examples of mapping object model classes to IDL are provided in Section 2.5.

Then, when the IDL description is completed and it should be implemented using SDL, an SDL stub is generated from it. Internal information that is not part of the interface definition can then be pasted as the SOMT method ordinarily describes.

Inherent in the IDL description (through modules and interfaces) is information about the system architecture, and this structure information is kept in the SDL system stub through the mapping rules that are described in Section 2.3. Once the system design has been performed, remaining activities such as providing the behaviour of the SDL system are performed as usual.

2.3 Mapping IDL to SDL

In order to create the implementation language stub it is necessary to provide mapping rules from IDL. In (OMG, 1996), several chapters are concerned with how different languages are mapped. For SDL, mapping rules have been proposed in (Born, 1996) and (Björkander, 1996).

Mapping Rules

The mapping rules from IDL to SDL described in (Björkander, 1996) can be summarised as follows:

- A module provides a namescope and a mechanism to group interfaces. As such it is mapped to a block type, where nested modules become nested block types.
- An interface is mapped to a process type. As the interface contains the attributes and operations that are available on an object, the corresponding process type contains the appropriate SDL definitions of these. Object references are mapped to PId values.
- An operation describes a service that is offered by an object, and it can be defined either as synchronous, where the client is blocked while the request is being handled by a server, or as asynchronous (using the keyword oneway), where the client making the request simply continues executing. In the former case, the operation is mapped to a remote procedure, while in the latter it is mapped to a signal. Raises and context expressions are still subject to be mapped, as are exceptions.
- An attribute is mapped to a declaration of a variable together with two remote procedures that are used to get and set the value of the variable. If the attribute is defined as readonly, the set operation is omitted.
- Interface inheritance must be flattened in SDL, where both operations and attributes defined in a base type have to be duplicated in a derived type, i.e., SDL inheritance cannot be used.
- Constants are mapped to synonyms.
- How basic types are mapped is shown in Table 1. Note that predefined types have the prefix 'CORBA_' in SDL. The type any currently have no suitable mapping in SDL.

Table 1 Mapping basic types

IDL Type	SDL Type	syntype of

long	CORBA_long	Integer
short	CORBA_short	Integer
unsigned long	CORBA_unsigned_long	Integer
unsigned short	CORBA_unsigned_short	Integer
double	CORBA_double	Real
float	CORBA_float	Real
char	CORBA_char	Character
boolean	CORBA_boolean	Boolean
octet	CORBA_octet	Octet

- Constructed types are mapped as follows:
  - An enum is mapped to a newtype with the appropriate literals.
  - A struct is mapped to a newtype with the corresponding struct.
  - A union is mapped to a newtype with a corresponding struct, where the first member of the struct is a tag matching the union's switch type.
- The template types are mapped as follows:
  - A sequence is mapped to the generator string.
  - A string is mapped to to the type CORBA_string, which is a syntype of Charstring.
- The complex declarators are handled as follows:
  - An array is either mapped to the generator array, together with an additional type to define its index range, or as the generator CORBA_array, which is defined to have a limited index range corresponding to that of the original IDL array.

Differences in the mapping proposals

The mapping proposals referred to above are very similar, but there are some minor differences. In (Born, 1996):

- All operations are mapped to remote procedures to avoid using channels and signal routes. This motive is questionable, and signals better convey the asynchronous nature of oneway operations.
- A context expression is mapped to an additional PId parameter of the operation for which it is defined.

- An exception is mapped to a struct value, which is stored in a new process instance in the server whenever an exception occurs. The client can then access the server to find out which exception has occurred. While the basic mapping is natural, it is not a good idea to pass a PId value (object reference) back to the client for it to examine. For example, it is not possible for the client to catch a system exception. A better mapping would probably be to pass the raised exception back in the form of a struct.
- All types are mapped as ASN.1 types. This is a good idea, but it should probably be possible to use either ASN.1 or SDL types in the mapping.
- Interfaces are mapped as process types on both the client and the server side. To include the process types on the client side seems strange and superfluous.

SDL System Structure

An SDL system that is based on the above mapping rules has a particular architecture; one package is used to define interface concepts, such as types, signals and remote procedures, whereas another is used to contain the structural information that can be derived from the IDL description. The interface package can be resused by other SDL clients that want to access services from this object implementation. Consider for example the SDL system that is shown in Figure 2.

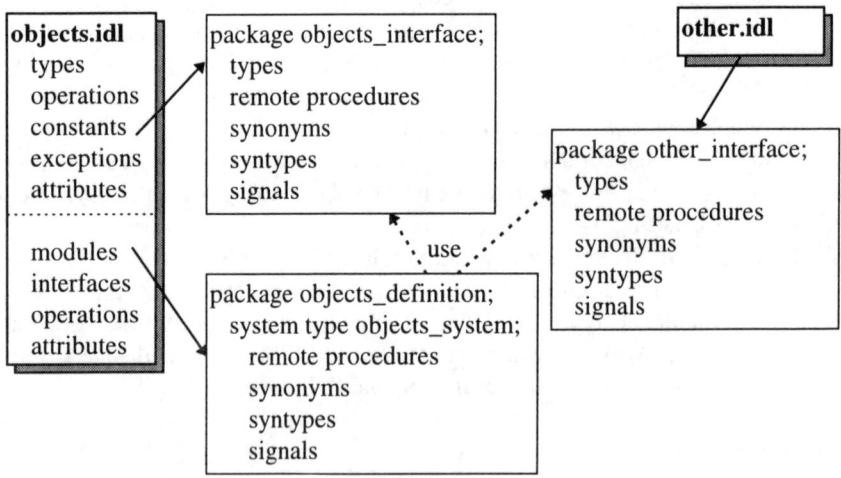

Figure 2 A schematic view of the SDL system structure.

In this example, an object implementation in SDL is created from the IDL file named *objects.idl*. The stub that is created will contain one package named *objects_interface*, and one package named *objects_definition*; this latter package also contains a system type named *objects_system*. The object implementation should make use of services that are defined in the IDL file *other.idl*. In order for the system implementing *objects* to access these services, the IDL description of *other* is also converted to SDL, but only the interface package (*other_interface*) needs to be used.

2.5 Extensions to SDL-92

The fact that it is allowed to call remote procedures over system boundaries is part of a larger standardisation effort to harmonise signals and remote procedures (and is included in an addendum to (ITU-T Recommendation Z.100, 1995) that is called SDL-96). This harmonisation also makes it allowed to include remote procedures in channels, signal routes, and gates.

Additional extensions in the above mappings are for example some new types. One type that has been added in SDT is the type Octet that is actually part of (ITU-T Recommendation Z.105), and which is used to implement the IDL type octet.

2.6 Mapping Object Models to IDL

In Section 2.2 it was hinted that it is possible to paste an object model as IDL; the transformation rules that are considered here are very simple and straightforward. The strength of this approach is that for each object that is pasted, an implink is created between the copied class and the IDL entity, thereby facilitating for example trace mechanisms.

Classes

An object model class can be mapped to either a module or an interface. In the former case, the resulting module is always empty, even if the class contained both attributes and operations. In the latter case, however, all attributes and operations are mirrored using IDL syntax as far as possible. This latter mapping is shown in Figure 3. The comment that is included in the mapping is used to contain the optional implink that is created between the class and the interface.

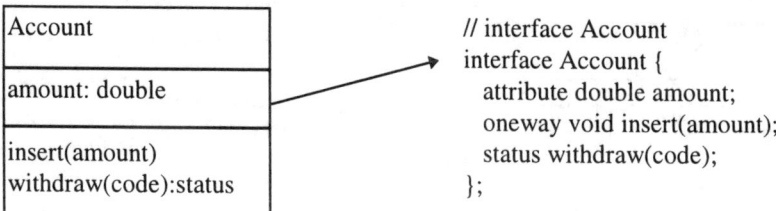

Account	
amount: double	
insert(amount) withdraw(code):status	

```
// interface Account
interface Account {
    attribute double amount;
    oneway void insert(amount);
    status withdraw(code);
};
```

Figure 3 Mapping an object model class to an IDL interface.

It is not possible to express all information that is needed in the IDL description simply by mapping a class like this; some manual additions must always be made, such as type definitions, constants, and exceptions, but also simple matters like defining for each operation parameter whether it is an in, inout or out parameter.

Furthermore, it might be necessary to define an attribute as readonly, or whether an operation is oneway or not. All operations are considered asynchronous by default, unless it can be determined that it should be synchronous. This is done by either defining a return type for the operation, or by inserting {sync} or {async} after it.

Inheritance

The mapping of object model class inheritance is only applicable when an object model class is mapped to an interface, and in that case the class inheritance is mapped as interface inheritance, as is shown in Figure 4.

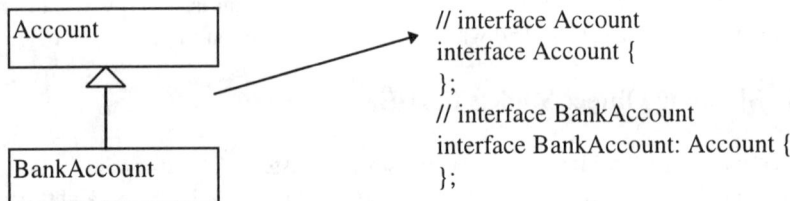

Figure 4 Mapping object model class inheritance to IDL inheritance.

Aggregation

When mapping aggregations, all object model classes that are not leaves are mapped as modules. However, the leaves themselves may be either modules or interfaces, and it is the responsibility of the user to decide which. Aggregations are thus mapped to nested modules, where the innermost layer may be either interfaces or modules. An example of aggregation mapping is shown in Figure 5.

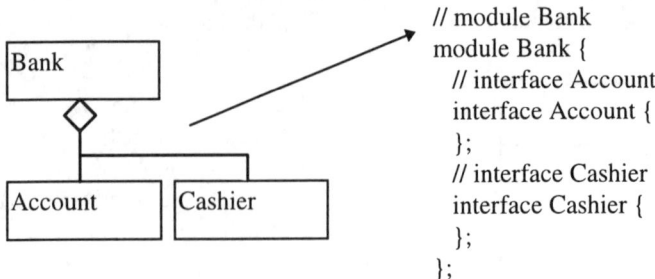

Figure 5 Mapping object model class aggregation to IDL.

3 IMPLEMENTING AN SDL APPLICATION

In this section the implementation of the CORBA-oriented approach is discussed in more detail. The created SDL application is supposed to be working as a client/server in a heterogeneous, distributed environment as only one of many such parts, i.e., the system has been decomposed into lesser parts that may or may not be implemented using different programming languages; the glue between all of these parts, however, is CORBA.

The tools that have been used in the implementation of the CORBA-oriented approach are SDT3.1 for modelling SDL and Orbix2.1mt (from IONA Technologies, Ltd.) to provide the CORBA implementation.

3.1 The system architecture

An SDL application can be used as both a server and a client, which is illustrated architecturally in Figure 6.

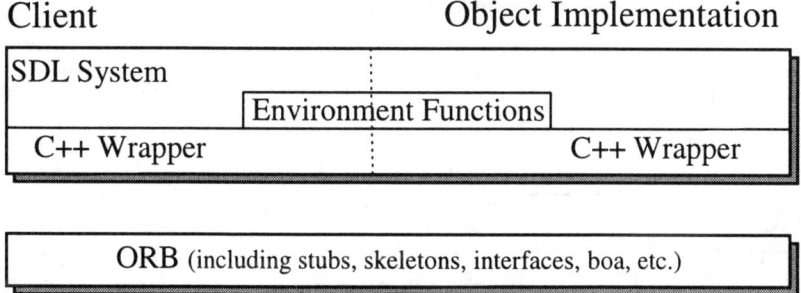

Figure 6 The system architecture of an SDL application.

An ordinary SDL application without the CORBA support only consists of the SDL system itself along with its environment functions that are used to connect the system to the outside world. In the CORBA-oriented approach, a wrapper is placed around the SDL system to imitate the behaviour of objects in a way that makes sense to the ORB. The C++ wrapper and its functionality is further discussed in Section 3.2. Orbix supports the C++ mapping of IDL, which is why the wrappers have been implemented using C++.

Based on the development process depicted in Figure 1, the development of the object implementation side starts with an IDL description. The complete process is then as follows:
1. The IDL description is used as the basis for an SDL stub, which is generated by an SDT specific IDL compiler. At the same time, a C++ wrapper for the object implementation side is also generated using the same IDL compiler.

2. The ORB also uses the IDL description to generate ORB specific C++ code, but has an ORB specific IDL compiler for this purpose.
3. A developer then provides the behaviour of the SDL system stub.
4. If any services from other servers are required (on the client side of the SDL application), their IDL descriptions should be converted to SDL using the SDT specific IDL compiler. Code is also generated for the C++ wrapper. As for the ORB, it uses its own IDL compiler to generate the necessary C++ code.
5. C code represented the designed SDL system is generated using the SDT C Code Generator.
6. The environment functions are predefined, and are always the same regardless of the SDL (CORBA) application; all system specific information is placed in the wrapper.

When the C and C++ code has been generated as above it is compiled and linked with a set of precompiled libraries (including an SDL kernel providing runtime support of the SDL system, as well as ORB specific libraries) to form an executable SDL application.

These steps are shown graphically in Figure 7.

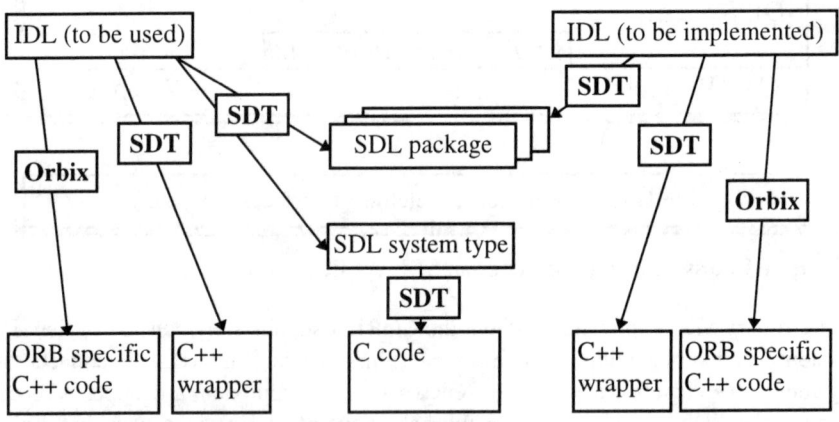

Figure 7 Generating code for the SDL application.

3.2 Wrapping the SDL system

In order to get a functioning SDL application, the behaviour of the SDL system wrapper is especially important. The wrapper can be said to consist of two parts: the environment functions and the class definitions that are generated from the IDL descriptions. The C++ wrapper on the client side has a behaviour that is very similar to the behaviour on the client side, even though there are some unique features, such as the need to be able to locate objects for particular requests.

Scheduling

Since the SDL system is capable of acting as a client, it has its own scheduling mechanism. Requests that made to other servers are sent through an environment function to the C++ wrapper, where the request is processed and then passed on through the ORB to the intended server.

At regular intervals, the scheduler calls another environment function to check whether there are events pending from the ORB. Should such an event be present, it is processed, and then sent to the appropriate process instance.

The entire SDL application is scheduled within a single UNIX process, and requests that are made internally in the SDL system are not managed by CORBA. Contrast this with the SDL-oriented approach, where all requests would be sent through CORBA, possibly from multiple UNIX processes.

Environment functions

The environment functions are responsible for managing the communication between the ORB and the SDL system. Two of these environment functions have already been mentioned above as responsible for passing requests to and from the SDL system.

Another environment is used to initialise the SDL system and its environment. In this case, it is particularly used to initialise the communication with the ORB.

The environment functions uses buffers to store all requests before they are treated, both for incoming and outgoing requests.

Objects versus process instances

When the ORB communicates with the object implementation, it only sees the C++ wrapper and the objects that are defined within it; the SDL system itself is completely hidden.

The SDL system, on the other hand, controls which objects that are present in the C++ wrapper by mirroring each process instance as an object. Whenever a new process instance is created in SDL a new object representing that process instance is created in the wrapper.

When a process instance is terminated, the object is also removed (when simulating the SDL system the object is preserved throughout the lifetime of the server in order to allow a trace of failed requests).

An object that receives a requests passes it along to its corresponding process instance (it first has to convert the request into a suitable format, i.e., a signal or remote procedure).

In SDL, all object references are represented using PId values. A C++ client, however, would still access the process instance of an SDL object implementation using its ordinary object references.

Multithreading

In order to handle requests correctly, it has been necessary to make use of a multithreaded ORB.

On the object implementation side, each new request from the ORB must be treated in a thread of its own. The reason for this is that a synchronous operation is mapped to a remote procedure in SDL, which in turn is implemented as the sending of two asynchronous signals (call and reply). When the request is received, a call signal is sent into the SDL system. While the operation is waiting for a reply, the SDL system must be able to execute other requests, which is not possible unless each request is executed in its own thread, while the SDL system is scheduled in a main thread.

On the client side, the situation is very similar. While waiting for a reply to a request made to the ORB, the SDL system must be able to continue to execute, which means that all such server requests must also be executed in separate threads.

The SDL system thus executes in one thread, and each request that is received by or sent from the SDL system results in a new, detached thread which disappears once the request has been handled.

The ORB is responsible for creating the appropriate threads for requests to the SDL system, while the C++ wrapper is responsible for creating threads for requests aimed at other servers. The C++ wrapper must also provide a sufficiently multithread-safe environment.

The C++ implementation classes

As part of the ORB specific code, a set of IDL C++ classes are generated which represents the IDL interfaces. It is then up to the developer to provide C++ implementation classes that implements these IDL C++ classes. The ORB provides mechanisms for connecting the IDL C++ classes and the C++ implementation classes to each other. In Orbix, either the BOAImpl approach or the TIE approach can be used for this task, and due to its greater flexibility the TIE approach is used in the SDT integration. Each IDL C++ class must have at least one corresponding C++ implementation class.

In the C++ wrapper, the C++ implementation classes are the most important part, as they provide the 'behaviour' of the SDL system. A request that is made to an object is passed to the appropriate process instance after having been processed. This processing is specific depending on whether the operation is asynchronous or synchronous. Due to the increased complexity when dealing with synchronous operations it is considerably easier to manage asynchronous operations.

For an asynchronous operation, the following steps have to be performed by an object:

1. Allocate memory for the signal.
2. Convert the parameters of the operation to SDL signal parameters.
3. Send the signal to the process instance corresponding to the current object.
4. Return, i.e., exit the request thread.

For a synchronous operation, on the other hand, some additional steps are necessary:

1. Allocate memory for the remote procedure call.
2. Convert the parameters of the operation to SDL remote procedure call parameters.
3. Send the remote procedure call to the process instance corresponding to the current object.
4. Wait for a reply, i.e., block the request thread until the appropriate remote procedure reply is received. It is necessary to pass information about the current thread's identity in the call/reply signals to ensure that the appropriate thread receives the correct reply (since the order in which replies are received is not guaranteed to be the same as the order in which they were sent).
5. Convert the SDL remote procedure reply parameters to C++ parameters.
6. Release the memory held by the reply.
7. Return, i.e., exit the request thread, and pass the obtained data back to the client.

Locating a server object
One particular problem that must be addressed is how an server object is located. In SDL, a request can be made either to a specific PId value, or without specifying a particular receiver. In the first case, the object reference of the receiver is already known, and there is no problem. In the second case, however, an appropriate receiver must first be located. There are two approaches to this problem:

- A bind concept can be introduced in SDL which would be responsible for finding a server implementing the required request. However, this solution does not remove the problem with requests that are made without specifying a receiver.
- When the C++ wrapper receives a request with no apparent receiver, an ORB specific bind command to find an appropriate server object can be performed. Once the request has been performed this object reference can then either be stored for subsequent requests of the same kind to reuse, or it can be released, thereby requiring a new bind command for each new request.

4   CONCLUSIONS AND FUTURE DIRECTIONS

The implementation that has been performed so far on the CORBA-oriented approach is by no means complete (as some IDL concepts are not yet supported), but still it must be considered an important contribution to the ways an interface can be implemented. Above all, this implementation shows that it is possible to use SDL in conjunction with CORBA using the CORBA-oriented approach. Together with (Olsen, 1995), where the SDL-oriented approach is demonstrated, this makes

SDL an important language to consider when dealing with distributed systems of different kinds.

One key issue of the CORBA-oriented approach is the that mapping rules between IDL and SDL need to be standardised. Currently, there are at least two different mapping proposals, and some work will have to be performed to gain a consensus on these, possibly through standardisation work in OMG or ITU.

As a continuation of the work performed hitherto, it is necessary to support all IDL concepts in SDL, and perhaps to update the methodology that has been presented in this paper.

A natural extension of this work would be to examine the SDL-oriented approach in more detail, providing methodology guidelines and implementations for when SDL should be targeted to CORBA. It is not expected that it will be possible to define general mapping rules for mapping SDL to IDL, as these will probably be implementation dependent.

5   REFERENCES

Björkander, M. (1996) Mapping IDL to SDL, Telelogic AB, Malmö.

Born, M. and Winkler, M. and Fischer, J. (1996) Formal Language Mapping from CORBA IDL to SDL'92 in Combination with ASN.1, Humboldt University, Berlin.

ITU-T Recommendation Z.100 (1995) CCITT Specification and Description Language (SDL). ITU-T, Geneva.

ITU-T Recommendation Z.105 (1995) SDL Combined with ASN.1 (SDL/ASN.1). ITU-T, Geneva.

Olsen, A. and Jorgensen, B.M. (1995) Using SDL for targeting services to CORBA, in *Bringing Telecommunication Services to the People, 3rd International Conference in Broadband Services and Networks* (ed. Clarke, A. and Campolargo, M. and Karatzas, N.), Springer-Verlag.

OMG (1996) The Common Object Request Broker: Architecture and Specification, Object Management Group, Framingham, MA (USA).

Telelogic (1996) The SOMT Method, in *SDT3.1 Methodology Guidelines: Part 1*, Telelogic, Malmö.

6   BIOGRAPHY

Morgan Björkander is a developer at Telelogic AB, and is mainly responsible for the CORBA support in SDL applications generated by SDT. He has also been involved in different RACE and ACTS projects focusing on service creation, for example SCORE and TOSCA.

13

Meta Information Management

S. Crawley, S. Davis, J. Indulska, S. McBride, K. Raymond
CRC for Distributed Systems Technology (DSTC)
University of Queensland, Qld 4072, Australia
crawley@dstc.edu.au

Abstract

The increasing openness and heterogeneity of distributed computing systems requires the representation of meta-information in distributed environments. This paper presents a conceptual framework for discussing meta-information, and describes the DSTC's design for a universal Meta-Information Manager (MIM) with the key properties of heterogeneity, extensibility and openness. The DSTC MIM design supports a wide range of applications for meta-information, from trading and binding services to design repositories and data warehousing.

Keywords

Distributed systems (C.2.4), programming environments (D.2.6).

1   INTRODUCTION

The increasing openness and heterogeneity of distributed computing systems requires the representation of meta-information in distributed environments, as evidenced by current standards activity in the OMG [1] and ISO [2][3]. Meta-information is information that describes other information. Everyday examples of meta-information include programming language data types, CORBA or DCE interface definitions, formal specifications and design diagrams, and schemas for databases.

A *Meta-Information Manager* (MIM) is a service that provides a repository for meta-information in a distributed environment. A MIM needs to support a wide range of clients, including infrastructure services, domain specific services and end user applications; for example, trader service types, interface types, database schemas, user profiles in a command interpreter.

Although low volume, meta-information is typically complex and highly inter-related; it is typically viewed as a navigable tree or graph. In addition, certain "paradigms" such as naming and scoping are common to many kinds of meta-information.

The DSTC MIM is the third generation of universal meta-information or type managers designed and developed by the DSTC [4][5][6]. The DSTC MIM described here differs from earlier DSTC work in that it represents types in structured form. This entails adding a well defined meta-meta-information level to the type management model to allow the representational aspects of each target type system to be modelled. One consequence is that the MIM can produce "specific" meta-information interfaces that are tailored to each target type system.

Apart from the DSTC work, there is little published material on universal type management. A few type managers have been designed to support a single interface type system; for example the CORBA Interface Repository [7] supports CORBA interface definitions, and the Commandos Type Manager [8] supports a canonical type model for the Comandos supported languages. The only other example of a universal meta-information manager is the UNISYS Universal Repository [9], which is designed primarily as a repository for software design tools.

2   OVERVIEW OF THE META-INFORMATION MANAGER

The terminology of this document frequently prefixes words with "meta-". In general, the *meta-* prefix indicates that the term denotes "a description of a something". In some cases the prefix is repeated; for example, a meta-meta-object is "an object that describes a meta-object".

2.1   Abstract Overview

The DSTC MIM is based on the abstract model of *meta-information* as illustrated in Figure 1. The universe of entities ("things") in a given domain of interest can be classified into *types*; i.e. sets of entities that share some property or properties of interest. The types for a given domain of interest are defined in some formal or informal *type language*. The types will have associated domain specific semantics, some of which can be expressed as *type relations*, consisting of tuples of related types (*type relationships*). For example, the notion of subtyping in a programming language can be modelled as a binary relation between types; i.e. the set of valid "subtype" / "supertype" pairs. A *type system* consists of a type language for a domain of types, and a definition of the type relations that are meaningful over those types. Finally, a *type schema* is a collection of types and relationships between them that jointly describe some "system" of interest.

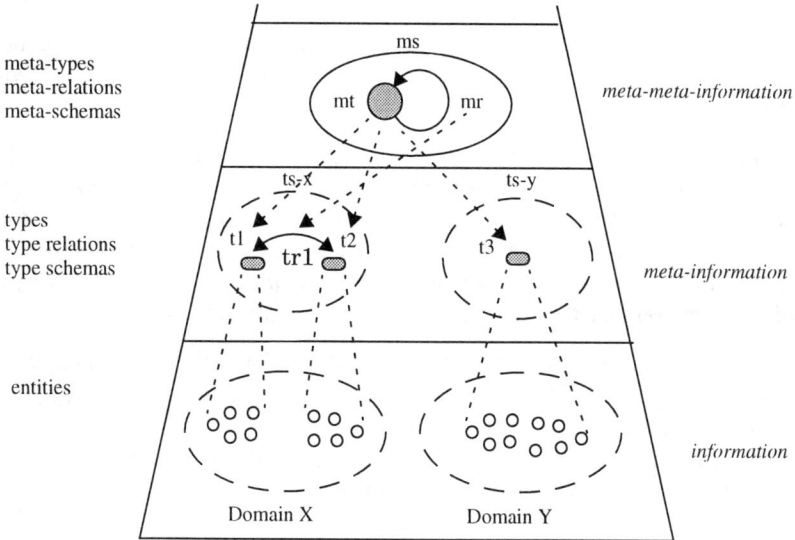

Figure 1   Meta-meta-information, meta-information and information.

Analogously, a type system can be said to define a domain of *meta-information* whose "values" are the types and relationships expressible in the type system. By applying the approach above, it can be reasoned the values in the meta-information domain can themselves have types. These types over the meta-information domain (i.e. the types of types, relationships) are known as *meta-types* and *relation* or *relationship types*, and are examples of *meta-meta-information*. The meta-meta-information domain is the one in which MIM type system definitions would be expressed. Further layers may be added to the abstract model, though three layers are generally sufficient for practical purposes.

The meaning of the abstract concepts in Figure 1 may be understood using some real world illustrations. If the meta-meta-information layer defined a type system for CORBA IDL, then the meta-information layer would contain schemas of CORBA IDL types and relationships between them, and the entities in the information layer would be CORBA objects. Alternatively, if the meta-meta-information layer defined a type system for design notations, then the meta-information layer would define specific notations, and the entities in the information layer would be design diagrams.

2.2   Concrete Overview

The DSTC MIM needs to represent types, relations and schemas in a form that is convenient for client programs. Since the MIM is intended to be "universal", it also needs to represent type system definitions and provide a mapping from these to meta-information interfaces. It needs to allow derived relationships between types;

e.g. based on analysis of the types or on inference from type system properties. Finally, the MIM needs to allow type system definitions to be augmented with semantics.

The DSTC MIM interfaces are all expressed in terms of the CORBA object model, and defined in CORBA IDL. Meta-information is represented as CORBA objects that are called *meta-objects*. Types and components of types are represented by meta-objects called *type objects* that have a fixed set of strongly typed properties. Relations are represented by meta-objects called *relation objects*. Collections of related types and relations are represented by *schema objects*.

Meta-meta-information is represented as CORBA objects that are called *meta-meta-objects*. Each kind of MIM type, relation and schema object in a type system is defined by a meta-meta-object; i.e. a *meta-type object*, *meta-relation object* or *meta-schema object* respectively. The closure of a meta-schema (i.e. the set of meta-meta-objects it depends on) constitute a *MIM type system definition*.

2.3   Type System Instantiation

The main reason that the DSTC MIM represents meta-meta-information is to allow programmers to define their own type systems. The programmer first defines a type system using a language and tools provided with the MIM. The type system definition is then *instantiated* to produce a group of CORBA services that manage the meta-objects which represent types, relationships and schemas. The programmer would then develop the tools to populate the meta-object services, and use the meta-information as required. Type system instantiation is illustrated in Figure 2.

The type system instantiation process could be automated using a *MIM toolkit*, or it could be entirely manual. This allows the DSTC MIM specification to be used for both "universal" meta-information repositories, and type system specific repositories. Assuming that it exists, a MIM toolkit could provide type system specification languages, compilers and IDL generators, along with tools for meta-object server implementation, and generic tools for browsing and managing meta- and meta-meta-objects. In either case, the steps in the type system instantiation process are as follows:

1) The programmer designs the conceptual structure of the types and relations in the new type system and causes the corresponding meta-meta-objects to be created. For example, the type system could be specified in a textual language (e.g. DSTC's "MODL" language [10]) and then compiled to produce meta-meta-objects.

2) The programmer translates (using a tool or by hand) the meta-meta-objects to the IDL for the new type system's meta-objects, following the mapping rules in the MIM specification. This IDL can then be compiled using a standard IDL compiler.

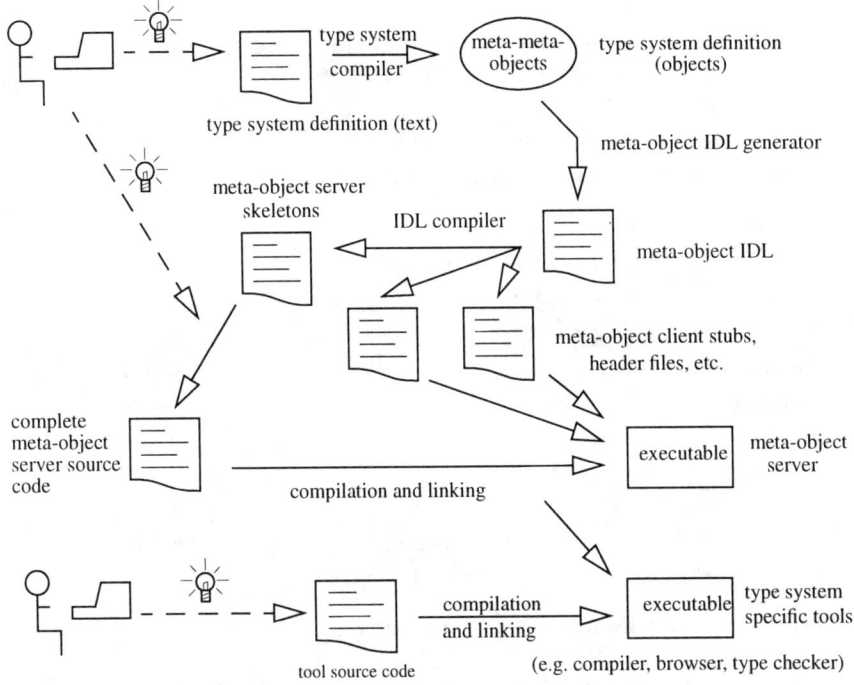

Figure 2 MIM type system instantiation and tool building.

3) The programmer implements the meta-objects services for the type system; e.g. by starting from server skeletons generated by the IDL compiler, or by using a meta-object implementation language and tools from the MIM toolkit.

When the type system has been instantiated, the programmer needs to implement any tools (e.g. browsers, compilers, type checkers, design tools, etc.) needed to create and use meta-information for the new type system. Once again, this could be automated using tools in the MIM toolkit.

3 META-OBJECTS

In this paper, the term *class* is used to refer to the interface type or signature of an object. In the CORBA / OMA world [7], object interfaces are typically expressed in the CORBA Interface Definition Language (IDL). Interfaces have operations with typed parameters, results and exceptions, and typed attributes. CORBA provides multiple interface inheritance. This paper uses the terms *super-class* and *sub-class* to denote the respective roles.

While a typical CORBA class can be instantiated as CORBA objects, some classes (known as *abstract classes*) are defined solely to be inherited by other interfaces. An *abstract base class* is a class defined to create a common super-class for group of classes. A *component class* is defined to provide a set of related features that will be reused in other classes.

The DSTC MIM represents meta-information using CORBA objects called *meta-objects*. There are many meta-object classes, all descended from the abstract base class "MO" (Meta-Object), as illustrated in Figure 3. The MO interface provides two attributes; one that links a meta-object to the meta-meta-object that describes it, and another that links it to the type schema that "owns" it.

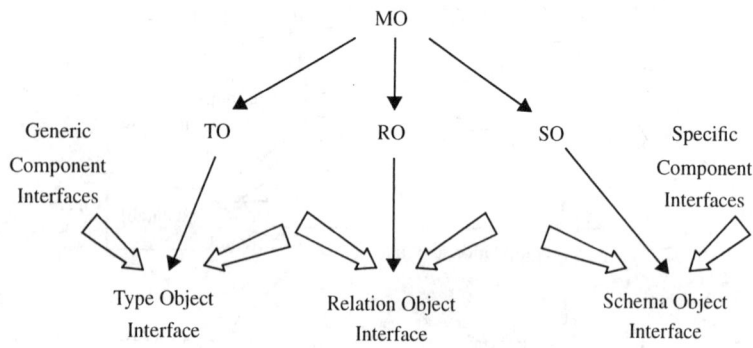

Figure 3   The MIM meta-object class hierarchy.

The full (i.e. most derived) interface for any MIM meta-object will inherit from two distinct kinds of interfaces. The *generic meta-object interfaces* provide a type system independent view of the meta-information, using textual names, unbounded sequences and the CORBA "any" type to achieve genericity. These interfaces are provided to support generic tools such as meta-information browsers that work with all kinds of meta-information.

The *specific meta-object interfaces* provide a view that is tailored to each specific type system. These interfaces provide specifically named operations and attributes with ordinary (static) CORBA types, and are designed to be more "user friendly" for applications working in a particular type system. The specific interfaces are derived from the type system's definition in the type system instantiation process (see Section 2), and should conform to the detailed rules for mapping meta-meta-objects to meta-object IDL as defined in [10].

3.1   Type Objects

A *type object* is a meta-object that represents a type or a component of a type. It has a fixed number of *properties*, where the property name and property type are specified in the type system definition. The type of a property may be any CORBA data type including a CORBA object type.

An instantiated MIM type system will define one or more type object classes. As Figure 3 shows, each type object class is descended from the "MO" class by way of the "TO" class. It also inherits a number of "generic" and "specific" component classes, depending on its properties, and on the relations that it may belong to. In the case of properties, the component classes provide operations to "get" and "set" the property values of the type object.

3.2   Relation Objects

A *relation object* is a MIM meta-object that is used to represent type relations. A type system definition will typically include a number of relation object classes. Some will serve to aggregate and connect component type objects to form complete types. Others may represent relationships between whole types; e.g. type equivalence or subtyping. Others still may be used to annotate types with other meta-information; e.g. descriptions or version stamps.

A MIM relation object represents a relation, and has roles with fixed role names and role types as explained in Section 3. The role types may be the classes of MIM type objects or general CORBA data types. A relation may be viewed as a "table" as shown in Figure 4. This example shows relationships between some type objects that are intended to represent record types and their component fields. Each row in the table represents a single relationship tuple, and each column represents a role in the relation.

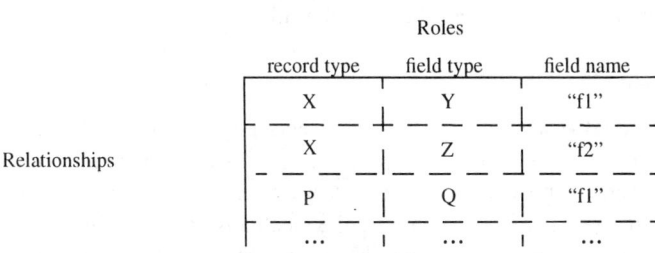

Figure 4   The tabular view of a relation.

A MIM relation may be defined to have one or more *unique key constraints*. A unique key constraint ensures that no two relationships can have the set of values in the unique key roles. (In the example in Figure 4, a unique key consisting of the "record type" and "field name" roles would ensure that field names are unique within a record type.) Every relation object implicitly has a *global uniqueness constraint* that ensures that no relationship tuple can appear more than once in the relation.

Relation object classes are descended from the "MO" class via the "RO" class as shown in Figure 3. They inherit component interfaces that define generic and specific operations for querying the relation (i.e. for finding all relationships with given values in particular roles), and for adding or removing relationships from the relation. When a relationship is added, the meta-object server for the relation must check that the new tuple does not violate the relation's uniqueness constraints.

3.3   Specialised Relations

Two kinds of type relationship occur very frequently in type systems. Types can often be modelled as *containers* for other types. For example, a CORBA interface type "contains" the types of the interface's operations. Types also commonly provide *name spaces* for other types. For example, a record type is a notional "name space" for the record fields.

The DSTC MIM defines three specialised kinds of relation object to support these uses. A *containment object* is a relation object in which two roles are designated the *container* and *contained* roles, with the contained role being a unique key, Thus a "contained" object can only be in one "container" at a time. A *naming object* has three roles designated the *namer, named* and *name* role respectively. The namer and name roles form a unique key for the relation. The naming and containment functions can be combined to give a *naming containment object* which has at least three roles, and two unique key constraints.

The interfaces for specialised relations (as defined in the IDL mapping rules) provide extra generic and specific operations to support the naming or containment paradigms. For example, a naming object has name lookup and reverse name lookup operations. The DSTC MIM implementation can be changed to support other specialised relations by extending the mapping rules.

3.4   Schema Objects

A *schema object* is a MIM meta-object that represents a type schema. A schema object is a container for a collection of types and relationships belonging to an instantiated type system. It provides "object factory" operations for creating type and relation objects in the schema, and provides the mechanisms for finding all type and relation objects in the schema. Schema objects inherit from "MO" by way of the "SO" class as shown in Figure 3.

A schema object contains a relation object for each of the type system's relation object classes, and may contain many type objects for each type object class. When these meta-objects are created using the schema's factory operations, they are automatically added to per-class attributes of the schema object. In addition, each meta-object's "my_SO" attribute is set to refer to schema object that created and owns it. Finally, a schema object may create only one relation object of each class.

3.5   Schema Factories

Each instantiated MIM type system defines a *schema factory* class; i.e. an object interface for creating new schemas for the type system. This interface consists of a single operation to create a schema object. The DSTC MIM specification does not say whether the factory object for a type system is unique, or how a client would go about locating it. Indeed, there are many potential uses of the MIM where no schema factory object would be required.

3.6   Known Relations

The DSTC MIM supports two distinct models of relationships between type objects. The *relational model* is characterised by general query operations over a table of relationships for many types. The *navigational model* is characterised by operations for navigating a graph of related type objects. While the relational model is a more powerful model, it is rather heavyweight if the client has no need to perform searches. For example, using queries to find the fields of a record type object in Figure 4 would be cumbersome and inefficient.

A type object interface can "know about" the relationships that the object participates in. For example, the record type objects in Figure 4 can "know about" their relationships with field names and type objects. The type object is said to have a *known relation* or simply a *known*. The effect of providing a type object with a known is that it acquires a virtual "link" to the objects it is related to as shown in Figure 5.

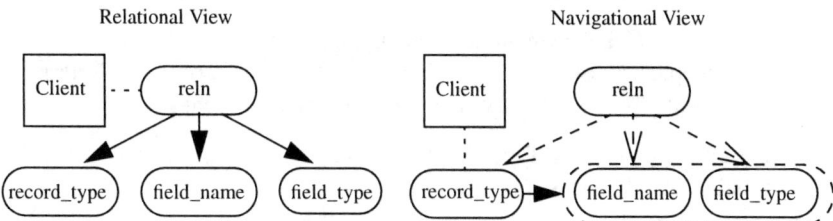

Figure 5   The relational and navigational views.

The definition of known relation of a type object specifies the role that the type object fills in the relation (the *known role*), and a *known name* by which the type object interface will refer to it. The known causes extra operations to be inherited by the type object class. These allow a client to find objects related to "this" object to create relationships between "this" object and others.

A complete and detailed description of these objects is available in [10].

4   CONCLUSIONS

In summary, this paper has introduced a conceptual framework for describing meta-information and meta-meta-information, and has described the design for a meta-information manager, which was successfully prototyped in CORBA. The key features that differentiate the DSTC MIM design from previous type management and repository research are as follows:

- Heterogeneity: the design supports multiple, arbitrary type systems.
- Extensibility: the design allows new type systems to be added at any time, either starting from scratch or by composing or extending existing type systems.
- Openness: the design allows multiple MIM servers to be federated into a seamless distributed meta-information service.

ACKNOWLEDGEMENTS

The work reported in this paper has been funded in part by the Cooperative Research Centres Program through the Department of the Prime Minister and Cabinet of the Commonwealth Government of Australia.

REFERENCES

[1] "Common Facilities RFP-5: Meta-Object Facility", OMG TC document cf/96-05-02, Object Management Group, June 1996.

[2] "Open Distributed Processing: Reference Model", ISO/IEC 10746, 1996.

[3] "ODP Type Repository Function" ISO/IEC CD 14769, 1997.

[4] "A Type Management System for an ODP Trader", by Jadwiga Indulska, Mirion Bearman and Kerry Raymond, in Proc. of the IFIP TC6/WG6.1 Int. Conf. on Open Distributed Processing (ICODP-93), Berlin, September 1993, North-Holland, pp. 169-180.

[5] "Types and Their Management in Open Distributed Systems" by Wayne Brookes, Stephen Crawley, Jadwiga Indulska, Douglas Kosovic and Andreas Vogel, Journal of Distributed Systems Engineering (to appear in 1997).

[6] "Type Management using Type Graphs" by Stephen Crawley, in Proc. DSTC Symposium '96, Brisbane. 11-12 July 1996.

[7] "The Common Object Request Broker: Architecture and Specification", Revision 2.0, Object Management Group, April 1995.

[8] "The COMANDOS Distributed Application Platform", Vinny Cahill, Roland Balter, Neville Harris and Xavier Rousset de Pina (eds.), ESPRIT Research Reports (Project 2071), Volume 1, Springer-Verlag, 1993.

[9] "Universal Repository (UREP)" (web pages), UNISYS.
http://www.unisys.com/marketplace/products/urep/

[10] "Meta-Object Facility", OMG TC document cf/97-01-01, Linnæus Project, DSTC, January 1997.

[11] "Object Oriented Modelling and Design", J.Rumbaugh et al, Prentice Hall, 1991.

BIOGRAPHY

Stephen Crawley is a Senior Research Scientist at the Defence Science Technology Organisation. Scott Davis is a Professional Officer at the Defence Science Technology Organisation. Jaga Indulska is a Senior Lecturer in the School of Information Technology, University of Queensland. Simon McBride is a Research Scientist in the CRC for Distributed Systems Technology. Kerry Raymond is a Senior Consultant at CiTR Pty Ltd.

14

Improving the Development and Validation of Viewpoint Specifications

N. Fischbeck, J. Fischer, E. Holz, O. Kath, M. Löwis, R. Schröder
Humboldt-Universität zu Berlin, Dept. of Computer Science
Axel-Springer-Str. 54a, 10099 Berlin, Germany
{holz/kath}@informatik.hu-berlin.de

Abstract

This paper presents an overview of the application of ITU-SDL for the development of distributed systems according to the RM-ODP. The use of the advanced SITE tools for the simulation and prototype code generation of/from SDL specifications of computational as well as engineering objects is illustrated. The advantages of this approach are shown by applying it on the ODP trading function as an example. Furthermore, concepts and principles for a smooth, semi-automatic SDL based transition between viewpoints, which are currently under development, are sketched out.

Keywords

RM-ODP, viewpoint, FDT, viewpoint languages, formal behaviour description, computational languages, TINA-ODL, CORBA-IDL, trading function, SDL-92

1   INTRODUCTION

Information technology systems are increasingly restructured as networked solutions to reap the benefit of cheaper and more flexible technology. Driven by the changes of the markets and the technology such systems are becoming more and more heterogeneous.

The development of the Reference Model for Open Distributed Processing (RM-ODP) is an ongoing joint standardization activity of ITU-T and ISO (ITU-T, 1995a). It aims at a framework to organize services within autonomous systems in order to facilitate interworking of software components distributed into larger and larger systems. The RM-ODP provides:

* general modelling approach and concepts and
* a method to divide a specification into viewpoints in order to simplify the specification process.

Although still in development, the ODP standard has already influenced major developments in the area of distributed and telecommunication systems. Two leading industrial consortia (OMG and TINA-C (TINA-C,1994)) have set up liaisons with the ODP standardization.

The viewpoint concept is essential for the whole RM-ODP. Each viewpoint is focusing on a subset of the properties of the system. The viewpoint concept can be applied at large at the whole system or at a different level of abstraction to individual components of the system. The RM-ODP does not define a general mapping or correspondence between every pair of viewpoint languages, however, special relationships between the computational and information resp. the computational and engineering viewpoint are identified. The RM-ODP does also not prescribe any chronological order for the development of specifications for the different viewpoints, on the contrary - it favours a parallel development of the specifications. Nevertheless, a development process starting with the enterprise specification, followed by (possibly interleaved) specification of information, computational and engineering viewpoints and concluded by the technology specification seems to be an adequate way for the development of ODP systems.

Part 3 of the RM-ODP (ITU-T, 1995c) contains the abstract definition of the viewpoint languages. A concrete syntax for these languages is not given, the language definitions contain only the concepts and rules for the specification from the selected viewpoint. This allows to use existing or evolving languages/notation techniques as concrete viewpoint languages. The languages envisaged here range from natural languages over programming languages to formal description techniques (FDT). Initial mappings between the computational and information languages and the standardized FDTs SDL, LOTOS, Estelle, Z and ACT.ONE are given in Part 4 of the RM-ODP (ITU-T, 1995d).

2   THE COMPUTATIONAL LANGUAGE

Seen from the computational viewpoint an ODP system can be viewed as a configuration of interacting (computational) objects which are supported by an infrastructure. The computational language is used to clearly identify in a distribution transparent manner the objects within the system, the internal activities of these objects and the interactions between these objects. The computational language imposes structuring rules on the objects and interfaces. These rules specify when and how two or more objects may interact, they give the set of actions a computational object can engage in, they define the set of failures and errors which can occur and finally they provide a base for portability. A computational specification is a specification whose objects all have environment constraints and own only computational interfaces.

The computational language distinguishes between two main kinds of (computational) interfaces, Operational interfaces and Stream interfaces. An operational interface is described by a set of operations where an operation is a single headed action which is part of the behaviour of the server object. The head of the action is the operation invocation according to the invocation template of the operation signature. The tail(s) of the action are terminations according to the termination templates of the operation signature.

In order for two computational objects to interact, first their interfaces have to be bound by an implicit or explicit binding. Implicit binding is available only for operational interfaces whereas explicit binding can be used for operational as well as for stream interfaces. A binding must satisfy the interface binding rules, which prescribe constraints on the types of interfaces which are permitted to be bound. One of these constraints for operational interfaces is the requirement that one interface must be a client interface and the other one a server interface. A successful explicit binding action results in a binding object, which controls then the execution of operations at the bound interfaces. This comprises also operations to change the set of interfaces involved in the binding.

The behaviour of a computational object is specified by the activities that occur within the object (internal) and the interactions at its interfaces (observable behaviour). Templates are specifications of the common features of a collection of objects, interfaces or actions and are the base for an instantiation.

3   CONCRETE VIEWPOINT LANGUAGES

With the application of FDTs as concrete viewpoint languages the full range of methodologies and tools for the development and analysis of specifications as well as for the implementation support is open to the designer of ODP systems and components. This comprises besides textual/graphical editors tools for the formal verification and for the simulation of specifications.Moreover, designers of distributed or telecommunication systems are often familiar with at least one of the FDTs which makes a transition to ODP more seamless.

Unfortunately none of the FDTs does support directly the ODP concepts by its semantic core. Therefore a modelling of the ODP concepts by composing basic concepts of the FDT is required, what may lead to a quite substantial overhead. In general it is also not possible to define a library of ODP concepts, which could be used to develop viewpoint specifications using the different FDTs. Instead Part 4 gives a set of patterns how to write down specification elements and how to combine those elements in order to obtain a complete viewpoint specification. A (formal) verification whether or not a specification is an ODP conform viewpoint specification is difficult or even impossible. Another shortcoming of the FDTs is their lack of object-orientation. Even SDL, which has adopted object-oriented concepts with the last language revision does not fulfil the requirements of ODP. This introduces again additional overhead to model ODP's object-oriented features.

Although in some cases the amount of specification to model basic ODP concepts may be large and therefore clutter the whole specification, the FDTs seem to be one appropriate means for the formal notation of viewpoint specifications. The best suiting language for the information viewpoint is Z, whereas SDL and LOTOS are expected to be key candidates for the computational viewpoint.

Interface and object description languages bridge the gap between (computational) specification and implementation. Different kinds of such languages have been developed by industry consortia, among them CORBA/ODP-IDL (ISO/IEC, 1996) and TINA-ODL (TINA-C, 1996). These languages all share the ability to describe the structure of objects and the signature of operations/services provided by objects in (programming) language independent terms.

IDL and ODL are well integrated in the application development process. Bindings to different programming languages (e.g. C++, Smalltalk, Eiffel) enable the automatic generation of stubs (client side) and implementation skeletons (server side). These skeletons have to be enriched manually (according to the viewpoint specifications) to obtain the complete object implementation.

ODL has been developed by TINA-C and is based on OMG-IDL. It extends IDL by providing support for those TINA concepts which are not part of the CORBA framework, e.g. streams, QoS-attributes, objects and object groups, objects with multiple interfaces. Moreover, an ODL specification describes not only the signatures of objects, interfaces and operations, it also specifies the required behaviour of an object or an interface and gives usage rules. However, these behaviour specifications can currently be given informally only. A detailed formal syntax for the behaviour specification is subject for further studies within TINA-C.

Although IDL and ODL both directly reflect most of the structural concepts of the computational language, they are not sufficient enough to check interoperability and conformance to standards. This is due to their missing abilities for an unambiguous formal behaviour specification. All conformance investigations are limited to a syntactical matching of object and interface signatures.

OMT is a further technique which has been applied for the development of viewpoint specifications, especially in the information viewpoint. However, similar to the interface/object description languages, only structural concepts are expressed with an OMT object model, the use of the dynamic and functional model (behavioural aspects) within the scope of ODP specifications occurs rather seldom.

4   COMBINED APPROACH

As it has been explained in the previous section, none of the languages and notation techniques currently applied is sufficient enough to be applied throughout different viewpoints. It is in general even not possible to express all the concepts of a single viewpoint (as they are defined in the RM-ODP) with one of the concrete techniques.

On the other side the application of existing techniques bears the advantage of the availability of design and validation/verification methodologies and supporting tools and the familiarity of the designers with these tools.

An approach to overcome the shortcomings of the different single techniques is therefore their combination. This increases not only the expressiveness of the specifications within on viewpoint but eases also the transition between different viewpoints (e.g. information to computational viewpoint, computational viewpoint to engineering viewpoint).

As FDT SDL has been selected as one of the most widely applied FDTs. The object-oriented features of SDL allow for a simple and appropriate modelling of entities. Within a joint project between Humboldt-University and the European Telecommunications Standardization Institute (ETSI) for the specification and simulation of large scale protocols like Broadband ISDN Signalling Protocol Q.2971 the advantages of object-oriented features in SDL have been proven (ETSI, 1997).

4.1 Combination of ODL/IDL and SDL

The main idea of this combination is the use of IDL or ODL as the initial computational language. The overall structure of the application is specified in terms of objects and their interfaces. An interface specification consist of the signatures of the operations provided by the interface, the exceptions which may be thrown and the attributes of the interface. Behavioural aspects are not included here (or may be given by informal text only). The next step is the transformation of these specifications into SDL. The result is an SDL-skeleton specification containing definitions of block types, (empty) process types and remote procedure specifications(signatures only).

In order to obtain a complete computational specification, the process types have to be complemented by the behaviour description (process graph specifying actions/ interactions). Such a computational specification serves then as input for the various tools available for SDL (simulation, symbolic execution, prototype generation).

The transformation of IDL and ODL to SDL is a rather straightforward mapping, the general principle is depicted in Figure 1, a detailed description is given in (Born, 1996). As it can be seen, the structure of the resulting SDL specification follows the rules defined in Part 4. A subset of the transformation rules concerning ODP-IDL was incorporated into the Part 4 standard. To automatize the transformation an prototype of an ODL2SDL translator has been implemented (Kolberg, 1997)

The main goal of the simulation of the final computational SDL specification is the validation of the functionality. The Integrated SDL environment SITE (SITE, 1997), which was developed at the Humboldt-University does support the analysis of an SDL specification by providing:

- Syntactical analysis
- Semantics analysis
- Generation of code (C++, Java)
- Simulation run-time library
- Prototype run-time library.

An overview of the SITE tools is given in Figure 2. The toolset does support the full range of the object-oriented features of SDL. The main difference of SITE

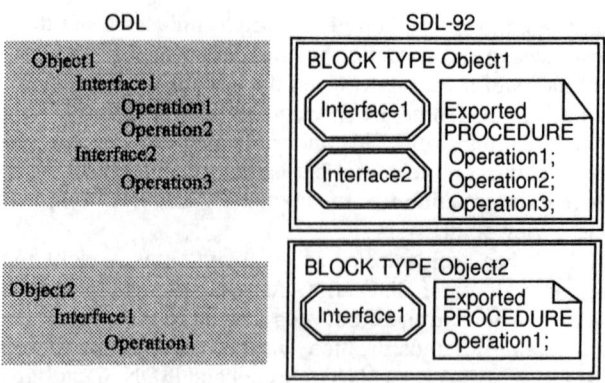

Figure 1 ODL to SDL mapping

compared to other SDL tools, is that the generated C++ code is not only structural equivalent to the SDL specification but can be used for simulation purposes as well as an implementation prototype. The purpose of the simulation is here the analysis

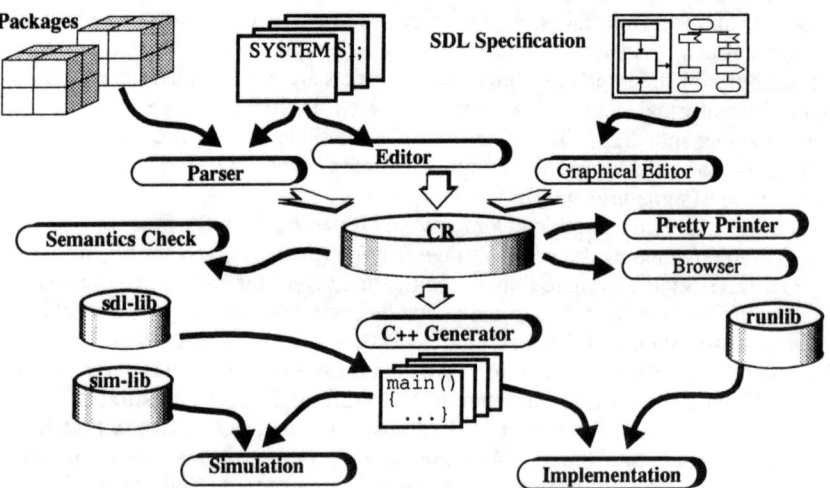

Figure 2 SITE structure

of the overall functionality of the specification at an computational level and independently from distribution aspects. The simulation of the computational specification is one aspect in the verification process. Because of the amount of data available from simulations, several techniques can be exploited to analyse these data, including analysis of a specific scenario, statistical analysis over a series of experiments, and the generation of message sequence charts. In order to support different analysis methods, the simulator is itself modular and provides interfaces to external analysis tools (Fischbeck, 1996). Since correctness and performance data depend partially on characteristics of engineering aspects (such as network delays), the simulation model can be extended with such information. As the simulated system is separated from the physical system, the simulation itself is not executed in a distributed fashion.

In contrast to the simulation the prototype implementation takes distribution aspects into account. The unit of distribution is an SDL system, i.e. each SDL system is translated into an executable object. The communication between the single systems is based on OMG CORBA. In the sense of TINA an SDL system corresponds to an engineering computational object (eCO).

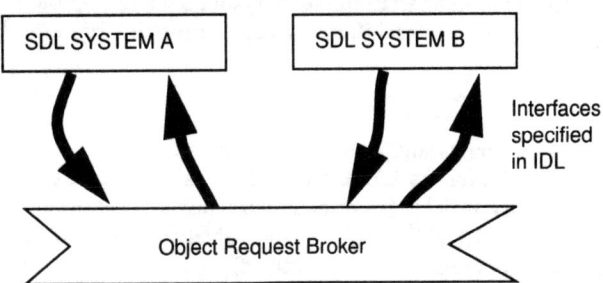

Figure 3 Corba based Prototype Implementation

Using the prototype the effects of the object distribution and of the underlying communication infrastructure on the functionality of the application can be studied at an computational level, which will also support the later development of an engineering specification. This approach has also been successfully used for the implementation of real applications (Fischbeck, 1996), using a proprietary communication platform instead of Corba.

4.2 Combination of OMT and SDL

The combination of SDL and OMT has been investigated within different projects (Holz, 1996b), (Guo, 1995) ranging from a translation of an OMT model to an SDL specification (including behaviour descriptions) to a pure linking mechanism between OMT class definitions to SDL data types. Especially the latter case is also supported by the commercial SDL tools, however caused by its restriction to data types it is limited to the information viewpoint.

Within the INSYDE (Holz, 1996b) project a three-stage methodology has been proposed. The first stage is the requirements capture and analysis. The Object Modelling Technique (OMT) as defined by Rumbaugh will be applied here. The second stage is the design phase, it is again split into two sub-phases: System Design and Detailed Design. For the System Design a special sub-set of OMT called OMT* has been defined. In contrast to OMT this sub-set has a well defined syntax and a transformational semantics. It is tailored towards the two target domains hardware and software. The Detailed Design uses the domain specific languages SDL for the specification of software. The initial specification for the detailed design is derived from the system design model by a semi-automatic translation of OMT* into SDL. In the final phase of the methodology a validation of the specifications of the hybrid system will be performed by simulation The methodology is supported by a prototype toolset consisting partly of off-the-shelf tools and partly by new developed components to bridge the gap between the different stages.

5   APPLICATION OF THE TECHNOLOGY

In order to examine the proposed combination technology for the development of computational specifications the specification of the Trading Function (ITU-T, 1995e) has been chosen as an example by taking the IDL specification of the Trader as starting point. The main specification steps performed for this task were (Niedrig, 1996):

*   Mapping
    The trader computational template described in OMG-IDL including its interfaces, attributes and operations (with their signatures and exceptions) is mapped to SDL by the IDL2SDL tool. As a result an SDL specification was obtained, which reflects the original structure as given in the ODP standard: A definition of an SDL block type (trading object template) including process type definitions (interface type definitions) with remote procedure definitions (operations) and their signatures and exceptions.
*   Addition of behaviour specification
    The generated SDL specification does not contain any behaviour description, i.e. all process type definitions are empty and all procedure definitions are virtual and empty definitions. The addition of the behaviour description follows the informal specification given in the standard and is expressed in terms of actions and interactions at interfaces. During this step also additional interfaces to other ODP services have been identified as necessary for the provision of the trading function (i.e. Type Management, Security Management). The result of this step is a computational specification of the Trader. A direct validation/verification of this specification in isolation is, however, not possible. This is due toe the missing specification of other ODP services, which appear here still in a generic form.
*   Adding an environment model & simulation
    The environment model completes the specification by importer and exporter objects and provides also access to the additional ODP services identified in the previous step. Within our work these services have been emulated by a very simple implementation or by a no-implementation (i.e. operations return a predefined value).

The final specification was used to derive conformance test cases for the trader. Part of this work was submitted to the trader standardization.

6   SUPPORTING DIFFERENT VIEWPOINTS

The positive results of the combination of different techniques within one viewpoint have encouraged further investigations on how to exploit such combinations also for a smooth transition between different viewpoints. The final goal of this development is to provide tool support for the stepwise transition from one viewpoint to another by keeping at the same time consistency between specification at different viewpoints. A schematic view of such a development methodology is given in Figure 4.

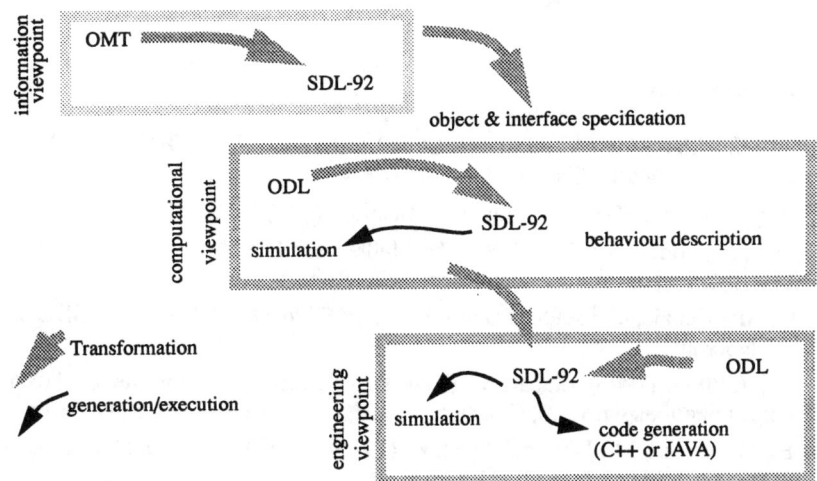

Figure 4 Transition steps between viewpoints

Because of the expected massive problems of transitions between computational and engineering specifications of objects and interfaces, the main focus of this development is the transformation of SDL computational descriptions into engineering ones. This approach will allow the simulation of engineering specifications under various distribution conditions as well as the generation of run-time code, done by the SITE tools. One main aspect of this transition is the semi-automatic generation and addition of engineering concepts to the computational SDL specification, e.g. stub, binder and object life cycle interface specifications for each object, done with ODL/SDL, too.

Another very important step in this development is to transit information specifications to analogous computational specifications, e.g. ODL descriptions with SDL skeletons.

Together these transformation concepts will bring an added value for the safe and fast development of software for large, distributed systems.

7   CONCLUSION

A combination of different specification and notation techniques has been shown as a feasible approach for the improvement of the development process. The main advantage lies in the availability of tools for existing techniques and the familiarity of the users with the techniques and tools. By combining the different techniques the strength of the single techniques can be exploited and there limitations can be evaded. A tool support is a necessary precondition.

The exploitation of the results for the support of the transition between different viewpoints will be investigated further and in more detail in the ongoing project CAMOUFLAGE (Holz, 1996; Kath 1996).

8   REFERENCES

Born, M.; Winkler, M. (1996) SDL-92, CORBA-IDL und TINA-ODL im ODP-Entwurfsprozeß, Master Thesis, Humboldt-Univ. Berlin

ETSI Project Team 87(1997) Formal specification (SDL) for B-ISDN DSS2 CS2

Fischbeck, N.; von Löwis, M. (1996) Abschlußbericht TERMO: Entwicklung einer Nutzerschnittstelle zur Simulation von SDL-Systemen

Guo, F.; MacKenzie, T. (1995) Translation of OMT to SDL-92, Proc. of SDL'95, Oslo Norway

Holz, E.; Kath,O. (1996) Einsatz von B-ISDN-Netzen im Rahmen einer TINA-Kommunikationsplattform, CAMOUFLAGE/S, Humboldt University Berlin

Holz, E.; Wasowski, M.; Witaszek D. et. al. (1996) INSYDE - The final Methodology Report, ESPRIT P 8641, D1.3

Kath, O.(1996) TINA-C und B-ISDN-Netze, Humboldt-Univ. Berlin

Kolberg, M. (1997) Compiler Frontend for TINA-ODL, Humboldt-Univ. Berlin

Niedrig, S.; Rademann, S. (1996) Eine Methodik für den ODP-Konformitätstest und deren Anwendung auf den ODP-Trader, Humboldt-Univ. Berlin

ISO/IEC (1996) DIS Open Distributed Processing - Interface Definition Language

ITU-T (1995) Rec. X.901 I ISO/IEC 10746-1: RM-ODP Part 1: Overview

ITU-T (1995) Rec. X.902 I ISO/IEC 10746-2: RM-ODP Part 2: Descriptive Model

ITU-T (1995) Rec. X.903 I ISO/IEC 10746-3: RM-ODP Part 3: Prescriptive Model

ITU-T (1995) Rec. X.904 I ISO/IEC 10746-4: RM-ODP Part 4: Architectural Semantics

ITU-T (1995) Rec. X.950 I ISO/IEC 13235:RM-ODP ODP Trading Function

SITE (1997) http://www.informatik.hu-berlin.de/Themen/SITE/SDL-tools.html

TINA-C (1994) Overall Concepts and Principles of TINA

TINA-C (1996) Object Definition Language - Manual Version 2.3

PART FIVE

OO Requirements Analysis and Design

15

Formal user-centred models

R. G. Clark
Department of Computing Science and Mathematics
University of Stirling, Stirling FK9 4LA, Scotland, UK
Tel: +44 1786 467427, Fax: +44 1786 464551, email: rgc@cs.stir.ac.uk

A. M. D. Moreira
Departamento de Informática, Faculdade de Ciências e Tecnologia
Universidade Nova de Lisboa, Portugal
Tel: +351 1295 4464, Fax: +351 1294 8541, email: amm@di.fct.unl.pt

Abstract

As informal requirements are usually expressed in terms of the behaviour which the environment expects from a system, we propose that the construction of a formal and executable user-centred model should be used as an intermediate step in the construction of a formal object-oriented specification. Rapid prototyping can be used to validate the user-centred model with respect to the requirements and the informal task of validating the object-oriented specification can be replaced by the formal task of verifying that it is equivalent to the user-centred model. As an example of this approach, we show its use within the Rigorous Object-Oriented Analysis (ROOA) method.

Keywords

Formal development, executable specification, object-oriented analysis, LOTOS, user-centred model

1 INTRODUCTION

Requirements analysis methods are concerned with understanding and structuring the information collected during requirements capture, with modelling the observable behaviour of the proposed system and with analysing the resulting model to ensure that the informal requirements are accurately represented. We have been investigating how formality can be introduced into the object-oriented analysis process and how an executable and formal object-oriented specification can be constructed. This is not an easy task as there is a wide gap between informal requirements and the structure and notation of a formal specification. It also cannot be a formal process.

Informal requirements are usually expressed, and are most easily understood, in terms of the behaviour which the environment expects from the system. We therefore propose that the initial formal specification should be a formal and executable user-centred model. As it is closer to the requirements than a system-centred object-oriented specification, the gap between the informal requirements and the initial formal specification is narrowed.

The user-centred model is concerned with events, i.e. interactions between the environment and the system. It is therefore best expressed in an executable event-oriented formal language such as LOTOS (Bolognesi *et al.* 1987). This allows rapid prototyping to be used both to validate the model against the requirements and to identify inconsistencies, contradictions, ambiguities and omissions in the requirements sufficiently early so that feedback can be given to the requirements capture process.

When the user-centred and object-oriented specifications are expressed in the same formal language, the validated user-centred model can be used as a step in the construction of the object-oriented specification. Furthermore, once we have an initial formal specification, it is possible, at least in theory, to verify that it is equivalent to subsequent specifications. Therefore, validating that the object-oriented specification satisfies the informal requirements can be achieved by verifying that it is equivalent to the user-centred model.

2   TWO FORMAL MODELS

A user-centred model is concerned with system behaviour as seen from the viewpoint of the environment. The environment includes human users, hardware devices and other software components which we refer to collectively as agents. The user-centred model is a set of agent views where each agent view describes, from the viewpoint of an agent, a way in which the system is to be used. An agent view is concerned with the role being played. An agent can play different roles, and therefore take part in more than one agent view, while different agents can play a similar role and therefore take part in separate instances of the same agent view. The focus is on the way agents use the system, not on the system itself. A system-centred model, on the other hand, is concerned with modelling an idealised view of the system and its interaction with the environment.

We describe, through a case study, the process to be followed in the building of a formal user-centred model and how it can be used as a step in the creation and validation of a formal object-oriented specification, i.e. a system-centred model. The case study shows how this approach can be integrated into the Rigorous Object-Oriented Analysis (ROOA) method which provides a process for the systematic construction of an executable formal object-oriented specification from a set of informal requirements (Moreira *et al.* 1994a, Moreira *et al.* 1994b, Moreira *et al.* 1996a, Moreira *et al.* 1996b). The ROOA model, which is expressed in LOTOS specifies the static, dynamic and functional aspects of a problem in terms of entities from the problem domain. As ROOA models a system as a set of communicating concurrent objects, it is ideally suited to the specification of distributed systems.

3 RELATED WORK

The creation of a formal user-centred model has also been proposed by Hsia *et al.* (1994), Glinz (1995) and Somé *et al.* (1996). However, they do not deal with how the development of a user-centred model can be integrated into the development of a formal system-centred model.

In the requirements analysis phase of the Object-Oriented Software Engineering (OOSE) method, Jacobson proposes the construction of a requirements model and an analysis model which correspond, in some respects, to our user-centred and system-centred models (Jacobson 1992). The OOSE requirements model is composed of a use case model and a simplified object model called a domain object model. The use case model is a complete set of use cases where each use case describes possible complete sequences of events which occur in response to some action initiated by an agent. The OOSE models are not specified in a formal language although Regnell *et al.* (1995) have shown how this can be achieved for the use case model.

LOTOS is often used in the specification of OSI systems where a service specification is produced initially to describe the system in terms of its external behaviour. This is then followed by a lower level protocol specification which includes internal structure (Clark *et al.* 1992, Turner *et al.* 1995). Although there is a similarity between a service specification and our user-centred model, the focus is different. A user-centred model is concerned with the roles played by the agents which constitute the environment and their view of the system while a service specification can be regarded as a very high-level system-centred model.

4 FORMALISING AGENT VIEWS

When an agent interacts with a system, a set of possible behaviours can occur. For example, when we lift a phone and dial a number, the call may, or may not, be answered. An agent view is the set of such possible behaviours. It can be represented by a tree which shows the possible alternative event sequences. We refer to a particular path through the tree, i.e. a sequence of external events, as an agent scenario.

The construction of an agent view helps us identify the objects which will provide the services required by the agent. Hence, a simple object model is constructed in conjunction with the agent views. Each agent view interacts with the objects which provide the external behaviour of the system as seen from that agent's point of view.

In general, an agent's interaction with the system will involve other agents. One agent may initiate a scenario, e.g. make a phone call, but that also involves a second (dependent) agent, namely the person being called. Initially, we concentrate on single-agent views as they are ideal in providing an initial understanding of a problem. In the phone call example we would have two single-agent views, one for the caller and one for the person being called. We show later how we can compose single-agent views to form a multi-agent view without modifying the single-agent views.

Creating the set of agent views which form the user-centred model is an iterative process. A major part of its construction is the resolution of inconsistencies and contradictions in the views and expectations of the different agents. We must also identify redundant or overlapping agent views. When, for example, several single-agent views have common initial behaviour, they can be amalgamated into one single-agent view.

5   THE CASE STUDY

In a road traffic pricing system, drivers of authorized vehicles can be charged at toll gates without having to stop. Drivers must install a device (a gizmo) in their vehicle. Authorized vehicles have a registration which includes vehicle details, the owner's personal data and an account number from where automatic debits are done monthly.

Each gizmo has an identifier which is read by the sensors installed in special lanes at the toll gates. With the information given by a sensor, the system can then store the necessary information so that the specified bank account can be debited. Different types of vehicles pay different rates.

When an authorized vehicle passes through a special lane, a green light is turned on and the amount being debited is displayed. If an unauthorized vehicle passes through a special lane, a yellow light turns on and a camera photographs the plate number. (Later, the owner of the vehicle will be fined.) There are two kinds of special lane: one where all vehicles of the same type pay the same fixed amount (e.g. at a toll bridge) and one where the amount to be paid depends on the distance travelled (e.g. on a motorway) and where the system has therefore to store the entrance point and the exit point for each vehicle.

We want to model the part of the system which receives the information from the gizmos, determines whether the vehicle is authorized, displays the amount to be debited, turns on the appropriate light, triggers a photograph when necessary, and deals with the account debits.

6 CREATING A USER-CENTRED MODEL

6.1 Describing agent views

By the end of this task we will have identified the agents which interact with the system, the views each one has of the system, and, as a result, the objects (which we refer to as interface objects) that allow those interactions.

The question to ask ourselves when identifying agents is: what interacts with the system? The first agent we may think of is the gizmo installed in a vehicle. A vehicle uses a toll gate and has a driver. We regard the composite vehicle-driver-gizmo to be a single agent which we refer to as *vehicle*. Another agent is the vehicle owner who has to buy a gizmo and be registered. Other agents include the bank and the system manager.

The second question to answer is: how does each agent view the system? To answer this, we imagine ourselves to be the agent and describe what we see when we use the system. As part of this process, we identify the interface objects such as the different types of toll gate with which we, the agents, interact. This information is used to create an initial object model. An object model using OMT notation (Rumbaugh *et al.* 1991) is shown in Figure 1.

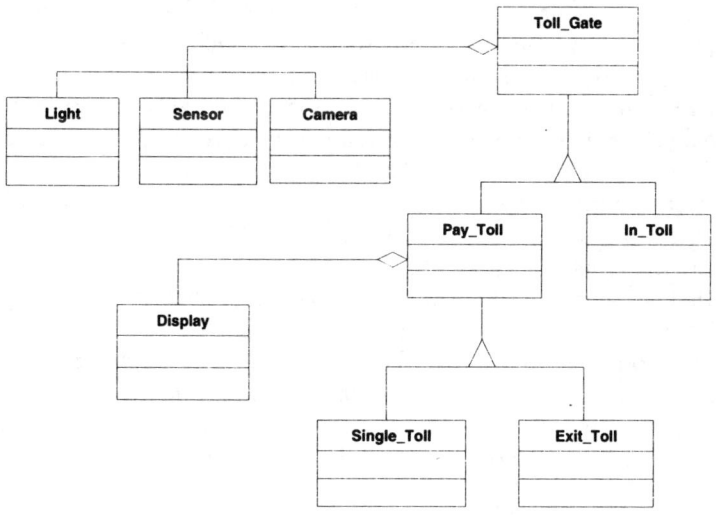

Figure 1 Initial object model.

The next step is to identify the ways in which the agents can interact with the system. The vehicle agent, for example, knows that it can use the system in two different ways: one where it always pays the same amount, the one-point vehicle view, and another where the amount paid depends on the distance travelled, the two-point vehicle view.

For each agent view, we construct an agent tree to give a graphical description of the possible event sequences which make up each agent view by:

- identifying the events which make up the interaction between the environment and the system;
- allocating the events to the appropriate interface objects;
- identifying the alternative event sequences which make up the agent view;
- adding the ordered events to the tree as nodes with each event being represented by the pair: ⟨class template⟩.⟨event⟩.

By focusing on one agent view at a time we can determine the different ways in which that agent can interact with the system without being sidetracked by other issues. Consider, for example, the two-point vehicle view. The vehicle passes an entry gate, a sensor reads its gizmo identifier and the system either responds by showing a green light or by showing a yellow light followed by triggering a camera.

The vehicle then proceeds onto the motorway. It may either leave by some unknown method (e.g. it could have an accident) or it passes through an exit gate. If a yellow light was shown in the entry gate, now the vehicle can only get another yellow, but, if a green light was shown, the vehicle may get another green light or a yellow light if the state of the system has changed (for example, the system now knows that that vehicle was stolen).

The agent view also describes the situation where a vehicle passes through an exit gate, but did not enter the motorway through an entry gate. This leads to a yellow light being shown. It should be noted that this situation was omitted from the original informal statement of requirements.

We must also consider the situation where a vehicle goes through a special lane either without a gizmo or with a gizmo which is broken and cannot be detected. A sensor must be able to detect that a vehicle has passed which means that the sensor must take part in two events; it first detects the presence of a vehicle and then attempts to read its gizmo.

The two-point vehicle view is shown in Figure 2. The other agent views are dealt with in a similar way. Based on the information in the agent views, we add any new interface objects, together with events which correspond to services offered by interface objects to the environment, to the object model.

6.2   Formalising agent views

Each agent view is formalised using LOTOS. The definition of an agent view is a LOTOS process which requires possible alternative series of services from the system. Each offered service is represented in LOTOS as:

<gate> <service name> <object identifier> <optional parameters>

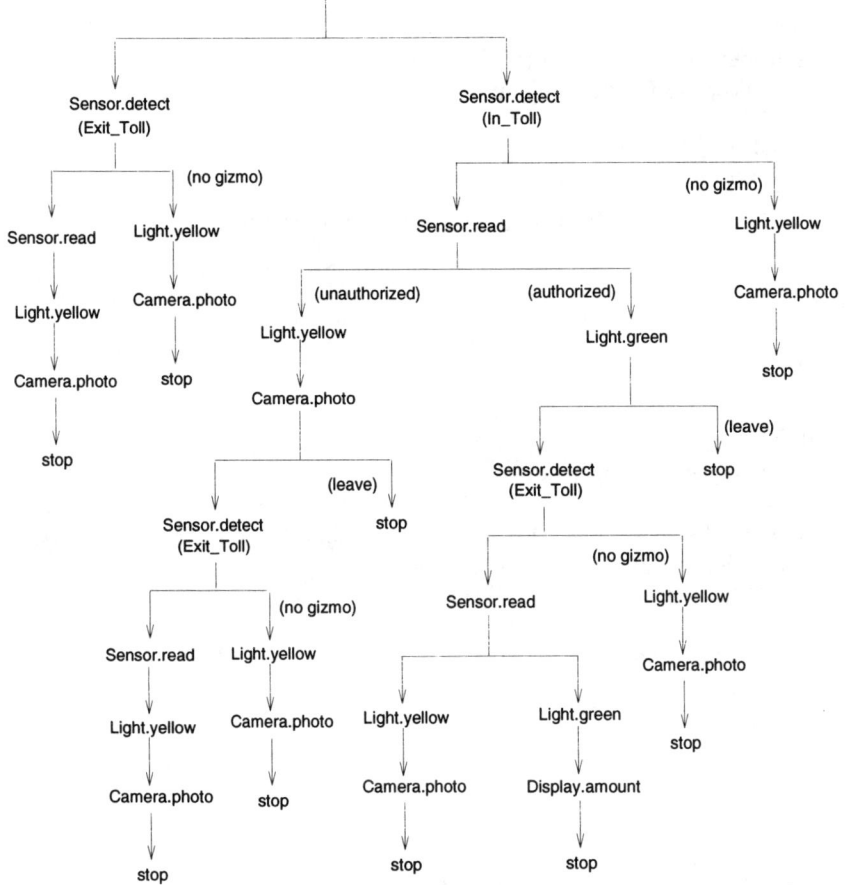

Figure 2 Two-point vehicle view.

It is straightforward to construct the outline of a LOTOS process definition from a tree by using the choice operator to offer each branch of the tree as a LOTOS choice expression. We can also factorise out those parts of the tree which are similar and define them in a separate process.

The LOTOS formal model contains more information than was held in the tree. We return to the requirements to determine the information which is to be passed between the agent and the system. The need for this extra detail is very useful in clarifying the requirements.

A LOTOS process definition for the view Two_Point_Vehicle shown in Figure 2 is given below. The process includes a LOTOS gate for communication with each toll gate component. Both the Gizmo and the Toll object identifier sorts are specified as ADTs.

```
process Two_Point_Vehicle[sen, lgt, dpl, cam]: exit :=
    sen !detect ?int: Toll_Id [Is_Enter(int)];
    ( Yellow_No_G[lgt, cam](int)
      []
      sen !read !int ?id_g: Gizmo_Id;
      ( ( Yellow[lgt, cam](id_g, int)
          >> Leave[sen, lgt, dpl, cam](id_g, was_yellow)
        )
        []
        lgt !green !int !id_g;
        Leave[sen, lgt, dpl, cam](id_g, was_green)
      )
    )
    []
    Missing[sen, lgt, dpl, cam]
where
    process Leave[sen, lgt, dpl, cam](id_g: Gizmo_Id, s: Entry_Val): exit :=
        sen !detect ?ext: Toll_Id [Is_Exit(ext)];
        ( Yellow_No_G[lgt, cam](ext)
          []
          sen !read !ext !id_g;
          ( [s eq was_green] → Green[lgt,dpl](id_g, ext)
            []
            Yellow[lgt, cam](id_g, ext)
          )
        )
        []
        exit (* leave in a non-special lane *)
    endproc (* Leave *)
endproc (* Two_Point_Vehicle *)
```

The LOTOS SMILE simulator (Eertink *et al.* 1993) allows the use of uninstantiated variables instead of values. As SMILE is able to determine when a combination of conditions can never be true, a single instantiation of an agent view can deal with a set of similar agent instances rather than be instantiated for a particular agent instance. Hence the **Toll** identifiers **int** and **ext** are introduced as variables; **int** has the constraint that it is an **In_Toll** identifier while **ext** has the constraint that it is an **Exit_Toll** identifier.

The behaviour at an exit gate is is factored out and defined in process **Leave** while processes **Green**, **Yellow** and **Yellow_No_G** specify the response to the user after a gizmo has been checked or a vehicle detected without a gizmo. Process **Missing** deals with the situation where a vehicle leaves through an exit gate without having entered through an entry gate. A complete LOTOS user-centred specification for the road pricing system is given in Clark *et al.* (1997).

6.3 Combining agent views

In general, individual agent views are not sufficient to specify external system behaviour fully; e.g. a gizmo must be registered before it can be used.

If two agent views are completely independent of one another, then they can be composed using the interleaving operator:

Agent_View_1[s] ||| Agent_View_2[t]

However, when two views are not independent, this composition does not work as it does not constrain the relative ordering of the events in **Agent_View_1** and **Agent_View_2**.

In the LOTOS specification of OSI systems, a service specification is produced using the constraint-oriented style in which behaviour is expressed as sets of constraints operating at external interfaces (Turner *et al.* 1995). A similar approach can be used to specify multi-agent views. Registration of gizmos occurs, for example, in the **Operator_Agent** view while gizmos are used in vehicle views. The processes defining the single-agent views are left unchanged. A constraint process, **Constrain_Gizmos** is defined to maintain the set of currently registered gizmo identifiers and the set which have entered, but which have not yet left a motorway.

process Constrain_Gizmos[rd, lgt](idgs, curr: Gizmo_Id_Set): **noexit** :=

   ...

endproc

This process interacts with **One_Point_Vehicle** or **Two_Point_Vehicle** on gate **lgt** and with **Operator_Agent** on gate **rd** and ensures, for example, that only a currently registered gizmo can lead to a green light being shown.

6.4 Validating the user-centred model

So that we can determine when an agent view has been successfully executed, the LOTOS process definition for an agent view is always instantiated from a process with the structure:

process X_Tree[gates, success](parameters): **noexit** :=
   X_Agent[gates](parameters) >> success; **stop**
endproc

Initially, we execute single-agent views independently. Execution using a simulator such as SMILE creates a tree showing all possible event traces. As this should have the same structure as the original agent tree, we have a simple check to verify that an agent tree has been correctly specified in LOTOS. We then define suitable constraints and interactively prototype multi-agent views.

To form the formal user-centred model, we first compose the LOTOS specifications of all the single-agent views using the interleaving operator. We then compose this behaviour expression in parallel with the constraint processes which define the interactions. When the resulting user-centred model is prototyped, the number of alternative events offered is too large to be handled conveniently. The user-centred model must therefore be composed in parallel with a **Test** process with the following structure (Quemada *et al.* 1995):

process Test[gates, success](parameters): **noexit** :=
 Agent_Scenario[gates](parameters) >> success; **stop**
endproc

The agent scenario in a test process is derived directly from a single or multi-agent view.

Validation is through a series of such test compositions. Initially, we prototype interactively to explore particular execution paths. Once we have shown that the composition is successful for at least some execution paths, we replace interactive prototyping with SMILE by the use of LOLA which automatically performs a complete state exploration of the composition of a specification and a test process (Quemada *et al.* 1995). LOLA reports one of three possible results:

Must where all possible execution paths lead to the **success** event.
May where at least one execution path leads to the **success** event.
Reject where no execution path leads to the **success** event.

This enables us to verify whether or not a particular test has been satisfied. A valid model must not only specify correct behaviour, it must rule out undesirable behaviour. Validating the user-centred model therefore involves rejection as well as acceptance tests.

The process of composing agent views and prototyping the subsequent user-centred model can show up inconsistencies, in which case we return to the agent view identification step, make the appropriate changes in one or more single-agent views and repeat the subsequent steps.

7 CREATING THE SYSTEM-CENTRED MODEL

We now construct a formal object-oriented specification which satisfies all the agent views. The ROOA method involves three main tasks: build an object model; refine the object model; build and validate the LOTOS ROOA model.

Task 1: Build an object model

With problems which are primarily data-oriented, an object model can be constructed without paying too much attention to dynamic behaviour. With process-oriented problems, on the other hand, the dynamic aspects of a problem must be explored to help identify the static object structure. Indeed, Rubin *et al.* (1992) suggest that the best way of identifying objects is to focus on their behaviour. When a method such as OMT is used to deal with process-oriented problems, we have found that agent views are a major help in identifying the required objects.

From the requirements we can identify some objects and their attributes which we add to the object model constructed during the creation of the user-centred model. However, as the problem we are analysing has a significant dynamic component, we cannot identify all the objects at this stage. The resulting incomplete object model is shown in Figure 3.

Task 2: Refine the object model

We first add those services, attributes, static relationships and message connections which can be easily derived from the requirements and from the user-centred model. We then examine the agent views and identify events which are sent to the environment and determine if the relevant interface object must offer a service to some internal object to trigger the external event. For example, in the two-point vehicle view, for `Camera` to generate the event `Camera.photo`, it must offer the service `take_photo`. Also, by analysing the object model in Figure 3, we can see that some of the static relationships are in fact message connections.

Agent views show the interactions between agents and interface objects. The consequent interactions between objects internal to the system are described in event trace diagrams (ETDs). An ETD shows the message passing between the objects in the initial object model and its construction may show new objects and services which are then added to the object model.

The scenario where a two-point vehicle view gets a green light results in the ETD shown in Figure 4. This ETD allowed us to identify new services and the new objects: `Gate_Processor`, `Current_Journey` and `Usage_Details`. As we construct the ETDs and identify new objects and services we add them to the object model.

We then collect the information in the ETDs and build an Object Communication Table (OCT) where we list the class templates; the services offered by each class template; and, for each service, the services that it requires and the

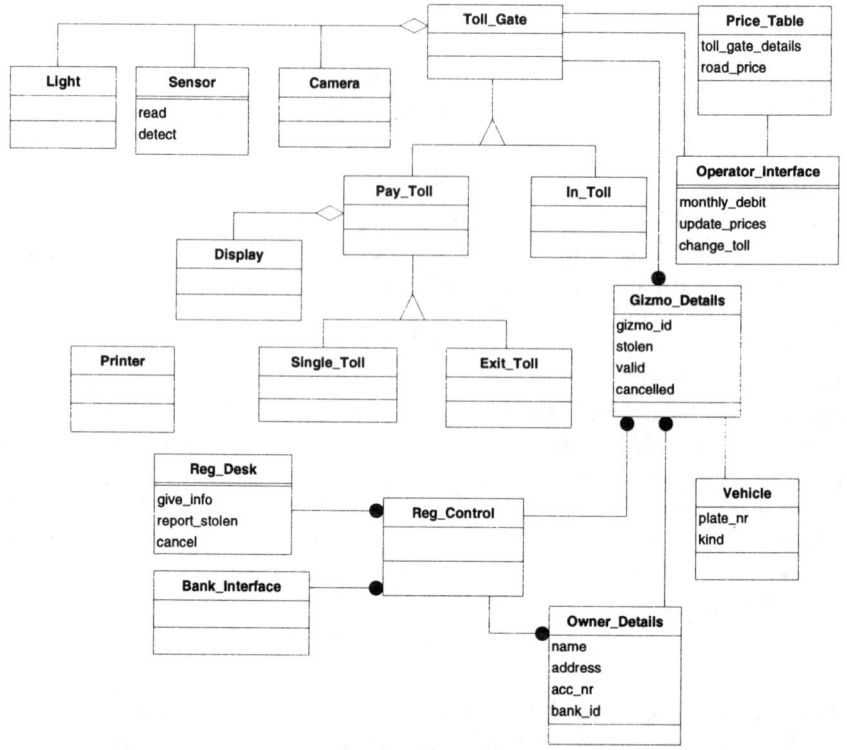

Figure 3 Initial OMT object model.

class templates (clients) which require it. The complete process is described in Clark *et al.* (1997).

Task 3: Build the LOTOS formal model

We first determine the LOTOS gates through which the objects are to communicate, add this information to the OCT and then construct a graphical representation of the ROOA model which we call the Object Communication Diagram (OCD).

To specify the behaviour of an object we place ourselves inside that object and act as if we were the centre of the system. By following this strategy, and using the information in the OCT and ETDs, we identify the events the object takes part in and their order. We start by specifying the interface objects and then proceed to the internal objects. Each event in the LOTOS process definition of an agent view, has a corresponding event in the LOTOS process definition of an interface object. Some events in the user-centred model show the initiation of an action to be carried out by the system together with the information which is required from the environment. The others show

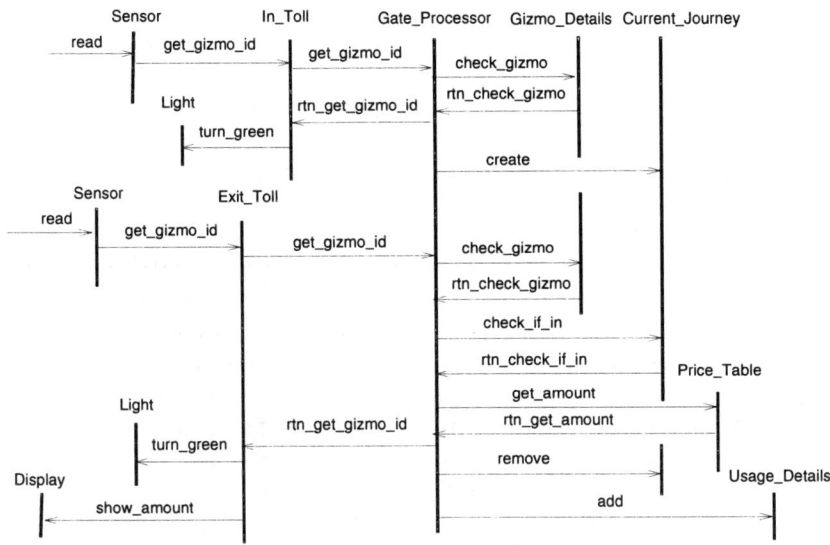

Figure 4 ETD for a vehicle passing a two toll gates and getting a green light.

responses from the system and the information which is to be passed to the environment. The formal user-centred model is therefore used directly in the construction of the LOTOS definition of the interface objects and determines the information which is to be passed during event synchronisation.

When we specify the system-centred model, it is not unusual for incompatibilities to appear in the expectations of different agents. Their resolution will often require us to go back to the users for further or revised information.

Once all the objects are specified, we follow the structure of the OCD and compose them, using the LOTOS parallel operators, into a behaviour expression which specifies the behaviour of the system-centred model.

8 VALIDATION AND VERIFICATION

Interactive prototyping with SMILE is used during the development of the ROOA model to increase our confidence that it satisfies the behaviour expected by the user-centred model. The ROOA model must now be validated. As the user-centred model has already been validated with respect to the informal requirements, validation of the ROOA specification is concerned with the formal task of verifying that it is equivalent to the user-centred model.

There are many different definitions of what it means for two specifications to be equivalent, but the only one which can be applied in practice to large specifications is testing equivalence (Quemada *et al.* 1995). In testing equivalence, the specifications under test are regarded as black boxes, i.e.

we only consider external behaviour, internal structure is ignored. The two specifications are composed with test processes in the same way as was described in the validation of the user-centred model in Section 6.4. However, we are not now concerned with the informal process of showing that the formal user-centred model satisfies the informal requirements, but with the formal process of demonstrating that two formal models cannot be distinguished by experiment.

Two specifications A and B are testing equivalent if every Must or May test on A gives the same result with B and vice versa. LOLA performs a complete state exploration of the composition of the user-centred and ROOA models with each test process and reports each of the results as Must, May or Reject. The drawback is, of course, that the set of test processes is not complete. However, by judicious choice of a sufficiently large number of tests we can have reasonable confidence when the specifications are indeed testing equivalent (Quemada *et al.* 1995).

The set of test cases used to validate the user-centred model with respect to the informal requirements were used to check that the two models specify the same behaviour. As the ROOA model has significant internal structure, the number of states generated, and which have to be explored, is very much greater than with the user-centred model. However, as testing equivalence is only concerned with external behaviour, the two specifications can still be equivalent. The much larger number of states generated with the ROOA model means that LOLA takes much longer to produce each result.

The complete specifications of the user-centred and ROOA models and a description of the use of LOLA in demonstrating testing equivalence is given in Clark *et al.* (1997).

9   CONCLUSIONS

Our approach has been to take ideas from several disparate areas, to combine them and to apply them in a novel setting.

From formal methods, we have taken the idea of a formal user-centred model (Hsia *et al.* 1994); from requirements engineering, the notion of expressing requirements as a set of multiple viewpoints (Kotonya *et al.* 1996, Nuseibeh *et al.* 1994) and from object-oriented analysis, the notion of expressing requirements as a set of use cases (Jacobson 1992).

We have adapted the LOTOS constraint-oriented style to the specification of the user-centred model, have demonstrated how LOTOS can be used in the specification of an object-oriented model and have used standard LOTOS techniques to validate the user-centred model and to show that the user and system-centred models are testing equivalent (Quemada *et al.* 1995).

Finally, although modelling the environment as part of the specification of a system is standard practice with embedded systems (Clark 1992), we believe that our modelling of the environment as a user-centred model, and its subsequent use in the creation and validation of a formal object-oriented specification, is new.

The problem that we have addressed is the large gap that has to be bridged during the transition from informal requirements to an initial formal requirements specification. We have proposed that a formal and executable user-centred model should be constructed to bridge this gap. The construction, and subsequent execution, of the user-centred model helps us to understand, structure and clarify the requirements. The user-centred model is then used to aid the construction of a formal object-oriented specification. Validation of the object-oriented specification is then concerned with verifying that it is equivalent to the user-centred model.

We have shown how the development of a formal user-centred model can be integrated into the ROOA method and have described how the formal user-centred model can complement the OCD and ETDs in the construction of the ROOA model. We have then shown how LOLA can be used to demonstrate that the user-centred and ROOA models cannot be distinguished by testing.

REFERENCES

Bolognesi, T. and Brinksma, E. (1987) Introduction to the ISO Specification Language LOTOS. *Computer Networks and ISDN Systems*, **14**, 25-59.

Clark, R.G. (1992) Using LOTOS in the Object-Based Development of Embedded Systems, in *Unified Computation Laboratory* (ed. C. Rattray and R. Clark), Oxford University Press, pp 307-319.

Clark, R.G. and Jones, V. (1992) The Use of LOTOS in the Formal Development of an OSI Protocol. *Computer Communications*, **15**, 86-92.

Clark, R.G. and Moreira, A.M.D. (1997) Using a formal user-centred model to build a formal system-centred model. *Computing Science and Mathematics TR 140*, University of Stirling.

Eertink H. and Wolz D. (1993) Symbolic Execution of LOTOS Specifications, in *Formal Description Techniques V* (ed. M. Diaz and R. Groz), North-Holland, Amsterdam, pp 295-310.

Glinz, P. (1995) An Integrated Formal Model of Scenarios Based on Statecharts, in *ESEC'95*, LNCS 989, Springer-Verlag, Berlin, pp 254-271.

Hsia, P., Samuel, J., Gao, J., Kung, D., Toyoshima, Y. and Chen, C. (1994) Formal Approach to Scenario Analysis. *IEEE Software*, **11** March, 33-41.

Jacobson, I. (1992) *Object-Oriented Software Engineering*. Addison-Wesley, Reading Massachusetts.

Kotonya, G. and Sommerville, I. (1996) Requirements Engineering with Viewpoints. *Software Engineering Journal*, **11**, 5-18.

Moreira, A.M.D. and Clark, R.G. (1994a) Combining Object-Oriented Analysis and Formal Description Techniques, in *ECOOP '94* (ed. M. Tokoro and R. Pareschi), LNCS 821, Springer-Verlag, Berlin, pp 344-364.

Moreira, A.M.D. and Clark, R.G. (1994b) Rigorous Object-Oriented Analysis, in *Object Oriented Methodologies and Systems* (ed. E. Bertino and S. Urban), LNCS 858, Springer-Verlag, Berlin, pp 65-78.

Moreira, A.M.D. and Clark, R.G. (1996a) LOTOS in the Object-Oriented Analysis Process, in *Formal Methods in Object Technology*, (ed. S. Goldsack and S. Kent), Springer-Verlag, Berlin, pp 33-46.

Moreira, A.M.D. and Clark, R.G. (1996b) Adding Rigour to Object-Oriented Analysis. *Software Engineering Journal* **11**, 270-280.

Nuseibeh, B., Kramer, J. and Finkelstein, A. (1994) A Framework for Expressing the Relationships between Multiple Views in Requirements Specification. *IEEE Transactions on Software Engineering*, **20**, 760-773.

Quemada, J., Azcorra, A. and Pavon, S. (1995) The Lotosphere Design Methodology, in *LOTOSphere: Software Development with LOTOS*, (ed. T. Bolognesi, J. van de Lagemaat and C. Vissers), Kluwer Academic Publishers, Dordrecht, The Netherlands, pp 29-58.

Regnell, B., Kimbler, K. and Wesslen, A. (1995) Improving the Use Case Driven Approach to Requirements Engineering, in *Second IEEE Int Symposium on Requirements Engineering*, IEEE Press, pp 40-47.

Rubin, K.S. and Goldberg, A. (1992) Object Behaviour Analysis. *Communications of the ACM*, **35**, 48-62.

Rumbaugh, J., Blaha, M., Premerlani, W., Eddy, F. and Lorensen, W. (1991) *Object-Oriented Modelling and Design*. Prentice-Hall, Englewood Cliffs.

Somé, S., Dssouli, R. and Vaucher, J. (1996) Towards an Automation of Requirements Engineering using Scenarios. *Journal of Computing and Information*, **2**, 1070-1092.

Turner, K.J. and van Sinderen, M. (1995) LOTOS Specification Style for OSI, in *LOTOSphere: Software Development with LOTOS*, (ed. T. Bolognesi, J. van de Lagemaat and C. Vissers), Kluwer Academic Publishers, Dordrecht, The Netherlands, pp 137-159.

10 BIOGRAPHY

Robert Clark is a senior lecturer in Computing Science at the University of Stirling in Scotland. He has a PhD from the University of Dundee and is a Chartered Engineer. He is the author of two books and his research is into the addition of formality to object-oriented development.

Ana Moreira is an assistant professor in Informatics at the New University of Lisbon. She received her MSc degree in 1991 from the New University of Lisbon and her PhD in Computing Science from the University of Stirling in 1994. Her research is on adding rigour to object-oriented analysis.

16

Approaches to the Specification of Object Associations

D. Ramazani and G. v. Bochmann
Département d'informatique et de recherche opérationnelle
Université de Montréal
C.P. 6128, Succursalle Centre-Ville
Montréal, Canada H3C 3J7
Phone: (514) 343-7484, fax: (514) 343-5834,
E-mails: {bochmann, ramazani}@iro.umontreal.ca

Abstract

Many practitioners agree on the key role of object associations during the requirements specification and analysis phases of application development, since they contribute to the definition of the semantics of applications. However, the literature shows that there are multiple semantics for associations, and confusion about how they should be represented. As a matter of fact, various interpretations of the concept of association exist, leading to a multiplicity of representations.

The contribution of this paper is an exposition of four practical approaches to the formal specification of associations. It also introduces a conceptual model for associations which is used as a baseline for comparing the four approaches to formal specification of associations. These four approaches are based on different constructs of the specification language Object-Z which can be used for formally describing associations. The way these approaches capture the requirements represented by associations is central to selecting the approach to be used for the application development.

Keywords

Associations, Formal specifications, Object-oriented modeling, Object interactions, Relationships

This work was funded by the Ministry of Industry, Commerce, Science and Technology, Quebec, and the Natural Sciences and Engineering Research Council of Canada under the IGLOO project organized by the Centre de Recherche Informatique de Montreal.

The cornerstone of convergence of object-oriented analysis and design methods is the explicit formulation of the semantics of concepts used in these methods. In this formulation, formal specifications play a key role. An important concept of these methods are associations. Based on the following arguments, we claim that special attention should be given to associations between objects. Mili et al. (1990) observed that a number of run-time interactions between objects correspond to enforcing relations. Among the weaknesses of object-oriented methods, Monarchi and Puhr, (1992) report the identification and representation of associations, and the maintenance of a consistent and correct semantics for associations. Associations contribute to the specification of the dynamic behavior of applications. Generally, the complexity of an application is due to complex interactions between its components. As some of these interactions are abstracted by associations, the explicit description of associations allows to manage complexity of applications.

In fact, it can be demonstrated that associations are key to the specification of applications, especially of complex applications (Kilov, 1993). A quick tour of existing object-oriented methods reveals that associations are neglected during the development of applications (Monarchi and Puhr, 1992). This results in applications which are difficult to enhance, to modify and to reuse (Tanzer, 1995). The problem is caused by the lack of traceability from requirements to the implementation code via the associations expressed in analysis specifications of applications. Many practitioners agree on the key role of associations during the requirements specification and the analysis phases of application development, since they contribute to the definition of the semantics of applications. As a result of this role, associations need to be represented and manipulated. Further, there are multiple semantics for associations, and confusion in how they should be represented. As a matter of fact, various interpretations of the concept of association exist (ANSI, 1995), leading to a multiplicity of representations. For instance, associations may be realized as attributes, as separate classes, as operations along with their results, as separate modeling construct, or not at all. Facing this problem in the context of the definition of standards, the ISO/IEC JTC1 SC21 Working Group 4 proposes a General Relationship Model (GRM) (ISO-2, 1993). GRM provides a framework for specifying semantic properties of relationships independent of how they are represented. It encourages the definition of generic, reusable relationship classes applicable to multiple management applications. To extend the usability of conceptual models for relationships to general applications, Kilov (1993) developed a generic concept of relationships.

The work described in this paper was done as part of an ongoing research dealing with the modeling of composite objects. In this research, it appears that object composition can not be successfully handled if we are unable to capture the semantic properties of object associations. This has led to the study of the formal approaches which can be used for specifying object associations. The contribution of this paper is an exposition of practical approaches to the formal specification of associations. Before this exposition, we need to agree on a common base for discussing the concept of association. For that purpose, in Section 2, we present a conceptual model for association. The remaining part of this paper reviews four approaches to the formal specification of associations. An example is used to demonstrate some of the semantic properties of associations. It illustrates the practical usability of the approaches. We close the paper with a review of the kind of requirements captured by each approach.

2   WHAT IS AN ASSOCIATION?

2.1   Review of associations in object-oriented methods

Object-oriented methods propose two distinct semantics for associations. Some methods describe an association as being a pair of attributes of the associated classes. Each attribute allows to find which objects are associated to the object to which it belongs. The other methods view an association as a relation which is equivalent to a relation in relational databases, i.e. a set of tuples with operations for its management. In addition, we may define association classes which allow associations to have attributes, operations, associations and subtypes. The difference between the two semantics lies in the fact that the first approach considers associations as properties of classes (i.e. attributes) while the second approach identifies associations as autonomous entities. This difference can be summarized by the question whether an association is allowed to exist on its own? In many cases, an association is merely a connection between objects, each participant object using the association for referencing a certain number of other objects. In other cases, we want to act on the association, e.g. by making some queries, adding tuples, etc. One association may imply these two kinds of situations. Therefore, the two approaches complement each other.

2.2   Evolution of associations from analysis to coding

The object-oriented development consists of three phases : analysis, design and implementation. Analysis serves to define the semantic properties of associations. These semantic properties determine the nature (semantics) of the connection between the objects. The semantics is expressed using abstract concepts, i.e. independent of any particular representation. An object-oriented method provides a notation which allows to capture the semantics of associations. In OMT-2, for instance, the semantic properties of associations are its degree, its roles including multiplicity and ordering, its attributes, operations and associations with other objects (Rumbaugh, 1996). The notation which is proposed consists of representing binary associations by lines between associated classes and ternary associations by diamonds with one line path to each participating class. Multiplicity and roles are indicated by text while attributes and operations are represented using the class construct.

During the design phase, the semantic properties of associations are interpreted in terms of object-oriented design artifacts. The interpretation depends on the design decisions taken by the designer. The design notation determines the repertoire of design artifacts. The designer may take five kinds of design decisions. He may choose to interpret associations as follows:

1 Not at all: associations are ignored by the designer.
2 Correlation of behaviors: associations are interpreted as object interactions. This is the case when object interactions may be linked to object associations.
3 Correlation of states: associations are interpreted as structural constraints.
4 Collection of correlated behaviors: In this case, not only the designer wants to interpret object associations as correlated behaviors like in 2, but he wants to act upon all the instances of the association at the application level.
5 Collection of correlated states: Here also the designer wants to interpret object associations as correlated object states like in 3, but he wants to act upon all the correlated states at the application level.

During the implementation phase, design artifacts are translated into constructs of some object-oriented programming language. This representation depends on the constructs provided by the language. For example, considering C++, design artifacts can be translated into object inclusion, pointer structures, operations along with their results, objects, macros, and templates.

In the evolution of associations from analysis to implementation, a formal basis for associations can play a significant role. Semantic properties of associations can be precisely defined allowing a rigorous interpretation of these semantic properties in terms of design artifacts. The formal representation contributes to the ease of understanding, to modifiability and reuse of association specifications (Kilov, 1993). The translation of design artifacts into programming language constructs can be mechanized since formal reasoning is possible and adequate translation rules can be devised. Further, using the formal basis, it becomes straightforward to produce tools, such as implementation generators.

2.3   A conceptual framework for associations

The model
An association is the abstraction of a set of constraints between classes of objects. It has a name and a set of instances. An instance of an association is a set of constraints on known object instances. These objects are the participants in the association. The participating objects in an association can be classified according to the type obligations (attributes, operations, associations and behavior) which define the responsibilities assumed by these objects. The set of type obligations which must be met by a participant is a role. Each role has a name. The number of roles of an association is the degree of this association. For each role, the number of objects which assume this specific role in an instance of the association is the role cardinality. Association cardinality is the number of instances of the association in which a given object may assume the same role. The participation of an object in an association is mandatory when the object can not exist without participating in the association with other objects. Otherwise, the participation of the object is optional. It is static when a participant can not be changed without destroying the instance of the association. Otherwise, it is dynamic.

An association may have mathematical properties including reflexivity, symmetricity and transitivity. In addition to its mathematical properties, an association may have application-specific properties. Among these properties we find constraints defined on states and behaviors of the participating objects. These properties are application-specific since they define the semantics of the application. Kilov uses the term business rules of an application to denote such properties (Kilov, 1993). Instances of an association may have state and behavior. The state is captured by attributes which characterize each instance of the association. The behavior consists of operations for querying and changing the state of an instance of the association. These operations should not be confused with the management operations of the associations which consists of operations for creating, deleting and testing the existence of an instance of the association. Similar to objects, associations can be mutually constrained. For example, we may define exclusive associations, derived associations, composed associations, etc.

Exemplifying the conceptual model
It is difficult to grasp a conceptual model without an illustration of its usability. We exemplify the model using the employment association. The informal description of this association follows. A person may work for many companies; she is an employee of these companies. A company may employ many persons; it is the employer of these persons. A person is characterized by her name, social security number (ssn), address and age. A company has a name, an address and a budget. When a person is employed by a company, she has a job in that company and she must be aged of at least 18. This job has a position and a salary which can be modified. The budget of the company includes all the salaries of its employees.

Association definition

name: employment		*degree:* 2
roles		
role-name:	*employee*	*employer*
type obligations:	with age ≥ 18	with budget
role cardinality:	one	one
participation:	optional	optional
association cardinality:	many	many
mathematical properties:	irreflexive, anti-symmetric, non-transitive, ordered	
application-specific properties:	company budget $\geq \sum$ salary	
attributes:	position, salary	
behavior:	change-position, change-salary	
management behavior:	hiring, firing, checkworking status	
constraints with other associations:	exclusive with the association "enrolled to unemployment"	

Figure 1 Employment association description according to the conceptual model

Hiring and firing an employee correspond to creating and deleting an instance of the employment association. We may also check if someone works for a given company (respectively if the company employs a given person). A person who is employed by a given company is not allowed to be enrolled to the unemployment insurance provided by the ministry of labor. Such an insurance has specific terms which can be modified by the ministry of labor. Using the conceptual model, we specify this association as described in Figure 1 which is self explanatory.

2.4   Comparison with other models for associations

We restrict this comparison to a few well-known models for associations, namely the ODMG object model (Loomis et al. 1993), the Unified method (Booch and Rumbaugh, 1995), GRM (ISO-1, 1995) and Kilov's model (Guttapale et al. 1992).

ODMG
In the ODMG object model an association is modeled by a pair of association signatures, each defining the type of the other object(s) involved in the association and the name of a traversal function used to refer to the related objects. Traversal functions are similar to roles. Traversal functions are defined as attributes (or operations) of the participating object types. Generic operations for adding, deleting and testing the existence of association instances as well as miscellaneous operations (traverse, create_iterator_for) are also provided. The ODMG model is grounded on the notion of a mathematical relation. Each tuple of the relation is represented by a pair of traversal functions. Considering the employment association, it can be described using ODMG as shown in Figure 2:

Employment
cardinality: many-to-many
traversal functions
signature: <employees: set of Person> *defined in the interface of:* Company
signature: <employers: set of Company>*defined in the interface of:* Person
generic operations: depend on the database system which is used

---

Figure 2 Employment association description using the ODMG model

GRM
In GRM, associations are defined using relationship classes. A relationship class consists of roles and behavior. Each role defines the participant type obligations, the role cardinality and the relationship cardinality. Cardinalities have minimum and maximum value sets. GRM also specifies static and dynamic participation of objects. The behavior defines the interactions between roles.

Relationship classes are defined independently of object classes. They define one or more roles, but do not specify the participants that can fulfill these roles. A relationship binding specifies the class of the objects that can participate in that relationship for each role. Relationship bindings are specified independently of the relationship classes. GRM is formalized using GDMO. To make our presentation concise, we shall assume that the reader has no background in GDMO. Instead, we use an informal notation which is intuitive (see Figure 3).

In GRM, attributes and operations for instances of the relationships can not be defined. Therefore, the constraint imposing that the company budget should includes its employee salaries can not be expressed.

Relationship class employment

roles

role-name:	*employee*	*employer*
type obligations:	age ≥ 18	none of interest
role cardinality:	[1, 1]	[1, 1]
relationship cardinality:	[1, *]	[1, *]
participation:	static	static

behavior

hiring an employee: creating a new instance of the relationship
firing an employee: deleting an instance of the relationship
mutual exclusion with "enrolled to unemployment" relationship class

Relationship binding for employment

role   employee   <u>binding</u>   Person
role   employer   <u>binding</u>   Company

Figure 3 Employment association description using GRM

Unified method

In the Unified method, an association is a structural relationship between objects. The instances of an association are called links. A link consists of a tuple of object references. Each end of the association is a named role. The role shows how its class is viewed by the other class. Multiplicity consists of minimum and maximum cardinalities of roles. Links may have attributes, operations and associations. These properties are regrouped under an association class which is shown in a class diagram by drawing a dashed line from the association line to a class box that holds the attributes, operations and associations for the link. Other semantic properties can be defined for a role, e.g. navigability and mutability of a role indicates that the link can be modified.

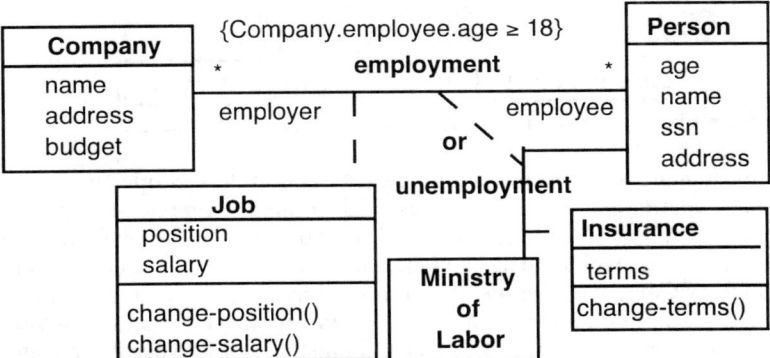

Figure 4 Employment association description using the Unified method

In Figure 4, we have two object classes Company and Person, and a binary association employment with roles employer and employee. Each role has a multiplicity of many indicated by the "*" symbol. In addition, there are attributes and operations for links. These properties are indicated by the association class

"Job" within the class diagram. Constraints between participating objects are indicated by text within braces free standing in the class diagram. It is not clear from (Booch and Rumbaugh, 1995) that we may express constraints involving the attributes of the association instances and the attributes of the associated objects. As a consequence, we drop the constraint imposing the budget of the company to includes all its employee salaries. Unemployment and employment associations are mutually exclusive. This is indicated by linking the two associations by means of a dashed line labelled "or".

Kilov's model

Kilov's approach tries to harmonize object-oriented modeling with entity-relationship modeling. Associations are described by its participants, cardinalities, and invariant, as well as the create, read, update and delete (CRUD) operations. These CRUD operations represent the management behavior of the association. The approach postulates the existence of primitive generic associations which are used (refined or combined) to define an association according to the application at hands. The employment association can be defined as a specialization of a relationship association, one of the generic associations provided by Kilov. A relationship association consists of relationship objects. Each relationship object associates several entities. It corresponds to exactly one instance of each of its participating entities. It has properties that provide information about itself, and not information about any of its participating entities. The employment association is pictured in Figure 5.

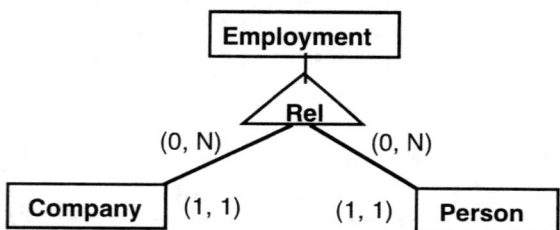

Figure 5 Employment association description using Kilov diagrams

The employment association is an entity since it has properties. In Kilov diagrams, rectangles indicate entities and triangles represent associations. Cardinality is specified using upper and lower bounds. In this diagram, for one part, an instance of employment is associated with only 1 instance of Company; an instance of Company is associated to a minimum of 0, and a maximum of N, instances of employment. On the other hand, an instance of employment is associated with only 1 instance of Person; an instance of Person is associated to a minimum of 0, and a maximum of N, instances of employment. Kilov diagrams do not specifically indicate relationship roles. Position, salary, change-position and change-salary are specified as properties of the employment entity. The invariant of the association constrains each person to be aged of at least 18. It also states that the company budget includes the salaries of the employees. Hiring and firing employees correspond to CRUD operations.

Summary of comparison

The main features of these models are summarized in the Table 1. The ODMG model is a simple one, but one with a strong mathematical basis since it is based on the concept of a mathematical relation. The model of the Unified method captures more semantic properties than the ODMG model. It considers association instances as objects which may have attributes, operations and associations. On the other hand, GRM subsumes the ODMG model. In GRM, specific modeling features, such as the management behavior of an association, are introduced to reflect the semantics of associations in the context of system management. In addition, GRM decouples associations from object classes to make the representation of associations independent of object classes.

Kilov's approach tries to reconcile GRM with the Unified method. This is done by proposing generic associations which can be specialized to represent associations between types, as well as object associations. The conceptual model proposed in Section 2.3 subsumes all these models. Its main objectives are to be more abstract and to clearly distinguish between the various semantic properties of associations. Our goal in devising this model was to present an intuitive, clear and precise model for associations which can be used as a baseline by practitioners.

Semantic properties	ODMG	GRM	Unified Method	Kilov's models
name	no	yes	yes	yes
degree	binary	yes	yes	yes
(role) name	yes	yes	yes	yes
(role) type obligations	yes	yes	yes	yes
(role) cardinality	one and many	yes	yes	yes
(role) participation	no	yes	yes	yes
association cardinality	no	yes	no	no
mathematical properties	no	yes	yes	yes
application-specific properties	no	no	yes	yes
attributes	no	no	yes	yes
behavior	no	no	yes	yes
management behavior	generic operations	yes	no	CRUD operations
constraints with other associations	no	yes	yes	no

Table 1 Semantic properties of some well-known models for associations

3   FORMAL SPECIFICATION OF ASSOCIATIONS

In this section, we describe four approaches to the formal specification of associations using the Object-Z language (Duke et al. 1994). The choice of Object-Z is motivated by the following reasons:

• It is an extension of Z, a well-known formal specification language;

- It has a strong mathematical base for defining relations;
- It is used for the formal specification of OMG, ODP and OSI standards;
- It is a general purpose specification language with built-in object-orientation;
- It is model-oriented, we may express correlations of states and behaviors;
- It has been proven to be expressive enough to specify industrial applications.

In this presentation, there is no emphasis on the formal specification of cardinalities. The reader is referred to (Liddle et al. 1993) which contains a formalization of various notions of cardinality. In addition, the formal representation of Kilov's models can be found in (Guttapalle et al. 1992).

Object-Z, as many other object-oriented formal specification languages, does not have a specific construct for representing object associations. In order to explicitly represent associations, the specifier needs to express the associations in terms of the constructs provided by the language. Among the constructs offered by Object-Z, the constructs of attributes, operations, classes and mathematical relations may help for representing associations. This gives rise to four basic approaches to the formal specification of associations. These approaches are described hereafter.

3.1   Approach using attributes (or operations)

This approach consists of representing associations by attributes (or operations) of the associated objects, as suggested by the ODMG model. The complexity added to class definitions may compromise modifiability and reuse of class definitions, as reported in (Tanzer, 1995).

In this approach, the definition of associations is distributed among its participating classes. Instances of the association are represented by pairs of attribute (or operation results) values. The degree of the association is the number of attributes required for representing a single instance of the association. Roles are captured by attributes. Type obligations, defined for each role of the association, are captured by the type of the attribute representing the role. Role and relationship cardinality are conjointly represented by using attributes which may be references or set of references to objects. Mandatory and optional participations are represented using assertions on the attribute value. To some extent, we may define application-specific properties of the association using assertions in the class definitions of the participating objects. While the other semantic properties of the association are only expressible in terms of properties of the participating classes.

We illustrate this approach using the employment association. In Object-Z specifications the definition of attributes, within a class definition, consists of two parts, one part is devoted to the types of the attributes and the other part to the constraints on the attributes. In the class Person (see below), the attribute employers represents a set of references to Company objects since the role cardinality is many. Theses references represent the companies the person is working for. In the class Company, the attribute employees represents a set of references to Person objects. These references represent the employees of the Company. The usage of sets implies that the duplication of instances of the association is not allowed. The constraints on these attributes are specified by assertions included in the class definitions of Person and Company. For example, in the Company class, the assertion states that each employee must be at least aged of 18.

[STRING] [Money] [Position]

```
┌── Person ──────────────        ┌── Company ──────────────
│ ┌────────────────────────      │ ┌────────────────────────
│ │ name: STRING                 │ │ name: STRING
│ │ ssn: STRING                  │ │ address: STRING
│ │ address: STRING              │ │ employees: ℙ Person
│ │ age: ℕ                       │ │ budget: Money
│ │ employers: ℙ Company         │ │ ∀ p ∈ employees • p.age ≥ 18
│ └────────────────────────      │ └────────────────────────
└────────────────────────        └────────────────────────
```

Salary and position are determined by a person and the company she is working for. They can be defined as partial functions which may be declared as global variables or localized within the classes Person or Company.

```
│ salary: (Company× Person) → Money
├─────────────────────────────────────────
│ ∀ p:Person, c: Company•
│ ((∃ m: Money•salary((c,p)) = m) ⇔ c ∈   p.employers∧ p ∈ c.employees)
```

Salary and position are defined by axiomatic descriptions (see above the definition of salary). An axiomatic description introduces one or more global variables in a specification, and optionally specifies a constraint on their values. We use surjections to denote the fact that two employees may hold the same position or they may have the same salary. Salary takes as input a tuple consisting of a Company and a Person. It returns the salary of the person within the company. An assertion links salary to the employment association. It states that for each tuple of Company and Person such that salary is defined, this tuple is an instance of the employment association. Forcing the domain of the partial surjections salary and position to be the employment association means that to each instance of the association corresponds a salary and a position.

Operations consisting of changing the salary and the position are defined as operations affecting the partial functions representing these attributes of the association. In Object-Z, the definition of an operation consists of two parts, the parameter declarations (input parameters are decorated with ? and output parameters with !), and the pre- and post-conditions for the operation. As illustrated below, the operation changesalary modifies the current value of the salary of an employee of a given company. As input, it takes the company, the person, and her newsalary within this company. Its pre-condition imposes that the salary of a person working for the given company is different from her new salary. Its post-condition guarantees that when the partial surjection salary is applied to the tuple formed by the company and the person, it returns the newsalary.

Operations for managing the association are represented by operations decoupled from the association representation. For instance, hiring an employee is an operation which adds an instance to the employment association. It is declared as an operation of the class Company as illustrated above. The operation hire is included in the definition of the class company, since intuitively a company hires one of its employees. We define the operation fire analogously.

```
┌─ ChangeSalary ──────────
│ Δ (salary)
│
│ employee?: Person
│ employer?: Company
│ newsalary?: Money
├─────────────────────────
│ ∃ m: Money •
│   ((salary(
│     (employer?, employee?))
│    = m)  ∧  m ≠ newsalary? )
│ salary'(
│ (employer?, employee?))
│   = newsalary?
└─────────────────────────
```

```
┌─ Company ────────────────────
│ ┌─ Hire ─────────────────────
│ │ employee?: Person
│ │ position?: Position
│ │ salary?: Money
│ ├─────────────────────────────
│ │ employee? ∉ employees
│ │ self ∉ employee?.employers
│ │ employees' = employees  ∪ {employee?}
│ │ employee?.employers' =
│ │         employee?.employer  ∪ {self}
│ │ position'((self, employee?)) = position?
│ │ salary'((self, employee?)) = salary?
│ └─────────────────────────────
```

3.2   Approach defining the entire association as an object

In this approach, one considers the association as an object having state and behavior. This allows to apply operations to the entire association. Many expressions over the entire association can be written concisely. The state of the object representing an association consists of the set of association instances and the semantic properties of the association. These elements are expressed in terms of class attributes and assertions on these class attributes. The behavior of the object includes update and query operations. Among these operations, there may be operations for adding or deleting association instances, testing the membership of given association instances, selecting a subset of the association instances according to some condition, and iterating over the set of association instances. In addition, the invariant of the class captures the association invariants and other constraints which need to be maintained between the participating objects.

Here, the definition of associations is localized in a single class definition. Instances of the association are stored using an attribute as a container (i.e. a set of tuples). The degree of the association is represented in the type definition of this attribute. This is also the case for roles, type obligations defined for each role, and role cardinalities. Association cardinality, mandatory and optional participations, and mathematical and application specific properties are represented by assertions in the class definition of the association. In this approach, like in the previous one, it is difficult to represent attributes and behavior of associations instances. Therefore, we also use the equivalence between association attributes and partial functions. These partial functions are attributes of the object representing the association. Association behavior is represented by operations affecting the partial functions. Management operations for the association are represented by operations of the object representing the association. Constraints with other associations are represented by constraints between objects representing these associations.

We define below a class Employment representing the employment association. The attributes employees and employers in the class definition allow to write concise assertions on the employment association. For instance, using these attributes we may act upon one kind of the particpants at a time. This is useful when constraining each employee to have an age of at least 18. The tuples of the association are stored in the attribute instances. The partial functions position and salary are introduced in the definition of the class Employment.

```
 ___ Employment _____
| instances: Company ↔ Person
| position: (Company× Person) ⇸ Position
| salary: (Company× Person) ⇸ Money
|_____
| ∀ p ∈ ran instances · p.age ≥ 18
| dom position = instances
| dom salary = instances
|_____
| <definition of changeposition, change salary, fire and hire> ___
```

In this case, the operations hire, fire, changesalary and changeposition are declared as operations of the class Employment.

3.3   Viewing each instance of the association as an object

In this approach, one considers each instance of the association as an object having state and behavior. The state of the object representing an instance of the association consists of the objects playing the roles along with other semantic properties. Application-specific properties and attributes of the association are expressed in terms of attributes and invariant of the object. The behavior of the object consists of application-specific operations which can be applied to each instance of the association.

In this approach, the instances of the association are objects. The degree of the association is represented by the number of the attributes required for representing the participant objects in an instance class. Roles, type obligations defined for each role, and role cardinality are captured by the type definition of the attributes representing the participant. Application-specific properties are represented by assertions in the class definition. Association attributes and behavior are made attributes and operations of the object representing the instance of the association. Management operations for the association are represented by operations for creating and deleting objects. However, using Object-Z, these operations can not be represented since Object-Z does not provide operations for creating and deleting objects. On the other hand, when the constraints with other associations are expressed in terms of association instances, they can be represented.

As an example, we define below the class Employment which represents a single instance of the employment association. Each role is represented by a specific attribute. With this approach, the representations of change-salary and change-position operations are intuitive.

```
┌─ Employment ──────────────────────────────────────────────────
│ employee: Person      ┌─ ChangePosition ──┐   ┌─ ChangeSalary ──┐
│ employer: Company     │ newposition?: Position │  │ newsalary?: Money │
│ position: Position    ├───────────────────┤   ├─────────────────┤
│ salary: Money         │ position ≠ newposition? │  │ salary ≠ newsalary? │
├───────────────────    │ position' = newposition? │  │ salary' = newsalary? │
│ employee.age ≥ 18     └───────────────────┘   └─────────────────┘
└─────────────────
```

3.4   Approach viewing associations as mathematical relations

This approach consists of reducing an association to a mathematical relation. Instances of the association are tuples of the relation. The degree of the association is the degree of the relation. The roles, and the type obligations are defined by definitional constraints on the domain and the range of the relation. The role cardinality defined for each role are specified as mathematical properties of the relation. Association cardinality, mandatory and optional participations, and mathematical and application specific properties are also defined in the same way. Association attributes can be represented using existential quantification while application specific operations can not be represented. Management operations for the association are represented as operations acting on the relation. Constraints with other associations are expressed byconstraints between mathematical relations.

The formal representation of associations with this approach is done using axiomatic descriptions. For instance, the employment association is defined as follows.

$$
\begin{array}{l}
\text{Employment: Company} \leftrightarrow \text{Person} \\
\hline
\forall \quad p: \text{Person} \cdot p \in \quad (\text{ran employment}) \Rightarrow p.age \geq 18
\end{array}
$$

All the semantic properties of an association can at some level be represented using mathematical relations. Such a description of an association becomes overloaded and for each semantic property there may be many representations. In order to avoid such a level of detail, we limit ourselves to the direct mapping between an association as a set of related objects and a mathematical relation understood as a cross product over domains.

4   OBSERVATIONS AND FINDINGS

As described in Section 3, we may use four approaches to the formal specification of associations. These approaches, due to the constructs of Object-Z on which they are based, can capture certain semantic properties directly, certain other properties only by using some specification artifacts, and may not be able to represent some other semantic properties. This is summarized in Table 2. In addition, the last line of the table indicates which of the well-known models for associations can be formally represented using each approach.

Semantic properties	Attribute	Association as an object	Instance as an object	Mathematical relation
name	no	yes	yes	yes
degree	yes	yes	yes	yes
(role) name	yes	yes	yes	yes
(role) type obligations	yes	yes	yes	yes
(role) cardinality	yes	yes	yes	yes
(role) participation	yes	yes	yes	yes
association cardinality	yes	yes	no	yes
mathematical properties	ordering	yes	no	yes
application specific properties	no	yes	yes	yes
attributes	no	no	yes	no
behavior	no	no	yes	no
management behavior	no	yes	no	no
constraints with other associations	no	yes	no	yes
Models captured	ODMG	GRM and Kilov	Unified method	GRM

Table 2 Semantic properties which can be formally captured by the approaches

These basic approaches for the specification of associations can be combined to overcome their respective disadvantages. A question which any requirements specifier may ask, is what are the criteria for selecting one approach over another? This raises the question of the relative importance of the different semantic properties. We believe that it is up to the requirements specifier to evaluate which requirements are more important, and to select the appropriate approach accordingly.

In general, we would suggest to use the instance of an association as an object approach if needed combined with the association as an object approach. This recommendation is based on our experience with these approaches as well as the experience of other practitioners. This approach avoids redundancy and it removes any direct couplings between the associated classes. In addition, it leads to complete and reusable requirement specifications .

5   CONCLUSION

The importance of associations in the development of applications, especially complex applications, and the key role played by formal specifications in the formulation of semantic properties of associations led us to review how we may formally specify associations. We identified four approaches. In order to ease the comparison of these approaches, we have introduced an abstract conceptual model for associations. Our experience with the well-known models presented in Section 2 and the different approaches for providing formal specifications reveals that there is still a confusion on the semantics of cardinalities (multiplicities) of associations. We note that (Liddle et al, 1993) give an excellent formal treatment of cardinalities for various models. Attributes of association instances can be represented by means of partial functions. In addition, there is sometimes a confusion between associations and attributes. Embley et al. (1992) have shown that attributes of objects may be represented by associations, and the converse is not necessarily true. In addition, management operations for associations are most of the time ignored, except in database and system management applications.

The main contribution of this paper is the demonstration that although there exists multiple well-known approaches for modelling object associations corresponding to different interpretations, as exemplified by the use of Object-Z, only four basic approaches suffice to formally specify object associations. These four basic formal approaches can be used individually or combined in order to capture various semantic properties of object associations. In addition, the four basic formal approaches can be applied in the context of any model-oriented formal specification language.

6   REFERENCES

ANSI-95 (1995) Object Data Management Reference Model. ANSI Accredited Standards Committee X3, Information Processing Systems.

Booch, G. and Rumbaugh, J. (1995) *The Unified method*, Rational Corporation.

Duke et al. (1994) Object-Z: a Specification Language Advocated for the Description of Standards. Technical Report No. 94-45, SVRC, The University of Queensland, Australia.

Embley et al. (1992) *Object-Oriented Systems Analysis, A Model Driven Approach*, Yourdon Press/Prentice Hall Englewood Cliffs, NJ.

Guttapalle et al. (1992) *The Materials: A Generic Object Class Library for Analysis*. Information Modeling Concepts and Guidelines, ST-OPT-002010, BellCore.

ISO-1 (1993) Information Technology - Open Systems Interconnection - Management Information Services - Structure of Management Information - Part 7: General Relationship Model, CDC ISO/IEC 10165-7.

Kilov, H. (1993) Information Modeling and Object-Z: Specifying generic reusable associations, Proceeedings of NGIT'93.

Liddle, S. et al. (1993) Cardinality Constraints in semantic data models, *Data & Knowledge Engineering*, **11**, 235-70.

Loomis, M. et al. (1993) The ODMG Object Model, *JOOP*, **6** (3):64-9.

Mili, H. et al. (1990) An Object-Oriented Model Based on Relations, *Journal of Systems Software,* **12**, 139-55.

Monarchi, D. and Gretchen, I., A Research Typology for Object-Oriented Analysis and Design, *CACM,* **35** (9), 35-47.

Rumbaugh, J. (1996) Models for design: Generating code for associations, *JOOP,* **8** (9), 13-7.

Tanzer, C. (1995) Remarks on object-oriented modeling of associations, *JOOP,* **7** (9), 43-6.

17

Exploring The Semantics of UML Type Structures with Z

R. B. France, J.-M. Bruel*, M. M. Larrondo-Petrie, and M. Shroff
Department of Computer Science & Engineering
Florida Atlantic University
Boca Raton, FL-33431, USA.
Email: {robert,maria}@cse.fau.edu

* Laboratoire IRIT/SIERA
F-31062 Toulouse Cedex, France

Abstract

The Unified Modeling Language (UML) builds upon some of the best object-oriented (OO) modeling concepts available, and is intended to serve as a common OO modeling notation. Given its intended role, it is important that the UML notation have a well-defined semantic base. In this paper we present some early results from our work on the systematic formalization of UML modeling constructs. The paper focuses on the formalization of UML Class Diagrams. The formal notation Z is used to express the semantics of Class Diagrams.

Keywords

Formal Specification Techniques, Object-Oriented Analysis and Modeling, Unified Modeling Language, Z.

1 INTRODUCTION

The *Unified Modeling Language* (UML) (Booch *et al.*, 1997) is a proposed common object-oriented (OO) modeling notation, currently being developed by some of the more experienced OO methodologists. The potential primary strengths of UML constructs, their simplicity and intuitive appeal, are also potential sources of problems. A significant problem is UML's reliance on informally defined semantics. This can lead to situations where models are interpreted differently because of differing viewpoints on what the semantics are. This is more likely to occur when complex structures (e.g., those involving recursive structures) are involved.

A formal semantic base for the notation allows one to rigorously reason about the models being built. Our work on formalizing other OO and structured notations indicates that the ability to rigorously analyze models strengthens validation and verification of the models. In our past work we have used formalized semantic bases for graphical techniques to animate requirements models, and to statically analyze properties (e.g., see (Bruel *et al.*, 1996; France *et al.*, 1997)).

In this paper we present a Z (Spivey, 1992) formalization of the UML constructs used to build Class Diagrams consisting only of types and their associations. Such diagrams can be used to model the static structure of systems at the requirements level. We assume that the reader is familiar with the Z notation. In section 2 we give the current form of our rules for transforming Class Diagram constructs to Z, and in section 3 we show how they can be used to formalize a non-trivial Class Diagram. We conclude in section 4 with an overview of our future work on the formalization of the UML notation.

2   FORMALIZING UML ANALYSIS-LEVEL CLASS DIAGRAMS

A Class Diagram is a model of the static structure of a system expressed in terms of classes, types, objects (class instances) and their associations. A UML *type* is a specification of concrete UML classes. In UML, classes implement types, that is, classes provide concrete implementations of the attributes and operations abstractly defined in types. We will refer to Class Diagrams that consist solely of types as *Type Diagrams*. Type Diagrams provide appropriate abstractions for modeling problems at the requirements analysis phase of software development, and is the focus of the formalization given in this paper.

In our formalization, a Type Diagram characterizes a set of instance structures, referred to as *valid instance structures* or *configurations*. A configuration is one that exhibits the properties expressed in the Type Diagram. One can view a configuration as a snapshot of a system's structure at some point in time, where the instances are those that have been created but not yet destroyed in the system. In this section we illustrate the rules for transforming UML type structures to Z specifications that characterize configurations.

2.1   A formalization of types

A type, like a UML class, has a name, and consists of a set of attributes and operation specifications. Graphically, it is depicted as a rectangular box with three compartments: the top compartment contains the type name, the middle compartment contains the set of attributes (with optional types and initial values), and the third compartment contains the list of operations (with optional argument lists and return types).

The set of all instances of a type in a configuration is called the *state* of the type. This set is to be distinguished from the set of all *possible* instances that satisfy the type properties. Such a set is called the *type space* of the type. A type state must be a subset of the type space. When interpreted in isolation, a type denotes its type space. When interpreted in the context of a Type Diagram, a type denotes a type state.

The state of an instance consists of two components: a data state and a set of operation states. The data state of an instance consists of attribute and association values. The attribute values are the values associated with type attributes, and association values are the instances that are linked to the instance under consideration. Associations will be discussed in the next section. In a state of an instance, each operation is associated with an operation state of the form $<$ *before_data_state, after_data_state(inputs)* $>$. The *before_data_state* is the (current) data state of the instance and the *after_data_state(inputs)* is the data state that is attained when the operation is performed to completion with inputs *inputs*.

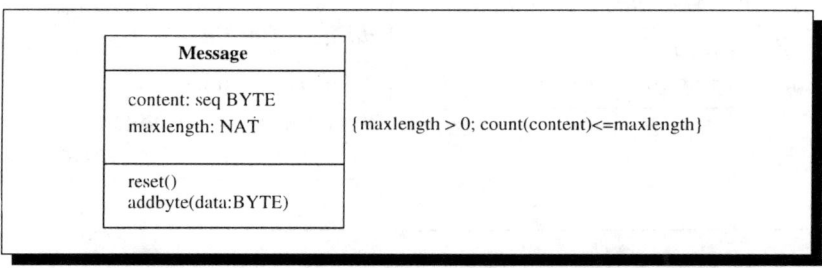

Figure 1   Message Type

In our formalization, a UML type is associated with a Z basic type consisting of elements representing unique instances of the UML type (they can be thought of as object identifiers). The attributes of a type are defined in a Z schema, referred to as an *attribute schema*. UML type invariants are expressed as predicates in the predicate part of attribute schemas. The attribute schema for the *Message* type shown in Fig. 1 is given in EXAMPLE 1. The type operation *reset* clears the message contents, and *addbyte* appends a byte of data to the message.

Example 1 Attribute schema for *Message* type

$[BYTE]$

─── *Message_Attributes* ──────────────────────────────
$content$: seq $BYTE$
$maxlength$: $\mathbb{N}$
──────────────────────────────
$maxlength > 0$
$\#content \leq maxlength$

Type operations are specified by Z schemas, called *operation schemas*, that relate before-data-states to after-data-states. Our operation schemas differ notationally from the traditional Z operation schemas in that we use a variable to represent the before-state and another to represent the after-state. The operation schemas for the *reset* and *addbyte* operations are given in EXAMPLE 2. In the schemas, m represents a before-state, and m' an after-state.

Example 2 Operation specifications for *Message* type

─ *Reset* ──────────────
m, m' : *Message_Attributes*
──────────────
$m'.content = \varnothing$
$m'.maxlength = m.maxlength$

─ *AddByte* ──────────────
m, m' : *Message_Attributes*
$data?$: $BYTE$
──────────────
$\#(m.content) < m.maxlength$
$m'.content = m.content ^\frown \langle data? \rangle$
$m'.maxlength = m.maxlength$

Semantically, an instance in a configuration can be viewed as a mapping of its object identifier to its data and operation states. This is captured formally by a Z schema that declares a variable representing the instances, and functions that map instances to their states. Such a schema is called a *type schema*. The type schema for the *Message* type is given in EXAMPLE 3 ($[MESSAGE]$ is supposed defined). The first three predicates of the *Message* type schema state that only configuration instances (elements of *instances*) are associated with data and operation states. The fourth predicate in the schema states that the before-state of an operation (m) must be the (current) data state of the instance (as determined by the function *attributes*).

Example 3 The type schema for *Message*

$\begin{array}{l}
\underline{\quad Message} \underline{\qquad\qquad\qquad\qquad\qquad\qquad\qquad\qquad\qquad} \\
instances : \mathbb{P}\, MESSAGE \qquad\qquad\qquad \text{[set of existing instances]} \\
attributes : MESSAGE \nrightarrow Message_Attributes \\
reset : MESSAGE \nrightarrow \mathbb{P}\, Reset \\
addbyte : MESSAGE \nrightarrow \mathbb{P}\, AddByte \\
\underline{\qquad\qquad\qquad\qquad\qquad\qquad\qquad\qquad\qquad\qquad\qquad\qquad\qquad} \\
\text{dom}\ attributes = instances \\
\text{dom}\ reset = instances \\
\text{dom}\ addbyte = instances \\
\forall\, p : instances;\ att1 : Reset;\ att2 : AddByte \\
\quad |\ att1 \in reset(p) \wedge \\
\quad att2 \in addbyte(p) \bullet att1.m = attributes(p) \\
\quad\quad \wedge\ att2.m = attributes(p)
\end{array}$

2.2 Formalization of UML associations

Semantically, an association is a set of *links,* where a link is a pair of instances of the form $(a \mapsto b)$, indicating that a and b are linked.

Multiplicity of an association constrains how many instances of a type can be associated with one instance of another (or the same) type. A range multiplicity is of the form $m..n$, where m is the lower bound and n is the upper bound of the range. The range $m..m$ can be simply written m. The multiplicity symbol '*' indicates many i.e. an unlimited number of objects. By itself, the symbol '*' is equivalent to '0..*' i.e. zero or more.

When associations are present, the data state of a type instance includes the instances that are related to the type. At the analysis level associations are bidirectional, that is, each linked instance knows about the instances it is linked to. This implies that each type instance includes information about its linked instances in its data state. Decisions related to restricting visibility of linked instances are best made during the design phase.

Consider the Type Diagram for a library system consisting of a *Copy* type and a *Borrower* type, with a many-to-one *Borrowed_by* association, shown in Fig. 2. The *Borrowed_by* association has an attribute *due_date* and a multiplicity that restricts a borrower to a maximum of 5 copies.

The formalization of the Type Diagram shown in Fig. 2 is given in EXAMPLE 4.

The association schema *Borrowed_by* defines the association as a set of pairs (*Rel*). The states of instances of the types *Borrower* and *Copy* are defined in

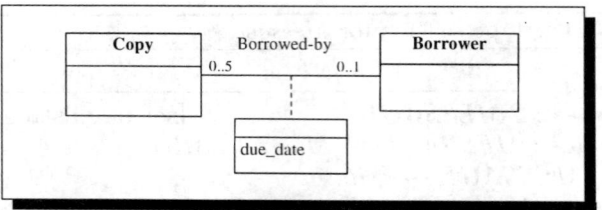

Figure 2 Example of an association with attributes

Example 4 The type schema for the *Borrowed_by* association

$[DATE, BORROWER, COPY]$ ┌─ *Borrowed_by_Attributes* ─────
 due_date : *DATE*

┌─ *Borrowed_by* ─────────────────────────
 Rel : $COPY \twoheadrightarrow BORROWER$
 Rel_Attributes : $(COPY \times BORROWER)$
   $\twoheadrightarrow$ *Borrowed_by_Attributes*
├──────────────────────────────────
 dom *Rel_Attributes* = *Rel*
 $\forall\, b :$ ran *Rel* $\bullet\ \#(Rel \triangleright \{b\}) \leq 5 \wedge \#(Rel \triangleright \{b\}) \geq 0$
                              [multiplicity constraint]

┌─ *Borrower* ────────────── ┌─ *Copy* ──────────────
 instances : $\mathbb{P}\ BORROWER$ *instances* : $\mathbb{P}\ COPY$
 borrowed_by : *Borrowed_by*   *borrowed_by* : *Borrowed_by*
├──────────────────── ├──────────────────
 ran *borrowed_by.Rel* $\subseteq$ *instances* dom *borrowed_by.Rel* $\subseteq$ *instances*

┌─ *AssocStruct* ─────────────────────────
 b : *Borrower*
 c : *Copy*
├──────────────────────────────────
 b.borrowed_by = *c.borrowed_by*

---

the type schemas with the respective type names. The schema *AssocStruct* is
a formalization of the Type Diagram shown in Fig. 2.

2.3   Formalization of Aggregation

An aggregate structure is a special type of association indicating a conceptual whole-part relationship. UML provides a weak and a strong form of aggregation. The strong form of aggregation is called a *composition*. An aggregate structure is depicted as an association of types in which a diamond is placed on the end of the association connected to the type that is the whole. If the diamond is filled then it is a composition, implying that each part can belong to only one whole, and that the lifetime of the parts are "coincident" with the lifetime of the whole (page 47, section 4.23 in (Booch *et al.*, 1997)). Such lifetime binding is not implied by the weak form of aggregation[†]. An unfilled diamond represents a weak aggregation, in which sharing of parts is allowed. For a composition, the multiplicity at the whole end must be no greater than 1 (restricting parts to belong to at most one whole). For the weak aggregation a multiplicity greater than one is allowed.

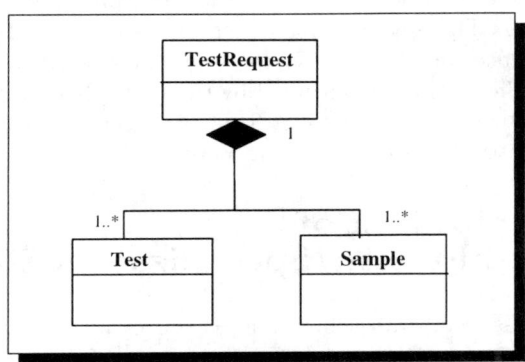

Figure 3   Example of a composition.

Below we formalize the composition shown in Fig. 3:

$$[TEST, SAMPLE, TESTREQUEST]$$

```
┌─ Test ─────────────────        ┌─ Sample ─────────────────
│ instances : ℙ TEST             │ instances : ℙ SAMPLE
```

[†]To be more precise, the UML manual does not indicate that there is a lifetime binding of parts to whole in a weak aggregation.

$$
\begin{array}{|l|}
\hline
\_\ TestRequest _____ \\
instances : \mathbb{P}\ TESTREQUEST \\
comp1 : TESTREQUEST \leftrightarrow \\
\quad TEST \\
comp2 : TESTREQUEST \\
\quad \leftrightarrow SAMPLE \\
\hline
\mathrm{dom}\ comp1 = instances \\
\mathrm{dom}\ comp2 = instances \\
\forall\, t : \mathrm{ran}\ comp1 \bullet \#(comp1 \rhd \{t\}) = 1 \\
\forall\, t : \mathrm{ran}\ comp2 \bullet \#(comp1 \rhd \{t\}) = 1 \\
\hline
\end{array}
\qquad
\begin{array}{|l|}
\hline
\_\ AggStruct _____ \\
tests : Test \\
samples : Sample \\
testrequests : TestRequest \\
\hline
\mathrm{ran}(testrequests.component1) \\
\quad = tests.instances \\
\mathrm{ran}(testrequests.component2) \\
\quad = sample.instances \\
\hline
\end{array}
$$

In the schema *TestRequest*, the components of the aggregate are specified as mappings from the instances of *TestRequest* to instances of the parts *Test* and *Sample*. The first and second predicates restrict part instances to instances in the configuration. The third and fourth predicates state that no two distinct instances of the type *TestRequest* can share parts. This restriction is not needed for weak aggregation in which the multiplicty at the whole-end is greater than one. The schema *AggStruct* formalizes the Type Diagram in Fig. 3. The instances of *Test* and *Sample* must be related to one *TestRequest* instance in a configuration (as specified by the multiplicty of 1 at the diamond end). The formalization of the weaker form of aggregation can be obtained by weakening the restrictions of the stronger form.

2.4   Formalization of Generalization/Specialization Hierarchies

A generalization-specialization hierarchy captures a supertype-subtype relationship between types. It is represented as a link from the subtype to the supertype, with a large hollow triangle at the supertype end.

The attribute structure of a generalization-specialization hierarchy is represented in Z by including the schemas defining the shared attributes in subtype attribute schemas. Our formalization covers the four combinations of generalization-specialization according to whether or not the supertype is abstract (i.e., all instances are instances of some subtype in the model) or the subtypes are disjoint (i.e., the subtypes do not share instances).

3   A FORMALIZATION EXAMPLE

In this section we illustrate the application of the UML-to-Z rules outlined in the previous section on a small, but non-trivial Type Diagram.

A Type Diagram for a Glyph structure is shown in Fig. 4. The complexity of recursive structures often causes modelers to underspecify their desired properties. Formalizing such structures forces modelers to consider and express required constraints that may have been glossed over in a less formal approach. Once a formal model is obtained, it can be used to demonstrate that the desired properties are present.

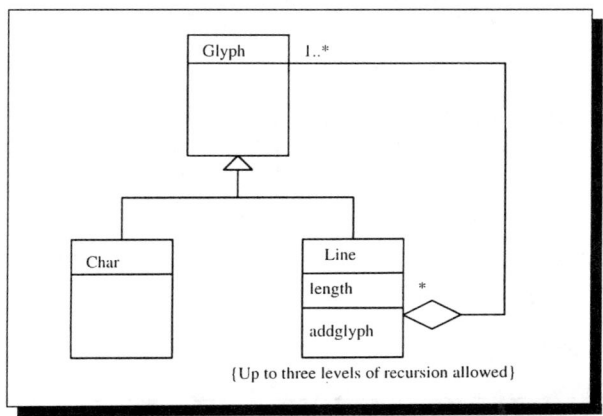

Figure 4 A recursive UML Type Diagram

A glyph is defined to be an abstract type for objects that can appear in a document structure. Its subtypes are the primitive graphical character elements (elements of type *Char*) and structural line elements (elements of type *Line*). This technique of composing increasingly complex elements out of simple ones in a hierarchical fashion is called *recursive composition*.

The type schemas for *Glyph* and *Char* are given below:

$$[GLYPH]$$

$$\begin{array}{|l}\hline Glyph\_\_\_\_\_ \\ instances : \mathbb{P}\,GLYPH \\\hline\end{array} \qquad \begin{array}{|l}\hline Char\_\_\_\_\_ \\ instances : \mathbb{P}\,GLYPH \\\hline\end{array}$$

The recursive composition is constrained as follows (the 'part-of' aggregation relationship is transitive):

- The aggregation is anti-symmetric (i.e., if line $l1$ is a part of line $l2$ then $l2$ cannot be a part of $l1$).
- The aggregation is irreflexive (i.e., a line cannot be a part of itself).
- No more than three levels of nesting is allowed (a line $l1$ can contain a line $l2$ that contains a line $l3$; $l3$ must consist only of characters). This is an application-specific constraint (see annotation on diagram).

The type schema for *line* is given below:

$$
\begin{array}{|l}
\hline
\_\,GlyphComp \rule{4cm}{0pt} \\
\;\; aggrel : GLYPH \leftrightarrow GLYPH \\
\hline
\;\; \forall\, l1, l2 : GLYPH \mid (l1, l2) \in aggrel^{+} \bullet (l2, l1) \notin aggrel^{+} \\
\qquad\qquad\qquad [aggrel^{+}\text{ is the transitive closure of } aggrel] \\
\;\; \forall\, l : GLYPH \bullet (l, l) \notin aggrel^{+} \\
\;\; \forall\, l1, l2, l3, l4 : GLYPH \mid (l1, l2) \in aggrel \,\wedge \\
\;\; (l2, l3) \in aggrel \bullet (l3, l4) \notin aggrel \\
\hline
\end{array}
$$

$$
\begin{array}{|l}
\hline
\_\,Line\_Attributes \rule{2cm}{0pt} \\
\;\; length : \mathbb{N} \\
\hline
\;\; length > 0 \\
\hline
\end{array}
\qquad
\begin{array}{|l}
\hline
\_\,AddGlyph \rule{2cm}{0pt} \\
\;\; l, l' : Line\_Attributes \\
\;\; c, c' : GlyphComp \\
\;\; g? : GLYPH \\
\hline
\;\; l'.length = l.length + 1 \\
\;\; c' = c \cup \{g?\} \\
\hline
\end{array}
$$

$$
\begin{array}{|l}
\hline
\_\,Line \rule{7cm}{0pt} \\
\;\; instances : \mathbb{P}\, GLYPH \\
\;\; attributes : GLYPH \nrightarrow Line\_Attributes \\
\;\; components : GlyphComp \\
\;\; addglyph : GLYPH \nrightarrow \mathbb{P}\, AddGlyph \\
\hline
\;\; \mathrm{dom}\; attributes = instances \wedge \mathrm{dom}(components.aggrel) = instances \\
\;\; \forall\, g : instances \bullet addglyph(g).l = attributes(g)\, \wedge \\
\;\; addglyph(g).c = components \\
\hline
\end{array}
$$

Using the schemas defined above, the following Z formalization of the Type Diagram in Fig. 4 is obtained:

$$
\begin{array}{|l}
\hline
\_\,GlyphStruct \rule{6cm}{0pt} \\
\;\; glyphs : \mathbb{P}\, Glyph \\
\;\; lines : \mathbb{P}\, Line \\
\;\; chars : \mathbb{P}\, Char \\
\hline
\;\; \langle lines.instances, chars.instances \rangle\ \text{partition } glyphs.instances \\
\;\; \forall\, l : lines \bullet \mathrm{ran}((l.components).aggrel) \subseteq glyphs.instances \\
\hline
\end{array}
$$

The first predicate in *GlyphStruct* states that the supertype is abstract and the subtypes are disjoint. The second predicate states that lines are composed of existing glyphs.

The benefit of having a formal model of the recursive structure is that one can prove properties that cannot be demonstrated simply be examining the diagram (e.g., the property that up to 3-levels of nesting is allowed). Building the formal model also forces one to consider, in detail, the constraints that are needed. Formalizing and analyzing UML structures can lead to a better understanding of the modeled structure.

4 CONCLUSION AND FUTURE WORK

The formalization of UML models can lead to a deeper understanding of modeled structure, and allows one to rigorously reason about modeled properties. In this paper we have illustrated how formalization can help clarify the meaning of non-trivial structures such as recursive definitions[‡]. Recursive structures have always been complex to model and to analyze. The lack of firm semantic bases for OO models compounds the complexity problem by increasing the chances of introducing ambiguous, incomplete, and imprecise statements of desired behavior and structure.

We have provided a semantic model that supports the formal interpretation of complex UML type structures, and the rigorous analysis of modeled properties. The Z specifications derived from the semantic model are tedious to produce by hand. Mechanical support for the generation of Z specifications from UML type structures is essential and possible. We are currently extending a tool we built for generating Z specifications from Fusion Object Models (see (France *et al.*, 1997)) to support the UML-to-Z transformation.

REFERENCES

Booch, Grady, Rumbaugh, James, & Jacobson, Ivar. 1997 (Jan.). *Unified Modeling Language*. Version 1.0. Rational Software Corporation, Santa Clara, CA-95051, USA.

Bowen, Jonathan P., & Hall, J. Anthony (eds). 1994. *Z User Workshop, Cambridge 1994*. Workshops in Computing. Springer-Verlag, New York.

Bruel, Jean-Michel, France, Robert B., & Benzekri, Abdelmalek. 1996 (21–25 Oct.). A Z-based Approach to Specifying and Analyzing Complex Systems. *In: Proceedings of the Second IEEE International Conference on Engineering of Complex Computer Systems (ICECCS'96), Montreal, Canada*.

France, Robert B., Bruel, Jean-Michel, & Larrondo-Petrie, Maria M. 1997. An Integrated Object-Oriented and Formal Modeling Environment. *To appear in the Journal of Object-Oriented Programming (JOOP)*.

Spivey, J. Michael. 1992. *The Z Notation: A Reference Manual*. Second edn. Englewood Cliffs, NJ: Prentice Hall.

[‡]The authors wish to thank the members of the Methods Integration Research Group (MIRG) for their participation on this project. For more information on MIRG, and integrated formal and OO modeling techniques see the WWW site at: *http://www.cse.fau.edu/research/MIRG/*. This work was partially funded by NSF grant CCR-9410396.

PART SIX

Formal Specification (I)

18

Expressive flexibility in constraint-oriented specification: LOTOS and Co-notation

Tommaso Bolognesi
CNR, Istituto IEI - Pisa - t.bolognesi@cnuce.cnr.it

Abstract - We illustrate and compare the flexibility of two specification techniques for concurrent systems, namely LOTOS and our recent constraint-oriented specification notation ('Co-notation'). *Flexibility* is intended here as the ability to match as closely as possible the structure of the initial, informal behavioural description of the system, and to directly formalize the conceptual links among data variables and events, and their groupings, as identified by that description. We show that the simple yet powerful composition operator of the co-notation (constraint conjunction), supporting both shared-action (rendez-vous) and shared-variable process interactions, achieves a higher expressive flexibility than LOTOS, and a stronger support to constraint-oriented reasoning.

Keywords - specification languages, formal methods, process algebra, constraints, parallel logic programming.

1. Introduction

This paper is concerned with expressive *flexibility* in formal specification, a somehow fuzzy concept which has nonetheless a great importance for the applicability and widespread acceptance of formal methods in system development. While on computation-theoretic grounds most non trivial specification languages offer maximum *expressive power*, that is, they can simulate Turing machines, not all of them exhibit the same *flexibility* in adapting to the way humans conceive the initial, informal description of a system and of its behaviour. The adoption of a flexible specification language guarantees a smooth transition from the informal to the formal description phase, by facilitating the mutual comprehension between 'non-technical people' (problem owner, customer) and the system development team, and by possibly increasing the involvement and contribution of the former in the early development steps. Let us now consider the *informal* behavioural specification of a generic concurrent system, which may exist as a document in natural language or simply as a set of ideas in the mind of the specifier. We maintain that, at this very early stage, one can readily recognize three types of elements, which may then find a more or less satisfactory formal counterpart in the features of the various specification languages. These are:

- actions and events;
- state/data variables;
- 'capsules', that is, means for encapsulation.

An *event* is the instantaneous occurrence (performance) of an *action* . An action may be pure -- just a name -- or it may involve data parameters, that can be atomically observed, offered to, or accepted by the environment when the action occurs. *State/data variables* are entities that describe the current configuration of the system and of its components; each variable holds a possibly structured value.

Various types of interdependencies among events and state/data variables are specified in the informal description, including temporal ordering of events, relations among (the values of) actions and state/data variables before and after an event, and state invariants. Groups of actions and/or variables, and their interdependencies, are encapsulated into specification fragments that are conveniently understood as unitary behavioural components. We use the generic term *'capsule'* for these chunks of *informal* specification. Of course encapsulation may be multi-level. For example, a capsule may describe a complex data structure, with its associated operations and properties, or a temporal pattern of actions, or a combination of the two. Encapsulation is a way to break and manage complexity, and is apparent from the textual structure of the informal description, which is organized into sentences, paragraphs, bullets, sections. Capsules do not describe independent fragments of behaviour: they are related to one another. For example two paragraphs may describe two phases of behaviour that must be performed in sequence. In this paper we focuse our attention on what we consider as the most elementary type of capsule interrelation mechanism, a somehow implicit mechanism which we are used to take for granted in any natural language description, and which is so primitive and pervasive as to become almost invisible. We refer to the *sharing of actions and state/data variables* among capsules. The very fact that the same action or variable name is used in k capsules implies a relation, a mutual influence between these descriptive components. The k capsules provide different, partial viewpoints about that action or variable, they constrain that item in different ways: all k viewpoints must then be traced back in the global picture of the system behaviour; all *constraints* must be *satisfied*.

We regard a formal specification language as *flexible* when it supports in a most direct way the transposition of actions, state/data variables, capsules, and their mutual links -- most notably, the sharing of actions and variables -- from the informal to the formal specification; a flexible specification language is convenient in that it supports a close matching between these two specification forms. A non flexible specification language forces one to play tricks, and to compensate for the absence of primitive expressive tools by creating artificial specification machinery which obscures the essential behavioural aspects.

In this paper we shall illustrate and compare the flexibility of two different specification techniques, namely LOTOS, and our constraint-oriented specification notation (co-notation), by applying these languages to a running example. The central point of the paper is that the co-notation achieves a higher expressive flexibility, due mainly to the fact that its composition operator encompasses both shared action (rendez-vous) and shared variable communication, while LOTOS supports only the former. Familiarity with LOTOS is assumed.

2. Jewellers

We provide now an informal description of our running example. We identify in italics the keywords that shall be used in the subsequent formalisations.

1. A jeweller shop is run by three friends: *Mike, Mary* and *Jane*. Together, they open the shop every morning and close it every evening (*openShop, closeShop*), perhaps via a three-key lock. Every night Mike and Mary go to the theater (*toTheater*), while Jane goes to a discoteque (*toDisco*). The following is a description of what may happen during the day in the shop.
2. *Mike* takes care of accepting new precious pieces (*pieceIn*) from external producers that need not be specified, and of storing them in the *safe* . At any time he may also update a table (*valTable*) that indicates the value per weight unit (*unitVal*) of each precious metal, namely *platinum, gold,* and *silver*. An update consists of a new (material, unitary-value) pair (*matValPair*). The input pieces are characterized by their weight, material, and value (*wei, mat, val*), the latter being computed based on the table entry for *mat*.
3. *Jane* is the window dresser; she may move pieces between the *safe* and the shop *window* (*pieceToSafe* and *pieceToWind*).
4. *Mary* is in charge of the cash deposit (*cash*), which she opens (*cashUnlock*) after entering the shop, and closes (*cashLock*) before leaving. Only the pieces in the *window* may be sold (*pieceOut*), at a *price* which is *1.3* times their value according to their material and weight. This money is put in the *cash* deposit. Note that the value of a piece when leaving the shop may differ from its value when entering it, in case the table was updated between these two events.

Two different customer behaviours are described.

5. *Peter* has a bag (*bag*) which he can fill with pieces that he buys during a shopping session. The current weight of the bag (*currWei*) cannot exceed a maximum weight capacity (*MaxWei*), but Peter is not happy until the money spent in the current shopping session (*currExp*) has reached a given threshold (*TargetExp*); at that point he empties his bag (*reset*) and is ready for a new shopping session.
6. *Paul* buys a piece only if he has enough funds (*funds*). However, he can *refill* his funds by adding some amount of money (*delta*), from time to time.

The next two sections provide alternative formalisations of the Jewellers system; for avoiding duplications, we have factored out type and variable definitions.

Data types

weight	= natural;
material	= {platinum, gold, silver};
value	= natural.

Data variables

wei, currWei:	weight;
mat:	material;

val, unitVal, price, cash, currExp, funds, delta: value;
valTable: material -> value   (a partial function);
safe, window: bag of (weight × material);
bag: bag of (weight × value).

Global variables
(These appear in italics also in the formal specifications.)
TargetExp: value;
MaxWei :   weight.

Actions
Some actions are not associated with data exchange, and therefore have no type. These are called 'pure actions'.

openShop, closeShop, toTheater, toDisco, cashUnlock, cashLock: pure actions;
pieceIn, pieceOut: weight × material × value;
pieceToWind, pieceToSafe: weight × material;
matValPair: material × value;
reset: bag of (weight × material);
refill: value.

3.   Using LOTOS

In LOTOS [B89, BB87] a system is typically described as a set of processes that interact with one another and with their environment via interaction points called *gates*. The interconnection structure is described by instances of the binary parallel composition operator, which is parameterised by the set of *synchronisation gates*.

In our subsequent specification we shall omit the 'process' and 'endproc' keywords, the process <functionality> attribute and the sort identifiers, implicitly assuming those factored out in Section 2. In cases of recursive processes, we have shortly indicated by '[=]' the gate list associated with a process instantiation, whenever it is the same list that appears in the header of the enclosing process definition. The LOTOS specification of the Jewellers system is given below. All persons mentioned in the informal specification are modelled as LOTOS processes. The first three are grouped into a process *Jewellers*.

System
[openShop, closeShop, cashUnlock, cashLock, pieceIn, pieceOut,
pieceToWindow, pieceToSafe, matValPair, toTheater, toDisco, reset, refill] :=
   Jewellers
   [openShop, closeShop, cashUnlock, cashLock, pieceIn, pieceOut,
   pieceToWindow, pieceToSafe, matValPair, toTheater, toDisco]
   ({}, {}, {(silver, 80), (gold, 100), (platinum, 120)}, 450000)
   |[pieceOut]|
   (**Peter** [pieceOut, reset] *({}, 0, 0)* ||| **Paul** [pieceOut, refill] *(100.000)*).

Jewellers
[openShop, pieceIn, pieceToWindow, pieceToSafe, pieceOut,
matValPair, closeShop, toTheater, toDisco]
(safe, window, valTable, cash) :=
   ((**Mike** [openShop, closeShop, pieceIn, matValPair, toTheater]
   |[openShop, closeShop, toTheater]|
   Mary [openShop, closeShop, cashUnlock, cashLock, pieceOut, toTheater]
) |[openShop, closeShop]|
   Jane [openShop, closeShop, pieceToWind, pieceToSafe, toDisco]
)
  |[pieceIn, pieceToWindow, pieceToSafe]|
   (**SafeWindow**[pieceIn, pieceOut, pieceToWind, pieceToSafe]*(safe, window)*
   |[pieceIn, pieceOut]|
   ValTable[pieceIn, pieceOut, matValPair]*(valTable)*
   |[pieceOut]|
   Cash[pieceOut]*(cash)*
).

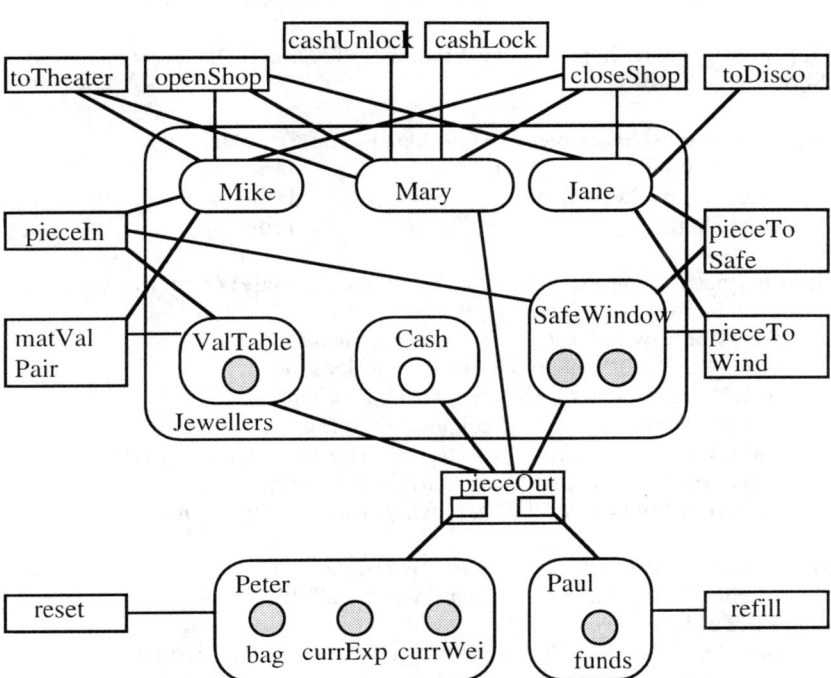

Figure 1 - Upper layers of the LOTOS specification

The structure of the upper layers of the LOTOS specification presented so far can be
represented graphically as in Figure 1, which shows the sharing of actions among

processes and the data variables that are encapsulated in some of the processes. The lower layers of the LOTOS specification are presented below.

Mike [openShop, closeShop, pieceIn, matValPair, toTheater] :=
   openShop;   Mike1 [=] .

Mike1[openShop, closeShop, pieceIn, matValPair, toTheater] :=
   pieceIn ?wei ?mat ?val;   **Mike1**[=]
   []   matValPair ?mat ?unitVal;   **Mike1**[=]
   []   closeShop;   toTheater;   **Mike**[=].

Mary [openShop, closeShop, cashUnlock, cashLock, pieceOut, toTheater] :=
   openShop;   cashUnlock;   **Mary2** [=].

Mary2 [openShop, closeShop, cashUnlock, cashLock, pieceOut, toTheater] :=
        pieceOut ?wei ?mat ?price; **Mary2** [=]
   []   cashLock;   closeShop;   toTheater; **Mary** [=].

  Jane [openShop, closeShop, pieceToWind, pieceToSafe, toDisco] :=
   openShop;   **Jane1**[=].

Jane1 [openShop, closeShop, pieceToWind, pieceToSafe, toDisco] :=
        pieceToWindow ?wei ?mat;   **Jane1**[=]
   []   pieceToSafe ?wei ?mat;   **Jane1**[=]
   []   closeShop;   toDisco;   **Jane** [=].

SafeWindow [pieceIn, pieceOut, pieceToWind, pieceToSafe] *(safe, window)* :=
        pieceIn ?wei ?mat ?val;
        SafeWindow [=]*(safe U {(wei, mat)}, window)*
   []   pieceOut ?wei ?mat ?price [(wei, mat) IN safe];
        SafeWindow [=]*(safe \{(wei, mat)}, window)*
   []   pieceToWind ?wei ?mat [(wei, mat) IN safe];
        SafeWindow[=]*(safe \ {(wei, mat)}, window U {(wei, mat)})*
   []   pieceToSafe ?wei ?mat [(wei, mat) IN window];
        SafeWindow[=]*(safe U{(wei, mat)}, window \ {(wei, mat)}).*

ValTable [pieceIn, pieceOut, matValPair] *(valTable)* :=
        pieceIn ?wei? mat ?val   [val = wei * valTable(mat)];
        ValTable [=] *(valTable)*
   []   pieceOut ?wei ?mat ?price [price = 1.3 * wei * valTable(mat)];
        ValTable [=] *(valTable)*
   []   matValPair ?mat ?unitVal;
        ValTable [=] *(valTable ⊕ (mat , unitVal)).*

Cash [pieceOut] *(cash)* :=
   pieceOut ?wei ?mat ?price;   **Cash** [=] *(cash + price).*

Peter [piece, reset] *(bag, currExp, currWei)* :=
    piece ?wei ?mat ?price [currExp<*TargetExpr* $\wedge$ (currWei + wei)$\leq$*MaxWei*];
           Peter [=] *(bag U (wei, price), currExp + price, currWei + wei)*
[]   reset;   **Peter** [=] *({}, 0, 0)* .

Paul [piece, refill] *(funds)* :=
    piece ?wei ?mat ?price [price $\leq$ funds];   **Paul** [=] *(funds - price)*
[]   refill ?delta;   **Paul** [=] *(funds + delta)*

The specification above is an example of the LOTOS constraint-oriented specification style [VSvS88], in which several processes cooperate in defining the temporal ordering and value of the actions that they share. For space reasons we omit a detailed illustration of the specification.

Assessment

In LOTOS, encapsulation is achieved by the *process* concept: the global behaviour is structured into processes, that represent different phases or aspects of behaviour. A process may be created for encapsulating a data structure and the actions (operations) that affect it, or a temporal pattern of events, or a combination of the two. In the specification above we have actually separated processes that describe the ordering of events from those that manage data structures, thus breaking the complexity of the global system behaviour. However, the freedom offered by LOTOS for creating the aggregates of data structures and actions that form the different processes is *not optimal*: only limited forms of aggregation are possible, since processes can only share actions, not data structures. For example, while the internal structure of process *Jewellers*, depicted in the upper box of Figure 1, reflects precisely the association person-action found in the informal system description (where some of the actions are shared), it fails to reflect the other association found in it, namely person-data structure (where some of the data structures are shared).

The limitation above is perhaps better clarified by an abstract example. Consider Figure 2, expressing the mutual influences between a set of three actions *a*, *b*, *c*, and a set of three data items *s*, *t*, *u*.

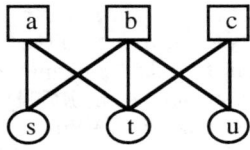

Figure 2 - Mutual influences between three actions and three data items

A link between an action, say *a*, and a data item, say *s*, indicates mutual dependency, which means either that *a* may or may not occur depending on the value of *s*, and/or that some value *x* established upon the occurrence of *a* depends on the value of *s*, and/or that the value of *s* is updated by the occurrence of *a* and, possibly, by the associated value *x*. LOTOS offers a few alternatives for describing this scenario. One extreme solution consists in encapsulating all actions and data parameters in a single process P[a, b, c](s, t, u). In the body of this process we would have complete access to the actions and data variables, and could freely express any relation among them. The obvious disadvantage is the lack of structure, and, related to that, the need to re-instantiate the whole, big process whenever one of the data items must be updated. We may then decompose the system into smaller, interacting processes, but the best that we can do is:

$$P[a, b](s) \quad |[a, b]| \quad Q[a, b, c](t) \quad |[b, c]| \quad R[b, c](u).$$

The three-process solution above corresponds to a covering of the gate set, namely {{a, b}, {a, b, c}, {b, c}}. No finer covering is possible because *a process that encapsulates a data item must necessarily insist on all the actions that affect that item.* On the other hand one might wish to consider an alternative aggregation, which perhaps matches more closely the informal description, and is in any case suggested by the symmetry in Figure 2. This would be the dual of the structure above, and would achieve maximal fragmentation of the action space; we represent it by the abstract expression 'X[a](s, t) * Y[b](s, t, u) * Z[c](t, u)', where the composition operator is left unspecified ('*'). The pattern above could only be formally specified by a language which admits *variable sharing*, but LOTOS does not support this feature.

4. Using the co-notation

We have designed the co-notation (Constraint-Oriented notation) [BA96] with the primary objective of supporting a constraint-oriented style for the specification of concurrent systems more effectively than in LOTOS. More generally, we felt the need to explore a possible integration of the good features of the process-algebraic approach, with its emphasis on event ordering and process composition, and those of a declarative approach such as Z [S90], with its emphasis on state variables and global state structuring.   The key idea was one of treating actions and state/data variables as uniformly as possible, and of supporting the specification and composition of constraints on these items in a most flexible way. Processes (or *blocks*, in our terminology) should cooperate both by shared variables and by shared actions, i.e. rendez-vous.   (A comprehensive introduction to interprocess communication mechanisms is found, for example, in [LL90].) A preliminary version of the co-notation was presented in [BC94], and a full introduction to the notation and its semantics is provided in [BA96].

The four types of object that constitute a specification in co-notation, called *co-specification* for short, are *actions*, state/data variables or, shortly, *state variables*, *elementary constraints* and *compound constraints* , also called *blocks* (see Figure 3). A co-specification describes networks of constraints on state variables and actions, where compound constraint are, recursively, networks of constraints, and may include local state variables. There exist two types of elementary constraint:

- *active predicates*, which insist on precisely one action, and zero or more state variables (there are two active predicates in Figure 3);
- *state invariants*, which insist on one or more state variables (there is one state invariant in Figure 3).

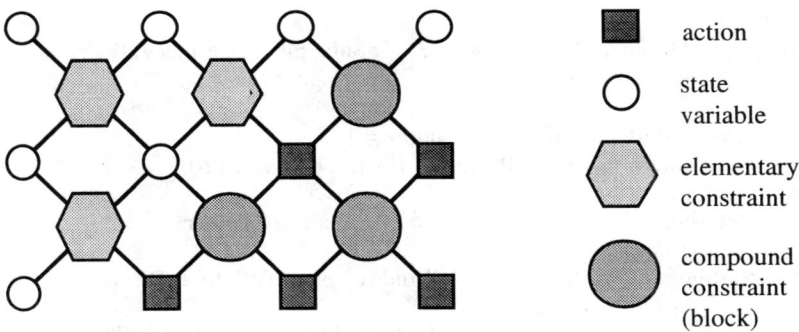

action

state variable

elementary constraint

compound constraint (block)

Figure 3 - A network of constraints of a co-specification

We shall not deal with state invariants in this paper, thus a constraint can only be a block or an active predicate. In essence, an action occurs when all the constraints that insist on it agree on its value and executability, also depending on the (possibly shared) variables upon which those constraints insist, and on the internal variables of the latter.

Similar to LOTOS, the co-notation is provided with an interleaving semantics based on labelled transition systems: a co-specification ultimately describes a possibly infinite labelled tree where arcs are labelled by action names and values, and nodes are labelled by assignments to the full set of state variables.

From the syntactic viewpoint, a co-specification is a list of definitions of types and of complex and elementary constraints, possibly enriched by a list of logical predicate definitions, for those predicates that are instantiated in the bodies of the elementary constraints. Compound constraints can be defined either graphically or textually. The block definitions are not nested, thus, for avoiding ambiguity, the constraint names must all be different. Compound constraint instantiations are not allowed to be (directly or indirectly) recursive, thus the structure of instantiations is a tree. Rather than further introducing the (few) syntactic and semantic features of the co-notation in abstract terms, we describe them as we comment the co-specification of the Jewellers system, which is provided below. Its reading should be facilitated by its similarity with the LOTOS specification of the previous section.

System
[openShop, closeShop, pieceIn, pieceOut, pieceToWindow, pieceToSafe, matValPair, toTheater, toDisco, reset, refill] :=
constraints

Jewellers
[openShop, closeShop, pieceIn, pieceOut, pieceToWindow, pieceToSafe,
matValPair, toTheater, toDisco] ,
(**Peter** [pieceOut, reset] ;   **Paul** [pieceOut, refill])

Jewellers
[openShop, pieceIn, pieceToWindow, pieceToSafe, pieceOut, matValPair,
closeShop, toTheater, toDisco] :=
variables

> cash = 450000, safe = { }, window = { },
> valTable = {(silver, 80), (gold, 100), (platinum, 120)}

constraints

  Mike [openShop, closeShop, pieceIn, matValPair, toTheater]
> *(safe, valTable)* ,
  Jane [openShop, closeShop, pieceToWindow, pieceToSafe, toDisco]
> *(safe, window)* ,
  Mary [openShop, closeShop, cashUnlock, cashLock, pieceOut, toTheater]
> *(cash, window, valtable)*.

We have provided above the definitions of two compound constraints, or blocks,
namely the *System*, which is the main block, and the *Jewellers*. The structure of
these two blocks can be represented graphically as in Figure 4, which shows the
sharing of actions *and* data variables among components, and the data variables that
are encapsulated in some of them.

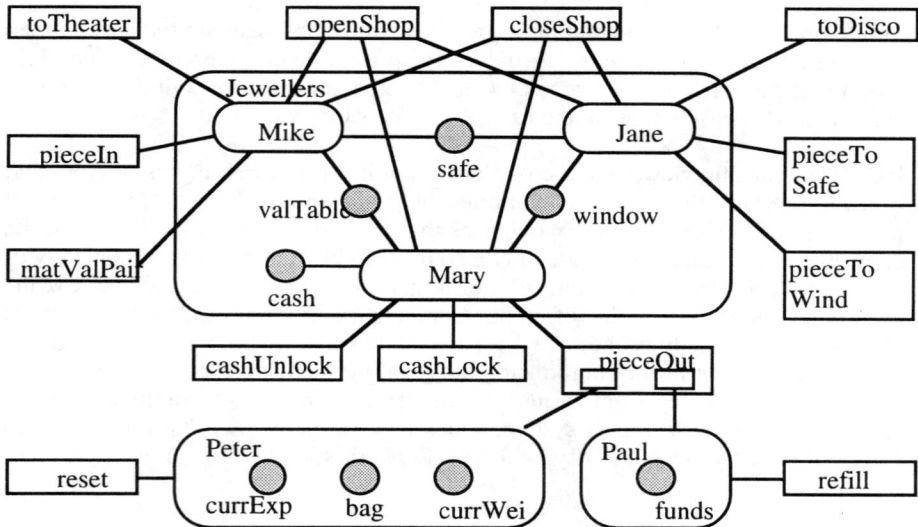

Figure 4 - Upper layers of the specification in conotation

The header of a constraint definition (be it a block or an active predicate) consists of a name, the list of actions, in square brackets, and, possibly, the list of state variables upon which the constraint insists. The items of the lists are understood as formal parameters, and should be associated with their sorts (e.g. 'x : nat'), but we omit the latter for conciseness, having provided them already in Section 2. Currently no internal actions are envisaged for the co-notation, thus all actions are listed in the header of the main block.

The body of a *block definition* consists of two sections:

- section *variables* defines the local state variables of the block and their initial values;

- section *constraints* contains a *co-expression*, which is formed by a set of *block* or *elementary constraint instantiations* connected by the binary composition operators of conjunction (',') and disjunction (';'). Formally:

<co-expr> ::=   <constraint-instantiation> | (<co-expr>)
          | <co-expr>, <co-expr> | <co-expr> ; <co-expr>.

Similar to the header of a constraint definition, a constraint instantiation consists of a name, a list of actions, in square brackets, and, possibly, a list of state variables upon which the constraint insists. Actions and variables, in this case, play the role of actual parameters: we are passing the actual names of actions and variables upon which the constraint must insist. Note that, unlike in LOTOS, we are not allowed to pass value expressions for binding formal variables to values: we pass actual state variable names, e.g. *safe* and *valTable* for instantiating formal state variable names (names may coincide).

A co-expression can only perform one of the actions that appear as parameters of its constraint instantiations, say action a. The disjunction operator ';' introduces nondeterminism, thus, in general, different, alternative subsets of the constraints that own that action may be involved in its execution, depending on the structure of the co-expression. Let E be a co-expression, and let $CC_a(E)$ denote the set of *sets of constraint* that may be involved in the execution of action a. Then:

- If $E = 'P[Acts](Vars)'$, that is, E consists of just a constraint instantiation, we have:

$$CC_a(E) = \{\{P[Acts](Vars)\}\} \qquad \text{if } a \in Acts;$$
$$CC_a(E) = \{\emptyset\} \qquad \text{if } a \notin Acts.$$

- If $E = 'E1 , E2'$, then the generic element of $CC_a(E)$ (this element is a set!) is the *union* of an element of $CC_a(E1)$ and one of $CC_a(E2)$.

- If $E = 'E1 ; E2'$, then the generic element of $CC_a(E)$ is an element of $CC_a(E1)$ *or* one of $CC_a(E2)$.

For example, if E_J is the co-expression defining the *Jewellers* block it is:

$CC_{toTheater}(E_J)$ = $\{\{Mike[...](...), Mary[...](...)\}\}$

while for the co-expression E_S defining the whole system it is:

$CC_{pieceOut}(E_S)$ = $\{\{Jewellers[...](...), Peter[...](...)\},$
                                         $\{Jewellers[...](...), Paul[...](...)\}$
                                         $\}.$

We provide now the definitions of the five blocks *Mike, Mary, Jane, Peter and Paul*, and of two blocks, *CycleA* and *CycleB*, that are used in the definitions of *Mike, Mary* and *Jane*.

Mike[openShop, closeShop, pieceIn, matValPair, toTheater] *(safe, valTable)* :=
constraints
   CycleA   [openShop, pieceIn, matValPair, closeShop, toTheater],
   Accept   [pieceIn]   *(valTable, safe)*,
   Revise   [matValPair]   *(valTable)*.

Jane[openShop, closeShop, pieceToWind, pieceToSafe, toDisco] *(safe, window)* :=
constraints
   CycleA   [openShop, pieceToWind, pieceToSafe, closeShop, toDisco],
   Move   [pieceToWind]   *(safe, window)*,
   Move   [pieceToSafe]   *(window, safe)*.

Mary [openShop, closeShop, cashUnlock, cashLock, pieceOut, toTheater]
   (cash, window, valtable) :=
constraints
   CycleB   [openShop,cashUnlock,pieceOut,cashLock,
                                         closeShop,toTheater],
   Sell   [pieceOut]   *(cash, window, valtable)*.

Peter [piece, reset] :=
variables   Bag = { }, currExp = 0, currWei = 0.
constraints
   GetPiece   [piece]   *(Bag, currExp, currWei)*,
   Reset   [reset]   *(Bag, currExp, currWei)*.

Paul [piece, refill] :=
variables   funds = 100
constraints
   Buy   [piece]   *(funds)*,
   Refill   [refill]   *(funds)*.

CycleA [openShop, loopAct1, loopAct2, closeShop, act3] :=
variables   s0 = 1;   s1, s2 = 0.

constraints

Arrow	[openShop]	(s0, s1),
Loop	[loopAct1]	(s1),
Loop	[loopAct2]	(s1),
Arrow	[closeShop]	(s1, s2),
Arrow	[act3]	(s2, s0).

CycleB [openShop, cashUnlock, pieceOut, cashLock, closeShop, toTheater] :=
variables   s0 = 1;   s1, s2, s3, s4 = 0.
constraints

Arrow	[openShop]	(s0, s1),
Arrow	[cashUnlock]	(s1, s2),
Loop	[pieceOut]	(s2),
Arrow	[cashLock]	(s2, s3),
Arrow	[closeShop]	(s3, s4),
Arrow	[act3]	(s4, s0).

The behaviour of *Mike* is described as the composition of one compound constraint (*CycleA*) dealing with the pure ordering of his actions, and two elementary constraints (*Revise* and *Accept*) dealing with the relations between some of his actions and the data structures that he handles. A similar separation of concerns is applied in the specifications of *Mary* and *Jane*.

Blocks *CycleA* and *CycleB* are specified in terms of elementary constraints *Arrow* and *Loop*, and can be readily understood as condition/event Petri nets, with variables $s0$-$s4$ playing the part of places, and actions *openShop* etc. playing the part of transitions. The *Arrow* and *Loop* active predicates, which express the 'token game' associated with transition firing, and the other active predicates of the co-specification, are defined below.

Arrow[trans](inPlace, outPlace) :- inPlace = 1,   outPlace = 0,
                                        inPlace' = 0,   outPlace' = 1.

Loop[trans](place) :-   place = 1, place' = 1.

Accept [pieceIn]*(valTable, safe)* :- pieceIn = (wei, mat, val),
                                            val   = wei * valTable(mat),
                                            safe'   = safe ∪ {(wei, mat)}.

Revise [matValPair] *(valTable)* :-   matValPair = (mat, unitVal),
                                           valTable'   = valTable ⊕ (mat, unitVal).

Move [piece] *(source, destination)* :- piece ∈ source,
                                              source'   = source - {piece},
                                              destination'= destination ∪ {piece}.

Sell [pieceOut]*(cash, window, valtable)* :-

```
pieceOut          = (wei, mat, price),
(wei, mat)        ∈ window,
price             = 1.3 * wei * valTable(mat),
window'           = window - {(wei, mat)},
cash'             = cash + price.
```

GetPiece [piece] *(Bag, currExp, currWei)* :-
```
piece             = (wei, mat, price),
currExp           < TargetExp,
currWei + wei     ≤ MaxWei,
Bag'              = Bag ∪ (wei, price),
currExp'          = currExp + price,
currWei'          = currWei + wei.
```

Reset   [reset] *(Bag, currExp, currWei)* :-
```
currExp           ≥ TargetExp,
Bag'              = {},
currExp'          = 0,
currWei'          = 0.
```

Buy [piece] *(funds)* :-
```
piece             = (_, _, price),
price             ≤ funds,
funds'      = funds - price.
```

Refill [delta] *(funds)* :-   funds' = funds + delta.

In co-notation we do not prescribe a specific syntax for the body of elementary predicate definitions; what we need is essentially first-order predicate calculus. We adopt a convention also used in the Z notation [S90], namely the primed decoration of state variables (e.g. s') for expressing the value of the variable after the occurrence of the (unique) action upon which the active predicate insists. In the definitions above we have also skipped quantifiers, as done in Prolog: for example, writing 'pieceIn = (wei, mat, val)' is equivalent to writing:

$\exists$ wei: weight, mat: material, val: value .
      pieceIn   = (wei, mat, val).

Note that in the body of active predicates, the conceptual distinction between action and state variables (in primed and unprimed forms) is lost: they are all treated as logical variables.

Assessment

In our opinion, the major advantage of the co-notation over LOTOS is that one can let the formal specification reflect very closely the conceptual associations among actions, state variables and 'agents' expressed in the original informal description. By comparing Figures 1 (LOTOS) and Figure 4 (co-notation) it is clear that the latter is closer to the informal description of Section 2: the five blocks *Mike, Mary,*

Jane, Peter and *Paul* of the co-specification correspond precisely, *up to referenced actions and state variables*, to as many bullets in the informal description, with the only exception that the actions *openShop, closeShop, toTheater* and *toDisco* were factored out, in the latter, in a kind of preliminary descriptive bullet.

We may strengthen this point by resuming the small abstract example of Figure 2, which depicted in a flat manner the mutual links between three actions and three data items. In co-notation we have maximum freedom in grouping these links: the grouping achieving the finest fragmentation of the data space would be described by the co-expression 'B1[a, b](s), B2[a, b, c](t), B3[b, c](u)', while the finest fragmentation of the data space would be represented by 'C1[a](s, t), C2[b](s, t, u), C3[c](t, u)'.

6. Conclusions

We have compared two specification techniques for concurrent systems, namely LOTOS, and our co-notation, both of which support a constraint-oriented specification style. Our comparison was focused on the flexibility offered by these techniques in expressing the links among actions and state/data variables, their groupings into constraints, or partial views on the global system behaviour, and constraint composition, as found in the informal description of the system. We have shown that only the co-notation fully supports this notion of flexibility, mainly due to the fact that its conjunction of constraints encompasses the sharing both of common actions and of common state variables, thus offering maximum freedom in fragmenting the global picture into partial system views.

If we now relax the emphasis on the links between actions and variables, and understand the term 'expressive flexibility' in a more general sense, then we should probably assign to LOTOS a higher flexibility score. Besides parallel composition, which corresponds both to conjunction (synchronisation) and to disjunction (interleaving) in co-notation, LOTOS behavioural operators include *choice, enabling* and *disabling*: these operators do correspond to elementary behavioural patterns which are often found in the informal description, and which the co-notation cannot describe directly. For enhancing the flexibility of the co-notation also in this respect one should investigate the extent to which these behavioural patterns can be built on top of the available constraint composition operators -- a subject for further research.

We have experimented with a few support tools for the co-notation, which include facilities for the graphical editing and the interpretation of co-specifications. Not surprisingly, it proved quite convenient to develop these tools in Prolog: the bodies of the elementary constraints can be expressed directly in that language, while those of the compound constraints can be expanded in a rather straightforward manner into Prolog code, with direct exploitation of the Prolog operators of conjunction (',') and disjunction (';'). A further, attractive track of research and development would be the investigate the convenience of other logic programming languages (e.g., parallel logic programming [Gr87]) as implementation supports for the co-notation.

References

[B89]   E.Brinksma (ed.), ISO - Information Processing Systems - Open Systems Interconnection - LOTOS - A formal description technique based on the temporal ordering of observational behaviour, ISO 4, February 1989, ISO, Geneva.

[BA96]   T. Bolognesi, F. Accordino, 'Constraint-oriented specification style and notation', Technical Rep. IEI B4-43, dec. 96.

[BB87]   T. Bolognesi, E. Brinksma, "Introduction to the ISO Specification Language LOTOS", *Computer Networks and ISDN Systems*, Vol. 14, No. 1, North-Holland, 1987, pp. 25-59.

[BC94]   T. Bolognesi, G. Ciaccio, 'Cumulating constraints on the *when* and the *what*', in R. L. Tenney, P. D. Amer, M. U. Uyar (editors), *Formal Description Techniques VI*, IFIP Transactions C-22, Proceedings of the IFIP TC6/WG6.1 Sixth International Conference on Formal Description Techniques - FORTE'93, North-Holland, 1994, pp. 433-450.

[G87]   H. J. Genrich, 'Predicate/Transition Nets', LNCS N. 254, Springer-Verlag, 1987, pp. 207-247.

[Gr87]   S. Gregory, *Parallel Logic Programming in PARLOG*, Addison-Wesley, 1987.

[LL90]   L. Lamport, N. Lynch, 'Distributed Computing: Models and Methods', Chapter 18 of J. van Leeuwen (ed.), *Handbook of Theor. Comp. Science, Vol. B - Formal Models and Semantics*, Elsevier - MIT Press, 1990.

[R85]   W. Reisig, *Petri Nets - An Introduction*, EATCS Monographs on Theoretical Computer Science, Vol. 4, Springer-Verlag 1985.

[S90]   M. Spivey, *The Z Notation - A Reference Manual,* Prentice Hall, 1990.

[VSvS88] C. A. Vissers, G. Scollo, M. Van Sinderen, 'Architecture and specification style in formal descriptions of distributed systems', in S. Aggarwal and K. Sabnani editors, *Protocol Specification, Testing, and Verification VIII*, North-Holland, 1988, pp. 189-204.

19

Objects as Abstract Machines

Simone Veglioni
Programming Research Group, Oxford University
Computing Laboratory, Wolfson building, Parks road, Oxford, U.K.
Tel: +44 1865 273838, Fax: +44 1865 273839
e-mail: veglioni@comlab.ox.ac.uk

Abstract
Most of the attempts aiming at providing a foundation to the object paradigm
are based on processes, sometimes with a high degree of success. However, so
far there is no standard model of objects as processes, thus justifying the exploration of other approaches. As an alternative model for objects behaviour,
we propose *Hidden Abstract Machine*, a framework which enriches the hidden
paradigm (Goguen & Diaconescu 1994, Malcolm & Goguen 1994, Goguen &
Malcolm 1997) by providing an observational semantic and interaction mechanisms.

Keywords
abstract machine, behavioural equivalence, hidden algebra, OBJ, algebraic
specification.

1 INTRODUCTION

The object paradigm, consisting of specification, programming and refinement, in the context of objects, classes, inheritance and concurrency, is the
most widely used paradigm in software engineering, even though, in spite of
the huge interest it raises, it still lacks of a standard model. For example, the
view of objects as processes, even though receives a wide consensus, has not
reached definitive results.

A homogeneous integration of static and dynamic aspects of the paradigm is
intrinsically difficult, but we think the choice of processes as model for objects
behaviour might not easy the problem. In fact, we realized that by founding
the approach on so-called Hidden Abstract Machines (HAM), abstract entities
which encapsulate the state and embody abstract data types, this integration
becomes relatively easier.

Roughly speaking, a HAM (or, simply, an abstract machines) can be viewed
as a black box having a hidden state, displays (i.e., attributes) giving details
of the state and buttons (i.e., methods) providing services, thus changing the
state:

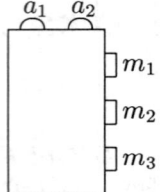

Figure 1 A graphical representation of a HAM with two attributes and three methods.

The notion of equality of two states is a consequence of their hiddenness: two states are equal if and only if they cannot be distinguished by any experiment, where an experiment is an execution of a series of methods followed by the request of an attribute.

Interaction between abstract machines is realized by means of a 'button' which, once pushed, pushes at the same time buttons of different machines. We will see how both synchronous and asynchronous communications can be expressed.

Abstract machines extend the hidden paradigm (Goguen & Diaconescu 1994, Malcolm & Goguen 1994, Goguen & Malcolm 1997), an algebraic and declarative specification framework supporting behavioural equivalence, to the extent that they advocate an *observational semantics*, which is obtained by combining behavioural semantics with reachability of states. As seen, interaction mechanisms are also defined.

In this paper, due to rigid space limits, we may only review the hidden paradigm in Section 2 and briefly introduce abstract machines in Section 3. In Section 4 we explain interaction of abstract machines only by example, without introducing the theory, which can be found in Veglioni (1997).

A little familiarity with Many-Sorted Algebra (Ehrig & Mahr 1985, Goguen 1997) can be of help in going through the technicalities of the theory, but is not needed to grasp the concepts.

ACKNOWLEDGEMENTS

I would like to express my gratitude to my supervisor Grant Malcolm for his constant help, and to Rocco de Nicola for many helpful discussions.

2   THE HIDDEN PARADIGM

One of the underlying concepts of the hidden paradigm is that in specifying objects we should distinguish data, visible and immutable, from states, hidden and mutable. Therefore, by continuing a long tradition on final algebra and behavioural equivalence (Reichel (1981) and Goguen & Meseguer

(1982), for example), data-values are dealt with visible sorts and equational logic, whereas states are dealt with hidden sorts and behavioural satisfaction. Attributes are visible-valued operations that are monadic on states, while methods by hidden-valued operations that are monadic on states. Two states are (behaviourally) equivalent if all their attributes have same values, and, after an application of any sequence of methods, their attributes still have same values.

More formally, if we assume that the data-values D have been decided upon, we can also assume that (V, Ψ) is a many sorted theory, and D is a fixed (V, Ψ)-algebra such that for each $d \in D_v$ with $v \in V$ there is some $\psi \in \Psi_{[],v}$ such that ψ is interpreted as d in D; for simplicity, we can assume that $D_v \subseteq \Psi_{[],v}$ for each $v \in V$. We call (V, Ψ, D) the UNIVERSE OF DATA VALUES.

Definition 1 A HIDDEN SORTED SIGNATURE (over (V, Ψ, D)) is a pair (H, Σ), where H is a set of hidden sorts, disjoint from V, and Σ is an $(H \cup V)$-sorted signature, such that:

- each $\sigma \in \Sigma_{w,s}$ with $w \in V^*$ and $s \in V$ lies in $\Psi_{w,s}$, and
- each $\sigma \in \Sigma_{w,s}$ has at most one element of w in H.

If $w \in S^*$ contains a hidden sort, then $\sigma \in \Sigma_{w,s}$ is called a METHOD if $s \in H$ and an ATTRIBUTE if $s \in V$. Then, call those hidden valued operations with no hidden sort in their domain (GENERALIZED) CONSTANTS.

A HIDDEN SORTED THEORY (or SPECIFICATION) is a tuple (H, Σ, E), where (H, Σ) is a hidden sorted signature and E is a set of Σ-equations; we may abbreviate this to (Σ, E) if the context permits.
□

The following example may help to clarify this definition. The code uses the syntax of OBJ3 (Goguen et al. 1996), where the semantics assumed is behavioural semantics, a loose semantics which we will soon explain formally.

Example 2 A 2-place buffer of natural numbers will have attributes to say whether it is empty (`empty`), whether it is full (`full`), and which value is in first position (`val`). It will have methods to insert in last position (`in`) and to extract in first position (`out`). If the buffer is full it will not accept any other element. If it is empty the attribute `val` gives a distinguished value (`#`). In the initial state the buffer is empty. The universe of data-values (`DATA`) will have a sort `Bool` for the booleans, a sort `Nat#` for the natural numbers plus a distinguished error element, and a (sub-)sort `Nat` for the natural numbers. In the following, `==` is an OBJ built-in boolean operation, checking (operationally) equality of terms. Notice that the data-universe is an initial algebra. This means that we could specify a buffer over whatever computable data type (Begstra & Tucker 1987).

```
obj 2BUFF is                    *** name of the buffer
  pr DATA .                     *** universe of data-values.
  sort h .                      *** (hidden) sort of states
  op init : -> h .              *** initial state

  op val : h -> Nat# .          *** at: value of first element
  op empty : h -> bool .        *** at: is the buffer empty ?
  op full : h -> bool .         *** at: is the buffer full ?

  op in : h Nat -> h .          *** me: put an element inside
  op out : h -> h .             *** me: take an element outside

  var H : h . var N, N' : Nat .

  eq val(init) = # .
  eq full(init) = false .
  eq empty(H) = val(H) == # .

  cq in(H,N) = H if full(H) .
  eq val(in(init,N)) = N .

  eq full(in(init,N)) = false .               *** one el. stored
  cq full(in(H,N)) = true if not empty(H) .
  eq val(in(H,N)) = val(H) if not empty(H) .

  eq out(H) = H if empty(H) .
  eq out(in(init,N)) = init .
  cq out(in(in(init,N),N')) = in(init,N') .  *** 2nd -> 1st
endo .
```

□

In order to understand which are the models of 2BUFF, we must formally define behavioural satisfaction. As seen, states are behaviourally equivalent if they "look the same" under every "experiment". Formally, such an "experiment" is given by a *context*, which is a term of visible sort having one free variable of hidden sort. This motivates the following:

Definition 3 Given a hidden sorted signature (H, Σ) and an $(H \cup V)$-sorted set X of variable symbols, then a Σ-CONTEXT is a visible sorted Σ-term having a single occurrence of a new variable symbol z. Call such a context APPRO-PRIATE for a term t iff the sort of t matches the sort of z, and write $c[t]$ for the result of substituting t for z in the context c. We let $T_\Sigma[z]$ denote the set of contexts containing z.

A Σ-algebra A BEHAVIOURALLY SATISFIES a Σ-equation $(\forall X)\ t = t'$ iff A satisfies each equation $(\forall X)\ c[t] = c[t']$ where c is an appropriate Σ-context; in this case, we may write $A \models (\forall X)\ t = t'$.

Similarly, A BEHAVIOURALLY SATISFIES a conditional equation e of the form $(\forall X)\ t = t'$ if $t_1 = t'_1, ..., t_m = t'_m$ iff for every interpretation $\theta : X \to A$, we have* $\theta^{\#}(c[t]) = \theta^{\#}(c[t'])$ for all appropriate contexts c whenever, for $j = 1, ..., m$, $\theta^{\#}(c_j[t_j]) = \theta^{\#}(c_j[t'_j])$ for all appropriate contexts c_j. As with unconditional equations, we write $A \models e$.
□

We have all the ingredients now to make the idea of model precise:

Definition 4 Given a hidden sorted theory (H, Σ, E), then a MODEL is a Σ-algebra A such that:

- $A|_{\Psi} = D$, and
- each equation in E is behaviourally satisfied.

A model of (H, Σ, E) is also called a (H, Σ, E)-ALGEBRA, a HIDDEN (Σ, E)-ALGEBRA, or even (Σ, E)-algebra if the behavioural context is clear. A Σ-algebra that satisfies the first condition is called a (H, Σ)-ALGEBRA, or a HIDDEN Σ-ALGEBRA. A HIDDEN Σ-HOMOMORPHISM is a Σ-homomorphism being an identity on visible carriers.
□

Notice that models of a hidden theory can be significantly different because, in a sense, they are constrained only the minimum necessary (in fact, their behaviour is constrained). For example, the equation out(H) = H if empty(H) does not constrain init and out(init) to be equal in all models, but only to appear the same under all experiments. In fact a history model, which is a model keeping complete history of interactions, thus distinguishing all states, is always allowed. This gives a high degree of abstraction to hidden theories.

Contrary to the visible case, a hidden theory may have no models. For example, if we add the equation empty(init) = false to 2BUFF, then we can prove that true = false which is in contrast with the first condition of Definition 4. Therefore, the theory would have no models.

Definition 5 A hidden theory is CONSISTENT iff it has at least one model.
□

The equation empty(init) = true in 2BUFF *constrains* the initial state to be empty, but sometimes, we can even leave a state completely unconstrained.

---

*Here, $\theta^{\#}$ is the unique Σ-homomorphic extension of θ.

For example, if we do not know, at the moment of writing 2BUFF, how is the initial state, we leave `init` unconstrained. We will see how this is related to non determinism (see Section 3.1).

Definition 6 A CONSTRAINT is a visible equation with at least a term (not in the condition) having a hidden subterm.
□

Notice that a state may be constrained by several equations, and that determining whether a set of constraints has a solution may be arbitrary difficult, even unsolvable. Therefore, also determining whether a hidden theory is consistent may be arbitrary difficult.

Proving behavioural equivalence requires, in general, context induction, which may be very awkward sometimes, but Malcolm and Goguen (Malcolm & Goguen 1994, Goguen & Malcolm 1997) have recently defined a powerful coinduction technique which neatly reduces the number of contexts to be checked and significatively simplify proofs. This result is based on the fact that behavioural equivalence is the greatest congruence (in much the same way as bisimilarity is the greatest bisimulation).

3   HIDDEN ABSTRACT MACHINES

In our opinion, hidden specifications do not really capture the notion of states, but rather the notion of hidden data. We believe, in fact, that states are *reachable*, i.e., they have been generated, thus having an origin. To this extent, streams (infinite lists) are not states because they do not have an origin and should be better defined as hidden data-values. Another crucial aspect in the specification of a state space is that models must consist of reachable entities, or equivalently, may contain unreachable entities, but then, they must all be considered as equivalent (because unobservable). This would allow to consider 2BUFF (Example 2) equivalent to the same specification but with `eq out(init) = init`   instead of   `ceq out(H) = H if empty(H)`. This is the very notion of observational equivalence, modeling indistinguishability, and this is why behavioural equivalence is not observational.

Hidden Abstract Machines model states by considering particular hidden specifications, and by combining behavioural equivalence with reachability, thus bringing the hidden paradigm closer to the automata tradition and to coalgebraic approaches like Jacobs (1996).

Definition 7 A (HIDDEN) ABSTRACT MACHINE — (H)AM — is a hidden-sorted theory (Σ, E) with only one hidden sort h and a hidden constant $init :\to h$, representing the initial state. Furthermore, every hidden constant, $c :\to h$, must be init-reachable: the equation $(\forall\emptyset)\ c = t$ must be in E, where t is a h-sorted ground term not-containing hidden-constants other than $init$.

Also parametric states, $p(V_1, ..., V_n)$, where $p : v_1...v_n \rightarrow h$, must be *init*-reachable: the equation $(\forall\{V_1:v_1, ..., V_n:v_n\})\ p(V_1, ..., V_n) = t$ must be in E, where t is a term not-containing hidden constants other than *init*.
□

It might be necessary to remark that, differently from processes, HAM are purely *reactive*: they continuously need stimuli from the environment to provide services.

As told, observational equivalence for states in abstract machines is obtained by combining behavioural equivalence with reachability of states: those hidden algebras whose reachable part verifies the specification must be considered as models.

We recall that:

Definition 8 An element $a \in A_s$ of a Σ-algebra A is REACHABLE iff $\exists\ t \in T_{\Sigma,s}$ such that $t_A = a$; otherwise it is UNREACHABLE. The REACHABLE SUB-ALGEBRA rA OF A consists of the carriers of A without unreachable elements, and the operations of A restricted to rA.
□

Notice that, in the context of abstract machines, reachability becomes *init*-reachability, thus capturing the automata notion of reachability.

Definition 9 Given a hidden signature Σ and a hidden Σ-algebra A, then OBSERVATIONAL-EQUIVALENCE on A, denoted $\equiv^{\mathcal{O}}$, is defined as follows, for $a, a' \in A_s$:

- $a \equiv^{\mathcal{O}}_s a'$ iff $c_A(a) = c_A(a')$, for all contexts c when both a and a' are reachable

- $a \equiv^{\mathcal{O}}_s a'$ otherwise.

Notice that when s is a visible sort then $a \equiv^{\mathcal{O}} a'$ iff $a = a'$.
□

Definition 10 A hidden Σ-algebra A OBSERVATIONALLY SATISFIES a (conditional) Σ-equation e of the form $(\forall X)\ t = t'$ if $t_1 = t'_1, ..., t_m = t'_m$, written $A \models^{\mathcal{O}} e$, iff for every interpretation $\theta : X \rightarrow A$, we have $\theta^\#(t)_A \equiv^{\mathcal{O}} \theta^\#(t')_A$ whenever, for $j = 1, ..., m$, $\theta^\#(t_j)_A \equiv^{\mathcal{O}} \theta^\#(t'_j)_A$.

Given an abstract machine $M = (\Sigma, E)$, a hidden Σ-algebra A is a (HIDDEN) CONCRETE MACHINE of M iff $A \models^{\mathcal{O}} E$.

Homomorphisms between concrete machines are hidden homomorphisms.
□

This definition models precisely the fact that concrete machines are those hidden algebras whose reachable part behaviourally satisfies the theory:

Proposition 11 Given a hidden theory (Σ, E) and a Σ-model A then $A \models^{\mathcal{O}}$ E iff* $rA \models E$.
□

For observational equivalence a coinduction technique has been defined (Veglioni 1997).

Let us see now how a coffee dispenser can be specified with an abstract machine. This example shows how to express "non availability" of services in some states, typical of process algebra:

Example 12 A coffee dispenser is a vending machine having attributes giving the number of coins inserted by the last client (money), the quantity of coffee available (level) and the total amount of money stored (amount); and methods to insert a coin (coin), to get a coffee (coffee), to fill in the coffee container (fill) and to collect all coins stored (collect). Only one coin is needed to get a coffee; and if there is no coffee available or the client has already inserted a coin (without getting a coffee) a coin is not accepted. The procedure of inserting a coin - getting a coffee cannot be interrupted, i.e., after having inserted a coin no service can be provided except that of giving a coffee. In the initial state the machine is empty.

```
obj COFFEEDIS is
  pr NAT .
  sort h .
  op init : -> h .

  op money : h -> Nat .
  op amount : h -> Nat .
  op level : h -> Nat .

  op coin : h -> h .
  op coffee : h -> h .
  op collect : h -> h .
  op fill : h -> h .

  var H : h . var N : Nat .
                              *** defn of init
  eq money(init) = 0 .
  eq amount(init) = 0 .
  eq level(init) = 0 .
```

---

*This proposition does not hold if E contains a conditional equation in which a quantified variable does not actually occur. However, this is not a restriction, because it is already required by term-rewriting.

```
                                        *** defn of coin(H)
eq coin(H) = H if level(H) == 0 or money(H) == 1 .
cq money(coin(H)) = 1 if level(H) > 0 .
eq amount(coin(H)) = amount(H) .
eq level(coin(H)) = level(H) .
                                        *** defn of coffee(H)
cq coffee(H) = H if money(H) == 0 .
eq money(coffee(H)) = 0 .
cq amount(coffee(H)) = amount(H) + 1 if money(H) == 1 .
cq level(coffee(H)) = level(H) - 1 if money(H) == 1 .
                                        *** defn of collect(H)
cq collect(H) = H if money(H) == 1 .
cq amount(collect(H)) = 0 if money(H) == 0 .
eq money(collect(H)) = money(H) .
eq level(collect(H)) = level(H) .
                                        *** defn of fill(H)
cq fill(H) = H if money(H) == 1 .
cq level(fill(H)) = 50 if money(H) == 0 .
eq money(fill(H)) = money(H) .
eq amount(fill(H)) = amount(H) .
endo .
```

Fourth axiom says that if there is no coffee in the machine or a coin has just been inserted, no coin is accepted (inserting a coin does not change the state), i.e., the service is not available, or in a process algebra terminology, the action coin cannot be performed.

□

After having explained how to express non availability of services, here is the straightforward definition of a terminal state:

Definition 13 A state $s \in T_{\Sigma,h}$ of an abstract machine $M = (\Sigma, E)$ is TER-MINAL iff for all methods m of Σ we have: $M \models^{\mathcal{O}} m(s) = s$.

□

3.1 Non-determinism

Non-determinism in the hidden paradigm is *underspecification*, and is faithful to software engineering practice of system development to the extent that all functions in a program are deterministic, but at a given stage of development there may still be many programs that can satisfy a given specification. This form of non-determinism is appropriate with refinement.

Definition 14 Given an abstract machine (Σ, E), a ground hidden Σ-term s (a state) is SPECIFIED iff for every attribute a (of appropriate sort), there exists exactly one $(d \in D)$ such that $E \models^{\mathcal{O}} a(s) = d$. It is UNDERSPECIFIED iff for some attribute a there is no d such that $E \models^{\mathcal{O}} a(s) = d$.
□

Definition 15 A consistent abstract machine (Σ, E) is DETERMINISTIC if every state $s \in T_{\Sigma,h}$ is specified; otherwise, it is NON-DETERMINISTIC.
□

COFFEEDIS is deterministic because all states are specified, but at an earlier stage of its development, the size of the coffee container could not be known, so neither the value of level(fill(init)) for example. This does not mean that it could take more than one value in a given model because each model is deterministic and will assign a particular value (of sort Nat) to an undefined state. At a later stage more information could be known: for example that its capacity was less than 100. This could be expressed by inserting the constraint level(fill(init)) < 100. It can be shown that adding this constraint, and then the constraint level(fill(init)) = 50 gives rise to two refinements (refinement for HAM is addressed in Veglioni (1997)).

Non-determinism in abstract machines also resembles that in constraint programming: a specification describes the states that an object could possibly have, but what states actually occur is co-determined with other objects through their interaction. For example a buffer X can receive whatever natural number between 1 and 5, but if the input is linked with the output of another buffer Y which only accepts even natural numbers, than X may actually receive only 2 and 4. Interaction is addressed in the following section.

Lemma 16 Given a deterministic abstract machine $M = (\Sigma, E)$ and a homomorphism between two concrete machines $h : A \to B$, we have that $A \models^{\mathcal{O}} e$ iff $B \models^{\mathcal{O}} e$ for all (conditional) Σ-equations e.
□

This theorem shows the aimed result that a deterministic abstract machine denotes a class of observationally equivalent models (though they can be significantly different, as seen).

We show now how internal choice can be expressed with underspecification:

Example 17 Let us consider an abstract machine having two methods:
```
op p : h -> h .               *** a method (i.e. an action)
op q : h -> h .               *** a method (i.e. an action)
```
If we add an operation (in in-fix notation), an attribute and two axioms:
```
op _ or _ : h h -> h .        *** internal choice
op left? : h -> bool .        *** underspecified attribute
```

```
ceq p(H) or q(H) = p(H) if left?(H) .
ceq p(H) or q(H) = q(H) if not left?(H) .
```

we perfectly express internal choice. In particular, every model will assign a boolean value to `left?(H)` for all states H. Different (deterministic !) models may assign different values, thus realizing different choices. Notice that the choice of executing p or q in a particular state is made by the models and does not appear at the specification level. This is the very notion of internal choice. The non-atomicity of _ or _ (cfr. Definition 1) can be accepted because the operation is *derived* (Veglioni 1997).

□

This gives us confidence that the hidden paradigm is not less expressive than process algebras. A comparative study will appear soon.

4   COMPOSITION OF ABSTRACT MACHINES

In the specification of distributed systems we need to compose the specifications of components. So far we have seen how to specify single components as abstract machines, now we see how to compose abstract machines. Unfortunately, there is no room here to show the theory underlying composition (which can be found in Veglioni (1997)), therefore we try to present all the concepts by means of examples.

In general, there are two kinds of composition: a composition in which components do not interact, which we call *Independent Composition*, and a composition in which the components interact somehow, which we call *Interactive Composition*.

Intuitively, if two abstract machines which do not interact are composed, they should be regarded as a unique machine, having the union of their attributes and methods:

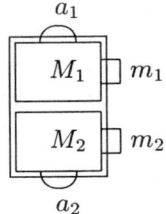

Its behaviour should be the product of the two behaviours, that is, if s_1 and s_2 are the current states of two machines M_1 and M_2, then the current state of their composition is $s = (s_1, s_2)$. And if a_1 and m_1 are respectively an attribute and a method of M_1, we will have: $m_1(s) = (m(s_1), s_2)$ and $a_1(s) = a_1(s_1)$ (symmetrically for M_2). This expresses non-interference of the two machines.

These two abstract machines interact if to their composition we add a "button" which, when pushed, pushes at the same time buttons of different machines:

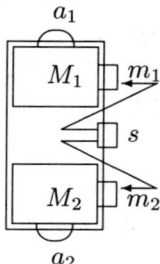

In this case s is an *interaction method* which establishes a communication line between M_1 and M_2 through m_1 and m_2. In principle, s realizes asynchronous communications because its execution does not depend on the availability of m_1 and m_2

However, a synchronous connection can be realized if the specification of s is as follows:

$s(H) = m_1(m_2(H))$ *if* "both m_1 and m_2 are available"
$s(H) = H$ *if* "at least one is not available"

The alternating bit protocol can be specified (Veglioni 1997) without explicitly build the channels, by defining asynchronous lossy interaction methods in a way like that below, where $TRAN$ is the interaction method which transmits a message from the sender to the receiver. The message can be lost, received or duplicated. *send* (respectively *rec*) is the method in the sender (receiver) to send (receive) the message. The attribute *choice* is an underspecified attribute (cfr. Example 17) which may return 0 (the message is lost), 1 (the message is received) and 2 (the message is duplicated).

$TRAN(H) = send(H)$ *if* $choice(H) = 0$
$TRAN(H) = rec(send(H))$ *if* $choice(H) = 1$
$TRAN(H) = rec(rec(send(H)))$ *if* $choice(H) = 2$

In a similar way the lossy transmission of acknowledgements can be realized.

Notice that interaction can be generalized to more than two machines. Notice also that methods forming a synchronization can still be executed separately (for example to form new synchronizations).

The simplicity of this interaction method gives us many advantages, such as distributivity of proofs (Veglioni 1997), with relevant computational benefits. We believe it is powerful enough as to express a broad class of interactive aspects of distributed systems.

4.1 Example of Independent Composition

Let us define two buffers of capacity one following same principles used for 2BUFF:

```
obj 1BUFF1 is
  pr DATA .
  sort h .
  op init : -> h .

  op val1 : h -> Nat# .
  op empty1 : h -> bool .
  op in1 : -> h Nat -> h .
  op out1 : h -> h .

  var H : h . var N : Nat .

  eq val1(init) = # .
  eq empty1(H) = val1(H) == # .
  cq in1(H,N) = H if empty1(H) .
  eq val1(in1(init,N)) = N .
  eq out1(init) = init .
  eq out1(in(init,N)) = init .
endo .
```

Let 1BUFF2 be like 1BUFF1 with 1 everywhere replaced by 2.

Their independent composition 1BUFF1 | 1BUFF2 will be a two-position buffer with in1, out1, val1 and empty1 to, respectively, insert, extract, read and check in one position, and with in2, out2, val2 and empty2 to operate in the other position. The initial state init represents the empty two-position buffer. 1BUFF1 | 1BUFF2 will consist of the union of the two signatures:

```
obj 1BUFF1 | 1BUFF2 is
  pr DATA .
  sort h .
  op init : -> h .

  op val1, val2 : h -> Nat# .
  op empty1, empty2 : h -> bool .
  op in1, in2 : -> h Nat -> h .
  op out1, out2 : h -> h .
```

and the union of their axioms, plus the following ones expressing independence of attributes and methods:

```
var H : h .   vars N, N' : Nat .

eq val1(in2(H,N)) = val1(H) .          *** methods do not
eq val2(in1(H,N)) = val2(H) .          *** interfere with
eq empty1(in2(H,N)) = empty1(H) .      *** attributes of
eq empty2(in1(H,N)) = empty2(H) .      *** other machines

eq in1(out2(H),N) = out2(in1(H,N)) .   *** methods of
eq in2(out1(H),N) = out1(in2(H,N)) .   *** different
eq in1(in2(H,N),N') = in2(in1(H,N'),N) . *** machines
eq out1(out2(H)) = out2(out1(H)) .     *** commute
endo .
```

Notice that the independent composition of abstract machines *is* an abstract machine. For example, it has only one hidden sort and an initial state.

This construction is profoundly inspired by *Independent Sum* (Goguen & Diaconescu 1994), where the main difference is that we use *vertical* signature morphisms instead of *hidden* ones (Veglioni 1997), which concern encapsulation and are not appropriate to express composition in general.

4.2   Example of Interactive Composition

A sender-receiver:

can be build as an interactive composition of 1BUFF1 (sender) and 1BUFF2 (receiver), where in1 is *in*; out2 is *out*, and in2-out1 forms the interaction *trans*:

```
obj SEN-REC is
  pr 1BUFF1|1BUFF2 .

  op trans : h -> h .            *** interaction method

  var H : h .

  cq  trans(H) = in2(out1(H),val1(H)) if empty2(H) .
  cq  trans(H) = H if not empty2(H) .
endo .
```

Here, **trans** extracts the value stored in the sender (val1(H)), and transmits it to the receiver, thus forming an interaction. Notice that the value is not transmitted if the receiver is not empty, because, otherwise, the value stored in the receiver would be lost before being extracted. It can be shown (Veglioni 1997) that SEN-REC is correct, that is, it *implements* a buffer of capacity 2 (2BUFF), where the methods in the buffer to insert and to extract are, respectively, trans;in, and out;trans, where _;_ expresses sequential execution.

Notice that process algebras generally involve 'silent' actions to specify a sender-receiver.

In Veglioni (1997) we show that both independent composition and interactive composition cannot generate inconsistency. This means that composing abstract machines is safe.

5 CONCLUSIONS

In this paper we support the view of "objects as abstract machines" rather the "objects as processes", and hope to have given evidence of its feasibility.

This hope is based on the fact that abstract machines are *reactive* entities which directly embody (abstract) data types and encapsulate states. States, in fact can be observed only by means of attributes, and can be changed only by using methods. An *observational* semantics for states is given: two states are equivalent iff their difference cannot be observed with any experiment (a series of methods followed by the observation of attributes).

Formally, this is realized within the hidden paradigm (Goguen & Diaconescu 1994, Malcolm & Goguen 1994, Goguen & Malcolm 1997), by using a restricted class of hidden specifications, and by combining behavioural equivalence with reachability of states.

In this paper we also show how simple but powerful interaction mechanisms can be defined for abstract machines, thus modeling interaction of components (objects) of a distributed system. In particular, we have seen how internal non-determinism and silent actions are intrinsic in this framework.

One of the main advantages of the view of objects as abstract machines, is the abstractness of specifications and effectiveness of proof techniques, besides that of having a very precise formal semantics.

All the theory underlying abstract machines, together with a coinduction proof technique, a formalization of refinement, and the specification of the Alternating Bit Protocol, can be found in Veglioni (1997), which also shows that in many cases proving global properties of abstract machines can be reduced to proving a number of local properties, with relevant advantages.

We wish to understand more deeply the relations holding between abstract machines and process algebra. We think to be able to encode process algebra within the hidden paradigm. In doing this we probably need to use Heterogeneous Unified Algebras (Veglioni & Parisi-Presicce 1997) as universe of data-values, in order to express bounded non-determinism.

REFERENCES

Begstra, J. & Tucker, J. (1987), 'Algebraic specifications of computable and semicomputable data types', *Theoretical Computer Science* **50**.

Ehrig, H. & Mahr, B. (1985), *Fundamentals of Algebraic Specification 1: Equations and Initial Semantics*, Springer.

Goguen, J. (1997), *Theorem Proving and Algebra*, MIT.

Goguen, J. & Diaconescu, R. (1994), Towards an algebraic semantics for the object paradigm, *in* H. Ehrig & F. Orejas, eds, 'Recent Trends in Data Type Specification', Springer-Verlag Lecture Notes in Computer Science 785.

Goguen, J. & Malcolm, G. (1997), A hidden agenda. To appear.

Goguen, J. & Meseguer, J. (1982), Universal Realization, Persistent Interconnection and Implementation of Abstract Modules, *in* M. Nielsen & E. Schmidt, eds, 'Proceedings, 9th International Conference on Automata, Languages and Programming', Springer-Verlag Lecture Notes in Mathematics 140.

Goguen, J., Winkler, T., Meseguer, J., Futatsugi, K. & Jouannaud, J.-P. (1996), Introducing OBJ, *in* J. A. Goguen & G. Malcolm, eds, 'Software Engineering with OBJ: Algebraic Specification in Practice', Cambridge University Press. Also available as a technical report from SRI International.

Jacobs, B. (1996), Behaviour-Refinement of Object-Oriented Specifications with Coinductive Correctness Proofs, Technical report, Computing Science Institute, Katholieke Universiteit Nijmegen.

Malcolm, G. & Goguen, J. (1994), Proving correctness of refinement and implementation, Technical Monograph PRG-114, Programming Research Group, Oxford University.

Reichel, H. (1981), Behavioural equivalence – a unifying concept for initial and final specifications, *in* 'Proceedings, Third Hungarian Computer Science Conference', Akademiai Kiado. Budapest.

Veglioni, S. (1997), Integrating Static and Dynamic aspects in the specification of concurrent, object-oriented and distributed systems, PhD thesis, Programming Research Group, Oxford University. To appear.

Veglioni, S. & Parisi-Presicce, F. (1997), Heterogeneous unified algebras: static and dynamic classifications, institutions, non-strictness and continuity. To appear.

6   BIOGRAPHY

Simone Veglioni is a *Training and Mobility of Researchers* funded Ph.D. student at the Programming Research Group of Oxford University. From June 1997 until September 1997 he will stay at the University of Amsterdam for the *Human Capital and Mobility* project EXPRESS.

20

State Based Service Description

Barbara Paech, Bernhard Rumpe[*]
Fakultät für Informatik,
Technische Universität München
80995 Munich, Germany,
http://www4.informatik.tu-muenchen.de/

Abstract

In this paper we propose I/O^*-state transition diagrams for service description. In contrast to other techniques like for example Statecharts we allow to model non-atomic services by sequences of transitions. This is especially important in a distributed system where concurrent service invocation cannot be prohibited. We give a mathematical model of object behaviour based on concurrent and sequential messages. Then we give a precise semantics of the service descriptions in terms of the mathematical model.

Keywords

Semantics, Formal Specification, Service Description, State Transition Diagrams, Mathematical System Model

1 INTRODUCTION

The object-oriented paradigm is based on the encapsulation of data within objects. This data can only be accessed by other objects through **service calls**. We use the term **service** as a synonym for **method**. Thus, services are the major constituent for object behavior. However, looking at the different object-oriented analysis and design methods, the abstract specification techniques of services and the interplay between different services within one object still lack a precise semantics. In most cases (e.g. OMT (Rumbaugh, Blaha, Premerlani, Eddy & Lorensen 1991), UML (Booch, Rumbaugh & Jacobson 1997), Syntropy (Cook & Daniels 1994)) state transition diagrams (STD) - inspired by

---

[*]This paper originated in the SYSLAB project, which is supported by the DFG under the Leibnizpreis and by Siemens-Nixdorf.

Harels' Statecharts (Harel 1987, Harel & Gery 1996) - are used to specify the object behavior. The STD determines the sequences of object states resulting from service executions. However, services are often not atomic, since even in sequential systems service execution may involve another service execution on the same object. In distributed systems, regarding complex services which involve calls to other objects as atomic, is, in general, a too strong restriction. Objects should react concurrently to as many service calls as possible, while preserving data consistency.

Therefore, we propose to use a whole state transition diagram for the description of one service. Transitions correspond to service steps between an input and an output. Object behavior is derived from the service description by interleaving of the service steps. The service description can also be marked to indicate at which execution states interleaving of other services is allowed.

Because the details of the object behavior are quite intricate, we give a mathematical semantics to object behavior based on the framework of stream processing functions (Broy, Dederichs, Dendorfer, Fuchs, Gritzner & Weber 1993, Klein, Rumpe & Broy 1996) and I/O^*-state machines (Rumpe & Klein 1996). In particular, we distinguish sequential and concurrent services calls. This allows to define multiple threads as in Java. As we will show, sequential and purely asynchronous systems are special cases of this model.

Altogether, the paper is structured as follows: First, we introduce the used formal foundation, in particular state machines for the modeling of object behavior. In the following section, we show how to adapt this model to the above sketched communication paradigm. Then we introduce I/O^*-state transition diagrams as the abstract description technique for services. We show how to give semantics to object behavior based on the service descriptions.

2   MATHEMATICAL SYSTEM MODEL

In (Klein et al. 1996) we developed a formal model of distributed systems, based on the theory of streams (Broy et al. 1993). This mathematical system model serves as a semantical basis for several description techniques, like object models, state transition diagrams, or process diagrams, as for example given in UML (Booch et al. 1997, Breu, Hinkel, Hofmann, Klein, Paech, Rumpe & Thurner 1997).

In this section, we extend the mathematical system model to service descriptions. The model emerged from (Grosu & Rumpe 1995, Rumpe & Klein 1996, Rumpe 1996) where the underlying theory of state machines is developed. In (Grosu & Rumpe 1995) a composition of object behavior is defined.

Basic assumptions
We make three basic assumptions about the kind of systems we take into account: First, objects can only read or modify parts of the state of another object through services, even those from the same class. Second, we do not

allow more than one service to be active at the same time (however, they may be interleaved). And third, communication between objects is asynchronous such that messages must be accepted, but may be delayed (sequential programming languages correspond to the special case where only one object is active at a time and activity is transferred with service calls).

I/O^*-State machines.

In the following, we introduce the mathematical basis for state based object behavior description. An I/O^*-**state machine**[*] (S, I, O, δ, S^0) consists of a nonempty set of object states S, a nonempty set of input messages I, a nonempty set of output messages O, a transition relation $\delta \subseteq S \times I \times S \times O^*$, and a nonempty set of initial states $S^0 \subseteq S$.

None of the above given sets need to be finite. The sets of input and output messages I contain service calls and return messages, possibly with arguments. The reaction to any input is attached to the same transition. This leads to a more compact notation compared to the well-known I/O-automata[*] (Lynch & Stark 1989). The transition relation δ is allowed to be **nondeterministic**. On one hand, this is adequate for the nondeterminism inherent in distributed systems. On the other hand, nondeterminism is important to cope with **underspecification** allowing refinement of such specifications. In (Rumpe 1996, Rumpe & Klein 1996), a refinement calculus for state machines is given which defines a set of development steps to be used for specialization of object behavior during development as well as for inheritance from superclass to subclass. Because of the basic assumptions about systems, an object cannot reject a message. This corresponds to **input enabledness** of the state machine: For each source state s and input message $i \in I$, there exists at least one destination state t and reaction $o \in O^*$ with $\delta(s, i, t, o)$.

Messages and States

Object states are composed of several parts that deal with the attribute state and active or suspended service states. We assume that local variables as well as arguments are private to the service invocation they belong to.

Let the set of variables VAR and the set of corresponding values VAL be given. We abstract from the fact that variables are typed, and regard each partial mapping $VAR \rightharpoonup VAL$ as **variable assignment**. We assume that each object has a fixed set of attributes and each service a fixed set of local variables, but do not formalize these constraints here. Given an abstract set PC of **program counters**, suspended service invocations are formalized as $SI = (VAR \rightharpoonup VAL) \times PC \times ID$, where the first component contains arguments and local variables. PC is used to denote special locations in the service code, where a message is awaited and therefore computation is suspended. The third component ID denotes the caller of the service. This is the object, where a

---

[*]We call them I/O^*-**state machines**, because each transition is labeled accordingly.
[*]In our classification I/O automata would be called $I \cup O$ automata.

(possible) response is to be delivered. To handle recursion of service calls, as usual, a **stack** of service invocations is used. We assume the mathematical datatype $stack(M)$ over set M with the services $push$, pop, top and $\emptyset$ for the empty stack to be given.

If considering multiple threads, one stack is not enough. Indeed, we need a separate stack for each thread. We abstract from actual threads by the set TAG, each tag denoting a thread identifier. We incorporate a mapping $TAG \rightarrow stack(SI)$ into each object state. Messages are tagged also with elements of TAG to indicate the thread they belong to. Thus, a message is a tuple

$$(sen, rec, tt, mn, ar) \in ID \times ID \times TAG \times MSG \times (VAR \rightharpoonup VAL),$$

where sen is the sender identifier, rec is the receiver identifier, tt is the (thread) tag, mn is the message name, and ar is the argument assignment.

The set MSG contains the service names, but also a special message ret that indicates return messages. The return value (if one exists) is encoded in the arguments of the return message. We use a pool for thread tags for each object, which is used whenever a new thread is started. Each two pools of different objects are disjoint. The states of objects are

$$(at, st, po, pt) \in (VAR \rightharpoonup VAL) \times (TAG \rightarrow stack(SI)) \times \mathbb{P}(TAG),$$

where at is the attribute assignment, st is a mapping, which assigns a stack to each thread, and pt is the pool for tags. This set of states is usually infinite. Note that one can easily extend this model to object creation with an additional pool for object identifiers such that object creation is just treated as a special message.

Transitions
To model data encapsulation, there are a number of restrictions on the state changes. We shortly repeat the most important restrictions here, without giving a formal definition. The set of attributes of an object and the value of attribute *self* are immutable. The tag pool may only be diminished. No tag may be used unless removed from the pool. Only one stack is changed in a transition. Either a service invocation is added, removed or the top invocation changed. If the top one is changed, the set of arguments and their values are immutable. Only call messages can add stack elements.

So each transition of the state machine resembles a part of a service execution. If a service calls other services, awaiting their answers, it is partitioned into several transitions.

3   MULTI-THREAD COMMUNICATION

In this section, we specialize the behavior model given above to a particular model of communication allowing for service calls where activity is transferred

mn_i	mn_o	$st'(tt_i)$
$sequ$	$= ret$	$= st(tt_i)$
$sequ, conc$	$sequ$	$= push(st(tt_i), (ar_i + loc, pc, snd_i))$
$conc$	$conc$	$= st(tt_i)$
$= ret$	$= ret$	$= pop(st)$
$= ret$	$sequ$	$= push(pop(st(tt_i)), (ar_i + loc, pc, snd_i))$
$= ret$	$conc$	$= st(tt_i)$

Figure 1 Restrictions on I/O^*-state machines

(**sequential**) as well as for service calls starting a new thread (**concurrent**). This model could be specialized to purely sequential calls, as in pure C++, or purely concurrent calls. The mixed style presented here is supported in Java, and also is the most flexible for modelling purposes.

Java allows different threads to simultaneously work on the same object and therefore allows to share data. It supports synchronization concepts, but the programmer is responsible to use them correctly. We prevent shared data access by interleaving the service executions. We therefore restrict the Java programming model at this point. However, this can easily be implemented in Java using semaphores. Altogether, we distinguish between **sequential call messages** where the caller awaits the return message, **return messages** that are answers to sequential calls, and **concurrent call messages** that invoke a new thread of computation.

We assume, that no service can compute internally for ever, such that each message is processed. As discussed in (Klein et al. 1996), the communication medium of the general system model ensures that the order of messages is preserved and that message contents are not changed.

Assume a transition $\delta(s, i, t, o)$. Let $s = (at, st, pt)$ be the source state, $t = (at', st', pt')$ the destination state, $i = (snd_i, rec_i, tt_i, mn_i, ar_i)$ the input message and $o = o_1 + + \langle (snd_o, rec_o, tt_o, mn_o, ar_o) \rangle$ the sequence of output messages, where the last message plays a special role. Only the stack of the input tag tt_i may be changed. Attribute assignments may change arbitrarily. For each concurrent output message in o_1 a new tag identifier is removed from pt. Sending a concurrent message does not interrupt the active service, but sending of a sequential one does. So only the last message emitted during a transition can be sequential. The tag of a possibly emitted sequential message has to be identical to the tag of the processed message. Is the processed service a concurrent one, the last message may be sequential, but only a call not a return message. All other conditions for state changes are shown in figure 1.

With $mn = ret$ we indicate return messages, with $sequ$ sequential and with $conc$ concurrent messages. The case of empty output is subsumed under the case of only concurrent output. In the simplest case (sequ-ret) an input call is immediately handled, the stack is not changed. If the output is sequential, the current service is suspended. A concurrent output does not change the stack. The other two cases deal with input return messages, where the stack has

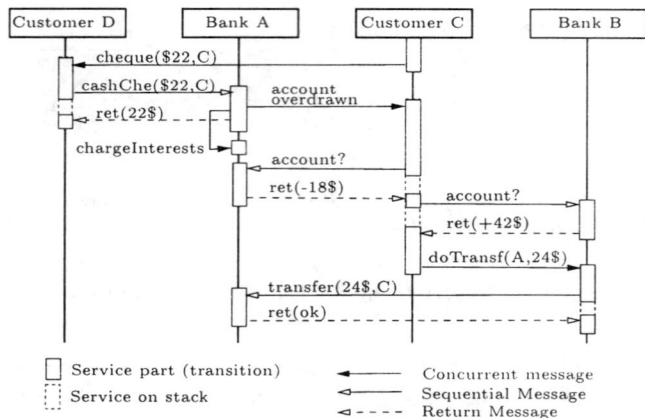

Figure 2 Bank scenario

to contain an according message invocation, which can be popped (ret-ret) or modified (ret-sequ). In case of modification an according program counter *pc* and an assignment *loc* of local variables denotes the internal state of the service invocation.

We illustrate this model by the following example (see figure 2). Assume we have two customers C and and D as well as two banks A and B. Customer C has one account per bank. B gives better interests, but A is used for payment transfers. Customer C uses a cheque for payment of customer D. In our concrete scenario, the account in bank A will be overdrawn, after D cashed the check and C gets an according request to balance. Now C is asking for the actual account at both banks and then placing an order to transfer \$24 from bank B to A. Bank B awaits the acknowledgment of A before completing the transfer.

4   SERVICE DESCRIPTION

In this section, we introduce a state based description technique for services and define object behaviour semantics in terms of I/O^*-state machines. We use an abstract version of I/O^*-state machines called I/O^*-**state transitions diagrams**. They allow for a finite description of the infinite state machines. We use state predicates to partition the state space. Similarly, we allow to abstract from the message parameters by using preconditions referring to attributes and input parameters and by using patterns for input messages. Also, we allow postconditions to describe the effect of data changes and patterns for output messages. The definition given below is a special case of the STD defined in (Grosu, Klein, Rumpe & Broy 1996), where input is restricted to a one-element sequence. Altogether, an I/O^*-state transition diagram $(att, I, O, S, \Lambda, \delta, S^0)$ consists of the set *att* of attributes, the nonempty

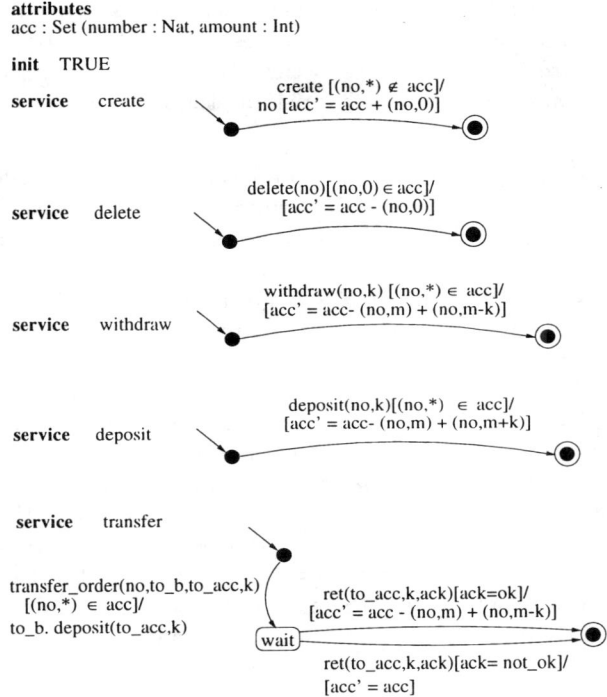

attributes
acc : Set (number : Nat, amount : Int)

init TRUE

service create

create [(no,*) ∉ acc]/
no [acc' = acc + (no,0)]

service delete

delete(no)[(no,0) ∈ acc]/
[acc' = acc - (no,0)]

service withdraw

withdraw(no,k) [(no,*) ∈ acc]/
[acc' = acc- (no,m) + (no,m-k)]

service deposit

deposit(no,k)[(no,*) ∈ acc]/
[acc' = acc- (no,m) + (no,m+k)]

service transfer

transfer_order(no,to_b,to_acc,k)
[(no,*) ∈ acc]/
to_b. deposit(to_acc,k)

ret(to_acc,k,ack)[ack=ok]/
[acc' = acc - (no,m) + (no,m-k)]

wait

ret(to_acc,k,ack)[ack= not_ok]/
[acc' = acc]

Figure 3 Bank description with I/O^*-STD for each service

set I of input messages, the nonempty set O of output messages, a nonempty, finite set of diagram states S, a mapping $\Lambda : S \to \langle Pred \rangle$ associating a predicate over the attributes att with the diagram states, a finite transition relation $\delta \subseteq S \times \langle Patt \rangle \times \langle Pred \rangle \times S \times \langle Expr \rangle \times \langle Pred \rangle$, where each transition is labelled with input pattern, precondition, output expression and postcondition, and a nonempty, finite set of initial diagram states S^0.

$\Lambda(s)$ must be satisfiable for all diagram states $s \in S$ and the predicates of two different diagram states exclude each other. Also the postcondition of a transition must be satisfiable, if the precondition is satisfied.

We call a set of diagrams describing one service each together with a predicate *init* characterizing the initial object states the **object behavior description**. As an example, consider a bank object. Figure 3 shows the object description with attributes defining the state space and with separate service diagrams for `create, delete, withdraw, deposit, transfer`.

The semantics of object behavior description is given in terms of I/O^*-state machines. Each diagram transition gives rise to a set of machine transitions satisfying the input pattern, the output pattern and the pre- and postconditions. In addition also the tags and stacks handling the interleaving of services have to be introduced. Thus, let $(att, loc_k, I_k, O, S_k, \Lambda_k, \delta_k, S_k^0)$, $k = 1, ..., n$, be a set of I/O^*-STD, where each STD describes a service over the object attributes *att* and the local service variables loc_k, and let *init* be a predicate over the attributes. The semantics of this object behavior description is any I/O^*-state machine $(\hat{S}, \hat{I}, \hat{O}, \hat{\delta}, \hat{S}^0)$ satisfying the following:

- $\hat{S} = \{\beta \in BEL : \beta.self = id$ and $\beta.tag$ associates with each $tag \in TAG$ a stack of service invocations $SI\}$, where BEL is the set of all variable assignments giving values to the attributes and some additional variables like $self$, tags and for the program counter of the currently active service. The set of service invocations SI_k is given by $(loc_k \rightharpoonup VAL, S_k, ID)$ and $SI = \bigcup_{k=1}^n SI_k$. Note that we use the states of the service STD as program counter values.
- $\hat{S}^0 = \{\beta \in \hat{S} : \beta \models init\}^*$
- $\hat{I}$ $(\hat{O})$ is derived from I (O) by using the appropriate message and parameter names and introducing the tag in the messages
- $(\beta_s, (snd_i, rec_i, tt_i, mn_i, ar_i), \beta_t, out ++ (snd_o, rec_o, tt_o, mn_o, ar_o)) \in \hat{\delta}$, if there exists $1 \leq k \leq n$, $T \in \delta_k, \beta \in BEL$ such that β satisfies the state predicates, pre- and postconditions, patterns and expressions of T (written as $\beta \models T$) and $\beta|_{att} = \beta_s$ and $\beta|_{att'} = \beta_t'$, where we use the slash notation to denote the values of the variables in the successor state, and either the stack of the tag is empty $(\beta.tag_i = \emptyset)$ and $\beta \models T$ and a new service execution is started $(\beta.pc = s \in S_k^0)$, or the stack is nonempty with program counter s on top $(first(\beta.tag_i) = (\gamma, s, id))$ and $\beta, \gamma \models T$ and the stack is handled according to section 3.

Note that with this semantics the labeling of the diagram states for the services carries a special weight: this labeling describes the set of all states the object may assume while the service is pending at that state. If the state predicate is not satisfied in a state where the pending service is to be continued, arbitrary behavior is possible (due to input enabledness). From a methodological point of view, it sometimes is necessary that services can be guarded from interleaving with other services. For example, account closure should not be possible while transfer is active. This could already be expressed using suitable preconditions and diagram state predicates such that the precondition for account closure is incompatible with the predicate labeling the wait-state of the transfer STD. However, we also allow a more direct way of specification, where diagram states may be labeled with service sets indicating the services which are not allowed to be interleaved at that state (called **exclusion sets**). With this extension, the semantics has to be adopted such that the transitions respect all exclusion sets of pending service invocations $(\pi_4(\pi_2(T)) \notin Ex(u)$ for all (γ, u, id) somewhere on some stack*).

---

*By $\beta \models init$ we denote that formula *init* is satisfied under variable assignment β
*By π_i we select the i-th component of a tuple.

5 CONCLUSIONS AND RELATED AND FUTURE WORK

We have discussed a semantic model for service execution in the context of multiple threads. We also have introduced a special kind of state transition diagrams for service description and shown how to this object behavior description can be given a precise semantics in terms of state machines taking care of different threads of activity through stacks.

Similar to SDL-92 (Braek & Haugen 1993), services are used to structure object (process) behaviour. In contrast to SDL services, the I/O^*-STD description of services makes explicit the state space of the object. This is necessary for an abstract description of service synchronization.

The major difference to Statechart-based description techniques is that we allow services to be distributed over several transitions, while usually only one transition per service is used. The latter kind of modeling is too restrictive, since not all services can be considered to be atomic (e.g. like the transfer service). In Syntropy and O-Mate, for a service additional internal events may be generated. However, a new external event may be treated only when the Statechart has stabilized, that means it has handled all the internal events generated in response to the last external event. Thus, internal events still do not allow e.g. two active transfer services.

Up to now, we have not treated nested states in I/O^*-STD. These states are very important for factoring object behavior over orthogonal sets of attributes. Since in our framework we do not allow internal events for communication between different substates, we avoid the usual difficulties of Statechart semantics (von der Beeck 1994). Thus, we do not expect any difficulties with incorporating nested states.

Another point we want to clarify in the near future is the use of refinement techniques as discussed in (Rumpe & Klein 1996). In that paper a calculus of refinement steps on STD is introduced which can be adapted to the framework here without difficulties. We will also explore this notion of refinement as a basis for an inheritance notion covering behavioral properties.

Acknowledgements
We thank our colleagues Ursula Hinkel, Peter Scholz and the anonymous referees for helpful comments.

REFERENCES

Booch, G., Rumbaugh, J. & Jacobson, I. (1997). *The Unified Modeling Language for Object-Oriented Development, Version 1.0*, RATIONAL Software Cooperation.

Braek, R. & Haugen, O. (1993). *Engineering Real Time Systems*, Prentice Hall.

Breu, R., Hinkel, U., Hofmann, C., Klein, C., Paech, B., Rumpe, B. & Thurner,

V. (1997). Towards a Formalization of the Unified Modeling Language, To appear at ECOOP'97.

Broy, M., Dederichs, F., Dendorfer, C., Fuchs, M., Gritzner, T. & Weber, R. (1993). The design of distributed systems - an introduction to FOCUS, *Technical Report TUM-I9202-2*, TU München.

Cook, S. & Daniels, J. (1994). *Designing Object Systems*, Prentice Hall.

Grosu, R., Klein, C., Rumpe, B. & Broy, M. (1996). State transition diagrams, *Technical Report TUM-I9630*, Technische Universität München.

Grosu, R. & Rumpe, B. (1995). Concurrent Timed Port Automata, *Technical Report TUM-I9533*, TU München.
    *http://www4.informatik.tu-muenchen.de/reports/TUM-I9533.html

Harel, D. (1987). Statecharts: A visual formalism for complex systems, *Science of Computer Programming* **8**: 231–274.

Harel, D. & Gery, E. (1996). Executable Object Modeling with Statecharts, *ICSE-18*, IEEE, pp. 246–257.

Klein, C., Rumpe, B. & Broy, M. (1996). A stream-based mathematical Model for distributed information processing Systems - SysLab system model, *in* E. Naijm & J. Stefani (eds), *FMOODS'96*.

Lynch, N. & Stark, E. (1989). A Proof of the Kahn Principle for Input/Output Automata, *Information and Computation* **82**: 81–92.

Rumbaugh, J., Blaha, M., Premerlani, W., Eddy, F. & Lorensen, W. (1991). *Object-oriented Modeling and Design*, Prentice-Hall.

Rumpe, B. (1996). *Formale Methodik des Entwurfs verteilter objektorientierter Systeme*, Ph.D. Thesis, Technische Universität München.
    *http://www.forsoft.de/~rumpe/Diss_Rumpe.html

Rumpe, B. & Klein, C. (1996). Automata Describing Object Behavior , *in* H. K. W. Harvey (ed.), *Specification of Behavioral Semantics in Object-Oriented Information Modeling*, Kluwer.

von der Beeck, M. (1994). A Comparison of Statecharts Variants, Vol. 863 of *LNCS*, Springer, pp. 128 – 148.

Bibliography

Dr. Barbara Paech studied Computer Science at the Technical University of Munich, Edinburgh University, and University of Pennsylvania. She received her Ph.D. in Computer Science from the Ludwig-Maximilians-University in Munich. Since 1993 she is a research assistant at the Technical University of Munich where she leads research projects on the formal foundation of software engineering as well as requirements and re-engineering.

    Dr. Bernhard Rumpe studied Computer Science and Mathematics at the Technical University of Munich. He received his Ph.D. in Computer Science from the Technical University of Munich. His research interests include the formal foundation of state-based behavioral specifications, and functional and OO programming concepts. Since 1997 he leads the SYSLAB research project on the formal foundation of OO software engineering techniques.

PART SEVEN

Subtyping and Inheritance

21

Subtyping for Distributed Object Stores

Jeannette M. Wing
Computer Science Department
Carnegie Mellon University
Pittsburgh, PA 15213 USA

1 INTRODUCTION

The programming language community has come up with many definitions of the subtype relation. The goal is to determine when this assignment

$$x\colon T := E$$

is legal in the presence of subtyping. Once the assignment has occurred, x will be used according to its "apparent" type T, with the expectation that if the program performs correctly when the actual type of x's object is T, it will also work correctly if the actual type of the object denoted by x is a subtype of T.

The question of when is one type a subtype of another is especially tricky to answer in the presence of shared mutable objects. For example, in the context of programming languages, we have the common situation where during the course of executing a program, two or more pointers reference the same object in the heap. A change to the object accessed by one pointer will be reflected in any further access to that object made through the other pointers. In Figure 1, for example, x and y are pointers of type T and subtype S, respectively, that refer to the same object; this aliasing means that a change made through y will be visible through x.

Relevant to this workshop's theme, the above aliasing situation is a special case of what happens in a distributed environment where objects are stored in persistent repositories. Generalize the notion of "pointer" to a notion of "handle," e.g., an index entry of a persistent database, a file name in a distributed file system, or a URL for the Web. Generalize the notion of a programming language's run-time heap to the notion of a persistent object store or

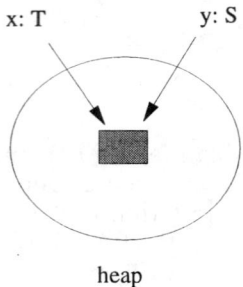

heap

Figure 1 Shared Access to Mutable Objects in a Heap

a distributed file system. Generalize the notion of multiple uses of an object during the execution of a single program to the notion of multiple users (or equivalently to multiple independent programs) that share access to the object. The situation in a distributed environment is more general since unlike a program's heap, objects live indefinitely, and do not disappear when the program terminates. Moreover, unexpected behavior that can result from one user changing an object with respect to another user's viewpoint is more likely since users may be unaware of each other's existence.

Programmers make two kinds of changes to a supertype definition when defining a subtype: they add new methods and they change old methods of the supertype. Unconstrained, however, both kinds of modifications can can lead to surprising behavior. Consider a type fat_set that has only *create*, *insert*, and *size* methods. If we were to define a subtype, set, by adding a new method, *delete*, then suddenly the fact that a fat_set object can only grow in size no longer is true, surprising users who think they have a fat_set object when it really is a set object. So, we cannot just add methods willy-nilly. Similarly, consider a plain_elephant type that has just one method, *get_color*, which always returns gray. If we were to define a subtype royal_elephant and correpondingly change the behavior (e.g., through overriding) of *get_color* to return blue, then users who think they have a plain_elephant object may see later that its color has changed. So, we cannot just change methods willy-nilly either.

What we need is a subtype requirement that constrains the behavior of subtypes so that users will not encounter any surprises:

No Surprises Requirement: Properties that users rely on to hold of an object of a type T should hold even if the object is actually a member of a subtype S of T.

The property users rely on for fat_sets is that they only grow and never shrink in size; the property users rely on for plain_elephants is that their color is always gray.

In their 1994 TOPLAS paper "A Behavioral Notion of Subtyping" Liskov and Wing [8] addressed the problem of what subtyping means, especially in the presence of shared mutable objects. They provide two alternative definitions in their paper. In this extended abstract, I summarize only one of these definitions, to highlight their main points. I then describe an object repository that the TinkerTeach Project built at Carnegie Mellon and use it to explain what relevance this subtyping notion has in practice.

2 REVIEW OF THE LISKOV AND WING NOTION OF SUBTYPING

Key to understanding the Liskov and Wing notion of subtyping is the use of the specification of an object's type. Determining when one type is a subtype of another is based on showing that certain properties hold between the two type specifications.

2.1 Type Specifications

A type specification contains the following information:

- The type's name.
- A description of the set of values over which objects of the type ranges.
- For each of the type's methods:

  - Its name.
  - Its signature, i.e., the types of its arguments (in order), result, and signaled exceptions.
  - Its behavior in terms of pre-conditions and post-conditions.

- A type constraint.

bag = **type**

uses BBag (bag **for** B)
for all b: bag
      invariant $\mid b_\rho.elems \mid \leq b_\rho.bound$
      constraint $b_\rho.bound = b_\psi.bound$

   $put =$ **proc** (i: int)
      requires $\mid b_{pre}.elems \mid < b_{pre}.bound$
      modifies b
      ensures $b_{post}.elems = b_{pre}.elems \cup \{i\} \;\wedge\; b_{post}.bound = b_{pre}.bound$

   $get =$ **proc** ( ) **returns** (int)
      requires $b_{pre}.elems \neq \{\}$
      modifies b
      ensures $b_{post}.elems = b_{pre}.elems - \{result\} \wedge result \in b_{pre}.elems \;\wedge$
             $b_{post}.bound = b_{pre}.bound$

   $card =$ **proc** ( ) **returns** (int)
      ensures $result = \mid b_{pre}.elems \mid$

   $equal =$ **proc** (a: bag) **returns** (bool)
      ensures $result = (a = b)$

end bag

Figure 2   A Type Specification for Bags

Figure 2 gives an example of a type specification for bounded bags. To the spirit of the theme of this workshop, I give formal specifications, written in the style of Larch [6], but I could just as easily have written informal specifications. Since these specifications are formal we can do formal proofs, possibly with machine assistance like with the Larch Prover [5], to show that a subtype relation holds [10].

The BBag Larch Shared Language trait and the **invariant** clause together describe the set of values over which bag objects can range. The **requires, modifies,** and **ensures** clauses specify the methods' pre- and post-conditions. The **constraint** clause specifies the type constraint.

A *type invariant* constrains the value space for a type's objects. In the bag example, the type invariant says that the number of integers stored in a bag in any state, ρ, is always less than or equal to the bag's (fixed) bound.

To ensure that the specification is *consistent*, the specifier must show that (1) each constructor of an object of the type establishes the invariant and (2) each of the type's methods preserves it. (For methodologically reasons, Liskov and Wing specify constructors separately from the other type's methods so none are shown in Figure 2.)

The inclusion of pre- and post-conditions in the specification of a type's methods allows us to relate the two types' behaviors; this is the main difference between the Liskov and Wing definition of subtyping and those that rely on just signature information (e.g., Cardelli [2]). For example, two methods with the same signature (e.g., *get* and *card* for bags) may have dramatically different behavior. Relying on just signature information identifies these methods that behave differently; thus, finer subtyping distinctions can be made when behavioral information is used in addition to signature information.

The inclusion of the *type constraint* is what distinguishes the Liskov and Wing work from all others (e.g., America [1], Cusack [3], Leavens [4, 7]) that also include some kind of behavioral information. To foreshadow what is coming in the next section: Not only must a supertype's type invariant and methods be preserved by the subtype's, but so must its type constraint.

The type constraint is intended to capture certain kinds of properties of an object that are required to remain invariant over the indefinite lifetime of the object. Liskov and Wing call them *history properties*. For example, if a window's size, an elephant's color, or a bag's bound is to remain constant, then these properties should be stated as type constraints; if the size of a set can only grow and never shrink or a counter's value only monotonically increases, then these properties should be stated as type constraints.

Stating history properties through type constraints is exactly how Liskov and Wing deal with mutable objects. Formally, a type constraint is a two-state predicate, $C(\rho, \psi)$, where ρ and ψ are any two successive states in any computation. A type constraint is similar to a type invariant except a type invariant is a single-state predicate–a property that holds in every state, ρ, of a computation. Since a type constraint is a property relating two successive states, it captures what behavior may *not* change from state to state; hence it captures additional "invariant" properties of mutable objects.

2.2   The Subtype Relation

The subtype relation is defined in terms of a checklist of properties that must hold between the specifications of the two types, S and T. Since in general the value space for objects of type S will be different from the value space for those of type T we need to relate the different value spaces; we use an *abstraction function*, A, to define this relationship. Also since in general the names of the methods of type S can be different from those of type T we need to relate which method of S corresponds to which method of T; we use a *renaming* map, R, to define this correspondence. (In a programming language like Java, this is just the identity map, as realized though method overloading.)

S is a *subtype* of T if the following three conditions hold (informally stated):

1. The abstraction function respects the invariants. If the subtype invariant holds for any subtype value, s, then the supertype invariant must hold for the abstracted supertype value $A(s)$.
2. Subtype methods preserve the supertype methods' behavior. If m is a subtype method then let n be the corresponding $R(m)$ method of the supertype.

   - Arguments to m are contravariant to the corresponding arguments to n; m's result is covariant to the result of n.
   - Any exception signaled by m is contained in the set of exceptions signaled by n.
   - n's pre-condition implies m's and m's post-condition implies n's (under the abstraction function).

3. Subtype constraints ensure supertype constraints. S's type constraint implies T's type constraint (under the abstraction function).

Why does this subtype relation guarantee that the No Surprises Requirement holds? Recall that the Requirement refers vaguely to "properties." What this definition of subtype guarantees is that certain properties of the supertype–those that are stated explicitly or provable from a type's specification–are preserved by the subtype. The first and third conditions directly relate the invariant and history properties; the second condition relates the behaviors of the individual methods, and thus preserves any observable behavioral property of any program that invokes those methods. Though the Liskov and Wing paper focuses on only these kinds of

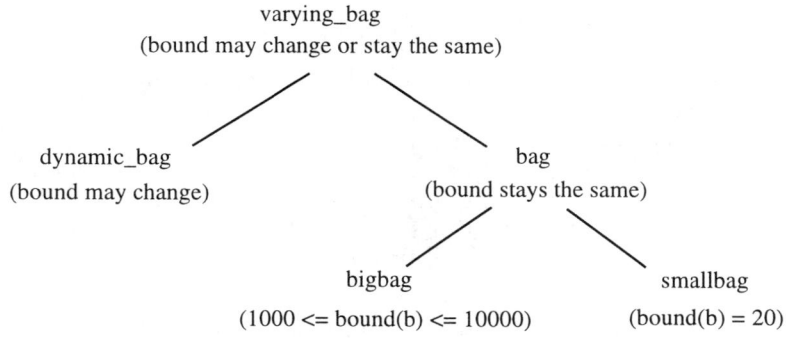

Figure 3 A Type Family for Bags

behavioral properties, it would be reasonable to extend this definition to other kinds of properties by extending the scope of what is included in a type specification.

Consider the type family of bags in Figure 3. The constraint for a supertype, varying_bag, of the bag type given in the previous section is that the bound may change or stay the same. Another subtype of varying_bag could be dynamic_bag, with the trivial contraint true (thus saying its bound may change). Dynamic_bag would have an additional method, *change_bound*,

$change\_bound =$ **proc** $(n:\text{int})$
    requires $n \geq |b_{pre}.elems|$
    modifies b
    ensures $b_{post}.elems = b_{pre}.elems \wedge b_{post}.bound = n$

which would not make sense for the bag type to have.

Further subtyping the bag type we can, for example, define bigbag with a constraint that its bound be within a certain range and smallbag with a constraint that its bound be fixed to be 20.

This definition of subtyping supports multiple supertypes. If S is a subtype of both T and U, then the designer is obligated to show the above checklist of conditions holds between S and T and between S and U. Implementation problems that arise because of multiple inheritance are irrelevant; subtyping is a relationship between specifications, not implementations.

3   AN EXAMPLE OBJECT REPOSITORY: THE TOM SERVER

How does understanding the subtype relation help the system designer? In the second half of this extended abstract I describe an object repository that the Carnegie Mellon TinkerTeach Project built and its key design principle. This project was done independently of the work on subtyping, but in retrospect two lessons can be learned from the implementor's design decision as related to the Liskov and Wing notion of subtyping:

- Avoiding mutability simplifies defining a subtype hierarchy.
- Mutability cannot be completely avoided in an open distributed environment.

Before I explain these seemingly inconsistent statements, I describe the object repository's functionality.

As part of his Ph.D. thesis, John Ockerbloom invented a Typed Object Model [9], a data model involving objects, types, and their associated metadata. He implemented an instance of this model, a TOM server, which currently supports the ability for users in a distributed environment to store data types and data conversion functions, to register new ones, and to find existing ones. The kinds of data types TOM supports today are different kinds of document types (e.g., Word, LaTex, PowerPoint, binhex, html) and "packages" of such document types (e.g., a mail message that has an embedded postscript file, a tar file, or a zip file). The kinds of data conversions TOM supports are off-the-shelf converters like *postscript2pdf* (i.e., AdobeDistiller^{TM}), off-the-Web ones like *latex2html*, and some home-grown ones like *powerpoint2html*.* As of mid-March, 200 sites from over 20 countries in 6 continents have accessed TOM.†

Today a specific application of TOM is to handle type conversion tasks, which includes a Web-based user interface built for the TinkerTeach Project. This user interface hides much of the complexity of type conversion from the user, in three ways:

---

*The conversion of a PowerPoint document to an .html file is actually done through the application of nine different intermediate steps going through different intermediate types like rtf, postscript, and ppm. These intermediate converters do things like converting postscript files to ppm files, resizing and rotating ppm files, and converting ppm files to gif files. Users see none of these intermediate steps.

†I use it on a daily basis. The Web site is: http://tom.cs.cmu.edu/.

- TOM is a system of *type brokers*. If a user makes a request to one instance of a TOM server, S, and S does not know about the data type or converter in question, but does know of another instance, T, that does, then completely transparently to the user, S will contact T to process the request. Thus, there can be multiple instances of a TOM server where each knows about a few types and converters; collectively all the TOM servers comprise a distributed object server.
- TOM can compose converters to do conversions. Given a source type and a target type (by the user), TOM can figure a plan of conversion steps to apply. It can make such plans on the fly, such as when it composes an *rtf2html* converter with an *html2text* converter.
- Given an object (e.g., a Word document) to convert, TOM uses heuristics to guess what the type of the source object is. It can also tell a user when a requested conversion is unsupported or meaningless.

3.1 Simplify Life: Avoid Mutability

When Ockerbloom originally designed TOM, he made the following critical design decision:

All objects are immutable.

The rationale behind the decision is that he wanted to treat arbitrary information in a distributed environment like the Web as objects. If objects can change in value, then issues of storage, update, and concurrency control must be resolved, perhaps using standard distributed file system or distributed database techniques. If objects cannot change in value, then TOM does not have to worry about how they are stored, where they are stored, how they are updated, if and how they are copied or replicated, and how to coordinate concurrent access to them. Rather, objects can live anywhere, be created by anyone, and be shipped around freely.

In principle, by deciding that no object can be mutable, Ockerbloom is able to avoid the problem of shared access to mutable objects. Thus, TOM does not have to worry about what subtyping means in the presence of mutability. As a corollary, showing the subtype relation holds of TOM's type hierarchy is simpler since showing constraints are preserved is trivial.

TOM does support an interesting subtype hierarchy. Figure 4 gives a subgraph of the TOM type hierarchy. For example, TOM

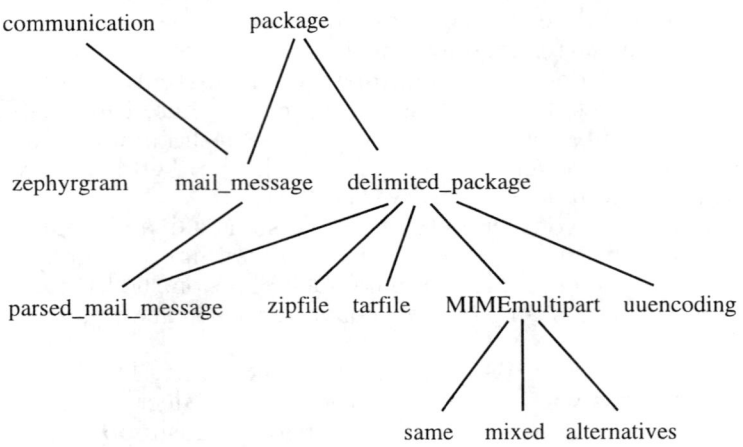

Figure 4 Part of TOM's Type Hierarchy

makes a distinction between a package type that has clear delim-
iters (delimited_package) and one which is just a mail_message,
which contains a mail header and some uninterpreted contents.
A parsed_mail_message is distinct from a mail_message because
the type of the message's contents has been determined (e.g., a
postscript file). Also TOM supports packages of packages, and so
for example, a mail message can contain a forwarded mail message which itself contains a MIMEmultipart file; TOM is "smart"
enough to unwrap these packages and present their contents in
a way that users can meaningfully interact with the individual
pieces.

Notice two examples of multiple supertypes in this subgraph.
The parsed_mail_message type is a subtype of both mail_message
and delimited_package, and a mail_message itself is a subtype of
both package and communication.

Ockerbloom carefully designed his type hierarchy so that each
subtype either only adds new methods or changes (by overriding)
old methods in a constrained way. Thus, proving that the Liskov
and Wing subtype conditions hold for the TOM hierarchy is relatively easy:

- If no changes to old methods are made then the proof is trivial. Since each object is immutable, neither old nor new methods can be mutators, so constraints are trivially preserved. Since no old method is overridden, invariants are preserved and the behavior of old methods is preserved. In the typical case, the subtype object simply has more state information, e.g, extra attributes, and the abstraction function is the obvious many-to-one function that throws away the extra state information. In their paper, Liskov and Wing call these *extension* subtypes since the subtype extends the supertype by providing additional methods and possibly additional state.
- If changes to old methods are made, then Part 2 of the sub-type definition applies: the contra/covariant rules, the exception rule, and most importantly the pre-/post-condition rules must be shown. If subtypes always only further constrain the behavior of the corresponding supertype methods, then it is easy to show that invariants and constraints are preserved (given that Part 2 holds and that the specifications are consistent). In their paper, Liskov and Wing call these *constrained subtypes* because the degree of nondeterminism is reduced in the subtype. The abstraction function is usually into rather than onto.

3.2   Life is Not So Simple: Context Matters

Since TOM's objects are all immutable, TOM can treat the values of these objects as the real "objects." In other words, from TOM's view there is a huge value space, where each value can be considered an "object" (an entity that provides a set of methods in the traditional sense). TOM also supports *handles* to objects, e.g., file names and URLs; handles are also TOM immutable objects and provide a *dereference* method. For example, the contents of a file is a TOM object, not the file itself; a file name is a TOM object, which when dereferenced, refers to the contents of a file.

By definition values of immutable objects cannot change. However, the *binding* between handles to values/TOM objects may change and *TOM has no control over this binding*. In particular, TOM's environment can change the binding between handles and values, and so from the user's viewpoint, it looks as if these objects are mutable. In other words, the dereference method on a handle might yield different results at different times, such as when someone has edited the file being referenced. And, two different handles, e.g., two different URLs, may dereference to the same file contents. TOM cannot control or even know about this binding. For example, it is common for many different URLs to refer to the same file on a given Web site and it is common for system administrators to export a URL for remote access but use an internal file name for local access. Thus, from a more global perspective, TOM objects are shared and these objects are mutable. We are back to our original problem: what does subtyping mean in the presence of shared mutable objects?

Unfortunately, as users of local and distributed file systems, the Web, or publicly accessible persistent object repositories, we have no control over the semantic guarantees that these different contexts provide. Unix-like file systems, for example, provide no consistency guarantees; a change by one user to a file may not be seen by another who has a replica or cached copy of that file. These weak consistency guarantees mean that while the subtyping relation may hold from TOM's internal viewpoint, it can be intentionally or inadvertently violated by someone who accesses a TOM object from outside of TOM, by implicitly changing the binding between some handle and TOM object.

This situation is neither new nor surprising. For any persistent object repository that does not sit in isolation, i.e., makes its objects available through means other than that repository's interface, the same situation will arise. Thus, this situation simply serves as a warning to the user of that persistent object repository and as a reminder to its designer: Context matters and must be taken into consideration when accessing the repository's objects.

4 SUMMARY

I reviewed the Liskov and Wing subtype definition that takes into consideration the problem of subtyping in the presence of mutable objects. The key ideas behind their notion of subtyping are (1) to consider the behavior of objects, as specified through pre-/postconditions for methods, and invariants and constraints for data types; and (2) to consider *history properties*, as captured by type constraints.

I also showed how this notion of subtyping is relevant to the design of the TOM object repository whose main application today is a data type conversion service. While TOM views all its objects as immutable, its environment may indeed provide alternative means of access to these objects and thus users may make changes to them. In TOM, this situation is realized by changing the binding between a handle and TOM object; TOM has no control this binding. If users make such changes then they need either to ensure that the changes are consistent with the behavior specified for the objects' types or to realize that the No Surprises Requirement can be violated.

5 ACKNOWLEDGMENTS

I thank Barbara Liskov, my co-author on previous papers on this subject. Parts of this abstract, in particular the opening paragraph, the statement of the No Surprises Requirement, the bag specification, and the bag type family, are taken from our TOPLAS paper. I would also like to thank John Ockerbloom for building TOM and his constructive comments on a draft of this abstract. Finally, I thank all the TinkerTeach Project members, in particular, Norm Papernick, our staff programmer, who maintains the TOM conversion service.

This research is sponsored in part by the Defense Advanced Research Projects Agency and the Wright Laboratory, Aeronautical Systems Center, Air Force Materiel Command, USAF, F33615-93-1-1330, and Rome Laboratory, Air Force Materiel Command, USAF, under agreement number F30602-97-2-0031 and in part by the National Science Foundation under Grant No. CCR-9523972. The U.S. Government is authorized to reproduce and distribute reprints for Governmental purposes notwithstanding any copyright annotation thereon. The views and conclusions contained herein are those of the authors and should not be interpreted as necessarily representing the official policies or endorsements, either expressed or implied, of the Defense Advanced Research Projects Agency Rome Laboratory or the U.S. Government.

REFERENCES

Pierre America. A parallel object-oriented language with inheritance and subtyping. *SIGPLAΓˉ*, 25(10):161–168, October 1990.

Luca Cardelli. A semantics of multiple inheritance. *Information and Computation*, 76:138–164, 1988.

Elspeth Cusack. Inheritance in object oriented Z. In *Proceedings of ECOOP '91*. Springer-Verlag, 1991.

Krishna Kishore Dhara and Gary T. Leavens. Subtyping for mutable types in object-oriented programming languages. Technical Report 92-36, Department of Computer Science, Iowa State University, Ames, Iowa, November 1992.

S.J. Garland and J.V. Guttag. An overview of LP, the Larch Prover. In *Proceedings of the Third International Conference on*

Rewriting Techniques and Applications, pages 137–151, Chapel Hill, NC, April 1989. Lecture Notes in Computer Science 355.

J.J. Horning, J.V. with S.J. Garland Guttag, K.D. Jones, A. Modet, and J.M. Wing. *Larch : Languages and Tools for Formal Specification*. Springer-Verlag, New York, 1993.

Gary T. Leavens and Krishna Kishore Dhara. A foundation for the model theory of abstract data types with mutation and aliasing (preliminary version). Technical Report 92-35, Department of Computer Science, Iowa State University, Ames, Iowa, November 1992.

Barbara Liskov and Jeannette M. Wing. A behavioral notion of subtyping. *ACM TOPLAS*, 16(6):1811–1841, November 1994.

John Ockerbloom. Exploiting structured data in wide-area information systems. Technical Report CMU-CS-95-184, Carnegie Mellon Computer Science Department, Pittsburgh, PA, 1995.

Amy M. Zaremski. Signature and specification matching. Technical Report CS-CMU-96-103, CMU Computer Science Department, January 1996. Ph.D. thesis.

22

Inheritance Anomaly
- A Formal Treatment

Lobel Crnogorac[†], Anand S. Rao[‡] and Kotagiri Ramamohanarao[†]
[†]*Dept. of Computer Science, The University of Melbourne, Parkville Vic. 3052, Australia, E-mail: {lobel,rao}@cs.mu.oz.au*
[‡]*Australian Artificial Intelligence Institute, Level 6, 171 La Trobe Street, Melbourne Vic. 3000, Australia, E-mail: anand@aaii.oz.au*

Abstract

Inheritance is one of the key concepts in object-oriented programming (OOP). However, the usefulness of inheritance in concurrent OOP is greatly reduced by the inheritance anomalies. These anomalies have been subjected to intense research, but they are still only vaguely defined and often misunderstood. In this paper we show that concurrency is not the real cause of inheritance anomalies. We formally define the inheritance anomaly as a relationship between inheritance mechanisms and behavioural hierarchies. Our framework can be used to analyse the occurrence of inheritance anomalies in many different paradigms. A formal definition of the problem and a clear exposition of its causes are pre-requisites for a successful integration of inheritance and concurrency.

Keywords

inheritance, behavioural subtyping, concurrent object-oriented programming

1 INTRODUCTION

Inheritance is one of the key concepts in object-oriented programming (OOP). It is a widely used methodology for code re-use in sequential object-oriented programming. In recent years, the concepts from OOP have been applied in a concurrent setting, leading to the emergence of concurrent object-oriented programming (COOP) [7, 11, 12]. In its full generality COOP paradigm allows inter-object concurrency (multiple objects existing concurrently) and intra-object concurrency (multiple threads inside an object). It was found that most OOP concepts (*e.g.*, encapsulation) could be naturally integrated into COOP. However, the integration of inheritance and COOP has not been smooth. One of the main problems with inheritance in COOP is the *inheritance anomaly* [8, 10, 11, 12, 13, 15]. Inheritance anomaly arises when additional methods of a subclass cause undesirable re-definitions of the methods in the superclass. Instead of being able to incrementally add code in a subclass the programmer may be required to re-define some inherited code, thus the benefits of inheritance are lost.

Inheritance anomalies have been researched extensively, but they are still only vaguely defined and often misunderstood. There is a wealth of language proposals in the literature trying to solve the problem of anomalies, but almost no formal work has been done. The main practical progress has been made in the area of languages that do not allow intra-object concurrency. For comparison purposes the proposals were usually evaluated on a set of standard benchmark examples introduced by Matsuoka and Yonezawa [11]. The aim was to successfully avoid the anomaly in the benchmark examples.

Inheritance anomaly has never been formally defined in its full generality, although a particular type of anomaly, *state-partitioning*, has been formally investigated by Matsuoka *et. al.* [10]. We feel that a general formal treatment of the problem is needed in order to precisely define the inheritance anomaly and to formally compare the different proposals. Without a formal treatment we cannot be certain that the set of benchmark examples is exhaustive (there could be other undiscovered types of anomalies). The formal treatment should also analyse the causes of the inheritance anomaly. It is widely believed that interference between concurrency and inheritance is the cause of inheritance anomalies. However, the appearance of inheritance anomalies in other paradigms based on object-oriented concepts points to the need for a much more thorough examination of their causes. For example, inheritance anomaly in real-time specification languages [1] could not be caused by an interference between inheritance and concurrency because these languages are purely sequential. The introduction of inheritance into agent-oriented programming (AOP) [14] also leads to the appearance of inheritance anomalies [5]. The requirements we put on a framework for inheritance anomaly are:

- The framework must be general enough to allow a uniform analysis of inheritance anomaly for any existing inheritance mechanism in any existing object-oriented paradigm. For example, inheritance anomaly in truly concurrent languages (allowing intra-object concurrency [3]) with active objects seems to be a much harder problem than the more constrained case (only inter-object concurrency). The framework must be able to capture the general problem of inheritance anomaly in COOP.
- The framework must formally pinpoint the cause of the anomaly. These results should be used to predict appearances of anomalies in new paradigms, and explain why the anomaly doesn't arise in sequential OOP.
- Are the known types of anomalies the only possible anomalies? Are they caused by different reasons or are they just different manifestations of a sin-

gle conflict? The framework must be able to derive the complete taxonomy of the types of inheritance anomalies across different paradigms.

- By using the framework it should be possible to explore whether an ideal solution actually exists. Also, the framework must be able to formally explain and justify the choices made in the development of better inheritance mechanisms (*e.g.*, the need to separate inheritance of synchronisation code from inheritance of functionality code should be formally justified [12]).
- The framework must present a formal definition of inheritance anomaly.

We share the views in [2, 13] that inheritance should be an unconstrained, flexible tool for code re-use. Inheritance should maximise the amount of code that can be re-used when defining a new specification from an existing specification. *Subtyping* is concerned with the use of objects, and is usually based on method signatures. Whenever we require an object, a subtype of that object would do equally well. Inheritance and subtyping are distinct concepts [2], since inheritance is concerned with the internal structure of objects (code sharing), while subtyping is concerned with the external behaviour of objects (the way objects are used). The work on *behavioural subtyping* [2, 9] extends the concept of subtyping to more general notions of behaviour.

The key insight of this paper is the connection between the notions of inheritance and subtyping: the inheritance mechanism should be powerful enough to incrementally mimic the behavioural (subtype) hierarchy. Hence, inheritance should be able to reach any "sub-behaviour" of a given specification without any need for re-definitions. Inheritance anomaly is defined as the failure of the inheritance mechanism to incrementally mimic the behavioural hierarchy. The aim of this paper is to provide a formal definition of inheritance anomaly and to address most of the previously stated requirements.

After a brief overview of the problem in COOP (Section 2), we introduce the domain of syntactic specifications of objects (Section 3). An inheritance mechanism is defined as a transition relation on the set of syntactic specifications. Defining inheritance in terms of a transition relation means we can avoid giving a particular semantics to inheritance. Thus, we avoid unnecessarily constraining our framework. In Section 4 we introduce the concept of behavioural hierarchy, which can be viewed as a generalised notion of subtyping. The relationship between syntactic specifications and behaviours is given by an abstraction function which maps specifications of objects into actual behaviours. The formal definition of inheritance anomaly is based on the relationship between the behavioural hierarchy and the inheritance mechanism. An inheritance mechanism is anomaly-free with respect to a given behavioural hierarchy if it can mimic that hierarchy in an incremental fashion. Inheritance anomaly is highly language dependent. This dependence is encoded in the transition relation. The framework is then applied in the context of sequential and concurrent OO languages (Section 5). We show that our formal definition matches informal examples of anomalies given in literature. We give fundamental results stating theoretical limitations of inheritance mechanisms in COOP. In particular, we show that an ideal solution does not exist.

The primary contribution of this paper is the use of the correspondence between an inheritance mechanism and a behavioural hierarchy to motivate a formal definition of the inheritance anomaly. Our framework can be used to analyse the occurrence of inheritance anomalies in many different paradigms. The analysis based on the formal definition provides a clearer understanding of the causes of the anomaly. We provide results that explain the recent directions of research into better inheritance mechanisms [3, 12] and show that an ideal solution does not exist. A formal definition of the problem and a clear exposition of its causes are the pre-requisites for a successful integration of inheritance into the existing object-oriented paradigms.

2   OVERVIEW OF INHERITANCE ANOMALY IN COOP

This section gives a brief overview of the problems caused by the inheritance anomaly in COOP. The examples used are due to Matsuoka and Yonezawa [11]. The anomaly results in the inability of COOP languages to inherit synchronisation code without re-definitions. Concurrent object-oriented programming languages have to provide facilities for expressing synchronisation constraints of objects. For example, the programmer needs to be able to express that adding an element into a full buffer or removing an element from an empty buffer is not allowed. Inheritance anomaly (in the context of COOP) is the conflict between concurrency and inheritance where extensive re-definitions of inherited methods are necessary in order to maintain the synchronisation constraints of concurrent objects. Matsuoka and Yonezawa [11] have distinguished three kinds of inheritance anomalies in COOP languages:

● *state-partitioning:*
Execution of a concurrent object can be thought of as a sequence of transitions between states. Each state is determined by the current values of all the state variables and the methods that are acceptable. The state-partitioning anomaly occurs when the addition of a new method further partitions the state-space. A new state is added to the set of states a class can be in. All the other states remain. The code changes in the superclass are caused by the difference in the number of states possible in the superclass and the subclass. The extra states possible in the subclass have to be accounted for in all the methods in the superclass. The classical example involves a language based on *accept sets* [7, 15]. The synchronisation scheme of these proposals uses explicit sets to determine which methods are acceptable at any time. The methods that are not currently acceptable are either rejected, or suspended and placed into a message queue. The keyword *become* denotes the explicit switching between the states. Consider the situation in Figure 1. Class *Buffer* implements a bounded buffer (storing at most MAX elements). It provides methods *put* and *get* which add and remove a single element from the buffer.

The synchronisation code, expressed by explicit accept sets, needs to describe the following constraints: "Method *put* is acceptable unless *Buffer* is full. Method *get* is acceptable unless *Buffer* is empty. Method *numOfElements* is always acceptable." We define a subclass of *Buffer*, *NewBuffer* with an additional method *get2* that removes two elements. Method *get2* can be used only if *NewBuffer* contains more than one element. This is an extra state that the object could be in. The anomaly appears when the *become* statements of *put* and *get* (as well as the most of the *behaviour* block) need to be re-defined to accommodate for the extra state.

State-partitioning has been circumvented by proposals that employ method guards instead of accept sets [11]. A method is accepted if its guard evaluates to *true*, otherwise it is suspended (placed into a message queue) or rejected. The syntax is "**method** *method_signature* **when** *guard*" (Figure 2). Here, *NewBuffer* is constructed from *Buffer* by adding the statement "**method** *get2* **when** *numOfElements > 1*" along with the code for *get2*. Unlike the situation in Figure 1, re-definitions of *put* and *get* are not necessary with method guards. The guards solve state-partitioning anomaly. However, method guards and accept sets can't prevent the next two types of anomalies.

● *state-modification:*
Additions of new methods to a class can introduce a finer-grained distinction for the set of states under which the methods can be invoked. Code re-definitions

```
class  Buffer{
int in=0, out=0;
behaviour:empty = {put(x),numOfElements}
          partial = {put(x),get,numOfElements}
          full = {get,numOfElements}
method   numOfElements
    code for numOfElements
method   put(x)
    code for put
    if (numOfElements==MAX) become full;
    else become partial;
method   get
    code for get
    if (numOfElements==0) become empty;
    else become partial;
}

class  NewBuffer: Buffer{
behaviour:/* the set empty is inherited cleanly */
          partial = {put(x),get,get2,numOfElements}
          full = {get,get2,numOfElements}
          one = {put(x),get,numOfElements}
method   get2
    code for get2
    if (numOfElements==0) become empty;
    else if (numOfElements==1) become one;
    else become partial;
method   put(x)
    code for put
    if (numOfElements==MAX) become full;
    else if (numOfElements==1) become one;
    else become partial;
method   get
    code for get
    if (numOfElements==0) become empty;
    else if (numOfElements==1) become one;
    else become partial;
}
```

Figure 1 State-partitioning

```
class  Buffer{
int in=0, out=0;
method   numOfElements /* always accepted */
    code for numOfElements
method   put(x) when (numOfElements < MAX)
    code for put
method   get when (numOfElements > 0)
    code for get
}
class  LockableBuffer: Buffer{
Bool locked = false;
method   lock{ locked=true;}      /* always accepted */
method   unlock{ locked=false;} /* always accepted */
method   put(x) when ( !locked &&
         (numOfElements < MAX))
    code for put
method   get when (!locked && (numOfElements > 0))
    code for get
```

Figure 2 State-modification

are caused by the need to account for this finer-grained distinction of states. The standard example involves adding a locking capability to the *Buffer* class. Methods *put* and *get* can be accepted only when the object is unlocked. The method *numOfElements* is not affected by the new locking operations. Method guards (Figure 2) are used to specify the conditions under which methods are accepted. State-modification arises from the need to add new variables to distinguish between states, *e.g.*, the variable *locked* in Figure 2. Methods *put* and *get* have to be re-defined to take *locked* into account.

- *history-only-sensitiveness*:

In COOP we often encounter situations that depend on history of an object. The need to re-define code arises because the methods in the superclass need to leave some trace of their execution for the future (usually, the methods are re-defined to update some new variables). The standard example involves extending the *Buffer* class with a new method *gget* which behaves exactly like *get*, except that it cannot be invoked immediately after an execution of *put*. To define this new class, *HistoryBuffer*, methods *put* and *get* are re-defined to use a new variable *after_put*. Thus, the benefits of inheritance are lost.

3   MODELLING INHERITANCE AS A TRANSITION RELATION

An inheritance mechanism defines the way a new class specification can be obtained by re-using code from an existing class specification. New services may be added, the inherited services re-defined or omitted. An inheritance mechanism of a language is usually given by defining the semantics of its inheritance operator [4]. We take a different approach. Here we formalise the inheritance mechanism of an arbitrary language as a transition relation on the set of syntactic specifications. A pair of syntactic specifications forms a transition if the second specification can be obtained from the first by using the inheritance rules. Expressing all inheritance mechanisms in terms of transition relations provides a uniform view of different languages. It avoids giving a particular semantics to inheritance, thus it does not constrain the formal framework to one paradigm. This is important since there is no clear agreement about the inheritance semantics in COOP or AOP (unlike in sequential OOP).

We capture a language with inheritance as a set of syntactic specifications that do not use inheritance, and a transition relation between them. Let *Spec* be the set of all possible syntactic specifications (without using inheritance) of classes in some language. In an OOP language, *Spec* is the countably infinite set of all classes that can be written without using the inheritance operator of the language. For example, the definition of class *Buffer* (Figure 2) is an element of the set *Spec* of the illustrated language. However, class *LockableBuffer* is not an element of *Spec* since it is defined by using the inheritance operator ":". We assume that an element $p \in Spec$ is a function $p : Keys \to Exp_\perp$ where *Keys* is a countably infinite set of names. The set $Exp_\perp$ is the set of expressions that can be written in the language. It is a partial order (actually, a flat cpo) under the $\preceq$ ordering. In OOP $p \in Spec$ maps method signatures and variables (the keys) to their respective bodies (the expressions).

Definition 1 A *preordering* on a set D is a binary relation that is reflexive and transitive. A *partial ordering* is an antisymmetric preordering. Let $e, f \in Exp_\perp$. If $e \preceq f$ then either $e = \perp$ or $e = f$. We extend $\preceq$ to functions. ∎

Definition 2 An *inheritance mechanism* is a pair $(Spec, \dashrightarrow)$ where $\dashrightarrow \subseteq Spec \times \Delta \times Spec$. An element of $\dashrightarrow$, (p, δ, q) is called a *transition* where $p, q \in Spec$ and $\delta \in \Delta$. Δ is the set of syntactic entities specifying the differences between p and q. We write $p \overset{\delta}{\dashrightarrow} q$ for $(p, \delta, q) \in \dashrightarrow$. Furthermore, $p \overset{\delta \in \Delta^*}{\dashrightarrow} q$ is used to denote the reflexive and transitive closure of $\dashrightarrow$ *i.e.*, a sequence of individual transitions. Overloading of the notion $p \overset{\delta}{\dashrightarrow} q$ is harmless. ∎

The transition relation $\dashrightarrow$ is a set of triples (p, δ, q). Transitions specify how inheritance can be used to move from a syntactic specification p to a new specification q by specifying the differences (*e.g.*, new methods) in δ^*. Transitions may simulate re-definitions, deletions or omissions of components of syntactic specifications. Hence, very general inheritance mechanisms can be defined. The sets *Spec* and Δ are determined by the language that is being analysed. Since $q \in Spec$, Definition 2 assumes that everything that can be defined by means of inheritance can also be defined without it.

Example 1 Figure 2 shows a single transition of the inheritance mechanism employing method guards. Here, p corresponds to *Buffer* and δ corresponds to *LockableBuffer*. The syntactic specification q is not shown however. It would correspond to a *fully expanded* version of *LockableBuffer* with the method *numOfElements* and the variables *in* and *out* explicitly defined. Thus, in practice, a language specifies transitions by giving the modification δ from p. ∎

Inheritance anomaly arises when instead of being able to incrementally add code in a subclass, the programmer is required to re-define some inherited code. In order to capture this property we need to formally make a distinction between "incrementally adding code" and "re-defining code".

Definition 3 Transition $p \overset{\delta}{\dashrightarrow} q$ is *incremental* if* $p \preceq q$. The subset of all incremental transitions is denoted $\dashrightarrow_I \subseteq \dashrightarrow$. ∎

An incremental modification means that new functionality is added to a syntactic specification without re-defining any of the existing services. Hence, if $p \overset{\delta}{\dashrightarrow}_I q$ then whenever p maps a key to a defined expression, q maps the same key to the same expression also. Furthermore, if p maps a key to $\perp$, then q maps the same key to $\perp$ or to a defined expression. The relation $\dashrightarrow_I$ corresponds to the elegant, incremental use of inheritance.

---

*The definition of transition relation can be extended to model multiple inheritance. We focus on single inheritance in this paper.

*In most contexts this is actually a double implication. However, in order to model inheritance in languages that use *self* and *super* [4] the definition of incremental transitions needs to be modified to include some additional transitions. Basically, re-instantiation of *self* and *super* is modelled by fully expanding all references to them and by including the transition in $\dashrightarrow_I$. Details are beyond the scope of this paper.

4   THE DEFINITION OF INHERITANCE ANOMALY

This section presents the formal definition of inheritance anomaly. As discussed earlier in Section 1 an anomaly-free inheritance mechanism needs to be powerful enough to simulate the behavioural hierarchy in an incremental fashion. If a subclass **preserves and extends the behaviour** of its parent then we would like to re-use the parent's specification as a whole. The subclass would be defined incrementally from the original specification without any need for re-definitions of the parent's methods. Alternatively, if a subclass **modifies the behaviour** of its parent (behaviour of the parent is not preserved) all we can expect to re-use are parts of the parent's specification (some methods are inherited cleanly while other methods are re-implemented).

We now introduce the notion of *behavioural hierarchy*. Behavioural hierarchy is a partially ordered set $(Beh, \leq)$. The set Beh is the set of all possible behaviours, ordered by a partial ordering $\leq$ which defines the meaning of behaviour preservation/extension. That is, if $\theta, \phi \in Beh$ and $\theta \leq \phi$ then ϕ somehow preserves and extends θ. Consider the relationship between the sets $Spec$ and Beh. The behaviour of an object specification is determined by the language semantics. The semantics of the chosen language defines an *abstraction* function α, which maps the set $Spec$ into the set Beh. The abstraction function maps a syntactic specification into its behaviour. Hence, $\alpha : Spec \to Beh$ is a function mapping a single element of $Spec$ into a single element of Beh. Syntactic specification p *implements* the behaviour θ if $\theta \equiv \alpha(p)$ (where $\equiv$ is induced by the partial order $\leq$). In general α is not injective since different syntactic specifications may be mapped to the same behaviour. Intuitively, we can implement a required behaviour in infinitely many ways, *e.g.*, by changing the names of variables. In general, α is not surjective ($\alpha(Spec) \subseteq Beh$, where $\alpha(Spec)$ is the image of α over $Spec$) since $Spec$ is countable, while Beh may be uncountable. The relationship between inheritance mechanism and behavioural hierarchy is the basis for the following definition.

Definition 4 Let $(Spec, \dashrightarrow)$ be an inheritance mechanism, with a behavioural hierarchy $(Beh, \leq)$ and an abstraction function α. Let $p, q \in Spec$ and $\theta \in Beh$. Define $G_p = \{q : p \xrightarrow{\delta \in \Delta^*} q\}$. Let $I_p = \{q : p \xrightarrow{\delta \in \Delta^*}_I q\}$, $B_p = \{q : \alpha(p) \leq \alpha(q)\}$. Finally, let $S_p = \{\theta : \alpha(p) \leq \theta\}$. ∎

Consider Figure 3. For each syntactic specification p we define sets G_p, I_p, B_p and S_p. The set G_p is the set of all syntactic specifications that can be obtained from p by repeated applications of the inheritance mechanism. Note that $G_p \subseteq Spec$, but commonly, $G_p = Spec$ since in many inheritance mechanisms we can obtain any given specification from p by repeated re-definitions, deletions and additions. The set I_p is a subset of G_p which allows only incremental transitions from p. The set of all *syntactic specifications* which preserve and extend the behaviour of p is denoted by B_p. In

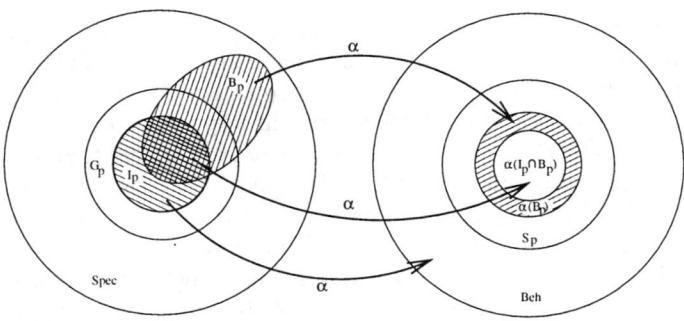

Figure 3 The definition of inheritance anomaly

general, I_p is not a subset of B_p because some incremental modifications can modify behaviour. This depends on the particular context, determined by α and $(Beh, \leq)$. For example, in AOP under a "natural" definition of agent behaviour it is possible to modify the behaviour of an agent by simple, incremental additions of new plans. The intersection $(I_p \cap B_p)$ is the set of all specifications which preserve and extend the behaviour of p and which can be incrementally (without re-definitions) obtained from p. The set S_p is the set of all possible *behaviours* that preserve and extend the behaviour of p. As shown in Figure 3, $S_p \subseteq Beh$. In general α may not be surjective, therefore the set B_p maps to $\alpha(B_p)$ which is a subset of S_p. In general $\alpha(I_p \cap B_p) \subseteq \alpha(B_p) \subseteq S_p$.

Definition 5 An inheritance mechanism $(Spec, \dashrightarrow)$ is *anomaly-free* with respect to $(Beh, \leq)$ iff $\forall p \in Spec$, $\alpha(I_p \cap B_p) = \alpha(B_p)$. ∎

Intuitively, consider the following scenario: a) The programmer has defined a specification p which implements the behaviour $\alpha(p)$; b) The programmer envisions a specialisation of $\alpha(p)$ which preserves and extends this behaviour; c) If this specialisation can be implemented in the language by some specification r then the programmer must be able to incrementally obtain a specification q from p, such that q also implements the required specialisation. Proposition 1 shows that this scenario is a consequence of Definition 5.

Proposition 1 *Let $p, r \in Spec$. If the inheritance mechanism $(Spec, \dashrightarrow)$ is anomaly-free with respect to $(Beh, \leq)$ and if $\alpha(p) \leq \alpha(r)$ then $\exists \delta \in \Delta^*, q \in Spec$ such that $p \overset{\delta}{\dashrightarrow}_I q$ and $\alpha(r) \equiv \alpha(q)$, where $\equiv$ is induced by $(Beh, \leq)$.*
Proof: *Since $\alpha(p) \leq \alpha(r)$ we have $r \in B_p$. Since $\alpha(I_p \cap B_p) = \alpha(B_p)$ by assumption, we can find $q \in I_p \cap B_p$ such that $\alpha(q) \equiv \alpha(r)$.* ∎

The concept of inheritance anomaly is defined with respect to a given behavioural hierarchy. An inheritance mechanism may be anomaly-free with respect to one hierarchy, while it may have anomalies with respect to another hierarchy. This observation leads to an explanation of the inconsistency in the current literature. Until now, the definition of behavioural hierarchy was only given informally by researchers, through examples. In the case of *HistoryBuffer* (Section 2) some researchers have claimed that it illustrates an anomaly [11, 12], while others have claimed that it actually *modifies* the behaviour of *Buffer* [13] and therefore should not be considered an anomaly. This inconsistency arises from different assumptions about the behavioural hierarchy. A formal definition of behaviour hierarchy is needed to unambiguously define the problem. The dependence of the inheritance anomaly on the chosen behavioural hierarchy gives rise to a new view of the most essential cause of the anomaly. The inheritance anomaly arises because the inheritance mechanisms suitable for one behavioural hierarchy may not be suitable for a different hierarchy. Of course, in the context of COOP it is still correct to view the anomaly as being caused by an interference between inheritance and concurrency. However, this view does not generalise well. For example, inheritance anomalies have been discovered in sequential real-time specification languages [1] (behaviour hierarchy would be based on temporal information).

5   EXAMPLE ANALYSIS

This section presents an example analysis of inheritance mechanisms in sequential and concurrent OOP. We examine two different behavioural hierarchies, which are based on the externally observable behaviour of objects. The object model that we adopt views objects as encapsulated message acceptors. The only way to communicate with such an object is by sending it messages. Thus, the externally observable behaviour of an object is defined by the way it reacts to messages. This behaviour is independent of the implementational details. The objects are assumed to be executing at most one service at a time. Thus, we do not deal with intra-object concurrency here. The formal definition of inheritance anomaly is illustrated by applying it to the examples of Section 2. Finally, we prove a general result which states the theoretical limitations in the search for anomaly-free inheritance mechanisms.

In the context of our chosen object model we define behaviour of an object to be the set of all possible sequences of messages that the object can accept. We use the concept of *traces* [6]. Suppose that an external observer notes down the message acceptances. A *trace* is a finite sequence of message acceptances by an object. The set of all such finite sequences of message acceptances of an object is called the *traces* of the object and is denoted by *traces(object)*. The set *Messg* $\subseteq$ *Keys* is the set of all keys which can be sent in messages and which are used to identify services. Thus, *Messg* is the set of all symbols that can appear in traces (*i.e.*, the *alphabet*), and the set of all possible traces is denoted *Messg**. An element of *Keys* is simply a method name or private variable name. Objects enforce encapsulation by using keys in *Keys* − *Messg* for their private services. The set of services (with their keys belonging to *Messg*) corresponds to the public interface of an object. A message can be accepted only if it matches a key of one of the services offered by the object. However, a message that matches one of the services may sometimes be not accepted. This case arises in COOP where messages are accepted according to the current state of the object. For example, a *get* message sent to a *Buffer* object may be accepted (if the object is non-empty) or not accepted (if the object is empty, in which case the message may be rejected or suspended).

Definition 6 Behaviour of an object O is *traces*(O). The set $\mathcal{P}(Messg^*)$ is the set of all subsets of *Messg**, *i.e.*, the set of all possible behaviours (hence, $Beh = \mathcal{P}(Messg^*)$ in this context). Behaviour $\theta \in \mathcal{P}(Messg^*)$ can also be viewed as a language over the alphabet *Messg*. Let $Reg \subset \mathcal{P}(Messg^*)$ be the set of all regular languages over *Messg*, that is, all languages (behaviours) that can be accepted by finite state machines. We use some standard operations on traces. Let $\theta \in \mathcal{P}(Messg^*)$ and let t be a trace in θ. The *restriction*, $t \restriction D$ denotes the trace t when restricted to symbols in the set D. The *length* of trace t is denoted $\#t$. *Catenation* constructs a trace from a pair of traces s and t by putting them together in that order. The result is denoted $s\hat{}t$. *Head* of a trace t gives the first symbol in t and is denoted t_0. The symbol m *occurs in* t iff $\#(t \restriction \{m\}) > 0$. Similarly, m *occurs in* θ iff $\exists t \in \theta \mid m$ *occurs in* t. ∎

Example 2 $\langle put, put, get \rangle$ is a trace of the behaviour of bounded *Buffer* from Figure 1 (assuming MAX is large), whereas $\langle put, get, get \rangle$ is not. $\alpha(Buffer) = \{t \in Messg^* \mid \forall s, v \in Messg^*, t = s\hat{}v \implies 0 \leq (\#(s \restriction \{put\}) - \#(s \restriction \{get\})) \leq MAX\}$. In other words, a trace of *Buffer* must have at least as many occurrences of *put* as it has of *get*, but the difference must be at most MAX. Additionally, every initial segment of this trace (*i.e.*, every prefix) must have the same property (*e.g.*, to disallow $\langle put, get, get, put, put \rangle$). It follows that every non-empty trace of *Buffer* starts with *put* (consider the initial segment of length 1) since *Buffer* starts off empty. The empty trace, $\langle \rangle$, is a trace of *Buffer*. To illustrate the operations on traces, we have: $\langle put, put, get \rangle \hat{} \langle put, get \rangle = \langle put, put, get, put, get \rangle$. $\#\langle put, put, get \rangle = 3$ and $\langle put, get, get \rangle_0 = put$. ∎

Behaviours are ordered by defining the relation $\leq$. Behaviour ϕ *preserves/extends the behaviour* θ if $\theta \leq \phi$. Intuitively, ϕ can behave like θ (can preserve θ), but it can also exhibit some additional behaviour (ϕ extends the behaviour θ). The first behavioural ordering (denoted $\leq_1$) is given by:

Definition 7 Let $\theta, \phi \in \mathcal{P}(Messg^*)$. Then, $\theta \leq_1 \phi$ iff $\forall m \in Messg$, m *occurs in* $\theta \implies m$ *occurs in* ϕ. Also, $\theta \equiv_1 \phi$ iff $\theta \leq_1 \phi$ and $\phi \leq_1 \theta$. ∎

Example 3 $\{\langle put, get \rangle, \langle put, put \rangle\} \equiv_1 \{\langle put \rangle, \langle get \rangle\}$ because the symbols *put* and *get* appear on both sides. Also, "the behaviour of *Buffer*" $\leq_1$ "the behaviour of *NewBuffer*" since *NewBuffer* may accept an extra message *get2*. ∎

It can be checked that $\leq_1$ is a partial order, and that $\equiv_1$ is an equivalence relation on $\mathcal{P}(Messg^*)$. Definition 7 states that $\theta \leq_1 \phi$ if the traces in ϕ contain more distinct symbols than the traces in θ. Hence, ϕ offers a larger set of services than θ. Behavioural equivalence $\equiv_1$ states that two behaviours are equivalent if they offer the same set of services (*i.e.* the same public interface). Definition 7 is equivalent to simple subtyping without covariant/contravariant rules [2] since messages contain only service names, and parameters are ignored (this can be extended). Sometimes, we need to distinguish between behaviours based on the actual sequencing of accepted messages. A stricter notion of behaviour preservation/extension, denoted by $\leq_2$, is given by:

Definition 8 Let $\theta, \phi \in \mathcal{P}(Messg^*)$ and consider $s, t, v \in Messg^*$. Then, $\theta \leq_2 \phi$ iff $\theta \subseteq \phi$ and $\forall s \in \phi, s = t\hat{\ }v$ for some $t \in \theta$ and for some v (possibly empty) such that the symbol v_0 (if it exists) never occurs in a trace of θ. The relation $\equiv_2$ is defined as in Definition 7. It follows that $\theta \equiv_2 \phi$ iff $\theta = \phi$. ∎

Again, we can check that $\leq_2$ and $\equiv_2$ define a partial order and an equivalence relation on $\mathcal{P}(Messg^*)$. Definition 8 states that two objects display equivalent behaviour if they can engage in exactly the same sequences of message acceptances (their traces must be identical). Furthermore, ϕ preserves/extends the behaviour θ if ϕ can engage in all the sequences that θ can engage in, and it may also engage in some additional sequences. However, such an additional trace of ϕ must start with a trace from θ until a new message (that never occurs in θ) is accepted by ϕ. Thus, ϕ and θ are identical until ϕ accepts a new message, after which ϕ produces some additional functionality.

Example 4 The behaviour of *LockableBuffer* (Figure 2) preserves/extends the behaviour of *Buffer* because it contains the same traces as *Buffer* (if the observer is not sending *lock* messages), but it also contains additional traces, all of which start with some trace from *Buffer*. For instance, the trace $\langle put, put, lock, numOfElements, unlock, get \rangle$ is such an additional trace which starts with the trace $\langle put, put \rangle$ from *Buffer*. Similarly, the trace $\langle lock, unlock, put \rangle$ is a trace of *LockableBuffer* which starts with the empty trace from *Buffer*. Hence, "the behaviour of *Buffer*" $\leq_2$ "the behaviour of *LockableBuffer*". Under Definition 8, *HistoryBuffer* (Section 2) also preserves/extends the behaviour of *Buffer*. For instance, it introduces new traces $\langle put, put, get, gget, put \rangle, \langle put, put, put, get, gget \rangle, \langle put, put, put, get, gget, gget \rangle$. Again, "the behaviour of *Buffer*" $\leq_2$ "the behaviour of *HistoryBuffer*". The behaviours of *HistoryBuffer* and *LockableBuffer* are incomparable. ∎

Definitions 7 and 8 give two different versions of behaviour preservation/extension. By using these definitions we can explain the set of standard informal examples that have been used as the "definition of inheritance anomaly" in literature. Different definitions of behaviour and behavioural hierarchy would be used to analyse inheritance mechanisms in other paradigms (*e.g.*, AOP, real-time specification, actor-based, COOP with internal concurrency). Note that we have not distinguished between the behaviour of classes and the behaviour of instances of classes in our definition of behaviour. This distinction is not necessary for the simple examples of Section 2, but it should be incorporated into a more thorough analysis.

Example 5 Consider Figure 1. We have, "the behaviour of *Buffer*" $\leq_2$ "the behaviour of *NewBuffer*". However, there is no incremental transition from the given specification of *Buffer* to any specification of *NewBuffer*. Hence, this inheritance mechanism is not anomaly-free with respect to $(\mathcal{P}(Messg^*), \leq_2)$. Similar arguments can be formulated for other examples in Section 2. Thus, the formal definition matches the informal examples. ∎

We present the results of our example analysis and discuss their implications.

Theorem 1 *Consider a typical sequential object-oriented language with a simple inheritance mechanism* $(Spec, \dashrightarrow)$ *where* $p \dashrightarrow q$ *iff* q *has additional methods, or* q *has re-defined some methods from* p. $(Spec, \dashrightarrow)$ *is anomaly-free with respect to both* $(\mathcal{P}(Messg^*), \leq_1)$ *and* $(\mathcal{P}(Messg^*), \leq_2)$.
Proof: *Firstly note that* $\forall p \in Spec, \alpha(p) = \{m \in Messg \mid p(m) \neq \perp\}^*$. *If* $\alpha(p) \leq_1 \theta$ *and if* θ *is of the above form then* $\exists q \in Spec$ *such that* $p \dashrightarrow_I q$ *and* $\alpha(q) \equiv_1 \theta$ *(*q *simply defines all* $m \in Messg$ *which occur in* θ, *and for which* $p(m) = \perp$*). The case for* $\leq_2$ *is identical.* ∎

Consider a sequential class *Buffer* (*e.g.*, Figure 2 with the method guards removed). The behaviour of this class is the set of all possible sequences of its services, *i.e.*, $\{put, get, numOfElements\}^*$. In other words, a sequential *Buffer* cannot refuse a message if the message key matches one of its services (of course, the message may return an error or fail, but the observer only observes message acceptances). Contrast this with the behaviour of bounded *Buffer* (in COOP), which is an element of *Reg*. Most COOP languages (that employ synchronisation code) can implement any regular language (at least) *i.e.*, $Reg \subseteq \alpha(Spec)$ where $\alpha(Spec)$ denotes the image of α over *Spec*. Inheritance mechanisms suitable for $\leq_2$ in sequential OOP may not be suitable for $\leq_2$ in COOP. Note that there is no anomaly under $\leq_1$ in COOP.

Definition 9 An inheritance mechanism is *behaviour preserving* under $(Beh, \leq)$ iff $p \overset{\delta}{\dashrightarrow}_I q \implies \alpha(p) \leq \alpha(q)$. ∎

We introduce the notion of *behaviour preserving inheritance mechanisms* thus classifying inheritance mechanisms into two types. Behaviour preserving inheritance mechanisms are based on the principles used in sequential OOP. An ideal inheritance mechanism in COOP would be behaviour preserving. Such a mechanism produces only extensions of behaviour, if used incrementally. A simple inheritance mechanism, as used in sequential OOP is behaviour preserving . It can be shown that the proposals based on method guards and accept sets, informally described in Section 2 are also behaviour preserving.

```
class Buffer{
int in=0, out=0;
method numOfElements{...}
pre:      numOfElements < MAX
method put(x){...}
pre:      numOfElements > 0
method get{...}
}
class LockableBuffer: Buffer{
Bool locked=false;
method lock{...}
method unlock{...}
pre:      (get) ∧ !locked
pre:      (put) ∧ !locked
}
```

Figure 4

Example 6 Figure 4 illustrates a non-behaviour-preserving inheritance mechanism that avoids the anomaly from Figure 2. Pre-conditions (denoted by **"pre"**) act as guards, but they can also be incrementally composed in the subclasses. The semantics of a set of pre-conditions is given by the conjunction of their conditions. Note that a syntactic specification would use two distinct keys for the two pre-conditions of the method *put* (hence, the transition is incremental). It can be seen that the incremental addition of pre-conditions is essentially "modifying" the synchronisation constraints of the object in an incremental manner. For example, if the pre-condition of *put* in *LockableBuffer* is changed to *locked* (instead of *!locked*) then the behaviour of *Buffer* is not preserved. A behaviour preserving mechanism is clearly preferable. ∎

Theorem 2 *Given an inheritance mechanism* $(Spec, \dashrightarrow)$ *and* α *such that* $Reg \subseteq \alpha(Spec)$, *if* $(Spec, \dashrightarrow)$ *is behaviour preserving under* $(\mathcal{P}(Messg^*), \leq_2)$ *then it is not anomaly-free.*

Proof: *Assume the mechanism is anomaly-free. We construct* $p \in Spec$ *such that* $\alpha(p) = \{m_1, m_2\}^*$ *for some* $m_1, m_2 \in Messg$. *Let* $s \in \alpha(p)$ *and* $m \in Messg$ *such that* m *doesn't occur in* $\alpha(p)$. *Then,* $t = s^\frown\langle m \rangle \notin \alpha(p)$. *Consider the behaviour* $\theta = \alpha(p) \cup \{t, t^\frown\langle m_1\rangle\}$. $\theta \in Reg$ *(by closure) and* $\alpha(p) \leq_2 \theta$, *hence* $\exists r \in Spec$ *such that* $\alpha(r) \equiv_2 \theta$. *By Proposition 1,* $\exists q \in Spec$ *such that* $p \dashrightarrow_I q$ *and* $\alpha(q) \equiv_2 \theta$. *At any instant, the state of* q *is determined by the values of its instance variables* (q *has a superset of the set of instance variables of* p *since* $p \dashrightarrow_I q$). *Suppose that after accepting* t, q *is in some state* S. *Construct* q' *such that* S *is the initial state of* q' *(by changing the initial values of variables). We have* $\alpha(q') = \{\langle \rangle, \langle m_1\rangle\}$. *Construct* p' *such that* S *(restricted to the variables of* p) *is the initial state of* p'. *We have* $\alpha(p') = \{m_1, m_2\}^*$ *since all states of* p *and* p' *can accept* m_1 *and* m_2. *Hence, we have* $p' \dashrightarrow_I q'$ *(since* $p \dashrightarrow_I q$ *and the same simple mapping of initial values was used to obtain* p' *from* p *and* q' *from* q). *However,* $\alpha(p') \not\leq_2 \alpha(q')$ *since* $\{m_1, m_2\}^* \not\leq_2 \{\langle \rangle, \langle m_1\rangle\}$. *Hence, the inheritance mechanism is not behaviour preserving.* ∎

Theorem 2 states that there is no behaviour preserving, anomaly-free inheritance mechanism in COOP. Hence, proposals employing method guards and accept sets must give an anomaly. Theorem 2 does not apply to sequential OOP (α does not map to *Reg*). Further research into this area should lead to more formal classifications of the different types of anomalies. The immediate implication of Theorem 2 is that there is no ideal solution to the problem of inheritance anomalies in COOP. Hence, it supports the recent direction of research [3, 12] which separates the actual inheritance mechanisms of the synchronisation code and the functionality code, leading to non-behaviour-preserving inheritance mechanisms. The trade-off expressed by Theorem 2 is that an inheritance mechanism is either not powerful enough (not anomaly-free), or it is too powerful (not behaviour preserving). In other words, the use of pre-conditions leads to the possibility of making mistakes (incrementally adding new code may not preserve the original behaviour).

6 CONCLUDING REMARKS

This paper investigated the problem of the inheritance anomaly. We claim that a formal approach is needed to provide a better understanding of the problem, which would lead towards the design of better inheritance mechanisms. The main contribution of the paper is the use of the correspondence between an inheritance mechanism and a behavioural hierarchy to motivate a formal definition of the inheritance anomaly in a general setting. The formal definition shows that the interference between inheritance and concurrency is not the most basic cause of the inheritance anomaly. Rather, the anomaly arises because the inheritance mechanisms suitable for one behavioural hierarchy may not be suitable for a different hierarchy.

We presented some theoretical limitations of inheritance mechanisms, thus justifying the recent trends in the search for a cleaner integration of inheritance and concurrency. In particular, we proved that an ideal solution is not possible in COOP. The given framework can be the basis for a formal comparison of the different inheritance mechanisms. It could also be used to formally construct the complete taxonomy of the types of inheritance anomalies. Further work in this area should concentrate on applying the formal treatment to languages that allow intra-object concurrency.

REFERENCES

[1] M. Aksit, J. Bosch, W. van der Sterren, and L. Bergmans. Real-time specification inheritance anomalies and real-time filters. In *ECOOP'94*, LNCS 821, pages 386–407. Springer-Verlag, 1994.

[2] P. America. Designing an object-oriented programming language with behavioural subtyping. LNCS 489, pages 60–90. Springer-Verlag, 1990.

[3] M.Y. Ben-Gershon and S.J. Goldsack. Using inheritance to build extendable synchronisation policies for concurrent and distributed systems. In *TOOLs Pacific '95*, pages 109–121, 1995.

[4] W. Cook and J. Palsberg. A denotational semantics of inheritance and its correctness. In *OOPSLA'89*, pages 433–443, 1989.

[5] L. Crnogorac and A. S. Rao. Inheritance by extensions and restrictions in agent systems. In *ACSC'97*, Sydney, Australia, February 1997.

[6] C.A.R. Hoare. *Communicating Sequential Processes*. Prentice-Hall International Series in Computer Science. Prentice-Hall, 1985.

[7] D. G. Kafura and K. H. Lee. Inheritance in Actor based concurrent object-oriented languages. In *ECOOP'89*, pages 131–145, UK, 1989.

[8] U. Lechner, C. Lengauer, F. Nickl, and M. Wirsing. How to overcome the inheritance anomaly. In *ECOOP'96*, LNCS 1098. Springer-Verlag.

[9] B. Liskov and J. M. Wing. A behavioral notion of subtyping. *TOPLAS*, 16(6):1811–1841, 1994.

[10] S. Matsuoka, K. Wakita, and A. Yonezawa. Synchronization constraints with inheritance: What is not possible - so what is? Technical Report 10, Dept. of Information Science, the University of Tokyo, 1990.

[11] S. Matsuoka and A. Yonezawa. Analysis of inheritance anomaly in object-oriented concurrent programming languages. In *Research Directions in COOP*, chapter 1, pages 107–150. MIT Press, 1993.

[12] C. McHale. *Synchronisation in COO Languages: Expressive Power, Genericity and Inheritance*. PhD dissertation, Trinity College, 1994.

[13] J. Meseguer. Solving the inheritance anomaly in concurrent object-oriented programming. In *ECOOP'93*, LNCS 707, pages 220–246.

[14] Y. Shoham. Agent-oriented programming. *Artificial Intelligence*, 60(1):51–92, 1993.

[15] C. Tomlinson and V. Singh. Inheritance and synchronization with enabled-sets. In *OOPSLA '89*, pages 103–112. ACM Press, 1989.

7   BIOGRAPHY

Dr Anand Rao is the Chief Research Scientist at the Australian AI Institute. He obtained his PhD from the University of Sydney in 1988, and spent a year at IBM's T.J. Watson Research Center. He has published a number of papers in reactive planning and recognition; families of Belief-Desire-Intention (BDI) logics and their properties; and agent-oriented languages and methodologies. **Prof. Kotagiri Ramamohanarao** received his PhD from Monash University in 1980. He is well known for his contributions in the areas of dynamic hash indexing, partial match retrieval and deductive database systems with over 100 refereed papers. He has been a program committee member for several prestigious international conferences including VLDB, ICDE, ICLP, EUROPAR and ISLP. **Lobel Crnogorac** is a PhD student at The University of Melbourne working on incorporation of inheritance into AOP.

23

On Behavioural Subtyping in LOTOS *

H. Bowman[1], C. Briscoe-Smith[1], J. Derrick[1] and B. Strulo[2]
[1] *Computing Lab., Univ. of Kent, Canterbury, Kent, CT2 7NF, UK*
[2] *BT Laboratories, Martlesham Heath, Ipswich, IP5 7RE, UK*
Email: {H.Bowman,cpb4,J.Derrick}@ukc.ac.uk & bstrulo@srd.bt.co.uk

Abstract

We consider how the OO notion of subtyping relates to LOTOS testing theory. In particular, we investigate which of the standard LOTOS preorders is a suitable instantiation of behavioural subtyping and argue that each of the main preorders, trace preorder, trace extension, reduction and extension, is in some way deficient. Then, in the light of pre and post condition based models of OO subtyping, we re-work the basic interpretation applied to LOTOS behaviour descriptions. We argue that this re-interpretation enables reduction to be used as an instantiation of behavioural subtyping.

1 INTRODUCTION

This paper investigates possible definitions of *behavioural subtyping* in the process algebra LOTOS. Interest in this topic is motivated from a number of important areas of current research. Behavioural subtyping impacts on research concerned with,

1. enhancing the specification and development capabilities of process algebra by incorporating features of object oriented methodologies [MC93];
2. providing a theoretical basis for *concurrent OO programming* and models of so called *active objects*, which are objects that exhibit non-uniform service availability [Nie95]; and
3. enhancing existing formal description techniques in order that they can be applied to the new generation of distributed systems, which are typically object oriented [BDLS95].

In all these areas subtyping plays a pivotal role in obtaining incremental system development, with its relationship to different inheritance mechanisms being crucial. The third of these areas has particularly motivated the work presented here. Central to object oriented programming platforms such as CORBA, the TINA DPE (Distributed Processing Environment) and the ODP

---

*The research presented here has been partially supported by British Telecommunications Research Labs, through their funding of Charles Briscoe-Smith's PhD studentship.

(Open Distributed Processing) Computational Model is the notion of *trading*. A trader is a distinguished object used in order to locate required services. It accepts service offers from objects, and maintains a database of currently available offers. When an object wishes to find a service, it performs an *import* operation on the trader, specifying what kind of service it wants, and receives copies of a number of service offers in reply.

When a client sends a description of the service it wants to the trader, the trader must somehow match this to the offers it has in its database. If it cannot find an offer exactly matching the requested service, it should look for offers of similar services, providing all the facilities that the client wanted, but possibly having other facilities that the client will not use. It is looking for a service which has a superset of the operations the client asked for, which the client could use without knowing that it was any different from the service type it requested. In fact, the relationship between the service requested and the service returned by the trader should be *subtyping*.

The concept of subtyping is familiar from object oriented programming languages[FM94], It is defined as substitutability: type A is a subtype of type B iff objects of type A may be used in any situation where an object of type B was expected, without the object's environment being able to tell the difference. Thus, an object of any particular type can *masquerade* as, or stand in for, an object of any of its supertypes. Subtyping is naturally a reflexive and transitive relation, i.e. a preorder.

However, the state of the art in service matching for trading is signature-based subtyping. Unfortunately, such matching is not rich enough to ensure the safety of object interactions in a heterogeneous distributed processing environment. For example, two object types may have methods with the same name but quite different meaning. To take a rather frivolous example, consider the analogy of an artist and a cowboy. Both are able to perform an operation "draw," but the results in each case will be rather different. Thus, it is possible that although signatures match, compatibility in terms of the behaviour of services is not obtained. The insufficiency of purely signature based approaches is witnessed by the increasing interest within OMG for adding behavioural properties to CORBA IDL.

What is actually required is a more powerful interpretation of matching based on (stronger) *behavioural* notions of subtyping (in ODP terms *behavioural compatibility*). Determining suitable interpretations of behavioural subtyping is the subject matter of this paper.

As our notation for describing the behaviour of service types we use the process algebra LOTOS. There are a number of reasons for this choice, not least the role of LOTOS as a formal description technique for open distributed systems and the accepted benefits of the process algebra approach [Mil89]. However, a further benefit of considering LOTOS is that a wealth of correctness relations exist, many of which are related to substitutibility and hence behavioural subtyping. From this domain the testing theories are of particular relevance. In such theories specifications are related if they pass the same tests.

Testing theory is an extremely rich field. In fact it is possible to place the spectrum of process algebra correctness relations (at least those based upon interleaving models of concurrency) in a hierachy of strength, i.e. in terms of their level of discrimination [vG93]. The relative strengths of particular correctness relations is tied to the intrusive capabilities of the tester to observe the specification. In this paper we will use a standard notion of testing in which the tester has the power of a standard LOTOS process (no additional operators are added to the testing language). Since clients in the OO setting will be LOTOS processes this seems a sensible choice. The testing theory induces a preorder that, for the moment, we will call *compatibility*:

S_1 *is compatible with* S_2 *iff for all finite sets of observable actions* G *and processes* P, $S_1 \ |[G]| \ P \stackrel{\sigma}{\Longrightarrow} \approx$ **stop** *implies* $S_2 \ |[G]| \ P \stackrel{\sigma}{\Longrightarrow} \approx$ **stop**.

where $|[G]|$ is the LOTOS parallel composition operator, $\approx$ is weak bisimulation equivalence, **stop** is the deadlock process, σ is a trace of observable actions and relation composition is denoted by juxtaposition*. This notation will be clarified shortly, but informally, the condition states that S_1 is compatible with S_2 if and only if, for all possible testers, if S_1 can perform a trace σ and then deadlock, then under the control of the same tester S_2 can perform σ and then deadlock. Thus, even more informally, S_1 does not add any new deadlocks to those that can arise from S_2.

In terms of OO and subtyping, in the above definition, G reflects the possible interfaces between S_1 and the tester, i.e. the actions that they can communicate via, and P reflects possible client specifications/programs. We argue that this condition is the basis for an intuitively sensible instantiation of subtyping in the process algebra setting. In OO terms the definition states that,

S_1 *is a subtype of* S_2 *if and only if any client (tester) using* S_1 *according to any interface (synchronisation set) can only observe a trace and then observe a deadlock if the client could observe the same trace and a deadlock if it was using* S_2 *(with the same interface)* *.

From amongst the LOTOS correctness relations, **red** (reduction) is the most important. In particular, modulo handling of divergence, **red** corresponds to failures divergences refinement [Hoa85] and testing preorder [Hen88], which are the principle notions of refinement used in CSP and CCS (respectively). However as it stands, reduction is not a sufficient definition of subtyping. This is because subtyping in the OO context allows *extension of functionality*, e.g. a subtype can offer more operations than its supertype.

---

*i.e. $S \ |[G]| \ P \stackrel{\sigma}{\Longrightarrow} \approx$ **stop** means $\exists Q \ . \ S \ |[G]| \ P \stackrel{\sigma}{\Longrightarrow} Q \ \wedge \ Q \approx$ **stop**
*Since we test against all possible clients (and not just those that have a subset of the operations of S_2) we get a strong notion of subtyping. We believe that this strength is necessary, e.g. when objects are being concurrently interacted with.

In the process algebra setting extending functionality implies addition of traces. However, reduction enforces a trace subsetting property and thus, does not allow functionality to be extended. In response to this observation a number of previous workers [Rud91] [CRS89] [Nie95] have based their interpretation of subtyping upon an alternative relation: the extension relation (**ext**) [BS86]. However, we will argue against using this relation; rather we will show how to re-interpret LOTOS specifications in order that reduction is the appropriate relation.

Section 2 presents background on LOTOS and outlines how aspects of LOTOS can be related to OO concepts. Section 3 relates the spectrum of LOTOS refinement relations to behavioural subtyping. Section 4 considers the characteristics of behavioural subtyping in OO specification and programming languages and then shows how LOTOS processes can be transformed in order to reflect these characteristics. Then section 5 highlights a simple technique for transforming LOTOS specifications according to this new interpretation. Finally, section 6 summarises and concludes the paper.

2   BACKGROUND

LOTOS. We use a subset of full LOTOS [BB88]:

$$P ::= \textbf{stop} \mid a; P \mid P\,[]\,P \mid P\,|[G]|\,P \mid \textbf{choice}\ a \in A\ []\ P \mid X$$

where $a \in \textbf{Act} \cup \{i\}$ (**Act** contains all observable actions and i is the distinguished hidden action). Thus, our notation has a deadlock process **stop**, action prefix $a; P$, binary choice $P\,[]\,P$, parallel composition $P\,|[G]|\,P$, generalised choice **choice** $a \in A\ []\ P$ and reference to a process variable X, through which recursion can be defined. Process definitions have the form, $X := P$.

We do not include the other basic LOTOS operators, hiding, relabelling, disabling and enabling. This is not because they bring any technical difficulties, but rather to simplify the presentation.

We also assume some semantic constructions. In the following P, P', Q, Q' stand for processes. $\mathcal{L}$ is the alphabet of observable actions associated with a certain process (we will write $\mathcal{L}(P)$ when we need to be explicit about the process we are referring to). The standard semantics for LOTOS [ISO87] map LOTOS processes to Labelled Transition Systems (LTSs) using a structured operational semantics. We will not repeat these inference rules. However, in standard fashion, we denote transitions as: $P \xrightarrow{a} P'$, meaning that P can perform an a and evolve to P'. Furthermore, $\mathcal{L}^*$ denotes traces over $\mathcal{L}$, $\epsilon \in \mathcal{L}^*$ denotes the empty trace and σ ranges over $\mathcal{L}^*$. We assume the following definitions:

$\xrightarrow{\epsilon}$; the reflexive and transitive closure of $\xrightarrow{i}$;

$P \xRightarrow{a\sigma} P'$ iff $\exists Q, Q' \cdot P \xRightarrow{\epsilon} Q \xrightarrow{a} Q' \xRightarrow{\sigma} P'$;

$P \stackrel{\sigma}{\Longrightarrow}$ iff $\exists P' \cdot P \stackrel{\sigma}{\Longrightarrow} P'$;

$P \stackrel{\sigma}{\not\Longrightarrow}$ iff $\neg(\exists P' \cdot P \stackrel{\sigma}{\Longrightarrow} P')$;

$Tr(P) = \{\sigma \in \mathcal{L}^* \mid P \stackrel{\sigma}{\Longrightarrow} \}$; the set of traces of P;

P *after* $\sigma = \{P' \mid P \stackrel{\sigma}{\Longrightarrow} P'\}$; the set of states reachable from P by σ;

$Ref(P, \sigma) = \{X \mid \exists P' \in (P \text{ after } \sigma) . \forall a \in X : P' \stackrel{a}{\not\Longrightarrow} \}$; refusals of P *after* σ.

$initials(P) = \{ a \in \mathcal{L} \mid a \in Tr(P) \}$.

Relating OO Concepts to LOTOS. Before we consider subtyping it is worth clarifying how LOTOS specifications relate to OO concepts. This section highlights some basic relationships.

Class. A class describes the common behaviour of a set of objects. As noted by a number of authors, e.g. [DEBS96] [Smi95] [Rud91], in LOTOS the natural counterpart to a class is a process definition. This describes the common behaviour of instantiations of the process definition.

Object. In OO programming objects are instantiations of a class. Thus, a simple interpretation of instantiation in LOTOS is as process instantiation.

However, more sophisticated interpretations of object instantiation can also be given. For example, [Rud91] [CRS89] interpret instantiation as the LOTOS implementation relation **conf** (which is the LOTOS conformance relation). Thus, any process that conforms to the specification of a class is seen as an instantiation of the class. Although **conf** has a number of undesirable properties as an implementation relation, in principle such an interpretation of instantiation is much richer and more flexible than simple process instantiation. In particular, when working in a behavioural setting it seems sensible to interpret instantiation in behavioural terms rather than as a purely syntactic instantiation. Although we will not need to consider this issue of instantiation further in this paper, implicitly instantiations in our setting will be related to their class definition much more strongly than by **conf**; perhaps by testing equivalence.

Operations. The basic units of interaction between objects are operations, also called method invocations, member function calls, or feature calls. In process algebra, the basic units of interaction between processes are actions. The affinity between these two concepts is witnessed by the number of workers in this area who have related the two: [Nie95] [Rud91] [CRS89] [DEBS96] [Smi95].

However, it should be pointed out that this similarity may not be exact, since process algebra actions are considered to be atomic, whereas in many OO models operations have duration. The assumption of atomicity is highly significant in the process algebra setting as it justifies the modelling of concurrency as interleaving. Non-atomic interpretations of actions lead to more complex semantic theories. A simplifying assumption that Nierstrasz makes [Nie95] is only to model method requests. Such an assumption effectively justifies an atomic interpretation of actions when modelling operations. In accordance with this majority of workers we will also enforce a simplifying atomic interpretation of actions/operations.

Finally, the parameters of operations may be modelled using LOTOS's data passing attributes, "!" and "?".

Interface. An object oriented class definition will usually contain a statement of the interface to objects of that class: usually a list of calls which may be made on the objects. The LOTOS equivalent is the set of all non-hidden actions in the process definition.

The above are only the most basic correspondences; there are many more which can be made. For example, Rudkin[Rud91] describes how inheritance and *self* might be introduced into LOTOS and Najm and Stefani [NSF94] consider how object mobility may be obtained. The interested reader is also referred to part IV of [ITU95] which relates OO modelling concepts to LOTOS constructs in the ODP setting.

3   RELATING LOTOS RELATIONS AND SUBTYPING

In this section, we attempt to locate an interpretation of behavioural subtyping from amongst the existing LOTOS correctness relations. Firstly, since subtyping is reflexive and transitive, but not symmetric (a symmetric relation would suggest substitutability in both directions, which is too strong), we will only consider the preorder relations. This choice rules out the equivalences weak bisimulation ($\approx$), strong bisimulation ($\sim$), testing equivalence (**te**) and testing congruence (**tc**) and the implementation relation **conf**, which is not transitive.

Trace Subsetting and Supersetting. We first consider trace preorder, one of the simplest correctness relations. The fact that P_1 is a trace refinement of P_2 is defined as (notice the order that we write refinement, this contrasts with some other workers), $P_1 \leq_{tr} P_2$ iff $\text{Tr}(P_1) \subseteq \text{Tr}(P_2)$. This relation is clearly inappropriate since it does not allow P_1 to have any more traces than P_2, which contradicts the extension of functionality involved in subtyping.

An alternative to $\leq_{tr}$ is trace extension: $P_1 \leq_{tre} P_2$ iff $\text{Tr}(P_1) \supseteq \text{Tr}(P_2)$. This *does* allow new operations to be added and, in fact, is the interpretation of subtyping used in [Pun96]. In Puntigam's work, trace extension serves as a valid check for type safety. Where in this context, type safety ensures that the subtype can understand all operations that the supertype can. However, the relation is not a suitable instantiation of the stronger notion of behavioural subtyping since it allows deadlocks to be added. For example, if X and Y are defined as,

$$X := a; \textbf{stop} \, [] \, b; \textbf{stop} \qquad Y := a; \textbf{stop} \, [] \, b; \textbf{stop} \, [] \, i; \textbf{stop}$$

then $Y \leq_{tre} X$. However, Y is not a behavioural subtype of X. When placed in synchronisation with the process "a; **stop**", X will do action a, but Y may do a, or may do an internal action and then deadlock. If Y deadlocks in a situation where X would not, Y is distinguishable from X and is therefore not a subtype of/compatible with X. In fact, the same criticsm can be levelled at all solely trace based correctness relations, including trace preorder.

Reduction. Reduction[BS86] is a more discriminating refinement relation that adds consideration of liveness properties to trace preorder. Its definition is,

$$P_1 \textbf{ red } P_2 \text{ iff } \text{Tr}(P_1) \subseteq \text{Tr}(P_2) \wedge \forall \sigma \in \text{Act}^* . \text{Ref}(P_1,\sigma) \subseteq \text{Ref}(P_2,\sigma)$$

(that is, P_1 reduces P_2 iff $P_1 \leq_{tr} P_2$ and, after any trace, P_1 does not refuse more than P_2). Interpreting refinement as reduction corresponds to viewing development as reduction of non-determinism. In addition, in terms of our general testing constraint, the property we called compatibility in section 1, we have the following result[*]:

Theorem 1 *For all processes P_1, P_2, P and $G \subseteq \text{Act}$, the following are equivalent:*

1. $P_1 \textbf{ red } P_2$
2. $P_1 \ |[G]|\ P \overset{\sigma}{\Longrightarrow} \approx \textbf{stop}$ *implies* $P_2 \ |[G]|\ P \overset{\sigma}{\Longrightarrow} \approx \textbf{stop}$.

Thus, reduction ensures the deadlock property we are seeking. However, as discussed in section 1, it fails to allow extension of functionality. So, as it stands, reduction is not a suitable instantiation of subtyping.

Extension. Since extension[BS86] is sensitive to deadlock properties and supports extension of functionality, it appears at first sight to be an ideal candidate for the subtyping relation. This is witnessed by the large number of workers who have used it as the basis for definitions of subtyping [CRS89] [Rud91] [Nie95]. Its definition is,

$$P_1 \textbf{ ext } P_2 \text{ iff } \text{Tr}(P_1) \supseteq \text{Tr}(P_2) \wedge \forall \sigma \in \text{Tr}(P_2) . \text{Ref}(P_1,\sigma) \subseteq \text{Ref}(P_2,\sigma)$$

(that is, P_1 extends P_2 iff $P_1 \leq_{tre} P_2$ and, after any trace that P_2 can do, P_1 does not refuse more than P_2). Consider two LOTOS processes, X and Y:

$$X := a; \textbf{stop} \ [] \ b; \textbf{stop} \qquad Y := a; \textbf{stop} \ [] \ b; \textbf{stop} \ [] \ c; \textbf{stop}$$

Referring to the definition of **ext**, we see that Y **ext** X. Y can do every trace

---

[*]This is actually a slightly stronger result than that proved in [BS86], since we do not require trace subsetting between P_1 and P_2. However, this stronger result can be verified with minor changes (involving taking a larger synchronization set) to Brinksma et al's proof.

that X does (and more), and, after any trace that X can do, X refuses at least everything that Y refuses. Conceptually Y defines a class which adds an operation to class X, viz. the action c. Thus, extension enables interface enlargement.

Unfortunately extension does not fulfil our requirements for behavioural subtyping. In particular, extension does not guarantee the definition of compatibility that we gave in section 1. For example, the tester c; **stop** with synchronization set $\{c\}$ serves as a counterexample since,

$$Y \ ||[c]|| \ c; \mathbf{stop} \stackrel{c}{\Longrightarrow} \approx \mathbf{stop} \quad \text{but,} \quad X \ ||[c]|| \ c; \mathbf{stop} \stackrel{c}{\nRightarrow}$$

Extension only satisfies the following more restrictive theorem, which is proved in [BS86],

Theorem 2 *For all processes* P_1, P_2, P; $G \supseteq \mathcal{L}(P_2)$ *and* $Tr(P_1) \supseteq Tr(P_2)$, *the following are equivalent:*

1. P_1 **ext** P_2
2. $\forall \sigma \in Tr(P_2),\ P_1 \ ||[G]|| \ P \stackrel{\sigma}{\Longrightarrow} \approx \mathbf{stop}$ *implies* $P_2 \ ||[G]|| \ P \stackrel{\sigma}{\Longrightarrow} \approx \mathbf{stop}$.

Thus, extension only ensures compatibility when restricting to traces of the supertype. However, we require the stronger compatibility property that was highlighted in section 1.

Another way of looking at this problem is that our definition of behavioural subtyping is based on the principle that a subtype must be usable in any situation where the supertype could be used, and not be seen to behave differently. If we have a process which may be a X or a Y, we can detect which it is by trying to perform the action c on the process. If the c is accepted, we have Y, but if c is refused, we must have X. Since it is possible to tell that we have a Y, our definition of behavioural subtyping tells us that Y is not a subtype of X.

Interestingly, this problem with extension is one that Nierstrasz has observed [Nie95]. His illustrative example is that of a one place buffer supertype and a deleting buffer subtype. We can express his example in LOTOS as follows:

$$Buf1 := put; get; Buf1 \quad \text{and} \quad DelBuf := put; (get; DelBuf \ [] \ del; \mathbf{stop})$$

Thus, *DelBuf* behaves as *Buf1* does but it adds the possibility to delete the element in the buffer and then evolve to deadlock. The tester/client which distinguishes the two is analogous to the LOTOS process:

$$T := Prod \ ||| \ Cons \ ||| \ del; \mathbf{stop} \quad \text{with} \quad Prod := put; Prod \quad Cons := get; Cons$$

which yields the composite behaviour shown in figure 1. Now *DelBuf* is clearly

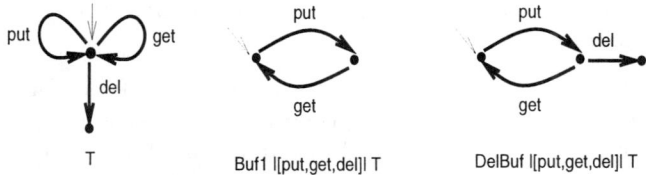

Figure 1 LOTOS Behaviours

an extension of *Buf1*, however, Nierstrasz observes that with the interface {*put, get, del*} and the tester *T*, *Buf1* cannot reach a deadlock state while *DelBuf* can. Specifically,

$$DelBuf\|[put, get, del]\| T \xrightarrow{put\ del} \approx \textbf{stop} \quad \text{but,} \quad Buf1\|[put, get, del]\| T \xrightarrow{put\ del} \nrightarrow$$

In fact, the problem here is exactly the same as that which we highlighted with behaviours *X* and *Y* above. Nierstrasz develops a number of concepts such as *request substitutability* and a notion of *restriction* in order to contain this problem. In contrast, our approach will be to reject extension as an interpretation of behavioural subtyping.

4 FUNCTIONALITY EXTENSION AND UNDEFINED

Undefined Operations in Object Oriented Methods. In order to inform this problem let us consider how functionality extension and particularly adding operations works in OO specification and programming methods.

- *OO Specification Techniques.* A relatively large number of OO specification notions now exist, for example, OO versions of Z, such as Object-Z [Ros92] and ZEST [CR92], OO versions of VDM, such as VDM++ [Lan95] and Liskov and Wing's notation [LW93]. Subtyping is not handled in a uniform way throughout these techniques, so, let us focus on the Liskov and Wing approach which has considered the topic in some depth. In [LW93] a number of conditions are highlighted which must all hold in order to ensure subtyping between a pair of specifications. However, the part of the definition that concerns us here is the pre and post condition relationship between operations. The definition requires that for every operation in the supertype there must exist a corresponding operation in the subtype (although, the subtype may contain extra operations) such that, for corresponding operations, the following holds,

    1. the precondition of the supertype operation implies the precondition of the subtype operation, and

2. the postcondition of the subtype operation implies the postcondition of the supertype operation.

Thus through subtyping, preconditions can be weakened and postconditions can be strengthened. In informal terms, weakening of preconditions enables operations to be applied (i.e. terminate) in more states, while strengthening of postconditions reduces non-determinism. This really does give us what we seek: addition of traces and reduction of refusals when we take subtypes. In spirit, subtyping behaves like refinement in state based specification notations such as Z.

Importantly though, this interpretation of subtyping only works because applying an operation outside its precondition has a very different meaning than the analogous occurrence in process algebra. In process algebras the analogue of applying an operation outside its precondition is the environment trying to perform an action when it is not currently offered, which has the result *deadlock*. In contrast, in state based specification notations such as Z or Liskov and Wing's notation, applying an operation outside its precondition is *undefined*, i.e. is completely unpredictable. In an "operational sense" anything could occur and the choice between these alternatives is non-deterministic.

- *OO Programming Methods.* In strongly typed object oriented systems, it is not possible to call an operation which is not offered by an object. However, other OO systems produce error messages when a program calls an undefined operation, or result in undefined behaviour (such as the program crashing or giving incorrect results), e.g. Smalltalk [GR83].

So, both these OO settings give justification for the argument that attempting to apply an operation that is not currently offered should result in undefined behaviour and not deadlock.

Undefinedness and LOTOS Specifications. What, then, would be the consequence of adapting LOTOS specifications to behave in an undefined fashion if an action that is not currently offered is performed? Unpredictable behaviour can be modelled in LOTOS using non-determinism. In fact, we can highlight the following process:

$$\Omega := (\textbf{choice } a \in \textbf{Act } [] \; i; \; a; \; \Omega) \; [] \; (i; \; \textbf{stop})$$

which offers a completely non-deterministic behaviour; at every point in its evolution it could offer any action and refuse any set of actions. Since $\text{Tr}(\Omega){=}\text{Act}^*$ and $\forall \sigma \in \text{Act}^*$, $\text{Ref}(\Omega, \sigma){=}\mathcal{P}(\text{Act})$, Ω is at the top of the reduction preorder; every behaviour is a reduction of it.

It turns out that we will be able to use reduction as the subtyping relation if the LOTOS definitions of our objects' behaviours are modified using Ω. We will show how the modification is done with two examples.

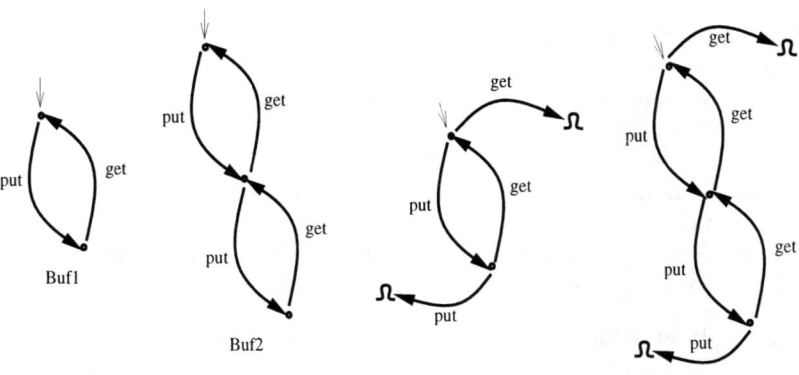

Figure 2 *Buf1* and *Buf2* without and with undefined added

Example 1. A one-place buffer, *Buf1*, was defined earlier. A two-place buffer may be defined:

$$Buf2 := put; Buf2a \qquad Buf2a := get; Buf2 \; [] \; put; get; Buf2a$$

The labelled transition systems corresponding to *Buf1* and *Buf2* are given in figure 2. For these definitions, $\mathcal{L} = \{put, get\}$.

We'd like the two-place buffer to be a subtype of the one-place buffer. Notice that, for the same reasons that we highlighted in our earlier example, as they stand, the two-place buffer is not compatible with the one-place buffer. Thus, to achieve this, we will modify the first two processes as shown in the right hand LTSs of figure 2.

We have added transitions such that every node has at least one transition leading away from it for every possible action in $\mathcal{L}$. Following any of the transitions we have added, the process evolves to Ω (this is in fact a relatively standard technique in process algebra which is used to enable parts of specifications to be extended when refining, see for example [LSW94]).

Using the fact that any behaviour reduces Ω, these two processes are now related in the way we wish; with the addition of undefined behaviour *Buf2* is both a reduction and a subtype of *Buf1*. To justify this, firstly observe that the traces of $\mathcal{T}(Buf2)$ and $\mathcal{T}(Buf1)$ (we will define the mapping $\mathcal{T}$, that adds undefined behaviour shortly) are the same, i.e. $\mathcal{L}^*$. This is because our transformation has ensured that at any state each process "may" perform any action in $\mathcal{L}$. Secondly, observe that for any trace in $\mathcal{L}^*$ the refusals of $\mathcal{T}(Buf2)$ are a subset of those of $\mathcal{T}(Buf1)$. Informally, $\mathcal{T}(Buf1)$ and $\mathcal{T}(Buf2)$ have identical refusals apart from those for traces of the form *put put* σ. For such traces, $\mathcal{T}(Buf1)$ will have evolved to undefined, and will thus refuse everything, while

$\mathcal{T}(Buf2)$ may still be performing defined behaviour, in which case it will refuse nothing. Thus, in addition, $\mathcal{T}(Buf1)$ is not a reduction/subtype of $\mathcal{T}(Buf2)$ since, for example, $\mathcal{T}(Buf1)$ can perform the trace *put put* and then refuse anything, while after the same trace $\mathcal{T}(Buf2)$ cannot refuse anything.

We introduce some terminology. The original LOTOS specification, i.e. before Ω's have been added, is called *the defined behaviour* of the specification, while the additional choices arising from the addition of Ω's is called the *undefined behaviour* of the specification. We call the LOTOS process resulting from the addition of undefined behaviour the *transformed process*, i.e. $\mathcal{T}$.

Example 2. Interestingly, using the label set $\{put, get, del\}$, when transformed *DelBuf* will be a subtype of *Buf1*. This is because in either of its defined states the transformed *Buf1* can perform a *del* and evolve to Ω. This contrasts with the approach taken in [Nie95], where Nierstrasz attempts to develop conditions that show that in their untransformed form *DelBuf* is not a subtype of *Buf1*.

5   ADDING UNDEFINEDNESS TO LOTOS SPECIFICATIONS

Transforming Specifications. Having introduced the concept of undefined behaviour we have to consider how to add this behaviour to LOTOS specifications in an automated way. There are three possible approaches; we could,

1. leave it in the hands of the specifier to explicitly include the undefined behaviour in their specifications;
2. develop a mapping which takes defined LOTOS specifications and maps them to LOTOS specifications with undefined behaviour; or
3. we could leave the LOTOS specifications unchanged, but rather add the undefined behaviour implicitly at the semantic stage.

Of these three, the first is not a feasible approach as it would make the specifier's task significantly more difficult. The second is feasible, however, defining the mapping is not straightforward. In particular, adding undefined behaviour through the parallel composition operator is quite subtle. Thus, it is the third of these alternatives that we select.

Our approach is to take the LTS of a LOTOS process and derive a new transition relation, which we denote $\vdash a \rightarrow$. This new transition relation will add states and transitions that reflect the required undefined behaviour. Where $\mathcal{L}$ is the label set of the specification we generate the smallest relation that satisfies the inference rules:

$$R1: \frac{P \xrightarrow{a} P'}{P \vdash a \rightarrow P'} \qquad R2: \frac{a \in \mathcal{L} \setminus initials(P)}{P \vdash a \rightarrow \Omega}$$

$$R3: \frac{a \in \mathbf{Act}}{\Omega \vdash i \rightarrow a; \Omega} \qquad R4: \frac{-}{a; \Omega \vdash a \rightarrow \Omega} \qquad R5: \frac{-}{\Omega \vdash i \rightarrow \mathbf{stop}}$$

$R1$ ensures that the new relation contains the relation $\rightarrow$. $R2$ adds the possibility to evolve to Ω when applying an action that is not currently offered. Rules $R3$, $R4$ and $R5$ code up the behaviour of the undefined process Ω.

Non-determinism. These inference rules define a simple means to add undefinedness to an LTS. For deterministic processes, the consequences of applying these rules are very straightforward. For example, the rules will map the first two LTSs of figure 2 to the second two LTSs. However, application of the rules is more subtle in the presence of non-determinism. Consider the following examples of the three archetypal forms of LOTOS non-determinism, with $\mathcal{L} = \{a, b, c\}$:

$$X := a; b; \mathbf{stop} \,[]\, a; c; \mathbf{stop} \quad Y := i; a; \mathbf{stop} \,[]\, i; b; \mathbf{stop}$$
$$Z := i; a; \mathbf{stop} \,[]\, b; c; \mathbf{stop}$$

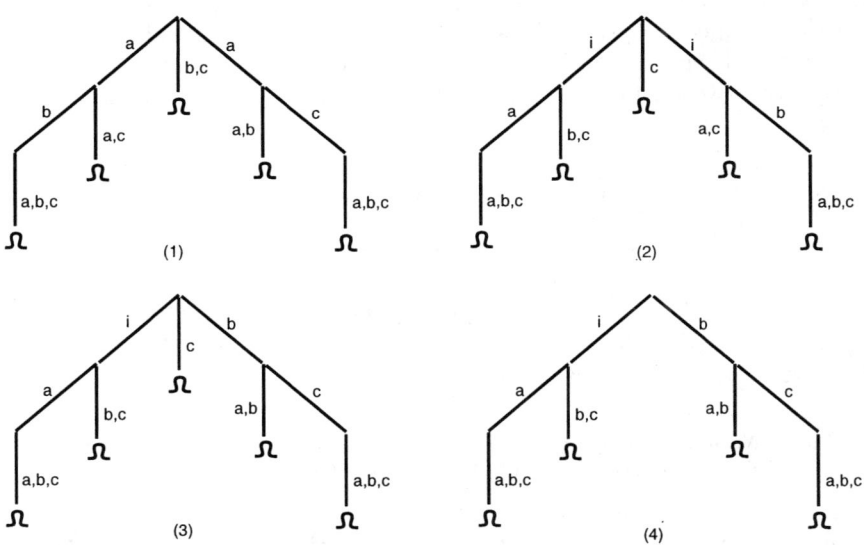

Figure 3 Transformed behaviours

LTSs resulting from adding undefinedness for each of these processes are shown as (1),(2) and (3) of figure 3. This transformation has the virtue of being extremely simple, however, it does not generate the minimum (in terms of least number of transitions) LTS. For example, in (3) the transition labelled c emanating from the start state is in fact redundant and (3) and (4) of figure 3 are testing equivalent.

A consequence of applying this transformation is that (modulo the addition of undefined behaviour) more processes are reductions than they would normally be. For example, once transformed all of the following behaviours would be reductions of Z.

a; **stop** $[]$ b; c; **stop** $[]$ c; **stop** a; **stop** $[]$ c; **stop** a; **stop** $[]$ b; a; **stop**

The last of these is perhaps the most suprising as the defined behaviour of the resulting specification requires an a action to be performed after the trace b, while Z requires a c action to be performed after the same trace. However, according to our intuition about subtyping in OO this is correct as the transformed Z can refuse the b action that leads to c, but cannot refuse the b action that leads to Ω. Thus, this situation is only odd if the processes are interpreted without undefinedness.

Further Examples. It is important to note though that while transforming LOTOS specifications in this way yields a more generous relationship between processes, which was after all our original intention, the resulting notion of subtying still remains sensitive to incompatible behaviour.

Consider two examples from [Nie95]: a variable and a non-deterministic stack:

$Var := put$; $Var2$ $Var2 := put$; $Var2$ $[]$ get; $Var2$
$NDstack := put$; $NDstack2$
$NDstack2 := put$; $NDstack2$ $[]$ get; $NDstack2$ $[]$ get; $NDstack$

Now, it can be checked that, $\mathcal{T}(Var)$ red $\mathcal{T}(NDstack)$ and $\mathcal{T}(NDstack)$ red $\mathcal{T}(Buf1)$, but $\neg(\mathcal{T}(NDstack)$ red $\mathcal{T}(Var))$ and $\neg(\mathcal{T}(Buf1)$ red $\mathcal{T}(NDstack))$. The latter two of these are because,

$Ref(\mathcal{T}(NDstack), put\ get\ get) = \mathcal{P}(\mathcal{L}) \not\subseteq \{\emptyset\} = Ref(\mathcal{T}(Var), put\ get\ get)$
$Ref(\mathcal{T}(Buf1), put\ put) = \mathcal{P}(\mathcal{L}) \not\subseteq \{\emptyset\} = Ref(\mathcal{T}(NDstack), put\ put)$

In addition, we can handle data passing processes by, in the usual way, expanding full LOTOS into basic LOTOS, using a richer action set and choice to model input alternatives *. The following two processes, an infinite stack and an infinite queue, are written in a pseudo full LOTOS.

$Stack(l : list) := put?x : nat$; $Stack(x\#l)$ $[]$ $[l \neq []] \rightarrow get!\ hd(l)$; $Stack(tl(l))$
$Queue(l : list) := put?x : nat$; $Queue(x\#l)$ $[]$ $[l \neq []] \rightarrow get!lst(l)$; $Queue(frnt(l))$

---

*There are actually some subtleties in how data passing has to be handled which we do not have space to discuss here. One issue is that mapping full LOTOS to basic LOTOS generates a deterministic modelling of output which is not always what is required. Ongoing research is currently seeking to resolve these issues.

where $x\#[x_1,..,x_n] = [x,x_1,..,x_n]$, $lst([x_1,..,x_n]) = x_n$, $frnt([x_1,..,x_n]) = [x_1,..,x_{n-1}]$ and hd, tl and empty lists, denoted $[\,]$, are treated in the usual way.

As would be expected these two behaviours are incomparable. The following trace/refusal properties demonstrate this (where $\sigma = put\_1\,put\_2\,get\_1$, $\sigma' = put\_1\,put\_2\,get\_2$ and a_v denotes the occurrence of an action at gate a with data value v):

$$Ref(\mathcal{T}(Stack([\,])),\sigma) = \mathcal{P}(\mathcal{L}) \not\subseteq \{\emptyset\} = Ref(\mathcal{T}(Queue([\,])),\sigma)$$
$$Ref(\mathcal{T}(Queue([\,])),\sigma') = \mathcal{P}(\mathcal{L}) \not\subseteq \{\emptyset\} = Ref(\mathcal{T}(Stack([\,])),\sigma')$$

As these examples demonstrate, transformed behaviours have a very precise trace/refusal character. Transformed specifications can perform any trace in $\mathcal{L}^*$ and after all traces either refuse nothing or refuse everything. One consequence of this is that for transformed specifications $red = ext = conf$. This is good news as it has previously been argued [BS86] that checking trace subsetting is a major hindrance to verifying reduction. In fact, this is one of the reasons that Brinksma considered $conf$ in the first place. In addition, the normal relationships between the LOTOS equivalences still hold, i.e. $\sim \subset \approx \subset te$.

A Note on Undefined Behaviours. Up to testing equivalence, there are actually several different processes that could be used as Ω. For example, both the following two processes have the same trace/refusal characterisation as Ω.

$$\Omega' := (\textbf{choice } a \in \textbf{Act } [\,]\ i;\ a;\ \Omega')\ [\,]\ (i;\ \textbf{stop})\ [\,]\ (i;\ \Omega')$$
$$\Omega'' := (\textbf{choice } a \in \textbf{Act } [\,]\ a;\ \Omega'')\ [\,]\ (i;\ \textbf{stop})$$

One of the reasons for this is that LOTOS trace/refusal semantics are not sensitive to divergence. Thus although from amongst these processes, Ω' is divergent and Ω and Ω'' are not, the three processes have the same semantic characterisation. The decision not to be sensitive to divergence concurs with the approach taken in bisimulation semantics [Mil89] and is tied to a subtle debate concerning *fair abstraction* [BBK87].

However, it should be pointed out that other models handle this issue differently. For example, in CSP, which employs a chaotic interpretation of divergence, only Ω' would give the most unpredictable behaviour.

6 CONCLUSIONS

A criticism of the approach to adding behaviour that we have presented here is that it is not very refined; a path to undefined is added at any state for any action that is not currently offerred. A more refined approach would allow the specifier to obtain refusal when (s)he wishes and undefined when (s)he wishes. This is an area of ongoing research.

To summarise, then, we have considered the spectrum of LOTOS correctness relations and argued that all fail in some respect to be a suitable instantiation of behavioural subtyping. Then through consideration of how subtyping behaves in OO specification and programming notations we have motivated a re-interpretation of LOTOS specifications in the OO setting. This involves adding undefined behaviour to LOTOS specifications. We have defined a simple LTS based mapping to add undefined behaviour to LOTOS specifications. The main consequence of applying this mapping is that the most well behaved of the LOTOS refinement relations, *reduction, really is behavioural subtyping*.

Acknowledgements. We would like to thank Tim Regan and Steve Rudkin for giving an initial stimulus to this research. In particular, Tim Regan was involved in preliminary discussions from which this paper has evolved. Also thanks to Benjamin Pierce for giving pointers to relevant literature and to Erik Poll and the anonymous referees for useful comments and suggestions.

REFERENCES

[BB88]   T. Bolognesi and E. Brinksma. Introduction to the ISO Specification Language LOTOS. *Computer Networks and ISDN Systems*, 14(1):25–29, 1988.

[BBK87]   J.C.M. Baeten, J.A. Bergstra, and J.W. Klop. On the consistency of koomen's fair abstraction rule. *Theoretical Computer Science*, 51:129–176, 1987.

[BDLS95]   H. Bowman, J. Derrick, P. Linington, and M. Steen. FDTs for ODP. *Computer Standards and Interfaces*, 17:457–479, September 1995.

[BS86]   E. Brinksma and G. Scollo. Formal notions of implementation and conformance in LOTOS. Technical Report INF-86-13, Dept of Informatics, Twente University of Technology, 1986.

[CR92]   E. Cusack and G. H. B. Rafsanjani. ZEST. In S. Stepney, R. Barden, and D. Cooper, editors, *Object Orientation in Z*, Workshops in Computing, pages 113–126. Springer-Verlag, 1992.

[CRS89]   E. Cusack, S. Rudkin, and C. Smith. An object oriented interpretation of LOTOS. In *Proceedings 2nd International Conference on Formal Description Techniques (FORTE'89)*. North-Holland, December 1989.

[DEBS96]   J. Derrick, E.A.Boiten, H. Bowman, and M. Steen. Supporting ODP - translating LOTOS to Z. In *First IFIP International workshop on Formal Methods for Open Object-based Distributed Systems*, Paris, March 1996. Chapman & Hall.

[FM94]   Kathleen Fisher and John C. Mitchell. Notes on typed object-oriented programming. In *Proceedings of Theoretical Aspects of Computer Software (TACS '94), Sendai, Japan*, volume 789 of

LNCS, pages 844–886. Springer, 1994.

[GR83]   A. Goldberg and D. Robson. *Smalltalk-80: The Language and its Implementation.* Addison-Wesley, 1983.

[Hen88]   M. Hennessy. *Algebraic Theory of Processes.* MIT Press, 1988.

[Hoa85]   C. A. R. Hoare. *Communicating Sequential Processes.* Prentice Hall, 1985.

[ISO87]   ISO 8807. *LOTOS: A Formal Description Technique based on the Temporal Ordering of Observational Behaviour,* July 1987.

[ITU95]   ITU Recommendation X.901-904 — ISO/IEC 10746 1-4. *Open Distributed Processing - Reference Model - Parts 1-4,* July 1995.

[Lan95]   K. Lano. Specification of distributed systems in VDM++. In *FORTE'95.* Chapman and Hall, 1995.

[LSW94]   K.G. Larsen, B. Steffen, and C. Weise. A constraint oriented proof methodology based on modal transition systems. Technical Report RS-94-47, University of Aarhus, 1994.

[LW93]   B. Liskov and J. M. Wing. A new definition of the subtype relation. In O. M. Nierstrasz, editor, *ECOOP '93 - Object-Oriented Programming,* LNCS 707, pages 118–141. Springer-Verlag, 1993.

[MC93]   A.M.D. Moreira and R.G. Clark. ROOA: Rigorous Object-Oriented Analysis. Technical Report TR 109, Computing Science Department, University of Stirling, Scotland, October 1993.

[Mil89]   R. Milner. *Communication and Concurrency.* Prentice-Hall, 1989.

[Nie95]   O. Nierstrasz. Regular types for active objects. In *Object-oriented Software Composition,* pages 99–120. prentice-Hall, 1995.

[NSF94]   E. Najm, J-B. Stefani, and A. Fevrier. *Introducing Mobility in LOTOS.* ISO/IEC JTC1/SC21/WG1 approved AFNOR contribution, July 1994.

[Pun96]   Franz Puntigam. Types for active objects based on trace semantics. In *First IFIP Workshop on Formal Methods for Open Object-Based Distributed Systems,* Paris, March 1996. Chapman & Hall.

[Ros92]   G.A. Rose. Object-Z. In S. Stepney, R. Barden, and D. Cooper, editors, *Object Orientation in Z,* Workshops in Computing, pages 59–78. Springer-Verlag, 1992.

[Rud91]   S. Rudkin. Inheritance in LOTOS. In K. R Parker and G. A. Rose, editors, *Formal Description Techniques, IV,* Sydney, Australia, November 1991. North-Holland.

[Smi95]   G. Smith. Extending $\mathcal{W}$ for Object-Z. In J. Bowen and M. Hinchey, editors, *9th International Conference of Z Users,* volume 967 of *Lecture Notes in Computer Science,* pages 276–295. Springer-Verlag, 1995.

[vG93]   R.J. van Glabbeek. The linear time - branching time spectrum (I and II). In *Concur'90 and Concur'93, LNCS 458 and LNCS 715.* Springer-Verlag, 1990 and 1993.

24

On the Specification, Inheritance, and Verification of Synchronization Constraints

Neelam Soundarajan
Computer and Information Science Depertment
Ohio State University
2015 Neil Avenue, Columbus, OH 43210, USA.
neelam@cis.ohio-state.edu

Abstract

Object-orientation and distributed systems are a natural match. Objects correspond to processes in a distributed program; the invocation of a method of one object by another object corresponds naturally to a message being passed between the corresponding processes in the distributed program. Despite this close correspondence, progress in developing an OO approach to concurrency has been limited. One important problem has been the so-called *inheritance anomaly* which is concerned with *how* and *how easily* synchronization constraints specified in a base class may be modified in a derived class. Our concern in the current paper is slightly different. We are interested in developing ways to abstractly *specify* these synchronization constraints, and ways to *verify* them. In other words we are interested in *what* these synchronization constraints do, and this is, of course, the critical question from the point of view of the users of these objects. We use the mechanism of *acceptance sets* in our specifications. We develop a proof method to verify that (base as well as derived) classes meet their specifications. We also consider the question of what kinds of modifications of synchronization constraints in the derived classes are easy for the clients of the class to deal with.

Keywords

Specification and verification, Synchronization constraints, Inheritance anomaly

1 INTRODUCTION

Object-orientation (OO) and distributed systems are a natural match. Active, autonomous objects correspond naturally to processes in a distributed program. Interactions between these objects in the form of messages, i.e., method invocations to carry out various tasks, correspond to interactions between the processes. Since individual objects do not have access to the internals of other objects, they can exist on separate machines with no shared memory. Despite this close correspondence, progress in using OO ideas in distributed systems has been slow.

Inheritance is one of the cornerstones of the object-oriented approach. It not only allows us to create classes in an incremental manner from existing classes, without having to repeat what the base class already provides. Although some authors have criticized inheritance, others like Meyer (1988) have presented persuasive arguments in its favor. One of the problems in applying OO ideas to concurrent programming is the *inheritance anomaly* (Matsuoka, Yonezawa 1993). To see the problem, consider the standard example of a *bounded-buffer* class. The problem arises because the code that is needed to ensure that appropriate synchronization constraints (such as not reading from an empty buffer, or not writing to a full buffer) is often interspersed with the code of the methods (such as *get* and *put*); as a result, if we attempt to develop a derived class, for example a *better-buffer* class that provides an additional operation *get2* that allows us to read two elements from the buffer, we may be required not only to provide the code for this additional operation, but also to rewrite the code of some of the existing operations to take account of appropriately synchronizing with the new operation. In other words, although the code for *performing* the existing operations, *get* and *put* in the case of the *bounded-buffer* class, has not changed –indeed that is the reason we are trying to use inheritance and *reuse* these operations defined in the base class– we are obliged to rewrite them anyway in order to encode the new synchronization conditions needed in view of the new operation. Various notations have been proposed to solve this problem, and we will consider some of them later in the paper.[*]

---

[*] Matsuoka, Yonezawa (1993) consider various schemes that have been proposed to specify the constraints and for each scheme present an example that forces us revisit the code in the base class because of the way in which the synchronization constraints are expressed.

Our interest in the current paper is somewhat different. From the point of view of the *client* of the class(es), the important question is not how the classes are constructed, but how they can be used. To serve this purpose, we need formal and abstract specifications of both the functional properties of the various operations provided by the classes, as well as their synchronization properties. Further, we need appropriate axioms and proof rules that can be used by the class designers to establish that the operations, either inherited from the base class, or defined or redefined in the derived class, do indeed meet these specifications.*
We develop a simple notation for expressing such specifications, and develop a method for showing that classes do meet their specifications. While object-oriented distributed systems can be expected to be easier to design and understand than non-OO systems, easy-to-use specification notations and verification methods of the kind developed in this paper are important to ensure that the resulting systems are reliable and behave as they are intended to.

Inheritance anomaly which is essentially a problem with how synchronization constraints are implemented in certain classes, has a counterpart in the specification/verification task. Indeed there it appears even in the absence of synchronization issues. The problem arises because an operation defined in a derived class may modify base class member variables. As a result, the derived class designer may be forced to reverify all of the base class operations, including those that are not redefined in the derived class, since these operations also use the same variables. We have developed a formal approach in (Soundarajan, Fridella 1997) to simplify this task in the sequential case. The key idea behind the approach in (Soundarajan, Fridella 1997) is to have *two* formal models of the (base) class. The first is the usual, abstract model for use by the client of the class. The second, for use by the designer of the derived class, is a *concrete* model in which operations of the (base) class are specified in terms of pre- and post-conditions on the *actual* data structures used in the representation, rather than in terms of an abstract math model. As a result, the derived class designer only needs to ensure that when an operation defined (or redefined) in the derived

---

*We will also need an appropriate formalism that clients of the class can use to show, given the specifications of various classes, that their own code that uses instances of these classes, is correct, i.e., is free of deadlock, or exhibits appropriate functional behavior, etc. We will not consider this formalism in the current paper. It should be possible to design such a formalism along the lines of existing systems such as (Soundarajan 1984), (Misra, Chandy 1981) that are applicable in the absence of user defined classes.

class finishes, the values of the various member variables including those inherited from the base class are acceptable to the operation that might be next invoked, according to the concrete pre-conditions of the various operations. The key point is that the concrete specification in terms of the actual data structures contains the additional information that a derived class designer needs, and that is missing from the abstract specification. With this additional information in hand, the derived class designer does not need to look into the actual code of the base class operation.

The formalism we develop in the current paper is a natural extension of that in (Soundarajan, Fridella 1997). Each class will have two specifications associated with it, an *abstract specification* for use by a client of the class, and a *concrete specification* for use by a derived class designer. Each specification will give us information about the functionality, i.e., input-output behavior of each operation in the class, as well as –and this is the extension being proposed in the current paper– about the synchronization properties, i.e., what operations may be invoked at various points during the execution of the program. As in (Soundarajan, Fridella 1997), the abstract specification will be in terms of an abstract mathematical model, whereas the concrete specification will be in terms of the actual data structures used in the class. We should note that we are *not* proposing a solution to the inheritance anomaly. Rather we are addressing the question of how behavior of classes, including synchronization behavior, can be specified and verified especially in the case when we use inheritance in defining some of the classes.

In the next section of the paper we introduce our specification notation by means of a few simple examples. In the third section we present our proof system using which we will be able to establish that a given class meets its specifications. One important point to note is that our specification notation as well as the general method of verifying that a given class meets its specifications is applicable to the various (programming) notations that have been proposed in the literature. Of course, the details of the proof method, and to an extent the primitives needed to express appropriate properties, do depend to an extent on the details of the programming mechanisms under consideration, but the overall approach is generally applicable. Thus in the third section we consider two different approaches (those of (Frolund 1992) and

(Kafura, Lee 1989)) to using inheritance in conjunction with concurrency and consider how our specification and verification method may be applied in each case. In the final section we briefly summarize the main ideas behind our approach, and consider some open problems.

2   ABSTRACT AND CONCRETE SPECIFICATIONS

Consider a class C with public methods $m_1, \ldots, m_n$. (We will generally use C++-like terminology and notation.) We will provide two specifications for C, an *abstract specification* for use by clients of C, and a *concrete specification* for use by designers of derived classes of C. Let us first consider the abstract specification. A client of C will need, first, the pre- and post-conditions of each m_i in terms of an abstract model of C. Second, the client must be able to tell at what points in the execution of the system the various methods of C will be enabled.* The most direct way of specifying this information is in terms of the value of the *acceptance set*, i.e., the set of methods that are enabled at the point in question. This specification could be provided in the form of a function over the *trace* or sequence of calls made so far to the various methods of C. In other words, we could specify a function $\mathcal{A}$ whose value $\mathcal{A}(t)$ for any given trace t is the subset of methods $\{m_1, \ldots, m_n\}$ that can be invoked at a given point if the sequence of methods that have been invoked thus far is t. While this approach would allow us to handle all types of synchronization schemes, it is usually more than is needed. A simpler approach would be to specify a *synchronization property* SP_C that imposes appropriate restrictions on the value of the acceptance set $\mathcal{A}$ in terms of the abstract model of C, and this is the approach we will use. (One disadvantage of the simpler approach is that we may not be able, as we will see in a later example, to deal with all possible types of synchronization behaviors.)

Let us consider the following class Bbuffer, the standard *bounded-buffer* example:

```
class Bbuffer {
// Buffer of size n
public:
```

---

* If an operation is invoked when it is not 'enabled', the call will be suspended, to be carried out at a later time when the operation is enabled. The call can become enabled at a later time because of calls from other concurrent clients who share this object with the current client.

```
        Bbuffer();
        void put(int k);
        int get( );
     protected:
        ⋮
};
```

A simple abstract model $M_{\texttt{Bbuffer}}$ for this class would just be a sequence of integers with initial value the empty sequence $\langle\rangle$. The put operation would append its argument to the current value of the sequence; the get operation would return the first element of the sequence as its return value and remove this element from the sequence. Thus the first part of the abstract specification, the pre- and post-conditions of these operations is:

$$\{\ \textit{true}\ \}\ \texttt{put}(\ \texttt{k}\ )\ \{\ \textit{self}_a = \#\textit{self}_a \,\hat{}\, \texttt{k}\ \}$$

$$\{\ \textit{true}\ \}\ \texttt{get}(\ )\ \{\ \textit{value} = \textit{head}(\#\textit{self}_a) \wedge \textit{self}_a = \textit{tail}(\#\textit{self}_a)\ \}$$

where $\textit{self}_a$ is the abstract bounded buffer object, $\textit{value}$ denotes the value returned by the operation under consideration; $\hat{}$ is the *append* operation; *head*, and *tail* are the usual functions on sequences; $\#$ denotes the value of the variable that follows the $\#$ at the start of execution of the operation in question.

Note that the pre-conditions of the operations are *true* rather than that there must be space in the buffer, or that the buffer must be non-empty respectively. That is because these conditions are part of the synchronization constraints, rather than the pre-conditions of the respective operations. If we had included these conditions as part of the pre-conditions, then that would mean that if the client were to invoke, say, the get operation when the buffer was empty, the (class) would be at liberty to do anything (such as returning a random value) since its pre-condition would not have been satisfied. By expressing the condition as part of the synchronization constraint, the client is assured that in this circumstance, the call will be suspended, to be resumed if and when the appropriate synchronization constraint is satisfied – as a result of calls by other concurrent clients who share this buffer. This will happen if the other client invokes the put operation, after the execution of which the buffer will no longer be empty.

So next we need to specify the abstract synchronization property, $ASP_{\texttt{Bbuffer}}$ which will capture these synchronization constraints:

$$ASP_{\text{Bbuffer}} \equiv [\,(self_a \neq \langle\rangle \Rightarrow \text{get} \in \mathcal{A}_{\text{Bbuffer}})$$
$$\wedge\,(|self| < n \Rightarrow \text{put} \in \mathcal{A}_{\text{Bbuffer}})\,]$$

where $\mathcal{A}_{\text{Bbuffer}}$ is the *acceptance set* of Bbuffer, i.e. the set of all methods of Bbuffer that can currently be invoked. ASP_{Bbuffer} essentially states that the method get can be invoked if the buffer is not empty, and that put can be invoked if the buffer is not full.

Next, consider the *concrete specification* of a class C. While the abstract specification of C gives us information about the functionality of each method and the synchronization properties of the various methods of C in terms of its abstract model, the concrete specification of the class will provide us with similar information in terms of the *internal* representation, i.e., the actual data members of C. The client of the class of course has no use for this information since she has no direct access to the data members of the class. (We are assuming here, following widely accepted principles of OO design, that no data members of a class are declared public.) But a *designer* of a class that inherits from C is in acute need of this information. Without this information that designer would be forced to study the actual code of the various operations of C to understand how they operate on the various data members, as well as the synchronization code to see what conditions the values of the data members must satisfy in order for specific operations to be enabled. It is only then that this designer will be able to ensure that the new operations that she introduces in the derived class, as well as redefinitions that she makes of operations inherited from the base class, are consistent with the design of the base class. Our concrete specification for C will provide this information in the form of pre- and post-conditions on the actual data members for each operation of C; and a synchronization property, again on the data members. We will term these pre- and post-conditions and the synchronization property 'concrete' in order to distinguish from their counterparts in the abstract specification. In addition, as in (Soundarajan, Fridella 1997), the concrete specification should include an invariant on the data members of the class; since our interest in this paper is on synchronization issues, we will generally ignore this invariant. Thus the concrete specification of C *will* contain more information than its abstract specification, but the details of the actual code bodies of the various operations of C will be abstracted away. That is the big advantage for the derived class designer; she needs to deal only with the concrete specification of C, not the actual code.[*]

Consider again the Bbuffer class. Suppose we use an array Elems[0:n-1] to store the elements currently in the buffer, and two variables in and out as pointers into this array to indicate where the next element added to the buffer must be stored, and where the next element read from the buffer should be fetched from. Thus the protected part of the class would look like:

```
protected:
  int Elems[n];
  int in, out;
```

The constructor function of the class will simply initialize in and out to 0. The operations put and get will store elements into the buffer and return elements from the buffer using the Elems array as a 'circular' array, i.e., increasing, modulo n, the appropriate in or out pointer by 1.

Thus the concrete pre- and post-conditions of these operations are:

$$\{ \textit{true} \} \, \texttt{put(k)} \, \{ \texttt{in} = \#\texttt{in} \oplus 1 \wedge \texttt{Elems} = \#\texttt{Elems}[\#\texttt{in} \leftarrow \texttt{t}] \}$$

$$\{ \textit{true} \} \, \texttt{get( )} \, \{ \textit{value} = \#\texttt{Elems}[\#\texttt{out}] \wedge \texttt{out} = \#\texttt{out} \oplus 1 \}$$

where the post-condition of put asserts that the value of Elems when put finishes is the same as when it started except the value of Elems[#in] is k. (The concrete invariant for this class would contain clauses like $0 \leq \texttt{in} < \texttt{n}$.)

As in the case of the abstract specification, the pre-conditions of the operations are just *true*. The requirements of a non-empty buffer for get and a non-full buffer for put are captured by the concrete synchronization property:

$$CSP_{\texttt{Bbuffer}} = [\, (\texttt{in} \neq \texttt{out} \oplus 1 \Rightarrow \texttt{put} \in CA_{\texttt{Bbuffer}}) \\ \wedge (\texttt{out} \neq \texttt{in} \Rightarrow \texttt{gett} \in CA_{\texttt{Bbuffer}}) \,]$$

where we have used $CA_{\texttt{Bbuffer}}$ to denote the concrete acceptance set of Bbuffer, although in fact there is no difference between the concrete and abstract acceptance sets.

So far we have only considered the specifications for the class. We still need to verify that the class, as actually implemented, meets these

---

[*] In this paper we will assume that all data members are protected, there being no private variables. Recall that private variables would not be accessible even in the derived class. In (Soundarajan, Fridella 1997) we explain how we can deal with both kinds of variables.

specifications. This verification method is the topic of the next section. Once that is done though, neither the client of the class nor the designer of the derived class would have to look at this actual implementation.

3   VERIFICATION OF CLASS BEHAVIOR

The verification task may be divided into two parts. First, we need to verify that the class operations and synchronization requirements as actually implemented, meet the concrete specification of the class. Second, that the concrete specification in some precise sense implies the abstract specification. The second part is the easier of the two and we deal with it in 3.1. How we handle the first part depends upon the details of the particular programming notation, specifically the notation used to express synchronization code. In 3.2 we consider two very different programming notations and explain how to deal with each one.

3.1   Verifying Abstract Specifications

In order to show that the concrete specification implies the abstract one, we must first define an abstraction function ε that maps the concrete state to the abstract state. Thus if ω is a concrete state, $\varepsilon(\omega)$ is the corresponding abstract state as seen by a client of the class. Note that since the abstract acceptance set $A\mathcal{A}$ is the same as the concrete one $C\mathcal{A}$, ε will map $C\mathcal{A}$ to $A\mathcal{A}$. Next let us introduce some conventions. Let f be any of the methods of the class. Let $a.pre_f$ and $a.post_f$ be its abstract pre- and post-conditions; similarly $c.pre_f$, $c.post_f$ are its concrete pre- and post-conditions. CSP be the concrete synchronization property, and ASP the abstract one. Then in order to show that the concrete specification implies the abstract one we need to establish the following $((1a)$ is for each f):

$$a.pre_f(\varepsilon(\omega)) \Rightarrow c.pre_f(\omega) \tag{1a}$$
$$c.post_f(\omega) \Rightarrow a.post_f(\varepsilon(\omega))) \tag{1b}$$
$$CSP(\omega) \Rightarrow ASP(\varepsilon(\omega)) \tag{2}$$

These implications simply represent the fact that the abstract and concrete specifications are directly related via the mapping function ε.

$(1a)$ and $(1b)$ are simplified versions of the corresponding rules in (Soundarajan, Fridella 1997); the rules in (Soundarajan, Fridella 1997) take account of the concrete invariant. We should also note here that while $(1a)$, $(1b)$, (2) are generally sufficient, they may have to be modified, as we will see in section 3.2, depending on the details of the programming notation.

3.2 Verifying Concrete Specifications

Now consider the first part of the verification task. This task consists of two subtasks. First we must verify that the bodies of the various methods meet their respective concrete specifications, i.e., their pre- and post-conditions. Second that the class as a whole meets CSP, the concrete synchronization property. So far we have been able to ignore the programming language notation in which these method bodies and synchronization code is written but now we must take account of this notation. The interesting question here is how the various notations that have been proposed in the literature (such as those of (Frolund 1992), (Kafura, Lee 1989), (Ishikawa 1992), (Matsuoka, Yonezawa 1993)) for specifying the synchronization conditions, may be handled.* We will consider, in turn, two of these notations – those of (Frolund 1992), and (Kafura, Lee 1989); we believe a similar approach will work with the others.

Frolund [(Frolund 1992)] proposes a natural approach to expressing synchronization code in classes. His proposal is to use *method guards* to express synchronization conditions. A method guard specifies a condition that *disables* a particular operation. Thus, for instance, in the case of the bounded buffer, one of the method guards would be:

$(\text{out} = \text{in})$ *prevents* get

This states that the get operation is *disabled* if the condition $(\text{out} = \text{in})$ is satisfied.

How do we capture, in our specification, the effect of such a method guard? Consider, for instance, the guard above. All we need to do, corresponding to this guard, is to include the following clause in our (concrete) synchronization property CSP:

---

* We will assume that the non-synchronization portion of the individual methods are written using a standard collection of constructs and will omit discussion of them.

$$(\texttt{out} = \texttt{in}) \;\Rightarrow\; \texttt{get} \notin C\mathcal{A}_{\texttt{Bbuffer}}$$

Thus each method guard g in the program would be represented in the *CSP* by the corresponding clause $CSP(g)$ as above. If a class has guards $g_1, \ldots, g_m$, the *CSP* for this class is be $CSP(g_1) \wedge \ldots \wedge CSP(g_m)$; this assumes that this class is a base class.

An important part of Frolund's proposal is that method guards become increasingly more restrictive in derived classes. In other words derived classes can only *add* new method guards; method guards of the base class will necessarily be inherited by the derived class and become part of its collection of method guards. Correspondingly all we need to do in our formalism, to obtain the CSP_D corresponding to a derived class is to conjunct to the base class CSP_B, the clauses $CSP(g_1) \wedge \ldots \wedge CSP(g_m)$ corresponding to the guards $g_1, \ldots, g_m$ of the derived class.

Two final points are worth noting: The requirement that synchronization conditions can only become more restrictive in derived classes does mean that Frolund's approach will not work in certain circumstances. For instance, if we wanted to define a derived class of the Bbuffer class which was of size $2n$ (where instances of Bbuffer class are of size n), we could not do so. Nevertheless, it is partly because of this restriction that our formalism corresponding to this approach turned out to be relatively simple. Second, Frolund has a construct called all-except which allows us to write guards that disable all methods except those that are named in the statement. To handle this construct, we would have to assume that we have a set called, say, *AllMethods* which is the set of all method names. The clause corresponding to an all-except guard would then simply say that none of the method names in *AllMethods* less those named in the construct are not in the set $C\mathcal{A}$ of the class (if the specified condition is satisfied).*

Next consider the approach of Kafura and Lee (Kafura, Lee 1989). They introduce the notion of *behavioral abstraction* which is more or less the same concept as our acceptance set. The value of the behavioral abstraction at any point is the set of method names that are *enabled* at that point. Each method body, as its last action, is required to execute a become command that essentially assigns a new value to the acceptance set. The reason for the name 'behavioral abstraction'

---

*Note that this does *not* say that the methods that are named in the construct *are* actually in $C\mathcal{A}$; some or all of them may well be prevented from being in $C\mathcal{A}$ because of other guards.

is that separately from the bodies of the individual methods, in a (protected) section that may be called the 'behavior' section, the class designer identifies a number of subsets of the set of all methods, and names these subsets with distinct names, say $B_1, \ldots, B_m$. Note that these subsets need not, and usually are not, disjoint from each other. The become statement at the end of each method is then of the form become B_j which says that when this method finishes, the set of enabled methods will be all (and only) those whose names appear in B_j. Thus the details of which methods are enabled are partly abstracted away in the definitions in the behavior section.

Further, when a class is defined by inheritance from another class, the derived class introduces its own behavior section. The corresponding behavior abstractions, let us call them, $B'_1, \ldots, B'_m$ are defined either from scratch or in terms of the $B_1, \ldots, B_m$ of the base class. An important construct that Kafura and Lee introduce is the redefines construct. This allows us them to redefine, in the derived class, the set of methods in a given behavior abstraction B_i of the base class. The advantage of this is that code that is inherited from the base class –although it has apparently not been rewritten– may do something different than it does in the base class! To see how this may happen, suppose that in a base class C the behavioral abstraction B_1 has been defined as the set of methods $\{g_1, g_3\}$. Suppose also that the method g_4, when it finishes, executes the command become B_1. Suppose also that in a derived class $C1$ of C we redefine B_1 to be the set $\{g_1, g_2\}$. Then, if we consider a call a.g_4 where a is an instance of C, at the end of the call the enabled methods will be $\{g_1, g_2\}$. If we consider a call b.g_4 where b is an instance of C_1, at the end of *this* call the enabled methods will be $\{g_1, g_2\}$! This is the important capability of their approach that allows Kafura and Lee to tackle several cases of inheritance anomaly – because with this mechanism, they are not forced to rewrite the code of operations like g_4 just in order to change, in the derived class, the set of methods enabled when it finishes.

How do we handle such a mechanism in our approach to specification and verification? We should first make an important point: The *client* of the classes C and C_1 has no interest in mechanisms like behavioral abstractions used by the designers of these classes to avoid the inheritance anomaly. What the client is interested in, as far as synchronization issues are concerned, is only information about the set of operations that are enabled at various points. Correspondingly, in our formalism, no information about behavioral abstractions will appear in the abstract specification of a class. This information must, however, appear in the concrete specification, for use by the derived class designer.

In order to deal with behavioral abstractions, we need to split the concrete synchronization property CSP_C of the class C into two parts. The first part, $CSP^d{}_C$ will give us the *definitions* of the various behavioral abstractions, i.e., for each B_i defined in the class, the set of method names corresponding to B_i. The second part, $CSP^c{}_C$ will specify the actual synchroniztion conditions, but in terms of the behavioral abstractions, rather than in terms of the actual method names. Moreover, we also need to include information about which behavioral abstraction will be enabled at the end of each method of the class, since this information is also of great importance to the derived class designer, as should be clear from the earlier discussion involving methods g_1, g_2 etc. In order to do this, $CSP^c{}_C$ will be expressed as a conjunction of a number of clauses, each of the form:

after g_i :: $(b \Rightarrow B_j)$

where b is a boolean (on the member variables of C). This clause means that following the execution of g_i, if the condition b is satisfied, then the set of methods in B_j will be enabled. The abstract synchronization property, ASP_C, will also consist of such clauses, except that rather than the behavioral abstractions B_j, the actual set of method names will appear in the corresponding clauses, and the conditions b will be on the abstract model of the class, rather than the concrete member variables.

$CSP^d{}_C$ will be established on the basis of the definition in behavior section of the class C. Essentially it is identical to this section if C is a base class. For a derived class $C1$ of C, we obtain $CSP^d{}_{C1}$ from $CSP^d{}_C$ by adding the new definitions provided in $C1$ and replacing the ones that are redefined in $C1$.

The individual clauses of $CSP^c{}_C$ will be established by considering the individual methods of C. If a method g_i is inherited without change from the base class, then we need do no work as far as the corresponding clauses in $CSP^c{}_C$ are concerned. This is the verification analog of the implementation reuse of Kafura and Lee's system. If the derived class implementor can reuse the base class operation although the behavioral abstraction that it enables when it finishes may mean a different set of methods in the derived class, then we, as verifiers, do not have to reverify the method body either. Of course, if the method g_i is redefined in the current class, or if it is entirely new, then we do have to verify the appropriate clause of CSP^c. To do this, we need a little bit of formal machinery to allow us to establish the appropriate clauses of CSP^c, i.e., the ones that begin *after* g_i ; we will omit the details of this machinery since it is fairly straightforward.

One final point is worth noting. When we apply the rule (2) of section 3.1 to the system of behavioral abstractions, we should note that CSP is the conjunction of CSP^d and CSP^c. The former gives us the sets of methods corresponding to the various B_j's that appear in the latter and it is these method names that appear in ASP. It is not clear to us whether information about different methods being enabled following calls to certain other specific methods is of interest to the client of the class or not. In any case, if we decide that it is, we will have to write not just the concrete synchronization property, but also the abstract one as a conjunction of after g_i clauses.

4 DISCUSSION

One school of OO thought (see, for example, (Liskov, Wing 1994)) has argued that inheritance should be used in a very limited way, specifically that it should not be used for purely implementation purposes. In other words, if D is a derived class of C, then it must be the case that instances of D behave in ways that are strictly consistent with behaviors that instances of C might exhibit. But other schools disagree. And indeed, it would seem that a better approach would be to develop ways of reasoning about different types of inheritance usage rather than disallow some usages by fiat. In the case of using OO ideas for concurrent or distributed systems, various proposals in the literature (Frolund (1992), Kafura, Lee (1989), Ishikawa (1992), Matsuoka, Yonezawa (1993)) for expressing synchronization constraints have been designed to allow derived classes to exhibit behavior, in particular synchronization behavior, different from that exhibited by the base class. An important consideration in all of these proposals is, of course, to try to limit the amount of work that needs to be done in the derived classes, and to take as much as possible from the base class.

Our goal in this paper has been to develop a general approach that can be used for specifying and reasoning about the behaviors of systems built using the notations proposed by these authors. The key idea underlying our approach is that each class needs *two* sets of specifications, one for use by clients of the class, the other for use by designers of classes that inherit from the given class. This is hardly surprising, given that these two groups of users need to know different things about the class. We applied this idea to the sequential case in (Soundarajan, Fridella 1997), and have tried to extend to concurrent languages in the current paper. For this purpose we had to associate with each class an abstract and a concrete synchronization property in addition to pre- and post-condition specifications of the various operations of the class. We sketched the proof techniques that would be used to show that a given implementation of a class meets both sets of its specifications. As we saw the details of the proof techniques do depend on the particular programming notation in question, but the basic approach seems general.

We wll conclude with a brief remark about future work. Kafura and Lee's (Kafura, Lee 1989) system depends heavily on their idea of *behavioral abstractions*. In a sense, this is a *second order* construct. In particular, commands like 'become' have a distinct 'second-order-flavor' about them. Nevertheless, we were able to deal with their system using our approach. This is very promising, and we believe that we will be able to extend our approach to other systems that use what look like second-order constructs even more extensively. That is not to say that we could deal with arbitrary second order operations on functions, but there is no need to do that. All we need to be able to do is to handle the types of opertaions that seem useful in building OO systems. Matsuoka and Yonezawa (Matsuoka, Yonezawa 1993) also remark on the importance of such limited second-order capabilities in building powerful systems. But it is not enough to *build* the systems. We need to be able to reason about them; that is why we believe our approach will prove useful.

REFERENCES

Frolund, S. (1992) Inheritance of synchronization constraints, *Proceedings of* ECOOP *1992*, 185–196.

Ishikawa, Y. (1992) Communication mechanisms on autonomous objects, *Prooceedings of* OOPSLA *1992*, 303–314.

Kafura, D., Lee, K. (1989) Inheritance in Actor based languages, *Proceedings of* ECOOP *1989*, 131–145.

Liskov, B., Wing, J. (1994) A behavioral notion of subtyping, *Transactions on Programming Languages and Systems*, **16**, 1811–1841.

Matsuoka, S., Yonezawa, A. (1993) Analysis of inheritance anomaly in concurrent OO languages, in *Research directions in concurrent OO programming*, ed. Agha, Wegner, Yonezawa, 107–150.

Meyer, B. (1988) *Object oriented software construction*, Prentice-Hall.

Misra, J., Chandy, K.M. (1981)Proofs of networks of processes, *IEEE Transactions on Software Engineering*, **7**.

Soundarajan, N. (1984) Axiomatic semantics of CSP, *ACM Transactions on Programming Languages and Systems*, **6**, 647–662.

Soundarajan, N., Fridella, S. (1997)Inheriting and modifying behavior, *technical report*, submitted to *TOOLS '97*, available at URL: http//www.cis.ohio-state.edu/~neelam.

5   BIOGRAPHY

Neelam Soundarajan is on the faculty at Ohio State University. His primary research interests are in programming languages, distributed and object-oriented systems, and specification and verification issues.

PART EIGHT

Distributed Systems:
ODP and CORBA (II)

25

Transformations and Consistent Semantics for ODP Viewpoints[*]

C. Bernardeschi[a], J. Dustzadeh[b], A. Fantechi[c], E. Najm[b], A. Nimour[b], F. Olsen[b,d]

(a) Dipartimento di Ingegneria della Informazione, Università di Pisa
Via Diotisalvi 2, 56100 Pisa, Italy

(b) Département Réseaux, Ecole Nationale Supérieure des Télécommunications
46, Rue Barrault, F–75013, Paris, France

(c) Dipartimento di Sistemi e Informatica, Università di Firenze
via S. Marta 3, 50139 Firenze, Italy

(d) PAA/TSA/TLR, France Telecom — CNET
38/40 rue Général Leclerc, 92794 ISSY MOULINEAUX CEDEX 9, FRANCE

Abstract

We discuss correctness preserving transformations of specifications in the passage from the Information viewpoint to the Computational viewpoint of the ODP ISO reference model. In our transformation exercise we use two relatively simple languages for each of these viewpoints; a class–oriented language for the Information viewpoint with the separation of actions from classes as the main distinguishing feature. For the Computational viewpoint the language chosen is a variant of the Actor language.

Keywords: ODP, Information Viewpoint, Computational Viewpoint, Correctness Preserving transformations, Actors, Objects, Object–Oriented Methods, Process Calculii.

[*]Supported by the Galileo project.

1   INTRODUCTION

The ISO reference model for Open Distributed Processing (ODP) introduces the concept of *viewpoint* to describe the system from a particular set of concerns (Farooqui et al. 1995). Five different viewpoints of the system have been introduced: "Enterprise" (requirement capture and early design of distributed systems), "Information" (conceptual design and information modeling), "Computational" (Software design and development), "Engineering" (system design and development) and "Technology" (technology identification, procurement and installation). Together, these different viewpoints allow a complete description of the system.

Formal methods provide a mathematical way to specify and analyze the behavior of concurrent and distributed systems. In the development of a system design trajectory supported by formal methods, transformations are usually defined in order to aid the development of correct specifications.

This is preliminary work in the direction of defining correctness preserving transformations from the Information viewpoint to the Computational viewpoint of the ODP reference model. In the Information viewpoint, a system is abstractly specified by giving the actions the system may perform; in the Computational viewpoint, a more concrete specification of the behavior of the system is given. In particular, the ODP computational model is an object–based framework. Its object–oriented nature emphasizes abstraction and encapsulation, while no assumption is made on the internal structure of objects or on linguistic support for objects. Since no constraint is placed on intra object activities, any type of intra object concurrency is allowed. Inter object concurrency is instead specified in the ODP computational model (Najm et al. 1995).

Due to their different nature, the passage from the Information viewpoint to the Computational viewpoint is a critical point in the system development methodology and it has been recognized as a possible area of intervention of transformations of specifications.

Information objects are mapped to computational object templates, and the abstract, collective behaviour of objects, as described by Information Actions, is mapped to message passing between autonomous interacting computational objects.

Structure of the paper. The core of the paper is section 4 where we present our proposed transformations from the Information viewpoint to the Computational viewpoint. Before that however, we describe the source language for the transformations in section 2 and the target language in section 3. The paper ends with an outlook for future work, comparisons with existing work and some concluding remarks.

2   INFORMATION VIEWPOINT: THE SOURCE LANGUAGE

2.1   Informal description

The language chosen for the Information viewpoint is a simple class–oriented language, where a system is defined as a collection of objects with attributes, but without methods, and a collection of actions that can be performed on the objects. The actions, contrary to methods, are perceived by the user as atomic, and operate on the object states as if acting on a single global state space.

2.2   Syntax

The syntax of the source language is given in syntax diagram 1 using a BNF notation, with some freedom for simplicity reasons. The keywords are given in bold.

$$\begin{aligned}
&\text{ClassDef} ::= \text{class\_id} = \text{Class} \\
&\text{Class} \quad ::= \textbf{class} \\
&\qquad\qquad\qquad \text{att}_1 : \text{Type}_1 ; \\
&\qquad\qquad\qquad \cdots \\
&\qquad\qquad\qquad \text{att}_n : \text{Type}_n ; \\
&\qquad\qquad \textbf{endclass} \\
&\text{ActionDef}:= \text{action\_id}(\text{var}_1 : \text{Type}_1, \ldots, \text{var}_n : \text{Type}_n) : \text{Type} = \text{Action} \\
&\text{Action} \quad ::= \textbf{action Decl Body endaction} \\
&\text{Decl} \quad ::= \text{var:Type;} \mid \text{Decl}_1\ \text{Decl}_2 \\
&\text{Body} \quad ::= \textbf{case}\ \text{Exp}_1 : \text{S}_1; \ \textbf{return}\ \text{Exp}_1'; \\
&\qquad\qquad\qquad \cdots \\
&\qquad\qquad \textbf{case}\ \text{Exp}_n : \text{S}_n; \ \textbf{return}\ \text{Exp}_n'; \\
&\text{S} \qquad ::= \text{var} := \text{Exp;} \mid \text{var.att} := \text{Exp;} \\
&\qquad\qquad \textbf{if}\ \text{Exp}\ \textbf{then}\ \text{S}_1\ \textbf{else}\ \text{S}_2\ \textbf{endif}; \\
&\qquad\qquad \text{S}_1\ \text{S}_2 \\
&\text{Exp} \quad ::= \text{c} \mid \text{var} \mid \text{var.att} \mid \textbf{nil} \\
&\qquad\quad\ \mid\ \ \text{op}(\text{Exp}_1, \ldots, \text{Exp}_n) \mid \text{action\_id}(\text{Exp}_1, \ldots, \text{Exp}_n) \\
&\qquad\quad\ \mid\ \ \textbf{new}\ \text{class\_id}(\text{Exp}_1, \ldots, \text{Exp}_n) \\
&\text{Type} \quad \mid\ \ \text{ground\_type} \mid \text{class\_id}
\end{aligned}$$

Syntax 1: The Source Language.

The terminals have the following meaning:
At this point it is important to make some comparisons between this language and other languages proposed for the Information viewpoint.

abbreviation	meaning
class_id	Class identifier
action_id	Action identifier
var	variable identifier
att	Attribute name
op	Operation on ground types
c	Constant

Of the formal languages, the most common of these is Z, see for ex. (Derrick et al. 1996) which discusses the use of Z in relation to transformations and viewpoint consistency. Our language is more concrete than Z — it is more close to an implementation language. We can imagine using it as the final refinement at the Information viewpoint. We could start with an abstract Z specification and end up with our language as the last description before going to the Computational viewpoint.

Of the object–modelling notations, we compare with OMT and Fusion. In comparison with OMT, our language is very close to OMT's static object model. The difference is first that actions (methods) are not encapsulated in classes at this level; this is what Fusion does. The second difference is that we do not provide explicit associations — instead associations are implicit via attributes referencing other classes.

2.3   Example

Let us consider the following simple example of an Information viewpoint specification of part of a bank account management system, made of a single object and a single action:

```
ACCOUNT = class owner: string;
                balance: integer;
           endclass

transfer(provider:account; recipient:account; amount:positive) =
(* the action transfers a sum equal to amount from the provider account *)
(* to the recipient account *)
action

case provider.balance < amount:      return NACK;

case provider.balance >= amount:
    provider.balance := provider.balance - amount;
    recipient.balance := recipient.balance + amount;
return ACK;
end action
```

2.4 Semantics

We give here the dynamic semantics of our source language. We suppose at this level that the programs are syntactically correct and type–checked. The dynamic semantics is given using the following judgments:

- $E \models (Decl \Rightarrow E')$ "In the environment E the declaration $Decl$ gives the environment E'" ;
- $E \models (S \Rightarrow E')$ "In the environment E the execution of the statement S gives the environment E'" ;
- $E \models (Exp \rightarrow v) \Rightarrow E'$ "In the environment E the evaluation of the expression Exp returns the value v and gives the new environment E'.

The values we consider here are predefined ground values and record references representing behavior–less objects. Each record is of the form:

$$< att_1 = v_1, \ldots, att_n = v_n >$$

where the v_i are also values. We consider also a nil value, noted $\bot$, that is polymorphic.

An environment is an ordered collection of pairs "variable/value" noted: $var = v$. A variable name may appear in more than one pair in an environment. The extension of the environment E with the pair $var = v$ gives a new environment denoted: $E, var = v$. We define a lookup function that gives for a given variable the most recent binding:

$$lookup((E, var = v), var') = \begin{cases} v & \text{if } var \equiv var' \\ lookup(E, var') & otherwise \end{cases}$$

Similarly, for the notation $E, var = v[var' \leftarrow v]$ we define the function:

$$E, var = v[var' \leftarrow v] = \begin{cases} E, var = v' & \text{if } var \equiv var' \\ E[var' \leftarrow v'] & otherwise \end{cases}$$

The dynamic semantics is then given by the following SOS rules:

Declarations:

$$E \models (var : Type) \Rightarrow E, var = \bot$$

A simple declaration extends its evaluation environment with the new declared variable bound to $\bot$.

$$\frac{E_1 \models Decl_1 \Rightarrow E_2 \ ; \ E_2 \models Decl_2 \Rightarrow E}{E_1 \models Decl_1 \, Decl_2 \Rightarrow E}$$

Two consecutive declarations extend twice the initial environment.

Action body:

$$E \models (Exp_j \to \text{FALSE}) \Rightarrow E \;,\; \forall j < i$$
$$E \models (Exp_i \to \text{TRUE}) \Rightarrow E$$
$$E \models S_i \Rightarrow E_i$$
$$\underline{E_i \models (Exp_i' \to v) \Rightarrow E_i'}$$

$$E \models \left(\left(\begin{array}{l} \text{case } Exp_1 : S_1 \;;\; \text{return } Exp_1'; \\ \qquad \cdots \\ \text{case } Exp_n : S_n \;;\; \text{return } Exp_n'; \end{array} \right) \to v \right) \Rightarrow E_i'$$

The conditions (Exp_i) of the case statement are evaluated. The branch corresponding to the first Exp_i evaluated to TRUE is executed. Note that the evaluation of the Exp_i has no side effects on the environment.

Assignments:

$$\frac{E \models (Exp \to v) \Rightarrow E'}{E \models (var := Exp) \Rightarrow E[var \leftarrow v]}$$

To assign an expression (Exp) to a variable (var), we first evaluate Exp and then change the binding of var to this value.

$$\frac{E \models (Exp \to v) \Rightarrow E'}{E \models (var.att := Exp) \Rightarrow E[var.att \leftarrow v]}$$

Similarly, to assign an expression to an object attribute, we evaluate this expression and then we update the attribute of the corresponding record.

IF statements:

$$\frac{E \models (Exp \to \text{TRUE}) \Rightarrow E' \;;\; E' \models S_1 \Rightarrow E''}{E \models \text{if } Exp \text{ then } S_1 \text{ else } S_2 \text{ endif } \Rightarrow E''}$$

$$\frac{E \models (Exp \to \text{FALSE}) \Rightarrow E' \;;\; E' \models S_2 \Rightarrow E''}{E \models \text{if } Exp \text{ then } S_1 \text{ else } S_2 \text{ endif } \Rightarrow E''}$$

If the condition expression is evaluated to TRUE (respectively, to FALSE) we execute the "then branch" (respectively, the "else branch").

Statement chaining:

$$\frac{E \models S_1 \Rightarrow E_1 \;;\; E_1 \models S_2 \Rightarrow E_2}{E \models S_1 S_2 \Rightarrow E_2}$$

The second statement is executed in the environment obtained after the execution of the first statement.

Constants:

$$E \models (c \to c) \Rightarrow E$$

A ground constant evaluates to itself.

Variables values:

$$E \models (var \rightarrow lookup(E, var)) \Rightarrow E$$

We have to search the environment for the last binding of var.

Function calls:

$$\frac{E \models (Exp_1 \rightarrow v_1) \Rightarrow E; \ldots; E \models (Exp_n \rightarrow v_2) \Rightarrow E}{E \models (op(Exp_1, \ldots, Exp_n) \rightarrow eval(op(v_1, \ldots, v_n))) \Rightarrow E}$$

We first evaluate the arguments of the function and then call the predefined evaluation function on ground expressions: eval. All the arguments are evaluated in the same environment.

Action calls:

$$\frac{\begin{array}{c} E \models (Exp_1 \rightarrow v_1) \Rightarrow E; \ldots; E \models (Exp_n \rightarrow v_n) \Rightarrow E \\ A(var_1 : Type_1, \ldots, var_n : Type_n) \stackrel{\triangle}{=} ActionB \\ E, var_1 = v_1, \ldots, var_n = v_n \models (ActionB \rightarrow v) \Rightarrow E' \end{array}}{E \models (A(Exp_1, \ldots, Exp_n) \rightarrow v) \Rightarrow E'}$$

As for ground functions calls, we first evaluate the arguments and then evaluate the body of the action in an environment where the formal parameters have been bound to the actual ones.

Attribute selection:

$$E \models (var.att \rightarrow lookup(E, var).att) \Rightarrow E$$

We look for the value of var in the environment and then select the appropriate attribute.

Object creation:

$$\frac{\begin{array}{c} E \models (Exp_1 \rightarrow v_1) \Rightarrow E; \ldots; E \models (Exp_n \rightarrow v_n) \Rightarrow E \\ Class \stackrel{\triangle}{=} class\ att_1 : Type_1, \ldots, att_n : Type_n\ endclass \end{array}}{E \models (\ new\ Class(Exp_1, \ldots, Exp_n) \rightarrow< att_1 = v_1, \ldots, att_n = v_n >) \Rightarrow E}$$

The expressions Exp_i are evaluated in the same environment then a new record is created according to the definition of $Class$ and initialized with the values of $Exp_1, \ldots, Exp_n$. The returned value is this record.

3 COMPUTATIONAL VIEWPOINT: THE TARGET LANGUAGE

The target language for the computational viewpoint is an object language related to the *Actor* paradigm (see (Agha 1986) for an introduction to Actors) and to mobile process algebras (see (Pierce et al. 1995) and (Milner et al. 1992) for information on mobile process algebras and concurrent objects). The syntax is given in syntax diagram 2 and some examples in section 4 (where we discuss the transformations).

```
ActorDef ::= actor_name(var₁, ... ,varₙ) = Behavior
Behavior ::= [m₁(ṽar₁) -> Behavior₁,
                ...,
              mₖ(ṽarₖ) -> Behaviorₖ]
           |   Actor.m(Val₁, ..., Valₙ); Behavior
           |   if val then Behavior₁ else Behavior₂
           |   become actor_name(Val₁, ..., Valₙ)
           |   create actor_name(Val₁, ..., Valₙ); Behavior
           |   let var = val in (Behavior)
Actor    ::= self | var
           |   new actor_id(Actor₁, ..., Actorₙ)
Val      ::= g_Val | var | nil | self
           |   new actor_name(Val₁, ..., Valₙ)
G_Val    ::= c | op(G_Val₁, ..., G_Valₙ)
```

Syntax 2: Target Language

It is not our purpose to completely and formally define the semantics of our language here (see (Agha et al. 1993) or (Agha 1986) for semantics of Actors). In short, the addressing part is based on Actors and the computational part on process algebras. Active objects (actors) have unique identity (mail address). Objects communicate using asynchronous message passing and each object has its own, private queue of incoming messages. These messages are treated in the order of arrival (i.e. the queue is FIFO).

Referring to the syntax diagram we explain the main distinguishing features of the language:

- *message reception*

$$[\ m_1(\tilde{var}_1) \text{ -> Behavior}_1,$$
$$\dots,$$
$$m_k(\tilde{var}_k) \text{ -> Behavior}_k]$$

  Wait for one of the messages $m_1, \dots, m_k$, and react by executing the corresponding *Behavior*.

- *message send*
    Actor.m(val₁, ..., valₙ); Behavior
  Send message m to *Actor* with arguments $val_1, \dots, val_n$ and become *Behavior*.

- *replacement behaviour*
    become actor_name(Val₁, ..., Valₙ)
  The current actor behaviour acts as specified in *actor_name*.

- *Actor creation*
    create actor_name(Val₁, ..., Valₙ); Behavior

Create a new actor that behaves as specified in *actor_ name* and then act according to *Behavior*.

4 TRANSFORMATION

4.1 Style of the Transformation

A class is transformed into an actor whose acquaintances (parameters) are the attributes of the class. For each attribute correspond *get* and *set* methods.
An action is transformed into an actor with only one method. This method is used to create a thread responsible for executing the action.
The main difference added by the transformation is that access to or modification of the object state, which is direct in the source language, is possible only through the *get* and *set* methods in the target language.

4.2 A transformation example

To illustrate the transformation we come back to the example given in section 2.3. The transformation would produce the following pair of actors, one corresponding to the ACCOUNT class and one corresponding to the transfer action.

```
Act_Account(owner, balance) =
   [get_owner(ret) ->
       ret.value(owner);
       become Account(owner, balance),
    get_balance(ret) ->
       ret.value(balance);
       become Account(owner, balance),
    set_owner(new_owner) ->
       become Account(new_owner, balance),
    set_balance(new_balance) ->
       become Account(owner, new_balance)
   ]
Act_Transfer() =
   [transfer(provider, recipient, amount, return) ->
       create Thread_Transfer(provider, recipient, amount, return);
       become Act_Transfer()
   ]
Thread_Transfer(provider, recipient, amount, return) =
  provider.get_balance(self);
  [value(p_balance) ->
     if p_balance < amount then return.value(NACK)
     else recipient.get_balance(self);
```

```
[value(r_balance) ->
    provider.set_balance(p_balance-amount);
    recipient.set_balance(r_balance+amount);
    return.value(ACK)
  ]
]
```

4.3   Definition and formalization of the transformation

We describe here a syntactic transformation from our information viewpoint language to the computational language. This transformation is a function, noted $\mathcal{T}$, that takes a well formed program text of the source language and returns a program text of the target language that has the same functionality. Function $\mathcal{T}$ is defined by induction on the structure of the source language.

Class definition: each class definition of the source language is translated into an actor definition in the target language.

$$\mathcal{T}\left[\begin{array}{l} C = \\ \text{class} \\ \text{att}_1 : \text{Type}_1; \\ \dots \\ \text{att}_n : \text{Type}_n; \\ \text{endclass} \end{array}\right] \longrightarrow \begin{array}{l} \text{Act\_C}(\text{att}_1,\dots,\text{att}_n) = \\ [\text{get\_att}_1(\text{ret}) -> \\ \quad \text{ret.value}(\text{att}_1); \\ \quad \text{become Act\_C}(\text{att}_1, \dots, \text{att}_n), \\ \text{set\_att}_1(\text{new\_value}) -> \\ \quad \text{become Act\_C}(\text{new\_value}, \dots, \text{att}_n), \\ \quad \dots \\ \text{get\_att}_n(\text{ret}) -> \\ \quad \text{ret.value}(\text{att}_1); \\ \quad \text{become Act\_C}(\text{att}_1, \dots, \text{att}_n), \\ \text{set\_att}_n(\text{new\_value}) -> \\ \quad \text{become Act\_C}(\text{att}_1, \dots, \text{new\_value})] \end{array}$$

A class C is transformed into an actor Act_C. The attributes of C become parameters of Act_C and for each attributes, att_i, there is a method named "get_att_i" that returns the value of att_i to the return address and a method named "set_att_i" that update the value of att_i with the specified value.

Action definition: each action becomes a stateless actor with a single method.

$$\mathcal{T}\left[\begin{array}{l} A(\text{var}_1 : \text{Type}_1, \\ \dots, \\ \text{var}_n : \text{Type}_n) \\ = \\ \text{action} \\ \quad \text{Decl; Body;} \\ \text{endaction} \end{array}\right] \longrightarrow \begin{array}{l} \text{Act\_A}() = \\ [\overline{\text{A}}(\text{var}_1, \dots, \text{var}_n, \text{ret}) -> \\ \quad \text{create Thread\_A}(\text{var}_1, \dots, \text{var}_n, \text{ret}); \\ \quad \text{become Act\_A}()] \\ \\ \text{Thread\_A}(\text{var}_1, \dots, \text{var}_n, \text{ret}) = \\ \quad \mathcal{T}\left[\text{Decl; Body}\right] \end{array}$$

An action A is transformed into an actor Act_C whose behavior is to repeatedly wait for reception of the message A and then creates a new actor (bound to "Thread_A") which is responsible for executing the action.

Action behavior: the translation is straightforward. Note that the body of an action is evaluated in a context where the declared variables have been introduced by a "let" statement.

$$\mathcal{T}\Big[\text{Decl Body}\Big] \longrightarrow \text{let } \mathcal{T}\Big[\text{Decl}\Big] \text{ in } \mathcal{T}\Big[\text{Body}\Big]$$

Declarations: Consecutive declarations are transformed into nested "let" statements.

$$\mathcal{T}\Big[\text{Decl}_1 \text{ Decl}_2\Big] \longrightarrow \mathcal{T}\Big[\text{Decl}_1\Big] \text{ in } (\text{let } \mathcal{T}\Big[\text{Decl}_2\Big])$$

A simple declaration binds the declared variable to nil.

$$\mathcal{T}\Big[\text{var:Type}\Big] \longrightarrow \text{var} = \text{nil}$$

Action Bodies: the "case" statement is transformed into a nesting of "if" statements.

$$\mathcal{T}\Big[\text{case Exp}_1\colon S_1; \quad \text{return Exp}_1'; \quad \ldots \quad \text{case Exp}_n\colon S_n; \quad \text{return Exp}_n';\Big]$$

$$\longrightarrow$$

$$\mathcal{T}\Big[\text{Exp}_1\Big]$$
$$[\text{value(cond)} -> $$
$$\text{if cond}$$
$$\text{then } \mathcal{T}\Big[S_1\Big]\Big[\text{return Exp}_1'\Big]$$
$$\text{else}$$
$$\mathcal{T}\Big[\text{case Exp}_2\colon S_2; \text{ return Exp}_2'; \quad \ldots \quad \text{case Exp}_n\colon S_n; \text{ return Exp}_n';\Big]$$
$$]$$

We first transform "Exp$_1$". We wait for the value of "Exp$_1$" (the method "value"). This value is bound to the cond. The "case" is transformed into an if statement such that if cond is "true", we execute the transformation of S_1, otherwise we execute the transformation of the rest of the case.

Return statement: a "return" statement simply evaluates its argument expression and then sends it to the return address "ret" which is an argument of the initial method.

$$\mathcal{T}\Big[\text{return Exp}\Big] \longrightarrow \mathcal{T}\Big[\text{Exp}\Big]$$
$$[\text{value(v)} -> \text{ret.value(v)}]$$

Statements: the statements of the source language can change the state of the configuration (object attributes or local variables). So we need an extra argument for our transformation function which is the continuation of the transformed statement. So that this continuation is executed in an environment where the state update has been taken into account.

$$\mathcal{T}\Big[var := Exp\Big]\Big[k\Big] \longrightarrow \boxed{\begin{array}{l} \mathcal{T}\Big[Exp\Big] \\ [value(v) \to let\ var = v\ in\ (k)\ ] \end{array}}$$

We first evaluate "Exp" and then execute "k" (the continuation of the assignment) in an context where "var" is bound to its new value.

$$\mathcal{T}\Big[var.att := Exp\Big]\Big[k\Big] \longrightarrow \boxed{\begin{array}{l} \mathcal{T}\Big[Exp\Big] \\ [value(v) \to var.set\_att(v);\ become\ k\ ] \end{array}}$$

For the assignment of a object attribute, we send to this object the message "set_att" with the new value as argument and then behave as specified by "k".

$$\mathcal{T}\Big[if\ Exp\ then\ S_1\ else\ S_2\Big]\Big[k\Big] \longrightarrow \boxed{\begin{array}{l} \mathcal{T}\Big[Exp\ \Big] \\ [value(v) \to \\ \quad if\ v\ then\ \mathcal{T}\Big[S_1\Big]\Big[k\Big]\ else\ \mathcal{T}\Big[S_2\Big]\Big[k\Big] \\ ] \end{array}}$$

An if statement is straightforwardly transformed into an if statement in the target language.

$$\mathcal{T}\Big[S_1\ S_2\Big]\Big[k\Big] \longrightarrow \boxed{\mathcal{T}\Big[S_1\Big]\Big[\mathcal{T}\Big[S_2\Big]\Big[k\Big]\Big]}$$

The transformation of the second statement will be executed in a context possibly changed by the first statement.

Expressions: the expressions are transformed such that they send their values by sending the message "value" with their value as argument.

$$\mathcal{T}\Big[c\ \Big] \longrightarrow \boxed{self.value(c);}$$

Constants are evaluated to themselves.

$$\mathcal{T}\Big[var\Big] \longrightarrow \boxed{self.value(var);}$$

As for constants, the transformation of an access to a variable (var) consists in invoking the method "value" of "self" with "var" as argument.

$$\mathcal{T}\Big[var.att\Big] \longrightarrow \boxed{var.get\_att(self);}$$

For attribute selection, we invoke the *get_att* method of actor *var* with *self* as return address.

$$\mathcal{T}\Big[\mathrm{op}(\mathrm{Value}_1, \dots, \mathrm{Value}_n)\Big] \longrightarrow \mathrm{self.value}(\mathrm{op}(\mathrm{Value}_1, \dots, \mathrm{Value}_n))$$

Ground expressions are evaluated to themselves.

$$\mathcal{T}\Big[A(\mathrm{Exp}_1, \dots, \mathrm{Exp}_n)\Big] \longrightarrow
\begin{array}{l}
\mathcal{T}\Big[\mathrm{Exp}_1\Big] \\
[\ \mathrm{value}(v_1) \to \\
\quad \dots \\
\quad \mathcal{T}\Big[\mathrm{Exp}_n\Big] \\
\quad [\mathrm{value}(v_n) \to \\
\qquad \mathrm{Act\_A}.A(v_1, \dots, v_n, \mathrm{self}) \\
\quad \dots \\
\quad] \\
]
\end{array}$$

Expressions Exp_1 to Exp_n are evaluated sequentially, then method A of actor Act_A is invoked with *self* as return address.

$$\mathcal{T}\Big[\mathrm{new}\ C(\mathrm{Exp}_1, \dots, \mathrm{Exp}_n)\Big] \longrightarrow
\begin{array}{l}
\mathcal{T}\Big[\mathrm{Exp}_1\Big] \\
[\ \mathrm{value}(v_1) \to \\
\quad \dots \\
\quad \mathcal{T}\Big[\mathrm{Exp}_n\Big] \\
\quad [\mathrm{value}(v_n) \to \\
\qquad \mathrm{self.value}(\mathrm{new\ Act\_C}(v_1, \dots, v_n)) \\
\quad \dots \\
\quad] \\
]
\end{array}$$

Expressions Exp_1 to Exp_n are evaluated sequentially, then method *value* of the current actor (*self*) with newly created actor behaving according to Act_C as parameter.

4.4   Proving the correctness of the transformation

Transformation $\mathcal{T}$ is one of the two steps that should be performed so that the semantics of the source information description is preserved in the target computational programs. Indeed, the target actors produced by $\mathcal{T}$ behave correctly with regard to their source information specification only when the executions of action translated actors are serialisable*. Thus, another transformation is needed in association with $\mathcal{T}$, and which generates the appropriate mechanism to control the concurrency of executions of actors (this is the role which is played by transaction object services in CORBA).

---

*Conditions looser than serialisable could be investigated, but this is out the scope of the present paper.

We discuss hereafter the semantics preservation properties of T only for the serialisable executions of actors; the concurrency control aspect is being researched and left for a second, forthcoming contribution. Let C be a collection of classes and actions of the source language and A the collection of actors resulting from the transformation of elements of C. We denote by $\mathbb{S}_C$ the set of instances of classes of C. Thus, for each instance c of $\mathbb{S}_C$, $\exists C \in C$ such that $c : C$. Similarly, we denote by $\mathbb{T}_A$ the set of instances of actors of A. Transformation T can be extended in order to be defined on instances of classes, i.e. $\widehat{T}$ maps instances of classes on instances of actors. The extension is given by:

for $< att_1 = v_1, \ldots, att_n = v_n >: C,$
$$\widehat{T}(< att_1 = v_1, \ldots, att_n = v_n >) = become\ T(C)(v_1 \ldots v_i)$$

Let $\overline{\mathbb{S}}_C$ denote the power–set of $\mathbb{S}_C$. Elements of $\overline{\mathbb{S}}_C$ denote configurations of class instances. Similarly, elements of $\overline{\mathbb{T}}_A$ denote configurations of Actor instances. For each action $A \in C$, the semantics of the source language defines a partial* function $[\![A]\!]_C$ which maps class instances on class instances: $[\![A]\!]_C : \overline{\mathbb{S}}_C \rightarrow \overline{\mathbb{S}}_C$. In a similar way, the semantics of the target language allows to associate to each actor A_c, a partial* function $[\![A_c]\!]_A : \overline{\mathbb{T}}_A \rightarrow \overline{\mathbb{T}}_A$

The semantics preservation of our transformation can now be stated: T is an information to computational semantics preserving transformation iff the diagram 1 is commutative.

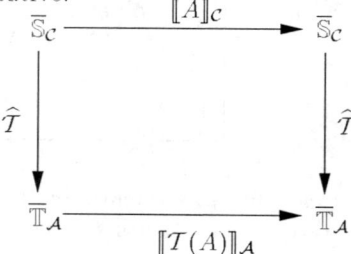

Figure 1 Transformation diagram.

The full proof of this property is too long to be given in this paper. It is achieved in two steps. In a first step, we define a labelled transition based semantics of the source language which we proof equivalent to the one given in section 2.4. In this setting, an information state of the source language is made of a configuration of class instances and of the current state of execution of the action under consideration. In the second step, we construct a simulation relation between a source description (a configuration made of class instances and one action) and its image in the transformation, whereby the information states are mapped on to computational states and elementary execution steps of the source language are mapped onto sequences of executions (possibly involving message exchanges between actors) in the target language.

---

*The partiality of this function reflects non terminating actions.

5 CONCLUSIONS

This paper has presented a correctness preserving transformation from a class–oriented Information viewpoint language to an Actor–like language for the Computational viewpoint. The language for the Information viewpoint has been given a formal semantics, whereas the language for the Computational viewpoint has been described informally only. We then defined the transformation in the form of a set of rules that perform syntactic transformations on the Information viewpoint grammar.

The main distinguishing feature of the Information viewpoint language is that actions (operations) are separated from class definitions. This will provide a lot of flexibility for composing objects at the Computational viewpoint. More work is needed to define rules for how objects can be composed.

The general issue of consistency in open distributed systems is a particularly challenging problem. So far, little work has been carried out in this area as pointed out by Bowman et.al. in (Bowman et al. 1996) and (Derrick et al. 1996). (Bowman et al. 1996) discusses this issue in the context of ODP and gives a general definition of consistency which covers several of the consistency definitions listed for ex. in the ISO ODP/RM (Farooqui et al. 1995). They emphasise the importance of unification of descriptions from several viewpoints into a final *implementation specification*. In (Derrick et al. 1996) the same authors propose a formal translation from LOTOS to Z. The emphasis is on the definition of a common semantic model to which specifications in the two languages can be translated and then compared for observational equivalence.

Apart from the work of Bowman et.al. we should also mention the work of the LOTOS community. The development of correctness preserving transformations carried out for the LOTOS specification language is a striking example (Cunha et al. 1995). In the development cycle of LOTOS, constraint oriented specifications are used in a first stage of the development, to pass then to resource oriented specifications for the late stage of the development. Transformations which allow this passage preserving the correctness of the specifications have been successfully defined and used.

In the near future we will extend this work to deal with transaction management as an integral part of the transformation. The motivation for this is to provide guarantees for the Information viewpoint's atomicity assumption. Furthermore, we consider to strengthen the links between the viewpoints by using reflection techniques. This might be particularly interesting in order to take into account design decisions made at the Computational viewpoint. The simple language we have defined for the Information viewpoint can only express functional requirements. There is a lot of other information that affects the final design. It is probably not feasible to provide formal languages for all of this information, but still, having more than one language is clearly imaginable. In that case it would be interesting to define transformations from each of the Information viewpoint languages and to have some way of combining these transformations into a coherent whole at the Computational viewpoint, with as much automatic support as possible.

REFERENCES

G. A. Agha, (1986), *Actors: A Model of Concurrent Computational in Distributed Systems*, MIT Press, Cambridge, Mass.

G. A. Agha, I. A. Mason, S. F. Smith and C. L. Talcott, (1993), *A Foundation for Actor Computation*, J. Functional Programming 1 (1).

C. Bernardeschi, A. Fantechi, F. Paternò, (1995), *Application of Correctness Preserving Transformations for Deriving Architectural Descriptions of Interactive Systems from User Interface Specifications*, Proc. *SEKE'95*, Rockville, Maryland.

H. Bowman, E. A. Boiten, J. Derrick, and M. W. A. Steen, (1996), *Viewpoint Consistency in ODP, a General Interpretation*. Proc. *First IFIP International Workshop on Formal Methods for Open Object–based Distributed Systems*, FMOODS'96, Paris, France. Chapman and Hall.

T. Bolognesi (Ed.), (1992). Catalogue of LOTOS Correctness Preserving Transformation. *Lo/WP1/T1.2/N0045/V03*, Final Deliverable, LOTOSphere ESPRIT project 2304.

P. Cunha, A. Fantechi, B. Mekhanet, E. Najm and J. Queiroz, (1995), *Correctness Preserving Transformations for the Late Phases of Development*. Chapter 9 of T. Bolognesi, J. van de Lagemaat and C. Vissers eds., LOTOSphere: Software Development with LOTOS, Kluiwer Academic Publishers, 181–199.

J. Derrick, E. A. Boiten, H. Bowman, and M. W. A. Steen, (1996). *Supporting ODP — Translating LOTOS to Z*. Proc. *First IFIP International Workshop on Formal Methods for Open Object–based Distributed Systems*, FMOODS'96, Paris, France. Chapman and Hall.

K. Farooqui, L. Logrippo, J. de Meer, (1995), *The ISO reference model for Open Distributed Processing: an introduction*, Computer Networks and ISDN Systems 27, 1215–1229.

H-M, Jarvinen and R. Kurki–Suonio, (1990), *The DisCo Language*, Report 8, Tampere University of Technology, Software Systems Laboratory, March.

Robin Milner, Joachim Parrow, David Walker, (1992), *A Calculus of Mobile Processes (Parts I and II)*, Information and Computation, 100:1-77.

E. Najm and J-B Stefani, (1995), *A formal semantics of the ODP computational model*, Computer Networks and ISDN Systems 27, 1305–1329.

B. C. Pierce and D. N. Turner, (1995), *Concurrent objects in a process calculus*, Proc. *Theory and Practice of Parallel Programming*, Sendai, Japan, 1994, Lecture Notes in Computer Science 907, 187–215.

26

Architectural Concepts for Agent Paradigm : A Way to Separate Concerns in Open Distributed Systems

A. Diagne
Université Pierre & Marie Curie
Laboratoire d'Informatique de Paris 6
Thème Systèmes Répartis et Coopératifs
4, place Jussieu 75252 Paris Cedex 05, France
Phone : (+33) (0)1 44 27 73 65, Fax : (+33) (0)1 44 27 62 86
E-mail : Alioune.Diagne@masi.ibp.fr

Abstract

The emerging agent paradigm is gaining legitimacy as a solution to the most and most complex needs in distributed systems. For instance, a new concept like remote programming is presented as an alternative to the limits of the classical client-server interaction modes including its derivates. Agent paradigm has also emphasized the concept of service as a set of functionalities with contractual constraints. Nevertheless, agent paradigm does not always deal with architectural concepts, it is mostly concerned with implementation of such advanced features in distributed systems. Meanwhile RM-ODP undertakes federation of open distributed systems with a generic architecture. This architecture is based on progressive structuration of systems within five viewpoints based upon object-oriented concepts. Viewpoints encompass representations from conceptual level to final implementation in a progressive way and supply a quite satisfactory basis for separation of the concerns. In this paper, we address the relevancy of such an architecture to the agent paradigm in order to separate the many concerns in open distributed systems (e.g. collaboration, cognition and reactivity).

Keywords

Objects, Agents, Service, Architecture for Open Distributed Systems, Separation of Concerns.

1   INTRODUCTION

Agent paradigm is gaining legitimacy as a solution to most and most complex needs in distributed systems (Magedanz 96). It brings concepts beyond the ones introduced by the object-oriented paradigm. These new concepts are presented as alternatives to limits of classical approaches in distributed systems like client-server and its various interworking modes (variations of the Remote Procedure Call, Message Passing, etc.).

The agent paradigm is nowadays supported efficiently by technological proposals (Sun 95), (Telescript 95). Most of the newly introduced concepts are supported efficiently at run-time and ongoing research is very active. Separation of Concerns is a new trend in software engineering which tries to formally separate and organize the many aspects one can have to handle through systems life-cycle (Hursch 95).

The RM-ODP is an international standard issued by ITU and ISO to overcome the necessity to put together heterogeneous systems in order to cooperate. Many vendors, each having a system (a set of services or a computing platform) with its own characteristics, must be able to cooperate by proposing or requesting some services. This can be made largely easier when systems share a common architectural basis. The RM-ODP is a proposal of a generic architecture for open distributed systems based upon concepts from the object-oriented paradigm. It proposes a prescriptive model and a descriptive one to structure such systems. As a generic standard, it is somewhat unprescriptive on some aspects which may vary with the application domain. However, it has achieved a first level of a federative approach for heterogeneous distributed systems.

Structuration and architecture are important research fields in agent-based systems. Many authors have already tackled the problem with separation of concerns as a major need to fullfil. Communication versus Knowledge Sharing is the basis of structuration in (Finin 94). The proposal in (Carle 95) focuses on the distributed artificial intelligence aspects like distributed problem solving and reduce reactive and collaborative aspects to a minimum. The proposal in (McKay 96) also stressed at the communication aspects for load sharing and knowledge sharing. Collaborative and reactive aspects are not considered. In (Pitt 95), the authors have proposed a satisfactory structuration - the Cooperative Services Framework - but the separation of concerns is not fully achieved. They fully consider the collaborative aspects with deontic logic but the cognitive aspects are forgotten and the reactive ones are not elaborated. In (Merz 96), the authors focus on the open infrastructure necessary to support openess of clients and servers cooperations. Petri net-based control flows which characterize a service allow to handle goal splitting and sub-agents creation with validation. A main remark from these works is that authors do not aim to apply formal engineering methods in their structuration. This aim needs a demarcation between the aspects that can be (and need to be) formally verified and those that can not. For these last ones, a validation by simulation can be helpful and sufficient.

The aim of this paper is to apply an ODP-like structuration to agent-based services within open distributed systems in order to separate the concerns in a way which enables formal engineering methods. The Section 2 is dedicated to a brief overview of ODP concepts while Section 3 presents some key ideas of the agent paradigm. Section 4 presents the new trend of service which nowadays influences the way to think about open distributed systems. In Section 5 we investigate the relevancy of an adaptation of ODP viewpoints to agent paradigm. In Section 6 we propose and discuss an architectural framework for ODP-like agent-based service in open distributed systems before conclusion in Section 7.

2   OVERVIEW OF THE **RM-ODP** CONCEPTS[(*)]

The RM-ODP defines first a consistent set of modeling concepts on which is based afterwards the definition of the proposed architecture. It therefore organizes the concepts into two main parts :

- a *general framework* which is the *descriptive model* of a distributed system (ITU X.902). It contains basic modeling, specification and architectural concepts.
- an *organization* and *structure* of a distributed system which is the *prescriptive model* (ITU X.903). It is organized into five progressive viewpoints, each one aims to capture part of the information processed in specification and design.

The three first viewpoints deal with the structuration of a system and can be made independent from the underlining technology. The two last ones cope more with realization aspects. Consistency between the many viewpoints of a system must be managed (Bowman 96). However, in the current trend of open distributed systems, we can not make the viewpoints as separated as in the ODP approach. Providing services in the «*open electronic market*» encompass problems which make the mixing of the many viewpoints and their representation at the processing level necessary.

3   OVERVIEW OF THE AGENT PARADIGM

Agent paradigm emerges from artificial intelligence works (Ferber 95). It aims to enhance software entities (data attached with behavior according to object orientation) with knowledge storage and inference capabilities, mobility and adaptability, advanced cooperation and collaboration mechanisms among other aspects. Agent paradigm covers a large set of application domains and can be considered from different points of view (reactivity, cognition, etc.).

3.1   Characteristics of Agents

What is an agent ? It is somewhat difficult to find in the literature a definition which makes a consensus. According to Ferber, «*an agent is an hardware of software entity able to act on itself and on its environment. It has a partial representation of its environment and is able to communicate with other agents. It aims an individual goal and its behavior is the result of its observations, knowledge, abilities and interactions it can have with other agents and with the environment*» (Ferber 95). Agents

---

[(*)]   For sake of place, the reader not familiar with the RM-ODP materials is asked to refer to published litterature.

are a self-contained entities which have several concerns to cope with. For this reason, agents are often tagged with the non-exhaustive following list of attributes :

- *intelligent* : an intelligent agent is able to have cognitive activities. An intelligent agent is a piece of software which behaves according to some attached knowledge and inference it can perform on it. There is a wide range of intelligence levels ranging from simple pre-defined rules to achieve indeterministic tasks to self-learning inference mechanisms. The knowledge of agents is often incomplete and they need to cooperate/collaborate with each other to enhance or refine it,

- *collaborative/cooperative* : agents are able to collaborate/cooperate with each other. Collaboration and cooperation are beyond communication and interaction because they encompass making agreement on the high level purpose of interactions, establishing the communication means and late supplying the context and making the binding necessary to the real execution. Cooperation is based on coordination of elementary activities in order to fullfil more elaborated ones. For instance, many servers can cooperate to enhance the quality of services they propose (e.g. fault tolerance by replication, best performances by competition, etc.). Collaboration is more a client/server notion based on contracts and proposal/request of services,

- *mobile* : a mobile agent is able to move code and data to a remote site and to adapt itself to the target run-time environment. Mobility includes remote execution and migration. Remote execution means moving data and code to a remote site which hosts the entire execution whilst migration means moving data and code during execution to perform progressively some tasks through several sites.

Nevertheless agents do have some basic characteristics :

- *autonomy* of an agent is twofold. First, an agent must be able to perform its tasks with a very loose coupling with others agents and users. Second, an agent must be able to initiate activities - like cognitive inference or cooperation/collaboration with other agents - on its own according to reached internal states and/or occurring events,

- *asynchronism* of behavior which means that triggering of the actions of an agent must be governed by its internal rules only. For instance, the invocation of a service provided by an agent can be delayed according to its current state and running activities,

- *communication* is the basis of the cooperation and collaboration between agents. An agent must be able to communicate with other agents which they know about the functionalities. It can also get in touch with other agents by trading mechanisms to know about their functionalities.

All these previous attributes make agents very relevant for the most and most complex needs in open distributed systems. Among these needs, a very difficult one to achieve is the slogan «*all information available at any time in any place*». The «*electronic market place*» metaphor is becoming a standard in the new electronic services. Services must be provided by electronic entities able to meet their consumers, achieve contracts with them and provide their services in a secure way. This need disclaims classical design and implementation solutions and asks for «*intelligent collaborative/cooperative mobile agents*».

3.2   Agents vs. Objects

Agent paradigm shares with object-oriented paradigm a basis of underlining concepts like *abstraction, modularity* and *encapsulation*. Agents can be viewed as elaborated entities which have capabilities such as reasoning and collaborating/cooperating that are not represented in classical object-oriented paradigm.

Nevertheless, some major differences exist between the two paradigms and agents can not be restricted to «*thinking objects*». In object orientation, one can distinguish between active and passive objects while agents are inherently active and autonomous entities. We can also notice that objects are essentially deterministic and have sequential methods often invocable one at a time. In agent-orientation concurrency (e.g. reasoning while collaborating/cooperating) and indeterminism (e.g. execution depending on context agreed on in collaboration/cooperation) are necessary. Method invocation in object -orientation is often free and only constrained by type-checking of the parameters while agents might interact according to well-defined protocols which can make some invocations allowed or not at a given context.

However, object-oriented paradigm can be a valuable basis to build agent-based systems because they offer a first structuring level of the system into entities that can be extended to agents. The concepts like modularity, encapsulation, elaborated interfaces can be applied to agents in a satisfactory way as a first structuration level.

4   SERVICE IN OPEN DISTRIBUTED SYSTEMS

Agent paradigm is expected to emphasize in emerging open distributed systems the concept of *service* (TINA-Consortium 95, Merz 96). A service is a collection of functionalities provided to be used under some conditions (Merz 96, Diagne 96a). The functionalities are supported by information stored and manipulated for providing the service. This information models the *resources* necessary to run the service. The conditions attached to a service can be of many kinds. A service can be attached with a protocol which specifies how to use it correctly e.g. by sequencing in a given way its many functionalities. It can also be submitted to access permissions or quality aspects which may vary according to the context in which it is used. Finally it can be submitted to any other restriction relevant to be considered in its application domain e.g. taxation, temporary (un)availability, progressive adaptation to users needs and contexts, etc.

Each functionality of a service is implemented by an operation the environment can invoke. The view the environment has on an operation consists of its quality(ies) of service, its signature and constraints from the service view (dependencies with other operations). A service is offered by a server and can be requested by clients. At the server side, the operation is implemented by a given behavior which can be run with different strategies according to the expected quality of service. For instance the same *behavior* which broadcasts a message is run with different *strategies* in order to achieve a reliable broadcast or an unreliable one. The service must therefore implement the behavior as well as its different strategies of application.

The concept of service must take into account the many aspects ranging from reactive properties to elaborated cognitive and collaborative/cooperative capabilities. Specification, realization, test and deployment of services need to be considered in a methodical way. Formal methods are being more or less applied to fullfil the needs of safety and reliability (Gervais 96).

Services encompass many aspects like collaboration with its users, reasoning on contextual knowledge and executing some functionalities to achieve the service. These aspects can be divided into three parts:

- the *collaborative/cooperative aspects* like negotiating and making contracts with users or other services. These contracts can be commitments on quality of service and/or access rights, consequent billing, etc.
- the *cognitive aspects* like making inference on the contextual knowledge attached to a service or to its execution. This contextual knowledge can determine the way the service is offered. It can also be associated to the profile of the service user,
- the *reactive* or *computational aspects* like modifying the resources and running some specific processing necessary to exhibit the right functionalities under the contractual constraints. Services are namely reactive because they must maintain a continuous interaction with their environment (users or other services).

These previous aspects are not independent from each other. We can notice that the contextual knowledge used in cognitive activities may depend on the previous collaborative/cooperative activities and may influence the reactive ones. These interdependencies must be considered while using formal methods in order to support verification and validation. For instance, formal methods can be considered on reactive aspects to ensure safety and reliability whatever are the cognitive and collaborative/cooperative aspects (Estraillier 96). We can so aim more normalization in the agent-based open distributed systems (Pitt 95). It appears therefore necessary to separate and manage the many concerns in order to avoid undesirable influences from each other (Hursch 95).

5   ADAPTING VIEWPOINTS TO THE AGENT PARADIGM

As agent-based systems need more architectural guidelines to achieve a first level of integration as well as a good separation of concerns, we propose to proceed like in the RM-ODP with some adaptations (McKay 96). The ODP viewpoints are well-suited to separate the many concerns in object-based open distributed systems. Nevertheless, we would not propose viewpoints for agent-based systems to be a progressive structuration like in ODP. We try, through our adaptation of viewpoints, to separate and organize the many concerns (collaborative/cooperative, cognitive and reactive) of such systems and to make their mutual dependencies more manageable. Viewpoints help us to separate concerns. We propose to apply the concepts underlining the ODP structuration to the agent-based system and we propose the following classes of agents. *For that purpose and for the understandability of the remainder of the paper, the reader is warned of the fact that in this proposal, all entities are processing ones unlike in RM-ODP where enterprise and information objects are not processing ones.*

5.1 Service Manager Agents

A Service Manager Agent is an entity managing the policies and rules attached to the availability and utilization of a collection of services. A Service Manager must be able to negotiate the offer of its services with consumers. It must also be able to cooperate with other Service Managers to use their services when needed. The two kinds of customer (humans and others Service Managers) put high requirements on the interface mechanism. High level trading capabilities and elaborated user interfaces are necessary in order to present services for the understandability of humans and availability for electronic entities.

A Service Manager must be able to represent the knowledge necessary to run the service and the relevant reasoning capabilities on that knowledge. Service Manager Agents can be mobile or fixed agents. They are responsible to negotiate with others agents or users in order to determine a context under which services will be provided (Finin 94). They match the collaborative/cooperative and cognitive aspects of a service.

Service Managers can cooperate between them to share the load of the service offer. Thus they do have distributed problem-solving capabilities to share their knowledge and the inference they perform on it (Carle 95). Load and knowledge sharing can be used by a set of Service Managers to present some kind of cooperation that supports «shared state» in the system.

Concerning the collaborative aspects, Service Managers must be able to establish contracts and to fullfil the subsequent obligations (Pitt 95). Service Managers must be able to accept or deny results of negotiation but once accepted, the subsequent contract must be carried out in a satisfactory way for the counterpart.

At this level, best effort must be put on separation of cognitive activities from the collaborative/cooperative ones. Collaboration and cooperation might need some cognition and cognition might depend on previous collaboration. The dependencies between the both aspects must be clearly identified and validated even in an informal way. The role of a Service Manager will be further clarified as we go along.

5.2 Resource Manager Agents

A Resource Manager Agent is an entity responsible to manage one or many resources on behalf of a Service Manager. It offers capabilities for access and modification of the managed resources. It defines the allowed access to the managed resources as well as integrity constraints that will be enforced.

Resource Managers are under control of a given Service Manager which can give - by authentication means - an access/modification permission to other entities (any other kind of agent) (Thirunavukkarasu 95). This permission will determine a subset of allowed accesses and modifications. Resource Manager can be made mobile in order to make the information available on remote site but the mobility is under control of the corresponding Service Manager. Resource Manager can encapsulate part of the reactive aspects of the service related with resource manipulation.

Exception processing must be enforced at this level to send some events back to the Service Managers because they might need some cognitive or collaborative actions. For instance, a temporary or definitive unavailability of some functionality(ies) must be signalled to the Service Manager to make it change its proposal to the environment relevantly. This allows to offer adaptable services. Access/modification attempts by agents which are not truthful or without the right permissions need also to be signalled back to Service Managers. These events are trapped by the Service Managers like exceptions, so the relevant activities can be performed. Exceptions allow one to have some kind of fault tolerance on Resource Managers.

Some Resource Manager may need to access functionalities offered by a counterpart in order to achieve its own ones. It must therefore ask its responsible Service Manager to undertake the necessary collaborative and cognitive activities in order to obtain the access/modification permission. This permission is delegated to the Resource Manager which therefore can send invocations. This problem can be solved by permanent contracts between Service Managers in order to access remote resources transparently. The access to the remote resource by the local Resource Manager can be achieved by creating ad-hoc Activity Managers (see next section).

We make the choice that Resource Managers do not receive any collaborative/cooperative or cognitive capabilities from their responsible Service Managers. They are therefore pure reactive agents which run according to the access permission attached to received invocations. We can then validate them formally and make sure that resources will not be corrupted independently from the indeterminism in collaboration/cooperation and cognition. Resources are critical to the correct operating of the service and therefore, all their concerns must be validated and verified formally.

5.3   Activity Manager Agents

An Activity Manager Agent is an entity able to perform a set of actions in order to fullfil a given goal. An Activity Manager is under control of a Service Manager and receives from it access permissions on its attached Resource Managers. It can also - according to its fixed goal - receive collaborative/cooperative and cognitive capabilities to address others Service Managers. Another possibility is that the responsible Service Manager carry on the collaborative/cooperative and cognitive activities with other Service Managers, then the Activity Manager only receives delegation on permission granted to its responsible Service Managers to access remote resources.

Activity Managers may divide their goals into sub-goals. Therefore, they clone themselves into others Activity Managers to handle these sub-goals. The global coherence must then be managed by the agent which initiates the goal splitting. Transaction-oriented facilities must be supported to manage this coherence. Activity

Managers can have collaborative, cognitive and reactive aspects according to their assigned goal. They are attached to a given Service Manager which will delegate them part or whole of its collaborative/cooperative and cognitive capabilities.

Activity Managers realize the reactive tasks necessary to Service Managers. This delegation is a way to isolate the reactive aspects from the others in order to formally verify their safety and reliability. Like for Resource Managers, we need to trap some exceptions for the behalf of Service Managers in order to process exceptional events that might happen to Activity Managers. For instance, a definitive failure on a (sub-)goal can eventually need new cognitive and collaborative/cooperative activities which the Activity Manager can not carry out on its own.

5.4 Engineering and Technological Agents

The two last viewpoints in the RM-ODP deal mostly with implementation aspects. In the agent paradigm also, we will consider such kind of agents as relevant to realize the three previous classes we have defined. Depending on an underlining technology, one must consider how the previous levels of abstraction can be realized. What can be called *engineering* and *technological agents* must therefore be defined to establish some correspondence between the needs in the previous levels with the concepts available in the underlining technology. We remain deliberately unprescriptive and refer to the RM-ODP for adaptation.

6 SERVICE-BASED ARCHITECTURE OF AN AGENT-BASED SYSTEM

Agent and Service are two valuable concepts to structure open distributed systems. They offer two levels of structuration which can be mixed to have a federative basis for such systems. Services can stand for structuration unit for which agents are used to represent their many concerns with a clear separation between them.

6.1 Overall Architecture

Given the set of agent classes we define above, we are going to issue an architectural proposal for structuring open distributed systems according to service and agent concepts. We will consider here the key idea of service as a main guideline in structuration of systems. We consider henceforth that a system is characterized by the set of services it can provide to its environment.

A service can be structured as follows:

- one or many Service Manager Agents responsible of the *policy of the service* encompassing its *use* by the environment and its eventual *collaboration* and *cooperation* with other services. They have the same lifetime than the service,
- one or many Resource Manager Agents responsible of the *local resources* necessary to the service and which it owns and manages. Their lifetime is up to the needs of the responsible Service Manager Agents,
- one or many Activity Manager Agents whose *goals* are determined by some Service Manager Agent. Then, they receive from that Service Manager eventual collaborative/cooperative and cognitive capabilities necessary to fulfill that goal. Their lifetime can end with the definitive success or failure for the as-

signed goal.

The relevancy of the service notion here is its reflexivity. In complex services, one can undertake decomposition and each Service Manager can be considered - with its attached Resource Managers and Activity Managers - as an elementary service. This decomposition allows to consider services at a granularity level which does not carry too much complexity. So test validation and verification can be achieved in a satisfactory way on elementary services before combining them into more elaborate ones. The feature interaction problematic deals with problems that might occurs like degradation of service functionalities caused by concurrency between instances of many services (or instances of the same service) is well known in telecommunication systems (Cameron 94). It can be tackled better at this low-level of granularity and it must be taken into account in the services composition procedure.

The delegation of reactive tasks to Activity Managers by Service Managers can be done one the fly and when needed. The Activity Managers do not need in that case any more collaborative or cognitive aspects. They are only created to perform a given task on behalf on an Service Manager. The Activity Managers can then be validated using formal methods in their reactive aspects. This possibility is left up to system designer and must be evaluated in front of the level of validation one can need in a given application domain.

6.2   Synthesis and Discussion

The ODP viewpoints are used to separate concerns in systems while managing consistency between them. We adapt them here in order to manage the dependencies between different aspects of a system (a collection of services). The progressiveness from Enterprise viewpoint to Technological one is lost because we have more operational associations to manage between a Service Manager and its attached Resource Managers and Activity Managers. Viewpoints are used to assign roles and activities to components in a service.

In figure 1 we show the correspondence between the ODP separation of concerns and the one we propose in this paper in order to structure services in an open distributed system by means of agents. Service Managers model the adaptation of the Enterprise viewpoint. Service Manager Agents need to represent the policy and general rules which govern service availability. They use these policy and rules to make contracts with users or other Service Manager Agents. Their representation of resources and activities is achieved through their attached Resource Manager and Activity Manager Agents. Resource Managers correspond to the Information viewpoint. Resources Manager Agents only represent the schemas that govern the management of resources. They can represent some computational activities in order to be able to invoke other local Resource Manager Agent or remote ones based on pre-established contracts between their akin Service Manager Agents. Activity Managers are the most complex entities because they must be flexible enough to allow the system designer to model different strategies. So they have basically computational aspects but can be enhanced with collaborative/cooperative and cognitive aspects as well as management of some temporary resources necessary for one execution at most This kind of resource, like the contextual information necessary to continue execution on remote sites in case of migration, does not enforce the need of a Resource Manager. Therefore, they can have capabilities of Resource Manager Agents and also receive delegation of e.g negotiation capabilities from their responsible Service Manager Agents. This allows them to make contracts with other Service Manager Agents.

An obvious advantage of this proposal is the fact that one can isolate the information managed (resources) in the Resource Manager Agents which can be validated formally. We therefore make sure that whatever the collaborative/cooperative and cognitive activities are, the resources will not be corrupted. Resources are critical for a service because their states determine the correctness, safeness and reliability of its behavior. For these reasons, it is important to verify and validate thoroughly their concerns.

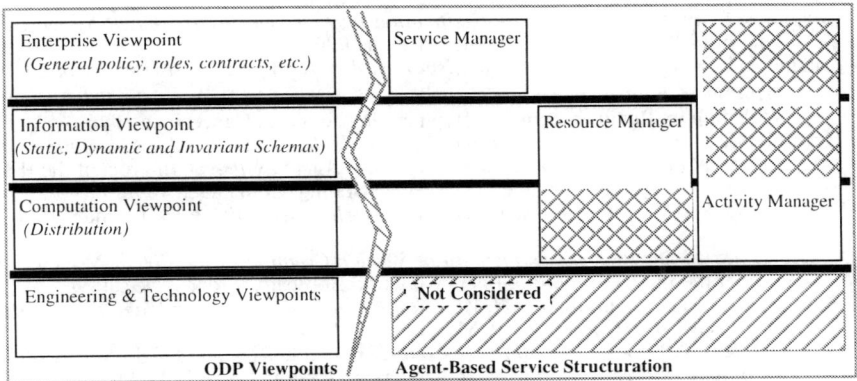

Figure 1   From ODP Viewpoints to Agent-based Service Structuration

7   CONCLUSION AND FUTURE WORK

We propose in this paper a way to mix the agent and service concepts and enhance them with architectural concepts which can fit for all purposes from the specification stage down to the final implementation. Agent-based service structuration of open distributed systems can therefore be emphasized with a satisfactory level of separation of concerns.

The proposed architecture first attempts a separation of concerns in order to better manage their own semantics and mutual dependencies. It also enables a service designer to apply formal methods when and where they can be needed to improve safety and reliability verification. This separation of concerns is difficult to achieve fully in some architecture because it depends on the intention and will of the service designer and the constraints of application domain. For instance the decision to make Activity Managers fully reactive in our proposal is up to the service designer under constraints enforced by the application domain.

In case of indeterminism in goals assigned to Activity Managers, it will be difficult to prevent them from having collaborative and/or cognitive capabilities in order to maintain a loose coupling with their responsible Service Managers. However, we give a valuable way to realize it more or less. We also claim that the architecture can be a first level of integration in the *«open electronic market»*. Services designed in a similar way are more easy to make interwork than otherwise. We have defined in (Diagne 96a) and (Diagne 96b) models which fit for the reactive aspects of our architecture proposal. We will henceforth stress our research effort on collaborative/ cooperative and cognitive models.

8   REFERENCES

H. Bowman, E.A. Boiten, J. Derrick & M.W.A. Steen, *«Viewpoint Consistency in ODP, a General Interpretation»*, In Proc. of FMOODS'96, Paris, France, March 1996.

E.J. Cameron, N.D. Griffith, Y.-J. Lin, M.E. Nelson, W.K. Shnure & H. Velthuijsen, *«A Feature Interaction Benchmark in IN and Beyond»*, In Feature Interactions in Telecommunications Systems, IOS Press, Amsterdam, Holland, 1994.

P. Carle, A. Collinot & K. Zeghal, *«Cassiopeia : a Method for Designing Computational Organizations»*, In Proc. of IJCAI'95, Montreal, Canada, Aug. 1995.

A. Diagne & P. Estraillier, *«Formal Specification and Design of Distributed Systems»* In Proc. of FMOODS'96, Paris, France, March 1996.

A. Diagne & F. Kordon, *«A Multi-Formalism Prototyping Approach from Conceptual Description to Implementation of Distributed Systems»*, In Proc. of the 7th IEEE Int. Workshop on Rapid System Prototyping, Greece, Porto Caras, June 1996.

P. Estraillier & F. Kordon, *«Structuration of Large Scale Petri Nets: An Association with High level Formalisms for the Design of Multi-Agent Systems»*, In Proc. of the IEEE Int. Conf. on System Man and Cybernetics, Beijing, China, Oct. 1996.

J. Ferber, *«Les Systèmes multi-agents : Vers une intelligence collective»* iia interEditions, 1995.

M.P. Gervais & A. Diagne, *«Formalization of Service Creation in Intelligent Network»* In Proc. of the Int. Conf. on Intelligent Network, Bordeaux, France, Dec. 1996.

T. Finin, Y. Labrou & J. Mayfield, *«KQML as an Agent Communication language»*, In Proceedings of ACM/CKIM'94, ACM Press Nov. 1994.

W.L. Hursch & C.V. Lopes, *«Separation of Concerns»*, Tech. Rep. NU-CCS-95-03, College of Computer Science, Northeastern University, Boston, USA, Feb. 1995.

T. Magedanz, K. Rothermel & S. Krause, *«Intelligent Agents : An Emerging Technology For Next Generation Telecommunications?»*, in Proc. of IEEE/INFOCOM'96, San Francisco, USA, March 1996.

M. Merz & W. Lamersdorf, *«Agents, Services and Electronic Markets : How do they integrate ?»*, In Proc. of the Int. Conf. on Dist. Platforms, Dresden, Germany, 1996.

D.P. McKay, J. Pastor, R. McEntire & T. Finin, *«An Architecture for Information Agents»*, In Advanced Planning Technology, AAAI Press, Menlo Park, CA, USA, May 1996.

J. Pitt, M. Anderton & J. Cunningham, *«Normalized Interactions Between Autonomous Agents : A Case Study in Inter-Organizational Project Management»*, In Proc. of COOP'95, Antibes-Juan-Les-Pins, France, Jan. 1995.

C. Thirunavukkarasu, T. Finin & J. Mayfield, *«Secret Agents - A Security Architecture for KQML»*, In Proc. of ACM/CKIM'95, Agent Workshop, Baltimore USA, Dec. 1995.

ITU X.902 & ISO/IEC 10746-2, *«Basic Reference Model of Open Distributed Processing, Part 2: Descriptive Model»*, 1995.

ITU X.903 & ISO/IEC 10746-3, *«Basic Reference Model of Open Distributed Processing, Part 3: Prescriptive Model»*, 1995.

TINA-Consortium, *«Service Architecture»*, TB_MDC.012_2.0_94, 31 March 1995.

Java, Sun Microsystems, *«The Java Language Environment : a White Paper»*, http://javasoft.com/, 1995.

Telescript, General Magic, *«Telescript Language Reference»*, http://www.genmagic.com/, 1995.

27

A Conformance Relationship for Stream Interfaces

F. Eliassen
University of Tromsø
Dept of Computer Science, 9037 Tromsø, Norway
frank@cs.uit.no

Abstract

In distributed object management systems, the notion of interface type is important for supporting type checking of interfaces during bind time. In particular, conformance rules for interface types express conditions for substitutability. Unfortunately, generally applicable conformance rules for stream interfaces have so far proven difficult to define. In this paper we propose a general definition of stream interface conformance satisfying the key requirement that it preserves computational compatibility. This ensures the continued service of existing clients when a stream interface is substituted for by a new conforming stream interface.

Keywords

stream interfaces, stream flows, type model, compatibility, conformances

1   INTRODUCTION

In an open distributed system, there is no guarantee that components are built using the same technology. In particular, future distributed multimedia applications will involve components supporting different audio/video formats, compression schemes, networking protocols, and hardware solutions.

It is therefore important to have in place object management facilities in support of distributed multimedia applications, capable of flexibly interconnecting stream endpoints despite the fact that they may exhibit various forms of heterogeneity with respect to media stream properties and coding formats. An important foundation on which to build this kind of support is a proper type model for streams - a model allowing stream endpoint conformance and compatibility to be checked and resolved. Informally, compatibility ensures that the properties of a source flow are as expected by the sink, while conformance express conditions for substitutability of flows and stream interfaces.

Generally applicable conformance rules for interfaces including continuous media flows, however, have so far proven difficult to define. This is partly due to the observation that the most appropriate basis for deciding compatibility (and subtyping) of multimedia flows is application-dependent.

In this paper we propose a definition of stream interface conformance based on an earlier proposed type model for stream flows (Eliassen & Nicol, 1996). In the model, a flow type specification is interpreted as a set of potential stream flow properties. This supports the definition of a variety of flow type relationships based on set theory. The proposed definition of conformance satisfies the key requirement of preserving compatibility of the flows. This ensures the satisfaction of existing clients when a stream interface is substituted for by a new conforming stream interface.

The remainder of the paper is organized as follows. In section 2, we offer further motivation for our work. Following that, in section 3 we then present the main ideas of a type model for continuous media flows. Section 4 introduces the notion of flow compatibility. Section 5 discusses requirements to a conformance relationship of stream interfaces, and offers a definition. Section 6 concludes.

2   TYPE CHECKING AND BINDING OF STREAM INTERFACES

The notion of stream interface have been proposed as the preferred means to convey media streams (Coulson et al, 1992). A stream interface consists of a collection of source and/or sink media flows. A continuous media flow (such as audio or video) is a time-based value interpreted as a (finite) sequence of temporally constrained data elements. The temporal dimension of the data elements refer to their required time and length of presentation. This is expressed as a QoS annotation specifying, for example, required sample rate. Stream interfaces have been adopted in the work on Open Distributed Processing (ITU-T Draft Recommendation X.901, 1995) and TINA-DPE (TINA-C, 1995) and more recently by OMG (Object Management Group, 1996).

Bindings are created between compatible stream interfaces for the purpose of exchanging continuous streams as dictated by the type and direction of the flows. During binding, the interfaces to be bound must be type-checked for compatibility.

Informally, type-checking means ensuring that the properties of each source flow are as expected by the corresponding sink flow.

Conformance rules for stream interfaces have been deemed outside the scope of the RM-ODP standard (ITU-T Draft Recommendation X.901, 1995); the proposed rationale for which is that the optimum form of subtyping will depend on the application. Although this is a valid observation, the advantages of detecting type errors at compile (or bind) time, rather than at run-time, are no less important for stream interfaces than they are for operational interfaces. In the work of TINA-C, the need for a compatibility relationship for stream flows that is more relaxed than equivalence, is recognized, but no definition is offered (TINA-C, 1995). Other works that recognize the need for type-checking of continuous media flows during binding but without providing generally applicable definitions, include (Bates & Bacon, 1995) and (Coulson et al, 1992).

3   FLOW TYPE SPECIFICATION

This section offers an overview of the type model for continuous media flows, and notation used as a basis for reasoning about type-checking of continuous media flows. Flows are used to model the behavior of continuous media devices or stored continuous media objects. In common with other proposals for multimedia information models, we distinguish between source and sink flows.

Our approach to a flow type declaration is to provide media descriptor information as additional parameters of the type declaration, i.e.,

> Flow fD [E_1;E_2;...;E_n] sC

The flow denoted above features multiple element types $E_1, E_2, ..., E_n$. The number of different element types we refer to as the *element cardinality* of the flow. Each element type specification includes a declaration of the generic element type such as Image or Audio, a label that within the flow type specification is a unique name for the element type, and for each generic element type, a specific set of attributes specifying quality properties of the element type. For images the set might include attributes such as width, height and depth, color-model, encoding, etc., while for audio the set could include sample-size, number of channels, and encoding, etc. The set of attributes denoted by fD may be used to specifiy temporal constraints on the elements of the flow. The symbol sC denotes a specification of the structural constraints of the flow. Structural constraints are specified using sequencing expressions similar to regular expression as follows:

A   a sequence consisting of single element of the type labeled A
p q   a sequence satisfying p followed by sequence satisfying q
p|q   the sequence satisfyoing p or the sequence satisfying q
p*   zero or more occurrences of sequences each satisfying p
p+   one or more occurrences of sequence each satsifying p

```
p[n]  exactly n occurrences of sequences each satisfying p
nil   the empty sequence
```

The purpose of sequencing expressions is to allow a source or sink to express "indvidual" requirements to the satisfaction of compatibility and conformance such as optional element types and element type dependencies.

Element and flow descriptors are collectively referred to as media descriptors. The type of a media descriptor is a tuple type $[a_1:T_1, \ldots, a_n:T_n]$ where $a_1, \ldots, a_n$ are attributes names and $T_1, \ldots, T_n$ are type expressions. The value of a media descriptor of type $[a_1:T_1, \ldots, a_n:T_n]$ is a tuple denoted $[a_1:v_1, \ldots, a_n:v_n]$ such that the type of v_i is a set of values each of type T_i. We assume that for each generic element type, a specific media descriptor type is defined by an appropriate authority.

The view of an attribute value as a set of "atomic" values, will in general enhance the chances of successful binding. For example when a sink flow type specifies a set of different names on the image encoding attribute, it actually declares that it can accept flows where the images can have any of the indicated formats.

Example 1: The following example is one of a partial specification of an MPEG video flow type featuring several optional playback rates.

```
type MPEG = Flow [playback_rate:{24,25,30,50}] [
        I:Image[encoding:mpeg_i];
        P:Image[encoding:mpeg_p];
        B:Image[encoding:mpeg_b] ]
            (I[1] (P+ B* | P*))*
```

This specification states that an MPEG flow consists of three different image element types labeled I, P, and B respectively. The element descriptors specify encodings, while the flow descriptor specifies playback rates. The structural constraints of the flow is specified as a sequencing expression indicating that each group of image elements must start with an I-frame, followed by one or more P-frames followed by zero or more B-frames, or by zero or more P-frames. ❑

Interoperability requires standard payload types be defined and uniquely named by appropriate authorities. This is the approach, for example, in the work of the Internet Engineering Task Force on a real time transport protocol (Schulzrinne et al, 1996) in which specific payload types such as different audio and video encodings, are assigned unique names by an appropriate internet authority. The value of the encoding attribute above could represent such globally unique names.

A media descriptor can be extended with a set of value constraints. A value constraint is a predicate that constrains the range of specified values for an attribute. This could be used, for example, to capture the aspect ratio property of visual media (in which the width dimension constrains the height dimension).

Definition 1 *(Interpretation of media descriptor)* The interpretation $I(e)$ of a media descriptor e is a relational algebra restriction of the (extended) cartesian product of the attribute values of the descriptor. ❑

The predicate of the relational restriction is derived from the value constraints of the flow type. Each tuple represents a possible combination of attribute values satisfying the predicate. This means that tuples not satisfying the value constraints of the flow, are removed (see (Eliassen & Nicol, 1996) for details).

Definition 2 *(Quality interpretation of flow type)* The quality interpretation QI(M) of a flow type M is the (extended) cartesian product of the interpretations of the media descriptors of its element types and its flow descriptor. ❏

Definition 3 *(Structural interpretation of flow type)* The structural interpretation SI(M) of a flow type M with sequencing expression p is the set of element sequences satisfying p. ❏

The above interpretations of a flow type allows us to define a variety of flow type relationships such as subtyping and compatibility based on set theory.

4   FLOW TYPE RELATIONSHIPS

When interpreting types as sets, the usual semantics of the subtype relationship is that of set inclusion. We may therefore define a flow type M to be a subtype of the flow type N if the interpretation of M is a subset of the interpretation of N. In the following let attr(T) denote the set of all attribute names of the flow type T, and ecard(T) the element cardinality of T.

Definition 4 *(Strict subtype relationship)* For flow types M and N with element types $D_1, D_2, \ldots, D_m$ and $E_1, E_2, \ldots, E_n$ respectively, M is a quality subtype of N, denoted $M <_q N$, if ecard(M) = ecard(N), and there exist a permutation p of 1..n such that D_i and $E_{p(i)}$ are of the same generic element type, and QI(M) $\subseteq$ QI(N). M is a structural subtype of N, denoted $M <_s N$, if SI(M) $\subseteq$ SI(N). Futhermore, M is a strict subtype of N, denoted M < N, if $M <_q N$ and $M <_s N$. ❏

From the above definition, we may, for example, derive that Flow[A;B] is a strict subtype of Flow[C;D] if I(A) $\subseteq$ I(C) and I(B) $\subseteq$ I(D).

Example 2: For the following two flow types M and N, M is a subtype of N

```
M = Flow [rate:4][I:Image
        [width:{640,320},height:480] ]
N = Flow [rate:{4,5}] [I:Image
        [width:{155,320,640},height:{240,480}]]  ❏
```

We also define a relaxed flow quality subtype relationship where the subtype may support fewer element types than the supertype such that, for example, Flow[A] is a relaxed subtype of Flow[B;C] when I(A) $\subseteq$ I(B).

Definition 5 *(Relaxed subtype relationship)* For flow types M and N with element types $D_1, D_2, \ldots, D_m$ and $E_1, E_2, \ldots, E_n$ respectively, M is a relaxed quality subtype of N, denoted $M <\sim_q N$, if ecard(M) $\leq$ ecard(N) and there exist a permutation p of 1..m such that D_i and $E_{p(i)}$ are of the same generic element type, and $QI(M) \subseteq \Pi_{attr(M)}QI(N)$, where Π denotes the relational algebra projection operator. Furthermore, M is a relaxed subtype of N, denoted $M <\sim N$, if $M <\sim_q N$ and $M <_s N$. ❑

Example 3: In this example, the flow type MPEG_V is a relaxed subtype of the flow type MPEG_AV.

```
type MPEG_V = Flow [] [
                 I:Image[encoding:mpeg_i];
                 P:Image[encoding:mpeg_p];
                 B:Image[encoding:mpeg_b]  ]
                      (I[1] P[1] B[2])*

type MPEG_AV = Flow [] [
                 I:Image[encoding:mpeg_i];
                 P:Image[encoding:mpeg_p];
                 B:Image[encoding:mpeg_b];
                 A:Audio[...]  ]
                      (I[1] P+ B* A*)* ❑
```

It seems, however, too restrictive to require that the source-supported set of flow qualities must be a subset of the sink-supported set of flow qualities to allow attempts to bind. A more appropriate approach would be to allow attempts to bind if they can at least support one common flow. This relationship we refer to as *quality compatibility*. If the two flow endpoints can support several common flow qualities, a flow quality negotiation protocol may be employed to choose the actual flow quality to be used.

Definition 6 *(Strict quality compatibility relationship)* For flow types M and N with element types types $D_1, D_2, \ldots, D_m$ and $E_1, E_2, \ldots, E_n$ respectively, M is strictly quality compatible with N, denoted $M <>_q N$, if $ecard(M) = ecard(N)$ and there exist a permutation p of $1..n$ such that D_i and $E_{p(i)}$ are of the same generic element type, and $QI(M) \cap QI(N) \neq \varnothing$. ❑

It follows from definition 6 that we may for example conclude that $Flow[A;B] <>_q Flow[C;D]$ if $I(A) \cap I(C) \neq \varnothing$ and $I(B) \cap I(D) \neq \varnothing$.

Definition 7 *(Relaxed quality compatibility relationship)* For flow types M and N with element types types $D_1, D_2, \ldots, D_m$ and $E_1, E_2, \ldots, E_n$ respectively, M is relaxed quality compatible with N, denoted $M <\sim>_q N$, if there exist subsets α of $1..m$ and β of $1..n$ of equal cardinality, and a bijection f from α to β such that for $i \in \alpha$, D_i and $E_{f(i)}$ are of the same generic element type, and $\Pi_{A\alpha(M)}QI(M) \cap \Pi_{A\alpha(M)}QI(N) \neq \varnothing$, where $A\alpha(M)$ denotes the union of the attribute names of the flow descriptor and the attribute names of the element types of M identified by α. ❑

A consequence of definition 7 is that two flow types M and N are relaxed flow quality compatible if a subset of the element types in each of the two flow types, are pair-wise compatible. Thus if $I(A) \cap I(C) \neq \varnothing$, $Flow[A;B] <\sim>_q Flow[C;D]$ even if B and D are incompatible, i.e. $I(B) \cap I(D) = \varnothing$.

Definition 8 *(Structural compatibility)* Two flows of type M and N respectively, are structurally compatible, denoted M $<>_s$ N, if their structural interpretations have non-empty intersections, i.e. SI(M) $\cap$ SI(N) $= \emptyset$. ❏

Different variants of the compatibility relationship are given by the following definition:

Definition 9 *(Flow compatibility)* A source flow of type M and a sink flow of type N are

i)   fully strict compatible, if M $<>_q$ N and M $<_s$ N .

ii) partially strict compatible, when M $<>_q$ N and M $<>_s$ N.

iii) partially relaxed compatible, when M $<\sim>_q$ N and M $<>_s$ N .

iv) fully relaxed compatible, when M $<\sim>_q$ N and M $<_s$ N. ❏

In a specific instance of a binding fully relaxed compatibility means that the sink might support element types not supported by the source, and the source may produce flow qualities not required by the sink. However, every flow structure that can be produced by the source is acceptable to the sink. Partially relaxed compatibility means that there might be element types supported by the source that is not supported by the sink and vice versa. Furthermore, partially (strict or relaxed) compatibility means that the source could produce sequences of element types that are not acceptable to the sink. In this case a binding could still be allowed if the flow structures produced by the source can be constrained to the common set of flow structures supported by both the source and the sink flow. This would make the resulting modified source flow type fully compatible with the sink flow type. Our approach is to allow a source to declare itself as constrainable.

Defintion 10 *(Constrainable source)* A source flow is constrainable if it in any instance of binding is able to constrain its generated flow structures to those it has in common with the sink of the binding. ❏

In general, the above definitions are meaningful only when the element types of M and N that are inclusion or compatibility related, have identical labels. We therefore assume this to be the case, possibly after an appropriate relabeling. The labels of the sequencing expressions must be relabeled accordingly.

5   STREAM INTERFACE CONFORMANCE

Conformance rules express conditions for substitutability. This means that a stream interface S may be replaced transparently by another stream interface T as long as clients that used to bind to flows of the S stream interface must be able to continue to do so for flows of the T stream interface without modification of their supported flow properties. In such a case, T is said to *conform* to S. It follows that an important requirement to steam interface conformance is that compatibility of flows must be preserved form S to T.

Conformance rules for distributed object-based systems were pioneered in the work on the Emerald programming system (Black et al, 1987). For operations they are based on the rule of contravariance. This means that an operation f conforms to an operation g if the argument type of f is a generalization (i.e., supertype) of that of g and the result type of f is a specialization (i.e., subtype) of that of g.

Contravariance may be applied to flows by viewing a source flow f : <<M as a function f : ( ) → M, and a sink flow h : >>X as a function h : (X) →. Then, according to the above rules, f may replaced in any context where f is expected by a function g : ( ) → N, if N < M, i.e. g does not produce concrete flow types that cannot be produced by f. Similar reasoning applies to sink flows.

The problem with the above definition of conformance is that the source flow g is allowed to support a smaller range of concrete flow qualities than the source flow f it replaces. Unfortunately, this means that compatibility is not always preserved from f to g. Hence we must require that g supports at least the flow qualities of f to preserve compatibility, i.e. $M <_q N$. For sink flows, however, contravariance seems sufficient as an underlying principle. On this background we propose the following conformance rules for stream interfaces.

Definition 11 *(stream interface conformance)* A stream interface type **T** conforms to a stream interface type **S** if and only if:

1. **T** provides at least the flows of **S**.
2. For each flow f of type M of **S**, let g of type N be the corresponding flow of **T**. Then the following conditions must be satisfied:
   i)   g has the same direction as f (source or sink);
   ii) if f is a sink flow then M is a (relaxed or strict) subtype of N;
   iii) if f is a constrainable source then g is constrainable;
   iv) if both f and g are non-constrainable sources, then $M <_q N$ and $N <_s M$.
   v)   if g is a constrainable source, then M <~ N. ❏

This definition of stream interface conformance preserves flow compatibility. A stronger requirement in item iv) would be to state $N =_s M$, i.e. g must be able to produce exactly the same flow structures as f, and no less.

Theorem *(compatibility preservation)* If a stream interface T conforms to a stream interface S, then computational flow compatibility is preserved from S to T such that if h is a flow compatible to some flow f in S, and g is the corresponding flow of f in T, then h is compatible to g. ❏

The proof can be made by showing that if f is a source(sink) to be replaced by the source(sink) g, and h is a sink(source) flow compatible with f, then g and h are compatible as well. For example if we assume case iv) of definition 11 and that h is a flow sink of type X such that X is fully compatible with M, then we can reason that $X <>_q M$ and $M <_q N$ giving $X <>_q N$, and that $M <_s X$ and $N <_s M$ giving $N <_s X$. This shows that X is fully compatible with N as well. Due to space limitations we omit the proof here.

Example 7: The following example illustrates a case where a video source of flow type MPEG is upgraded to support additional playback rates, video encodings and audio, all encapsulated in a single flow of type AMJMPEG. We see that MPEG <~ AMJMPEG. Hence the new video source must be constrainable.

```
type MPEG = Flow [playback_rate:{25,30}] [
                    I:Image [encoding:mpeg_i];
                    P:Image[encoding:mpeg_p];
                    B:Image[encoding:mpeg_b]]
                        (I P+ B*)*

type AMJMPEG = Flow [playback_rate:{20,25,30}] [
                    J:Image [encoding:jpeg];
                    I:Image [encoding:mpeg_i];
                    P:Image[encoding:mpeg_p];
                    B:Image[encoding:mpeg_b];
                    A:Audio[encoding:{PCMA,GSM}] ]
                        (I P+ B* A*)* | (J A*)* ❏
```

6 CONCLUSIONS AND FUTURE WORK

In this paper we have proposed definitions of compatibility and conformance of stream interfaces. These definitions rely on a proposed type model for stream flows. A salient feature of the model we propose concerns the interpretation of a type specification as a set of potential flow qualities and flow element sequences. This supports the definition of a variety of flow type relationships based on set theory. In this paper we defined subtyping and compatibility rules as a basis for defining conformance rules expressing conditions for substitutability of stream interfaces.

We concluded that the proposed notion of conformance preserves compatibility with existing clients. This ensures the continued service of these clients when a stream interface is substituted for by a new conforming interface.

In our current work work we focus on gathering practical experience using the type model in distributed multimedia applications. In order to accommodate continuous quality attribute types (such as float) in the flow type model we are reworking (some of) the definitions of flow type interpretation, flow subtype, and flow compatibility to a more intentional style. The definitions of this paper have an extentional style based on relational algebra that assumes discrete attribute values.

7   REFERENCES

Bates, J., Bacon, J.(1995) Supporting Interactive Presentations for Distributed Multimedia Applications. *Multimedia Tools and Applications*, Kluwer 1(1), 47-78.

Black, A. N. et al.(1987) Distribution and Abstract Types in Emerald. *IEEE Transactions on Software Engineering*, 13(1), 1987, 65-76.

Coulson, G., Blair G. S., Stefani, J. B., , Horn, F., Hazard, L.(1992) Supporting the Real-Time Requirements of Continuous Media in Open Distributed Processing. *Technical Report MPG-92-35*, Lancaster University.

Eliassen, F., Nicol, J. R.(1996) Supporting Interoperation of Continuous Media Objects. *Theory and Practice of Object Systems: special issue on Distributed Object Management* (ed. G. Mitchell), Vol.2, No.2, Wiley, 1996, 95-117.

ITU-T X.901 I ISO/IEC 10746-1 (1995) ODP Reference Model Part 1: Overview. *Draft International Standard.*

Schulzrinne, H., Casner, R., Frederick, R., Jacobson, V.(1996), RTP: A transport protocol for real-time applications. *IETF*, rfc 1889.

TINA-C (1995) TINA Object-Definition Language, Version 1.3. *TINA-C Deliverable.*

Object Management Group (1996) Control and Management of A/V Streams Request for Proposal. *OMG Document*: telecom/96-08-01.

8   BIOGRAPHY

Frank Eliassen is professor of computer science at the University of Tromsø, where he leads the Open Distributed Systems research group. His research interests are generally oriented towards issues of distributed systems and distributed databases caused by requirements of heterogeneity and autonomy. Particular research topics of interest include distributed object management systems, extended transaction models, and interoperability of distributed multimedia applications. Also see http://www.cs.uit.no/ODS/index.eng.html.

28

Test execution of telecommunications services using CORBA

L. Paula Lima Jr.[1] *and A. R. Cavalli*
Institut National des Télécommunications
LOR - 9, rue Charles Fourier - 91011 Evry Cedex - France
Email: lima{ana}@hugo.int-evry.fr

Abstract

Even though there are many testing methods for telecommunication systems very little has been said about test execution in distributed environments. This paper addresses this problem and proposes the implementation and test of telecommunication services on a CORBA-like platform. We present a test architecture for distributed systems and describe system components and test execution phases. An implementation solution is also proposed. The application is illustrated by the implementation of a telecommunication service: the BCS (Basic Call Service).

Keywords

Test execution, Formal Description Techniques, CORBA, telecommunications services.

1   INTRODUCTION

Intelligent Networks (IN) represented a remarkable step forward in the domain of telecommunications systems. The ability to add 'intelligent' nodes (i.e. computers) into switching networks made it easier (or possible) to introduce new services like

---

1. This work was funded by CNPq.

Call Forwarding Unconditional (CFU), Originating Call Screening (OCS), and many others. However, the service implementation was still dependent on the hardware and operating system upon which it was built. Nowadays, telecommunications services are migrating to open distributed environments where a layer called *middleware*[1] assures transparency, interoperability and cooperation among the several service components in spite of the (possibly) heterogeneous underlying systems. For instance, TINA[2] architectures specify the DPE[3] layer which is responsible for providing facilities to service programmers, so that they do not need to cope with communication details and may concentrate on the service implementation itself. Portability and re-utilization are other advantages that become apparent.

DPE layers can be implemented using CORBA-like platforms[4]. Although CORBA may not be adequate to provide all service quality requirements of some multimedia applications, people still use CORBA because it is easy to program and to set up a distributed application in this environment. There are more and more commercial tools for verification which are capable of generating CORBA code from FDT[5] specifications (for instance, SDT/ITEX from Telelogic and a recent version of GEODE from Verilog [2]).

It is widely known that testing is the most important (and the most expensive) phase in the life-cycle of communicating systems. This is especially true for telecommunications services due to the well-known potential problem of feature interactions. Failures of the network may have a catastrophic effect on the operator's image as a reliable service provider.

There are many different testing methods for telecommunications systems, ranging from traditional black-box testing in isolation [3] [4] to component testing (embedded testing [5], or testing in context [6]) and there are many proposals concerning feature interaction detection (e.g. [7] [8]). All these methods correspond to what is called the *Testing Generation Phase*, which starts with the system specification and ends with the generation of test suites. Of course, the testing generation phase is highly influenced by the test execution architecture. However, very little has been said about *test execution*, specifically, regarding distributed environments like CORBA. Thus, on the one hand we see a worldwide move to using distributed platforms to implement telecommunications services, and on the other hand the need for a suitable architecture to test these services.

---

1. The distributed environment must provide the semblance of a single system image across potentially millions of hybrid client/server machines. The *middleware* is the layer that creates this illusion by making all servers on the global network appear to behave like a single computer system.
2. TINA: Telecommunications Information Networking Architectures.
3. DPE: Distributed Processing Environment.
4. CORBA (Common Object Request Broker Architecture) is the Object Management Group's architecture that allows clients to invoke methods on remote objects either statically or dynamically hiding all communication details from the system programmers [1].
5. FDT: Formal Description Techniques (e.g. SDL, LOTOS, Estelle, Z, etc.).

This paper addresses precisely this problem suggesting a general CORBA-based test architecture for telecommunications services. In doing so, we will be able to come up with a practical view on, amongst others:

1. distributed test execution and generation (in embedded systems);
2. how to structure the test architecture for CORBA-like systems/objects;
3. object-oriented aspects of conformance testing; and
4. performance analysis aspects.

All these items are discussed in this paper, except item 4 which is left for future work. Methods for test generation from formal specifications have already been suggested in some of our previous works [3][15], therefore they will not be detailed here. Section 2 briefly describes the environment upon which the distributed system is implemented. Section 3 presents a test architecture for distributed systems built upon an ORB platform. Section 4 introduces the system components and goes through the test execution phases discussing some particularities of object-oriented conformance testing. Conclusions are drawn in section 5.

2   PLATFORM DESCRIPTION

The distributed platform chosen was Orbix™ (Iona Technologies) which is a full implementation of the Object Management Group's (OMG) Common Object Request Broker Architecture (CORBA) [9]. The key advantage of such platforms is that they make application construction and integration much easier, since software interfaces are defined using a standard language (IDL: Interface Definition Language) which allows objects to be accessed from anywhere in a distributed system.

Basically, all interactions (i.e. service requests) are performed by means of '*channels*.' Each channel is composed of a *stub* (on the client side) and a *skeleton* (on the server side) which are connected to the Object Request Broker (ORB). Channels allow a transparent communication between the client and the object implementation (Figure 1) providing the illusion that all procedure calls happen locally. The ORB in turn provides the functionality required to communicate across (possibly heterogeneous) platforms.

A service request consists of the following information:

- a target object;
- an operation;
- a list of parameters; and
- an optional request context.

'One way' operations (called *announcements* in the terminology of the Reference Model for Open Distributed Processing - RM-ODP [10]) allow asynchronous message exchanges amongst the various system components.

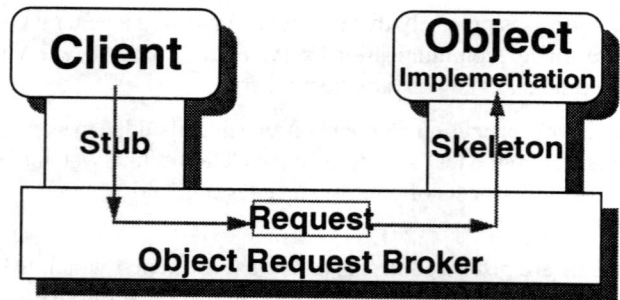

Figure 1 A request passing from client to object implementation.

Object can simultaneously be clients *and* servers, and must cooperate (or *interoperate*) in a useful way to achieve a given objective. This is exactly what CORBA provides.

3   TEST ARCHITECTURE FOR DISTRIBUTED SYSTEMS

Following the guidelines of classical test architectures for protocols as described in [11], we suggest that a test architecture for an open distributed object-oriented platform must have two sets of components: *tester components* and *components under test* that are (possibly) distributed over several sites (Figure 2). The tester compo-

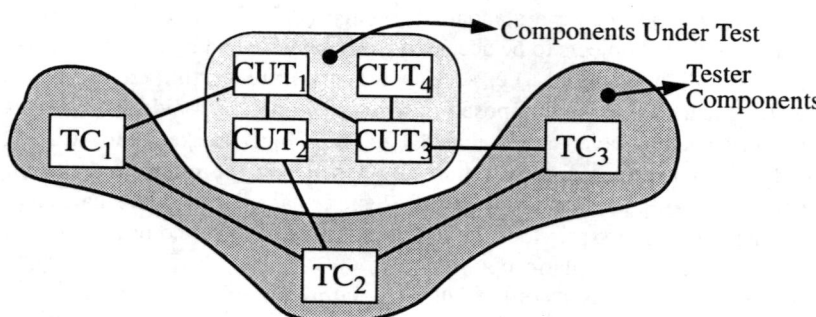

Figure 2 Components of the test architecture.

nents receive information about the object configuration (i.e. the collection of objects able to interact at their interfaces) and the test suite which is obtained through some test generation method from a formal specification.

Simple systems (with local architectures) are usually implemented using one single tester component and a single component under test.

Example 1. For simplicity, let us assume that we have just one tester component and two components under test, as shown in Figure 3.

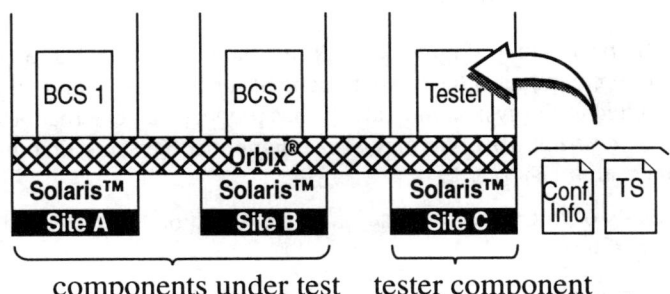

components under test   tester component

BCS = telephone Basic Call Service

Figure 3 Test architecture example.

These components are distributed over different sites, but the tester itself is centralized. Information about the configuration of objects to be tested ('*Conf. Info*' file in Figure 3) and the tests sequences (ideally in TTCN notation) [11] are the data that the tester needs to conduct tests through the ORB platform and to produce a verdict. Test sequences to some particular scenarios may also be expressed in terms of Message Sequence Charts [12] (MSC, Figure 4) ❏

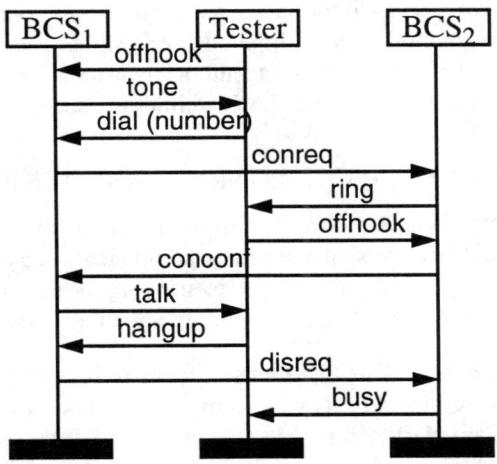

Figure 4 Example of a scenario to test.

After defining the components of the architecture and their configuration, we must establish some rules concerning the communication among the various components, because the tester must interact with the system under test in a well-defined manner. We assume that all communication is done through message exchanges and that each component has an input queue of a given finite length (infinite queues lead to infinite test sequences in the test generation phase).

Tester components are equipped with timers in order to detect deadlocks and livelocks. A different time-out value may be assigned to each transition we intend to test. In practice, a default time-out value is given to all transitions (some QoS requirements, e.g. performance requirements, can be expressed using different time-out values for particular transitions).

Example 2. In the system of Figure 3, let us say that the communication rule established is that all components have input queue of length 1 and that a new external input is submitted to the system only after it has produced an external output action (*I/O ordering constraint* [6]). We may also assume that the same time-out value is assigned to test all transitions. ❑

Given this set of rules, in order to test a single transition, the tester needs the following information:

- signal to be sent;
- target component under test to send the signal to;
- expected response;
- expected origin of the response;
- a time-out value.

Notice that information about the target component (or the expected origin of the response) may be implicit in the signal name. If after a certain period of time (specified by the time-out value), the tester has not yet received a response to its previous stimulus, then a 'FAIL' verdict is issued.

4   SYSTEM COMPONENTS AND TEST EXECUTION PHASES

In practice, we realize that, in addition to all the architecture components, we need a module that sets up all of these components in a predefined initial configuration and that triggers the test process by handing the control over to the tester (or tester components). We call this auxiliary module the '*install module.*'

Example 3. Our distributed system is basically composed of three main objects (two *service objects* - the components under test - and one *tester component*) and one auxiliary component ('*install module*') that sets up the initial configuration of the system and triggers the test process. All but the last are hybrid objects (they are simultaneously clients and servers). The *install* module is a client to all of the others. ❑

In order to assure interaction among the various components, we propose a common basic interface type (called *'Prot'*) that all components must inherit their own interfaces from. It has the following attributes/methods[1] (written in IDL):

```
interface Prot {
        readonly attribute long id;
        readonly attribute long state;
        oneway void SendMsg(in Prot sender, in string signal);
};
```

The semantics of each attribute/method is as follows:

- *long id* - Each object has an unique identification within the system. This attribute holds the object's identifier.

- *long state* - Each object has a defined behaviour with defined internal states. This attribute contains the current internal state of the object.

- *void SendMsg(sender, 'Message')* - Through this method, the components interoperate. All message exchanges are asynchronous ('oneway').

The main objects share these operations while reacting differently to message exchanges (each one has its own behaviour).

In fact, the *'Prot'* interface described above is a suggestion on an interface type that implements a common message exchange mechanism amongst the several components. Other interface types implementing similar message exchange mechanisms may be used instead.

In addition to the functionality of this basic interface, *components under test* offer to special clients the possibility of determining (loading) their behaviour at run-time. For this, their *'Prot'* interface is extended to include *LoadBehaviour* method that 'instructs' the object about its expected behaviour. It is true that this feature is pointless in 'normal' systems where each component has a specific predefined behaviour. However, it is extremely useful in order to study the impact of a component's wrong behaviour on the whole system or in the final verdict.

Example 4. Both our service objects exhibit the behaviour of the Basic Call Service (BCS, Figure 5). ❑

The tester has a similar function (*LoadTS*) that tells it what the expected behaviour of the whole system is. This corresponds to the test suite plus information about the components' configuration.

Test execution in this context can basically be divided into three phases:

---

1. Actually, in CORBA read-only attributes become procedure calls expressed in the target programming language.

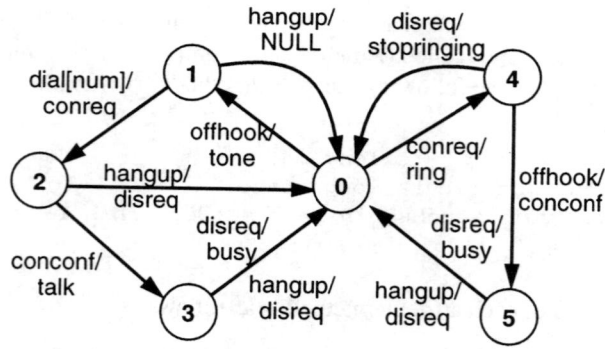

Figure 5 Basic Call Service automaton.

1. <u>Connection establishment phase</u>: The *install* module tries to connect to all objects that are concerned with the testing and the tester (to make sure they exist);

2. <u>Configuration set-up</u>: Again it is the *install* module that does all the job introducing all modules to each other, determining their behaviour and triggering the test process just before leaving the scene;

3. <u>Test execution</u>: The tester takes over and starts sending signals to the various components under test while collecting and checking their responses. This phase terminates with a verdict ('PASS' or 'FAIL').

In the following sections, we will detail each test execution phase using the example treated so far.

4.1 Connection establishment and configuration set-up

First of all, the *install* module tries to connect to the tester and to all objects that compose the system under test (❶ in Figure 6). By doing this, it verifies that all modules exist and that they can then become part of the testing system (SUT + tester). In order to establish a connection among the several components in a given configuration, the install module instructs them to introduce themselves to each other (❷ in Figure 6) using *SetUpPeer* and *SetUpNewModule* operations at the components' and tester's interfaces, respectively. In this way, the tester becomes aware of the existence and initial configuration of the components under test. All components then establish bidirectional channels between themselves (following the specified configuration) after having been introduced to each other (❸ and ❹ in Figure 6). The tester behaviour (defined by the test suite) is then loaded (*LoadTS*) and finally the test process is triggered (❺ in Figure 6).

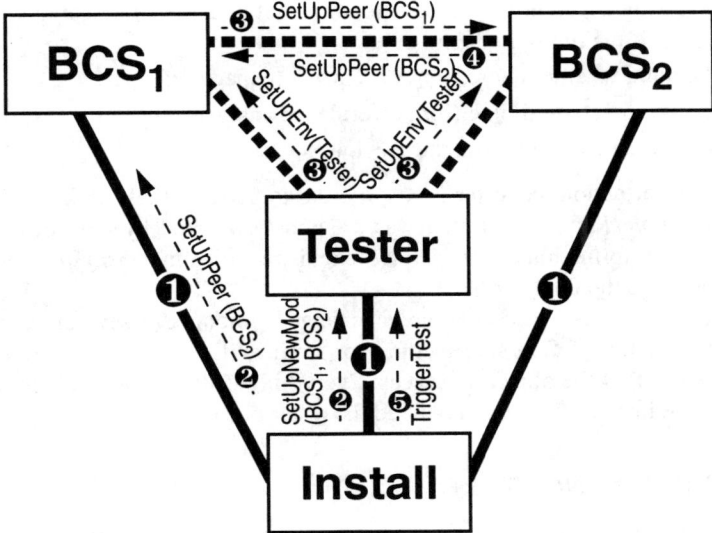

Figure 6 Connection establishment + Configuration set-up.

4.2 Test execution phase

After having set up the system configuration, the *install* module quits and the tester takes over sending signals to the components under test and receiving messages from them (Figure 7). This is done in several *test steps*. Each step consists of:

- Sending a signal to one of the components under test;
- Triggering a time-out which is specific to that particular signal (or transition being tested);

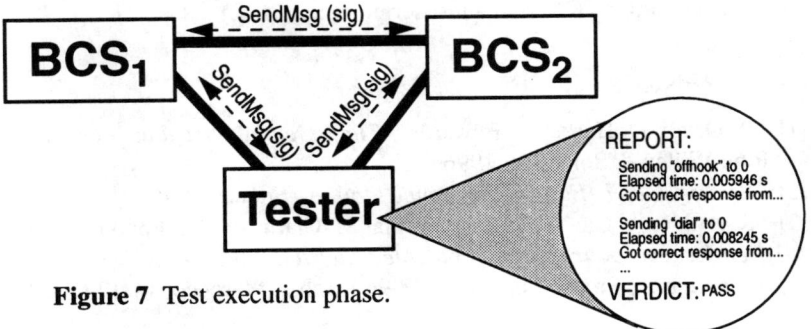

Figure 7 Test execution phase.

- Receiving a response from some component under test or issuing a 'FAIL' verdict if the time-out expires;
- Checking whether the response is correct (expected);
- Computing elapsed time (for performance analysis only);
- Generating a report about the previous steps.

All the information the tester needs in order to perform these tasks is given to it through the *LoadTS* operation that was previously invoked by the *install* module (Section 4.1). Information about each test step is recorded in a *test report file* whose last line presents the final verdict.

During this phase, messages are exchanged among the several components through invocation of *SendMsg* method (or similar, if a basic interface other than *'Prot'* was used) at the object interfaces. Everything is done automatically once we have built the install module and defined the tester behaviour.

5   CONCLUSION/FUTURE WORK

Although, in our case, the implemented tester is centralized, we intend, in the future, to implement it in a distributed way. To do this, the configuration information file will include details of how to interoperate amongst the various tester components in order to produce the final verdict. Possibly, different test suite files must also be produced for each component according to the distributed test generation technique used (for instance, synchronizable sequences).

The results of our work are encouraging, particularly taking into account that the majority of our colleagues working on the implementation of telecommunications services on a CORBA-TINA platform do testing manually [13]. Once the module representing a service has been implemented, they manually define a test script. They repeatedly express the need for a test environment where all the phases of the testing procedure are automated. The testing environment that we have presented in this paper is an effort in this direction: all phases are automated and the test results obtained are more complete. Another advantage is the possibility of using existing formal description techniques to describe: the system (using SDL) [14], the test suite obtained (using TTCN) and a test script (using MSC).

6   REFERENCES

[1]   R. Orfali, D. Harkey, J. Edwards, *"The Essential Distributed Objects"*, John Willey & Sons, Inc. 1996.
[2]   Verilog, *Geode Editor - Reference Manual,* France, 1996.
[3]   R. Anido, A. Cavalli, L. Paula Lima, M. Clatin, M. Phalippou, *"Engendrer des tests pour un vrai protocole grâce à des techniques éprouvées de vérification"*, Proceedings of CFIP'96 - Rabat, Marocco, 14-17 Oct. 1996.

[4]   D.P. Sidhu and T.K. Leung, *"Formal methods for protocol testing: a detailed study"*, IEEE Transaction on Software Engineering, 15(4), April 1989.

[5]   L. Paula Lima, A. Cavalli, *"Service Validation - An Embedded Testing Approach"*, Proceedings of EUNICE Summer School - Lausanne, Switzerland, 23-27 Jun. 96.

[6]   A. Petrenko, N. Yevtushenko, G. v. Bochman, *"Fault models for testing in context"*, Proceedings of FORTE'96 - Kaiserslautern, Germany - 8-11 Oct. 96.

[7]   SCORE Methods and Tools, *"The SCORE Approach to Formal Methods in FI Detection"*, Report on Methods and Tools for Service Creation (Second Version Volume I: Service Interaction Analysis) - R2017/SCO/WP2/DS/P/028/b2.

[8]   P.Combes and S. Pickin, *"Formalization of a User View of Network and Services for Feature Interaction Detection"*, In L. G. Bouma and H. Velthuijsen, editors, Feature Interactions in Telecommunications Services, Amsterdam, May 1994, IOS Press, pp. 123-135.

[9]   IONA Technologies Ltd. *Orbix® 2 Programming Guide,* November 1995 P1.

[10] ISO/ITU-T, *Open Distributed Processing - Reference Model - Part 3: Architecture - International Standard 10746-3,* ITU-T Recommendation X.903 - 1995.

[11] ISO/IEC, *Information Technology - OSI - Conformance Testing Methodology and Framework,* International ISO/IEC multipart standard No 9646, 1994.

[12] ITU-T Rec. Z.120 (1996), *Message Sequence Chart (MSC),* Geneva, 1996.

[13] Alcatel-Alsthom Recherche and UNISOURCE, *Service Pilot on TINA (SPOT),* Jan. 96.

[14] CCITT, *Specification and Description Language,* CCITT Z.100, International Consultative Committee on Telegraphy and Telephony, Geneva, 1992.

[15] A. Cavalli and B. M. Chin, *"Testing methods for SDL systems"*, Computer Networks and ISDN Systems, 28:1669-1683, 1996.

PART NINE

Formal Specification (II)

29

CSP-OZ: A Combination of Object-Z and CSP

Clemens Fischer
University of Oldenburg, Department of Computer Science
P.O. Box 2503, 26111 Oldenburg, Germany.
E-mail: Fischer@Informatik.Uni-Oldenburg.de

Abstract

In this paper we define a combination of Object-Z and CSP called CSP-OZ. The basic idea is to define a CSP-semantics for every Object-Z class. Special care is taken to capture the characteristics of input and output parameters properly and to preserve the expected refinement rules.

CSP-OZ is well suited for the specification and development of communicating distributed systems. It provides powerful techniques to model data- and control-aspects in a common framework. The language is easy to use for Z and Object-Z users.

Keywords

Z, Object-Z, CSP, concurrent systems, combining FDTs, refinement.

1 INTRODUCTION

For the definition of the semantics of parallel, object-oriented languages, two things must be considered: On the one hand structuring features like inheritance are very important (*internal view*) and on the other hand the dynamic behaviour of objects executed in parallel has to be defined (*external view*). It is convenient to separate these aspects because an external observer of the dynamic behaviour of an object should not be able to look into the internal structure. This paper solely deals with the external view of an object; i.e. the dynamic semantics of objects.

Object-Z [6] is an object-oriented extension of Z [22] for the predicative specification of objects. An Object-Z class consists of the specification of a

---

*This research is supported by the German Ministry for Education and Research (BMBF) as part of the project UniForM under grant No. FKZ 01 IS 521 B2.

state space and operations on this state space. The process algebra CSP [12] has been developed to describe the dynamic behaviour of a system and covers aspects like parallel composition, hiding and divergence.

The basic idea we follow in this paper was suggested by Smith [21]: We define the dynamic semantics of an Object-Z class using the semantic model of CSP. Thereby all CSP-operators like parallel-composition and hiding can be applied to objects.

This work is in the same line as [7], where CSP and Z are combined in a similar fashion. A major difference between [21] and [7] is the handling of input and output parameters of an operation. In [21] no semantic difference is made between input and output parameters, whereas in [7] the difference between input and output is captured semantically, but the mixture of input and output parameters is not considered yet.

The contribution of this paper is to define a CSP failure-divergence semantics of Object-Z integrating the views of [21] and [7] and extending them to mixtures of input and output parameters. This combination of Z, Object-Z and CSP is called CSP-OZ.

The rest of this paper is organised as follows. The ideas of the semantics are presented in the next section with an example. The version of Object-Z used here and the semantic model of CSP are introduced in the subsequent sections. In section 5 the failure-divergence semantics of CSP-OZ is defined. Finally, a conclusion is drawn and connections to related work are discussed.

2   ILLUSTRATING EXAMPLE

To illustrate the approach, we present a system for the management of free and used process identifiers (PIDs): $PID == 1 .. maxP$ where $maxP : \mathbb{N}_1$ is a global variable. The Object-Z specification of the system is the class $PIDmngr$.

$$
\begin{array}{l}
\underline{\quad PIDmngr \quad\rule{6cm}{0pt}} \\[4pt]
\left|\begin{array}{ll}
\underline{\qquad\qquad\qquad\qquad} & \underline{\text{\_Init}\rule{3cm}{0pt}} \\
used : \mathbb{P}\, PID & used = \varnothing \\[12pt]
\underline{\text{\_}reqP\rule{2.5cm}{0pt}} & \underline{\text{\_}relP\rule{2.5cm}{0pt}} \\
\Delta(used) & \Delta(used) \\
nu? : \mathbb{N} & rel? : \mathbb{P}\, PID \\
fr! : \mathbb{P}\, PID & \\
\underline{\qquad\qquad\qquad} & used' = used \setminus rel? \\
fr! \cap used = \varnothing & \\
used' = used \cup fr! & \\
\#fr! = nu? \vee (nu? > \#(PID \setminus used) \wedge fr! = \varnothing) & \\
\quad [\#S \text{ is the number of elements of the set } S.] &
\end{array}\right|
\end{array}
$$

The unnamed schema in the left corner defines the state space of the class. We use the keyword **State** to refer to this schema. The variable *used* stores the set of PIDs that are used by some process in the environment of *PIDmngr*. The schema Init specifies the initial values of *used*. The class has two operations: *reqP* and *relP*. The variables declared in the delta list ($\Delta(used)$) can be changed by the operation. The parameters of *reqP* are *nu*? and *fr*!. The decorations ? and ! are used for inputs and outputs, respectively. The schema *reqP* describes how the state of *PIDmngr* changes when *reqP* is applied. The undecorated variable *used* corresponds to the state before the operation and the dashed variable *used'* is the final value. Hence *reqP* computes a set of unused PIDs containing *nu*? elements. If there are not enough free PIDs available, the empty set is returned.

The basic idea of CSP-OZ is to define the semantics of a class using the semantic model of CSP. Thereby all CSP-operators – especially hiding and parallel composition – can be used to combine objects, and CSP and Object-Z syntax can be mixed.

The CSP-semantics is based on the alphabet of a process; i. e. the set of events the environment can observe. The observable events of a class are the operations with their parameters [20, 21, 7]. For example, ($reqP, \{nu \mapsto 3, fr \mapsto \{2, 4, 5\}\}$) and ($relP, \{rel \mapsto \{2, 5, 6\}\}$) are possible events of *PIDmngr* (provided $maxP \geq 6$). The set $\{rel \mapsto \{2, 5, 6\}\}$ denotes the function where *rel* is mapped to the set $\{2, 5, 6\}$.

To make these events explicit, we extend Object-Z with the declarations of channels. The schema names of the operations corresponding to a channel are prefixed with the keyword com (more keywords will be introduced later). The CSP-OZ specification of *PIDmngr* is the following class:

$$
\begin{array}{l}
\rule{0pt}{2ex}S\text{-}PID_1 \\
\hline
\text{channel } reqP : [\, nu : \mathbb{N};\ fr : \mathbb{P}\,PID\,] \\
\text{channel } relP : [\, rel : \mathbb{P}\,PID\,] \\
PIDmngr[\textsf{com\_}reqP\,/\,reqP, \textsf{com\_}relP\,/\,relP]
\end{array}
$$

The specification $S\text{-}PID_1$ inherits all schemas of *PIDmngr*, but *reqP* and *relP* are renamed to com_*reqP* and com_*relP*, respectively. The declaration of *PIDmngr* in $S\text{-}PID_1$ can textually be replaced by the definition of *PIDmngr*.

The semantics of a CSP-process is given in terms of the *failures* and the *divergences*. A failure is a tuple (tr, X) consisting of a trace *tr* of events the process may perform and a set of events X the process can refuse to engage in after *tr*. The divergences are a set of traces after which future behaviour of the process is unpredictable.

A class $\mathcal{C}$ can engage in the trace $\langle e_1, \ldots, e_n \rangle$ of events $e_i = (op_i, p_i)$ if the successive application of the operation op_i with parameter p_i transfers some initial state of $\mathcal{C}$ into some reachable state. The trace

$$
\begin{aligned}
tr_1 = \langle\ & (reqP, \{nu \mapsto 3, fr \mapsto \{2, 4, 5\}\}), \\
& (reqP, \{nu \mapsto 1, fr \mapsto \{8\}\}),\quad (relP, \{rel \mapsto \{2, 4, 8\}\})\rangle
\end{aligned}
\tag{1}
$$

transfers the initial state of $S\text{-}PID_1$ into the final state $used = \{5\}$.

The crucial question we deal with in this paper, is the definition of the refusals of a class. To understand the problem, we have to look at the CSP-refinement relation. A process P_1 *refines* (or correctly implements) a process P_2 if $\mathcal{F}(P_1) \subseteq \mathcal{F}(P_2)$ (P_1 is more deterministic) and $\mathcal{D}(P_1) \subseteq \mathcal{D}(P_2)$ (P_1 is less divergent; i.e. more defined), where $\mathcal{F}$ and $\mathcal{D}$ denote the set of failures and divergences of a process, respectively. CSP-refinement is compositional, i.e. whenever P_1 refines P_2 then $C(P_1)$ refines $C(P_2)$ for any system context $C(\cdot)$. We expect that the following class is an implementation of $S\text{-}PID_1$.

```
┌─ I-PID ─────────────────────────────────────────────────────────────
│  channel reqP : [ nu : ℕ;  fr : ℙ PID ]
│  channel relP : [ rel : ℙ PID ]        Init  ≙ [ ∀ p : PID • u(p) = 0 ]
│  ┌──────────────────────────────────────────────────────────────────
│  │ u : PID → {0, 1}        [u is used to store the status of every PID.]
│  │ next : 1 .. maxP + 1    [next is a pointer to the smallest free PID.]
│  ├──────────────────────────────────────────────────────────────────
│  │ next = min({p : PID | u(p) = 0} ∪ {maxP + 1})
│  └──────────────────────────────────────────────────────────────────
│
│  ┌─ com_reqP ──────────────────────────────────────────────────────
│  │ Δ(u, next)
│  │ nu? : ℕ
│  │ fr! : ℙ PID             [fr! is the set of free PIDs starting with next.]
│  ├──────────────────────────────────────────────────────────────────
│  │ nu? ≤ maxP + 1 − next ⇒ (#fr! = nu? ∧
│  │     min fr! = next ∧ ∀ p : fr! • u(p) = 0 ∧ u'(p) = 1 ∧
│  │     max fr! = next' − 1 ∧ ∀ p : PID \ fr! • u'(p) = u(p))
│  │ nu? > maxP + 1 − next ⇒ (#fr! = ∅ ∧ u' = u)
│  └──────────────────────────────────────────────────────────────────
│
│  ┌─ com_relP ──────────────────────────────────────────────────────
│  │ Δ(u, next)
│  │ rel? : ℙ PID
│  ├──────────────────────────────────────────────────────────────────
│  │ ∀ p : rel? • u'(p) = 0 ∧ ∀ p : PID \ rel? • u'(p) = u(p)
│  └──────────────────────────────────────────────────────────────────
└─────────────────────────────────────────────────────────────────────
```

The behaviour of *I-PID* differs from $S\text{-}PID_1$ in one aspect: A request for new PIDs is answered with the set of PIDs starting with *next*, not with an arbitrary set of free PIDs. E. g. if no PIDs are used, any request for three PIDs is answered with the set $\{1, 2, 3\}$, whereas $S\text{-}PID_1$ could send back any set of three numbers.

This is an acceptable behaviour of an implementation because *fr* is declared as an output parameter. The specification $S\text{-}PID_1$ makes a nondeterministic choice between different values of *fr*. Any implementation can reduce this nondeterminism. Following this idea, $S\text{-}PID_1$ can refuse the following sets of events initially (i. e. before engaging in any event):

$$R_1 = \{X : \mathbb{P} \; Event \quad | \quad (\forall \, n : 0 \, .. \, maxP \bullet \exists f : \mathbb{P} \; PID \bullet \#f = n \, \wedge$$
$$(reqP, \{nu \mapsto n, fr \mapsto f\}) \notin X) \, \wedge \qquad (2)$$
$$(\forall \, n > maxP \bullet (reqP, \{nu \mapsto n, fr \mapsto \varnothing\}) \notin X) \, \wedge \quad (3)$$
$$(\forall \, e : X \bullet first(e) \neq relP) \, \} \qquad (4)$$

To understand this definition, think of the events $S\text{-}PID_1$ must engage in. Consequently (2) - (4) says something about events that cannot be refused. In (2) the property is captured that $S\text{-}PID_1$ must deliver some set of PIDs if n is not larger than $maxP$. The existential quantification $\exists f : \mathbb{P} \; PID$ captures the nondeterministic choice of possible output parameters. Line (3) deals with n exceeding $maxP$. Then $S\text{-}PID_1$ does not have a choice for fr; the empty set is always returned. Finally, $S\text{-}PID_1$ can never refuse a communication on $relP$ for every choice of the parameter rel (4). The function $first$ computes the first component of the tuple e (i. e. the channel of e).

Recall that we just considered initial refusals, but the calculation of the refusals after some trace tr is similar. The initial refusals of $I\text{-}PID$ are

$$R_I = \{X : \mathbb{P} \; Event \quad | \quad (\forall \, n : 0 \, .. \, maxP \bullet$$
$$(reqP, \{nu \mapsto n, fr \mapsto \{1, \ldots, n\}\}) \notin X) \, \wedge \quad (5)$$
$$(\forall \, n > maxP \bullet (reqP, \{nu \mapsto n, fr \mapsto \varnothing\}) \notin X) \, \wedge \quad (6)$$
$$(\forall \, e : X \bullet first(e) \neq relP) \, \}. \qquad (7)$$

Line (6) and (7) are the same as (3) and (4), but the existential quantification in (2) is removed in (5). This models the fact that the nondeterminism is removed in $I\text{-}PID$. It initially always returns $\{1, \ldots, n\}$ when n PIDs are requested. The initial refusals of $I\text{-}PID$ are a subset of the initial refusals of $S\text{-}PID_1$ as expected. This is also true for the refusals after engaging in any possible trace. Thus $I\text{-}PID$ is an implementation of $S\text{-}PID_1$ in the failure-divergence semantics.

However, this is not the case if we follow the standard Object-Z semantics where no difference is made between input and output parameters [6]: Any parameter can be controlled by the environment *and* the system. Consequently, $I\text{-}PID$ is not an implementation of $S\text{-}PID_1$ according to [21], because the choice between the different values for $fr!$ would not be made nondeterministically.

Nevertheless, the blocking view of an operation can be useful for other examples and we integrate it into CSP-OZ. In the following version of $PIDmngr$ the parameter nu of $reqP$ is used without decoration. We call this a *simple* parameter. Its value is under the control of the object *and* the environment.

┌─ *S-PID₂* ────────────────────────────────────
│ $S\text{-}PID_1[\texttt{redef com\_reqP}]$

$$\begin{array}{|l}
\hline
\text{\_\_com\_}reqP \text{_____} \\
\Delta(used) \\
nu : \mathbb{N} \\
fr! : \mathbb{P}\, PID \\
\hline
\#fr! = nu \wedge fr! \cap used = \varnothing \wedge used' = used \cup fr! \\
\hline
\end{array}$$

The keyword **redef** indicates that $S\text{-}PID_2$ inherits $S\text{-}PID_1$ except for the definition of com_$reqP$. Instead of sending an empty set of free process identifiers, any unrealizable request for PIDs is blocked in $S\text{-}PID_2$. The initial refusals are the following sets of events.

$$R_2 = \{X : \mathbb{P}\, Event \mid \quad (\forall\, n : 0 \mathinner{\ldotp\ldotp} maxP \bullet \exists f : \mathbb{P}\, PID \bullet \#f = n \wedge$$
$$(reqP, \{nu \mapsto n, fr \mapsto f\}) \notin X) \wedge$$
$$(\forall\, e : X \bullet first(e) \neq relP)\ \}$$

The last question to address in this section is the guard of an operation, i. e. the condition when the operation is enabled. In Z operations are always enabled, but in Object-Z and a lot of interpretations of Z other views are adopted. To integrate these different views and to avoid confusion about the guard, we use an extra schema to determine the guard of an operation. This is neither in the tradition of Z nor of Object-Z, but it is in line with many different languages like action systems, B, VDM and MIX. The following specification can refuse to release PIDs that are not used.

$$\begin{array}{|l}
\hline
\text{\_\_}S\text{-}PID_3 \text{_____} \\
S\text{-}PID_2[\textbf{redef com\_}relP] \\
\hline
\begin{array}{|l}
\hline
\text{\_\_enable\_}relP \text{_____} \\
rel : \mathbb{P}\, PID \\
\hline
rel \subseteq used \\
\hline
\end{array}
\quad
\begin{array}{|l}
\hline
\text{\_\_effect\_}relP \text{_____} \\
\Delta(used) \\
rel : \mathbb{P}\, PID \\
\hline
rel \neq \varnothing \wedge used' = used \setminus rel \\
\hline
\end{array} \\
\hline
\end{array}$$

We use the prefixed keywords **enable** for the guard and **effect** for the schema describing the state change of an operation. Communications outside **enable** are blocked. A communication that is not blocked, but that is applied outside the precondition of **effect** leads to divergence. Thus $S\text{-}PID_3$ refuses any communication on $relP$ initially. If the empty set is released the system will diverge.

Note that an operation **effect_**c without simple parameters and without an enable-schema (abbreviates the guard *true*) models the Z-view of an operation. Hence CSP-OZ integrates the non-blocking view traditionally adopted in Z and the blocking view of Object-Z. It even allows arbitrary mixtures of these aspects. Only in this special case the enable schema is necessary.

3 SYNTAX OF CSP-OZ

Object-Z [6] extends Z with the notion of a class, specifying a state space and operations on this state space. In this paper we consider only a restricted version of Object-Z. The parallel and sequential composition operators are no longer used inside Object-Z (the corresponding CSP operators can be used instead); and we do not allow history invariants as they cannot be modelled properly in our semantic model.

CSP-OZ extends Object-Z with the notion of communication channels and CSP-syntax. A CSP-OZ class has the following basic structure.

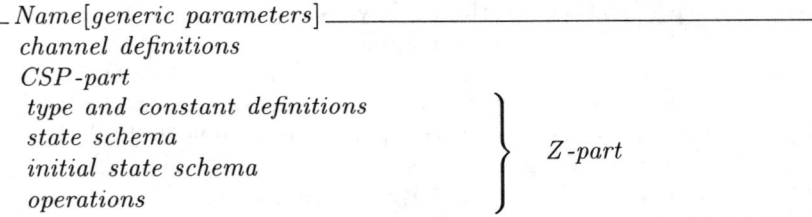

The *channel definitions* declare the *interface* of the class. Every declaration has the form channel $c : [p_1 : ty_1; \ldots; p_n : ty_n]$ where c is the channel name, $p_1, \ldots, p_n$ is the possibly empty list of undecorated parameter names and $ty_1, \ldots, ty_n$ are the type declarations of the parameters.

The CSP-part is a set of equations of the form *CSP-name = CSP-process* where *CSP-process* is defined using the syntax of the CSP model checker FDR [9] extended with Z-types. The channels of every process in the CSP-part must be a subset of the channels declared in the interface. If the CSP-part is not empty a process with the keyword main must be defined. This process is used to determine the semantics of the CSP-part. Other CSP-processes can be used to enhance readability of the CSP-part.

The type and constant definitions and the state and initial state schemas of the *Z-part* are the same as in Object-Z [20]. The schema, type and variable names used in the Z-part must be disjoint from the names used in the CSP-part.

For every channel c declared in the interface there must be either the schema effect_c and optional the schema enable_c or the schema named com_c, which is an abbreviation for

$$\text{effect\_}c \mathrel{\widehat{=}} \text{com\_}c$$
$$\text{enable\_}c \mathrel{\widehat{=}} \exists \, \text{State}'; \; p_{s+1}? : ty_{s+1}; \; \ldots; \; p_n! : ty_n \bullet \text{com\_}c.$$

All parameters of c must be declared in effect_c together with the possible decorations ? (input) and ! (output). An undecorated parameter stands for a simple parameter. The schema enable_c specifies the states and simple parameters in which c cannot refuse to communicate on c. Hence we assume that `enable_c` and `effect_c` can be normalised to

```
┌─enable_c──────────────────
│ p₁ : ty₁; ...; pₛ : tyₛ
│ ──────────────────────────
│ P_ena
└───────────────────────────
```

$$\begin{array}{l}\hline \text{effect\_c} \\ \hline \Delta(v_1, \ldots, v_m) \\ p_1 : ty_1; \ldots; p_s : ty_s \\ p_{s+1}? : ty_{s+1}; \ldots; p_{s+i}? : ty_{s+i} \\ p_{s+i+1}! : ty_{s+i+1}; \ldots; p_n! : ty_n \\ \hline P_{\mathit{eff}} \\ \hline \end{array}$$

Any of the parameter lists can be empty. If `enable_c` is omitted P_{ena} is set to true. Other schemas – not prefixed with a keyword – may freely be used to structure the specification, but there is always a class without further schemas with the same failure-divergence semantics.

As the Z-part of a CSP-OZ class corresponds directly to an Object-Z class, features like inheritance or instantiation of Object-Z classes can be used to structure a CSP-OZ class. But in this paper we assume that all classes can be normalised to the form given above.

CSP-OZ classes can be combined using the following CSP-operators:

- $\mathcal{C}_1 \, [\,|\, A\,|\,] \, \mathcal{C}_2$ is the parallel composition of $\mathcal{C}_1$ and $\mathcal{C}_2$ synchronising on the set of events A.
- $\mathcal{C} \setminus A$ is the class where all events in the set A are hidden; no $e \in A$ can be observed by the environment.
- $\mathcal{C}_1 \, |||\, \mathcal{C}_2$, $\mathcal{C}_1 \sqcap \mathcal{C}_2$ and $\mathcal{C}_1 \,\square\, \mathcal{C}_2$ are the interleaving, internal choice and external choice of $\mathcal{C}_1$ and $\mathcal{C}_2$.

The notation $\{|\ relP\ |\}$ denotes the set of events of the channel *relP*. As an example, we specify the registration desk of a hotel. A connection diagram [12] can be found in Figure 1. Every box stands for a CSP-OZ class. The lines between boxes denote hidden channels and the four dangling lines are the observable interface of the specification. An event on the channel *request* indicates that a guest wants to register. Then a new PID is requested and a card with the PID is printed. The printer delivers the card (*card*) or fails to produce a card (*nocard*).

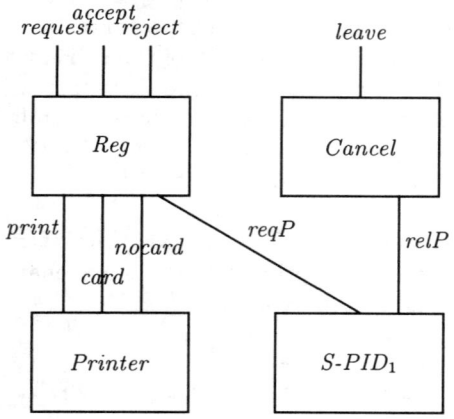

Figure 1 The connection diagram for *RegDesk*

A successful registration is indicated by *accept* and any error in this procedure results in a communication on *reject*. Leaving the hotel is indicated by a communication on *leave*. Then the PID is released. We omit a detailed specification of *Reg* and *Cancel*. The printer is modelled by the following specification.

Printer
channel $print : [\, p : PID \,]$
channel $card, nocard : [\,]$
main $= print?x \to (card \to$ main $\sqcap nocard \to$ main$)$

The precise semantics of the connection diagram in fig. 1 is given by the following CSP-expression.

$$RegDesk = \left(((Reg\,[|\,\{|\,print, card, nocard\,|\}\,|]\,Printer) \right.$$

$$\left. [|\,\{|\,reqP\,|\}\,|]\,\,S\text{-}PID_1)\,\,[|\,\{|\,relP\,|\}\,|]\,\,Cancel \right)$$

$$\setminus\{|\,relP, reqP, print, card, nocard\,|\}$$

4 THE SEMANTIC MODEL OF CSP

The standard semantic model of CSP is the failure-divergence model. The semantics given in [12] is restricted to finite alphabets. Straight forward extensions of this model to infinite alphabets are restricted to bounded nondeterminism. To allow unbounded nondeterminism, two models were developed: A new order for the definition of fix points is introduced in [17] and in [18] the failure-divergence model is extended with the set of infinite traces a process may engage in. Generally speaking, both models can be used for the semantics of CSP-OZ. The model in [18] avoids the anomaly that hiding of an event e leads to divergence if it can occur finitely many times without a finite bound for the number of possible occurrences of e. But we prefer here the model of [17] as it is similar to the standard model of CSP [12] and is the basis of a CSP-encoding in the theorem prover Isabelle [24] that is used in the project UniForM [15] to develop tool support for CSP-OZ. Nevertheless, the definitions presented here are also applicable to the infinite trace model and the differences completely vanish if only bounded nondeterminism is considered.

Definition 1 *Let $\mathcal{A}$ be a possibly infinite alphabet of events. Then the semantics of a CSP-OZ specification is a tuple $(\mathcal{F}, \mathcal{D})$ where*

1. *the failure set $\mathcal{F} : seq\,\mathcal{A} \leftrightarrow \mathbb{P}\,\mathcal{A}$ is a relation between traces on $\mathcal{A}$ and subsets of $\mathcal{A}$. A tuple (tr, X) is element of $\mathcal{F}$ iff tr is a possible trace and all communications from X can be refused after engaging in tr.*

2. *the divergence set* $\mathcal{D} : \mathbb{P} \, seq \, A$ *with* $\mathcal{D} \subseteq dom \, \mathcal{F}$ *is the set of traces after which the system may diverge,*

The set of $(\mathcal{F}, \mathcal{D})$ *fulfilling the healthiness conditions of [17]*[*] *is called* $\mathcal{N}'$. □

5   A FAILURE-DIVERGENCE SEMANTICS OF CSP-OZ

The semantics for CSP is defined in [17]. Thus our main task is the definition of the failure-divergence semantics of CSP-OZ classes with an empty CSP-part. The combination of the CSP-part and the Z-part is defined at the end of this section using the parallel operator.

Let Id_c denote the set of all channel identifiers. Let Id_v denote the set of undecorated variable identifiers and let Id_p denote the set of undecorated parameter names. Finally, let *Value* denote the set of possible values that can be assigned to variables and parameters. The exact definition of Id_c, Id_v, Id_p and *Value* is not formalised here. A state is a finite partial function assigning values to variable-identifiers.

$$State == Id_v \nrightarrow Value$$

An event is a tuple consisting of a channel-identifier and a parameter, which is a finite partial function assigning values to parameter-names.

$$Parameter == Id_p \nrightarrow Value$$
$$Event == Id_c \times Parameter$$

A CSP-OZ class with an empty CSP-part induces a set of channels, states, and initial states; a transition relation; a set of input, output, and simple parameters for every channel and a set of enabled parameters for every state and channel. This is modelled by the schema type *ClassStruct* with the following signature.

```
┌─ ClassStruct ──────────────────────────────
│  states, initial : ℙ State
│  channels : ℙ Id_c
│  trans : Event ⇸ (State ⇸ ℙ State)
│  simple_p : Id_c ⇸ ℙ Id_p
│  in_p, out_p : Id_c ⇸ ℙ Parameter
│  enable : Id_c × State ⇸ ℙ Parameter
└────────────────────────────────────────────
```

We only give an informal description of the translation of CSP-OZ syntax to the induced *ClassStruct*. The construction of *states*, *initial* and *channels* for

---

[*]The healthiness conditions of [17] rule out miraculous specifications; require a prefix closed set of traces, subset closed refusal sets, and any set of communications that are impossible in the next step are refused. Furthermore, the divergence set must be a suffix closed and the chaotic closure is included in $\mathcal{F}$ in case of divergence.

a given class C is obvious. We only consider classes where $initial \neq \varnothing$. The set $simple_p(c)$ consists of all simple parameter names declared for the channel c. The sets $in_p(c)$ and $out_p(c)$ contain all functions assigning values to the input and output parameters. The possible values of the parameters are only affected by the declaration of the channel, not by the corresponding schemas. If a channel c does not have input parameters we assume $in_p(c) = \{\varnothing\}$ (output parameters analogously). The set $trans(e)(st)$ consists of all states reachable by applying the event e to the state st. E. g. for $S\text{-}PID_1$ we have

$$trans(reqP, \{nu \mapsto 1, fr \mapsto \{1\}\})(\{used \mapsto \{4\}\}) = \{\{used \mapsto \{1,4\}\}\}$$
$$trans(reqP, \{nu \mapsto 3, fr \mapsto \{1\}\})(\{used \mapsto \{4\}\}) = \{\}$$

The function $enable$ computes for a given channel and state the enabled simple parameters. E. g. for $S\text{-}PID_3$ we have

$$enable(relP, \{used \mapsto \{1,3,4\}\}) = \{\{rel \mapsto X\} \mid X \subseteq \{1,3,4\}\}$$

and for $S\text{-}PID_1$ we have $enable(relP, st) = \{\varnothing\}$ for all states st because $relP$ does not have a simple parameter. If $S\text{-}PID_1$ could refuse a communication on $relP$ for some st we would have $enable(relP, st) = \{\}$. This trick of notation simplifies the following formal definitions significantly because we do not have to consider separate cases for channels without all three kinds of parameters.

We define the failure-divergence semantics of CSP-OZ based on *ClassStruct*. The function *reachable* computes the set of reachable states for a trace.

$$
\begin{array}{l}
\hline
reachable : ClassStruct \rightarrow (\text{seq } Event \rightarrow \mathbb{P}\, State) \\
\hline
\forall C : ClassStruct \bullet \\
\quad reachable(C)(\langle\rangle) = C.initial \,\wedge \\
\quad \forall\, tr : \text{seq } Event \bullet reachable(C)(tr \,^\frown \langle e\rangle) = \\
\quad\quad \{st : State \mid (\exists\, st_1 : reachable(C)(tr) \bullet st \in C.trans(e)(st_1))\} \\
\hline
\end{array}
$$

Using tr_1 (1) we have $reachable(S\text{-}PID_1)(tr_1) = \{\{used \mapsto \{5\}\}\}$ if we identify $S\text{-}PID_1$ with the induced *ClassStruct*.

The function $\mathcal{D}$ computes the set of traces after which a class may diverge. A trace tr is in $\mathcal{D}(C)$ if it can be divided into $s \,^\frown \langle c, v\rangle \,^\frown t$ where $\langle c, v\rangle$ is the event causing the divergence which means $\langle c, v\rangle$ is enabled after s but no transition is defined for $\langle c, v\rangle$. Because of the healthiness conditions any trace t must be possible after $s \,^\frown \langle c, v\rangle$.

$$
\begin{array}{l}
\hline
\mathcal{D} : ClassStruct \rightarrow \mathbb{P} \text{ seq } Event \\
\hline
\forall C : ClassStruct \bullet \mathcal{D}(C) = \\
\quad \{s, t : \text{seq } Event;\ c : Id_c;\ v : Parameter \mid \\
\quad\quad \exists\, st : reachable(C)(s) \bullet ((C.simple_p(c) \lhd v) \in C.enable(c, st) \wedge \\
\quad\quad\quad\quad [v \text{ (restricted to the simple parameters) is enabled in } st.] \\
\quad\quad\quad \forall\, o : C.out_p(c) \bullet C.trans(c, v \oplus o)(st) = \varnothing) \\
\quad\quad\quad [\text{No transition is defined for any choice of output parameters.}] \\
\quad\quad \bullet (s \,^\frown \langle c, v\rangle \,^\frown t)\} \\
\hline
\end{array}
$$

Recall that $\{D \mid P \bullet E\}$ is the set of all evaluations of the expression E in the context of all variables declared in D and restricted in P.

The failures of an object are defined in two steps: We start with the definition of the refusals – sets of events that can be refused – for a given state. Then we combine this definition with *reachable* to define the failures of an object.

A set of events X is a refusal of a class C in the state st iff for all enabled values of simple parameters s_p and for all input values i_p there exists an output value o_p such that the event $(c, s_p \cup o_p \cup i_p)$ is *not* in X. The transition for $(c, s_p \cup o_p \cup i_p)$ can either be taken or there exists no output value such that a transition is defined for s_p and i_p.

$$refusal : ClassStruct \to (State \nrightarrow \mathbb{P}\,\mathbb{P}\,Event)$$

$$\forall C : ClassStruct \bullet \Big(\operatorname{dom} refusal(C) = C.states \,\wedge$$
$$\forall st : C.states \bullet \forall X : refusal(C)(st) \bullet$$
$$\forall c : C.channels \bullet \forall s_p : C.enable(c, st) \bullet \forall i_p : C.in_p(c) \bullet$$
$$\exists o_p : C.out_p(c) \bullet ((c, i_p \cup o_p \cup s_p) \notin X \,\wedge$$
$$\text{[The event } e = (c, i_p \cup o_p \cup s_p) \text{ cannot be refused if]}$$
$$(C.trans(c, i_p \cup o_p \cup s_p)(st) \neq \varnothing \,\vee \qquad \text{[}\ldots e \text{ can be taken next or]}$$
$$\forall o_p : C.out_p(c) \bullet C.trans(c, i_p \cup o_p \cup s_p)(st) = \varnothing))\Big)$$
$$\text{[}\ldots \text{the system diverges after } e.\text{]}$$

E.g. taking R_1 and R_I from section 2 we have $refusal(S\text{-}PID_1)(\{used \mapsto \varnothing\}) = R_1$ and $refusal(I\text{-}PID)(\{used \mapsto \varnothing\}) = R_I$. The definitions of $\mathcal{D}$ and *refusal* even make sense if c has no parameters of a particular type [8].

The definition of the failures of a CSP-OZ class is now straight forward. The class C can engage in the trace tr if there exists a reachable state st for tr. The set of events X can be refused after tr if it can be refused in the state st.

$$\mathcal{F} : ClassStruct \to \mathbb{P}(\text{seq } Event \times \mathbb{P}\,Event)$$

$$\forall C : ClassStruct \bullet \mathcal{F}(C) =$$
$$\{(tr, X) \mid \exists st : reachable(C)(tr) \bullet X \in refusal(C)(st)\}$$
$$\text{[no divergence]}$$
$$\cup \{(tr, X) : \mathcal{D}(C) \times \mathbb{P}\,Event\} \quad \text{[chaotic closure in case of divergence]}$$

It is proven in [8] that the definitions fulfill the healthiness conditions of the model $\mathcal{N}'$.

Note that declarations of channels with the same name in different classes must not introduce type conflicts. Otherwise the alphabet $\mathcal{A}$ in definition 1 would not be defined.

The semantics of a CSP-OZ class having only a CSP-part is the semantics of the process **main**. This is defined [17] where also the semantics of the CSP-operators can be found. The combination of CSP- and Z-syntax is simply done by parallel composition.

S	S_C	S_Z
Interface	Interface	Interface
CSP-part	CSP-part	Z-part
Z-part		

Thus the semantics of S is given by $S_C \, [| \, \{ | \, c_1, \ldots, c_n \, | \} \, |] \, S_Z$ where c_i are the channels declared in *Interface*.

This approach of combining (Object-)Z and CSP reuses an enormous part of existing theory. We get a lot of theorems for free. For example, the monotonicity of the CSP-parallel composition allows the separate refinement of the CSP- and the Z-part and all CSP-laws are valid laws of CSP-OZ.

6 RELATED WORK

Roscoe, Woodcock, and Wulf [19] give an informal translation from Z to CSP by separating input and output communications. The application of a Z-operation is modelled by two CSP events. CSP-OZ is much more general than this approach.

He [11], Josephs [13], and Woodcock and Morgan [25] translate state based specifications to CSP and prove various refinement results. Butler [4] combines action systems and CSP similar to our approach. Our definition of the semantics is simpler as we do not redefine CSP operators in Z; we do not separate input, output, and non-value passing communications; and we can present specifications in a better structure by using Object-Z. Also the combination of CSP and Z syntax is new and to the best of our knowledge the mixture of input and output parameters in one event has not been defined before.

LOTOS [3] combines CSP- and CCS-like operators with an algebraic specification language for abstract data types (Act One). Work on combining LOTOS and Z is in progress [2, 5]. The semantics is defined by a translation of LOTOS to ZEST (an extension of Z similar to Object-Z; also with a blocking view of operations). By contrast, we define the Object-Z semantics in the CSP model. We can apply parallel composition to classes and hide operations which is impossible in [5]. Furthermore, we do faithfully model Z-operations instead of only using blocking operations, and we precisely capture the characteristics of input and output parameters. Our integration of Z, Object-Z and CSP is much deeper than the integration of ZEST and LOTOS.

Strulo [23] investigates the difference between Z and ZEST operations. It is shown that blocking operations are good for specifying active behaviour while non blocking operations are good for modelling passive behaviour. CSP-OZ can be seen as the formalisation of the hybrid approach [23] where specifications of active and passive behaviour can be mixed. An operation com_c corresponds to active behaviour and an operation effect_c with the empty schema as precondition corresponds to passive behaviour. We even extend the hybrid approach as the special properties of input and output parameters are not captured in [23].

The relations to the work of Smith [21] and a previous paper of the author [7] have already been stated in the introduction.

7   CONCLUSION AND FUTURE WORK

The development of CSP-OZ is driven by the idea of combining successful existing formal languages based on a well defined semantics. An important aspect is that CSP and to some extent also Object-Z and Z are proper sub languages of CSP-OZ. Hence users of any of these three languages can start using CSP-OZ without having to learn many new things. The additional features of CSP-OZ can be explored step by step. A first example of the application of this idea is the use of CSP-Z – the predecessor of CSP-OZ – as part of a Brazilian project [1], where the specification for the SACI-1 micro satellite is developed using CSP-Z.

The strength of CSP-OZ lies in distributed systems where both data and control aspects must be modelled. We tried to demonstrate this idea in section 3 with the example of a registration desk. The printer is modelled by a pure CSP-process as no internal data structure is relevant in this system. The PID manager is a mainly data driven application which we modelled without using any CSP. The combination of the different components was done using CSP-operators. Paraphrasing the title of [6], CSP-OZ is even better suited for the description of standards than Object-Z.

Another advantage of CSP-OZ is its compositional semantics. For example, the implementation of the PID manager developed in section 2 can be used in the registration desk without affecting the correctness of the design. This could be done for all components of the system. Reusing a CSP-semantics for CSP-OZ gives this important advantage for free. Thus CSP-OZ is especially well suited to build up a library of specifications together with their implementations.

The semantics of CSP-OZ might look a bit complicated at first look. We nevertheless believe that the intuition behind it is simple, and the step from (Object-)Z to CSP-OZ is fairly easy.

A manual for CSP-OZ with a detailed definition of the syntax and the semantics is under development. Refinement rules already developed in [16, 4] can easily be rephrased for CSP-OZ as done in [7, 10]. Z-data refinement is a proper refinement in CSP-OZ for channels modelling Z-operations (analogously for Object-Z data refinement). Small case studies that translate straightforwardly to CSP-OZ can be found in [21] (game of life), [7] (telecommunications protocol) and [10] (pocket calculator).

The success of a formal method crucially depends on tool support. A workbench to integrate different tools (FDR [9], Z and CSP encodings in Isabelle

[14, 24], an hierarchical editor for Z and a graphical editor for CSP) is under development in the project UniForM [15].

ACKNOWLEDGEMENTS.

I thank G. Smith for spotting my interests towards Object-Z. His cooperation and helpful comments are gratefully acknowledged. The Semantik-group in Oldenburg and the cooperation with the University of Bremen in the UniForM project provide a stimulating background for my work. G. Smith, H. Dierks, S. Kleuker, C. Dietz, S. Hallerstede, H. Fleischhack and E.-R. Olderog read draft versions of this paper and made valuable comments.

REFERENCES

[1] E. Barros and A. Sampaio. Towards provably correct hardware/software codesign using OCCAM. In *Proccedings of the Third International Workshop on Codesign.* IEEE Computer Society Press, 1994.

[2] E. Boiten, H. Bowman, J. Derrick, and M. Steen. Viewpoint consistency in Z and LOTOS: A case study. submitted for publication, 1997.

[3] T. Bolognesi, J. van de Lagemaat, and C. Vissers, editors. *LOTOSphere: Software Development with LOTOS.* Kluwer Academic, 1995.

[4] M. J. Butler. *A CSP Approach To Action Systems.* PhD thesis, University of Oxford, 1992.

[5] J. Derrick, E.A. Boiten, H. Bowman, and M.W.A. Steen. Supporting ODP – translating LOTOS to Z. In *First IFIP International workshop on Formal Methods for Open Object-based Distributed Systems.* Chapmann & Hall, 1996.

[6] R. Duke, G. Rose, and G. Smith. Object-Z: A specification language advocated for the description of standards. *Computer Standards and Interfaces*, 17:511–533, 1995.

[7] C. Fischer. Combining CSP and Z. submitted for publication, 1997.

[8] C. Fischer. Combining Object-Z and CSP. Technical report, University of Oldenburg, April 1997.

[9] Formal Systems (Europe) Ltd. *Failures-Divergence Refinement: FDR 2,* Dec 1995. Preliminary Manual.

[10] S. Hallerstede. Die semantische Fundierung von CSP-Z. Master's thesis, University of Oldenburg, January 1997. in german.

[11] J. He. Process simulation and refinement. *Formal Aspects of Computing*, 1(3):229–241, 1989.

[12] C.A.R. Hoare. *Communicating Sequential Processes.* Prentice-Hall International, 1985.

[13] M.B. Josephs. A state-based approach to communicating processes. *Distributed Computing*, 3:9–18, 1988.

[14] Kolyang, T. Santen, and B. Wolff. A structure preserving encoding of Z in Isabelle/HOL. In J. von Wright, J. Grundy, and J. Harrison, editors, *Theorem Proving in Higher Order Logics*, LNCS 1125, pages 283–298. Springer Verlag, 1996.

[15] B. Krieg-Brückner, J. Peleska, E.-R. Olderog, D. Balzer, and A. Baer. UniForM — Universal Formal Methods Workbench. In U. Grote and G. Wolf, editors, *Statusseminar des BMBF Softwaretechnologie*, pages 357–378. BMBF, Berlin, March 1996.

[16] C. Morgan. *Programming from Specifications*. Prentice Hall, 1990.

[17] A. W. Roscoe. An alternative order for the failures model. In *Two papers on CSP*, Technical Monograph PRG-67, pages 1–26. Oxford University, 1988.

[18] A. W. Roscoe and G. Barrett. Unbounded nondeterminism in CSP. In *Mathematical Foundations of Programming Semantics*, volume 442 of *LNCS*, pages 160–193. Springer-Verlag, 1989.

[19] A. W. Roscoe, J. C. P. Woodcock, and L. Wulf. Non-interference through determinism. In D. Gollmann, editor, *ESORICS 94*, volume 875 of *LNCS*, pages 33–54. Springer-Verlag, 1994.

[20] G. Smith. *An Object-Oriented Approach to Formal Specification*. PhD thesis, Department of Computer Science, University of Queensland, St. Lucia 4072, Australia, October 1992.

[21] G. Smith. A semantic integration of Object-Z and CSP for the specification of concurrent systems. submitted for publication, 1997.

[22] J. M. Spivey. *The Z Notation: A Reference Manual*. Prentice-Hall International Series in Computer Science, 2nd edition, 1992.

[23] B. Strulo. How firing conditions help inheritance. In J. P. Bowen and M. G. Hinchey, editors, *ZUM '95: The Z Formal Specification Notation*, volume 967 of *LNCS*, pages 264–275, 1995.

[24] H. Tej and B. Wolff. A corrected failure-divergence-model for CSP in Isabelle/HOL. submitted for publication, 1997.

[25] J. C. P. Woodcock and C. C. Morgan. Refinement of state-based concurrent systems. In *Proceedings of VDM Symposium 1990*, volume 428 of *LNCS*, pages 340–351. Springer-Verlag, 1990.

BIOGRAPHY

Clemens Fischer received a master degree in computer science from the University Oldenburg, Germany, in 1995. Since then he works in the research group headed by Prof. E.-R. Olderog at the same university.

His current research interests are combining FDTs, practical applications of formal methods and integration of fully automatic verification techniques with a transformational approach to the design of correct systems.

30

Towards the Refinement of Executable Temporal Objects

Michael Fisher
Department of Computing, Manchester Metropolitan University
Manchester M1 5GD, U.K.   EMAIL: M.Fisher@doc.mmu.ac.uk

Abstract

Concurrent METATEM is a high-level language in which the behaviour of an individual reactive component is represented by a temporal logic formula and is animated by direct execution. The combination of this executable temporal formalism, together with an operational model providing asynchronous concurrency and broadcast message-passing, presents a powerful and flexible framework in which to develop concurrent object-based, particularly agent-based, applications.

While Concurrent METATEM has been applied in a variety of scenarios, and techniques for the verification of properties of Concurrent METATEM systems have been developed, little work has been carried out on the basis for refining such systems. Here, we introduce simple mechanisms for the refinement both of an object's internal behaviour and interface, and of individual objects into new systems of communicating objects.

Keywords

Executable specifications, refinement, object-based systems, transformation

1   INTRODUCTION

Concurrent METATEM is a simple programming language developed for reactive systems (Fisher 1993) that has been shown to be particularly useful in representing and developing multi-agent systems (Fisher 1995a). It is based on the combination of two complementary elements: the direct execution of temporal logic specifications providing the behaviour of an individual object (Fisher 1996); and a concurrent operational model in which such objects execute asynchronously, communicate via broadcast message-passing, and are organised using a powerful grouping mechanism (Fisher 1994). While both the operational model and object representation technique are simple, they together provide a framework in which a variety of concurrent object-based systems can be specified and implemented. It is important to note that *object-based*, rather than *object-oriented*, systems are developed here, and so we are not directly concerned with features such as inheritance and classes; the only attributes these objects have are encapsulation and message-based communication. Note, however, that many 'standard' object-oriented features can be built on top of this basic framework, if required. In this sense, the closest related work stems from the development of the Actor paradigm (Agha 1986).

In this paper we consider the refinement of Concurrent METATEM, incorporating

1. the refinement of an individual object's behaviour, which corresponds to standard refinement of temporal specifications (Manna & Pnueli 1992);
2. the refinement of an object into a collection of new objects that together implement the original behaviour under appropriate communication constraints; and
3. the use of a fixed set of transformation rules, rather than arbitrary refinements, allowing a "pick and mix" approach to program development.

For simplicity, we consider propositional, rather than first-order, temporal specifications. While this is obviously a restriction, many of the techniques we discuss can be transferred, with a little work, to the first-order framework.

The structure of this paper is as follows. In §2, we provide a definition of the temporal logic we use, followed, in §3, by a brief review of the Concurrent METATEM language. In §4, we present the framework for refinement of Concurrent METATEM objects, and consider a range of simple examples. We also derive *fixed* transformations which are behaviour preserving, and present a larger example of system refinement. Finally, in §5, we present conclusions and identify future work.

2   TEMPORAL LOGIC

Temporal logic can be seen as classical logic extended with various modalities representing temporal aspects of logical formulae (Emerson 1990). The propositional temporal logic we use (called PTL) is based on a linear, discrete model of time (Gabbay, Pnueli, Shelah & Stavi 1980). Thus, time is modelled as an infinite sequence of discrete states, with an identified starting point, called 'the beginning of time'. Classical formulae are used to represent constraints *within* states, while temporal formulae represent constraints *between* states. This temporal logic can be seen as classical logic extended with various modalities, for example '$\Diamond$', '$\Box$', and '$\bigcirc$'. The intuitive meaning of these connectives is as follows: $\Diamond A$ is true now if A is true *sometime* in the future; $\Box A$ is true now if A is true *always* in the future; and $\bigcirc A$ is true now if A is true at the *next* moment in time. In this presentation, similar connectives are introduced to enable reasoning about the *past* (Lichtenstein, Pnueli & Zuck 1985).

2.1   Syntax

We begin with the formal syntax of the language. Formulae of PTL are constructed using the following symbols.

- A set, $\mathcal{L}_p$, of *propositional symbols* represented by strings of lower-case alphabetic characters.
- Classical connectives, $\neg$, $\vee$, $\wedge$, **true**, **false** and $\Rightarrow$.
- Future-time temporal operators, categorised as

  - nullary operators: **start**,
  - unary operators: $\bigcirc$, $\Diamond$, $\square$,
  - binary operators: $\mathcal{U}$, and $\mathcal{W}$.

- Past-time temporal operators, categorised as

  - unary operators: $\bullet$, $\blacklozenge$, $\blacksquare$,
  - binary operators: $\mathcal{S}$, and $\mathcal{Z}$.

The set of *well-formed formulae* of PTL (WFF$_p$) is defined as follows.

- Any element of $\mathcal{L}_p$ is in WFF$_p$.
- If A and B are in WFF$_p$, then so are

$$
\begin{array}{lllll}
\neg A & A \vee B & A \wedge B & A \Rightarrow B & \textbf{start} \\
\Diamond A & \square A & A\,\mathcal{U}\,B & A\,\mathcal{W}\,B & \bigcirc A \\
\blacklozenge A & \blacksquare A & A\,\mathcal{S}\,B & A\,\mathcal{Z}\,B & \bullet A
\end{array}
$$

2.2 Semantics

Intuitively, the models for PTL formulae are based on discrete, linear structures having a finite past and infinite future, i.e., sequences such as

$$s_0,\ s_1,\ s_2,\ s_3,\ \ldots$$

where each s_i, called a *state*, provides a propositional valuation. However, rather than representing the model structure in this way, we will define a model, σ, as

$$\sigma = \langle \mathbb{N}, \pi_p \rangle$$

where $\mathbb{N}$ is used to represent the sequence of states $s_0, s_1, s_2, s_3, \ldots$, and, π_p is a map from $\mathbb{N} \times \mathcal{L}_p$ to $\{\mathsf{T}, \mathsf{F}\}$, giving a propositional valuation for each state in the sequence.

An interpretation for this logic is defined as a pair $\langle \sigma, i \rangle$, where σ is the model and i the index of the state at which the temporal statement is to be interpreted.

A semantics for well-formed temporal formulae is a relation between interpretations and formulae, and is defined inductively as follows, with the (infix) semantic relation being represented by '$\models$'. The semantics of a proposition is defined by the valuation given to that proposition at a particular state:

$$\langle \sigma, i \rangle \models p \quad \text{iff} \quad \pi_p(i, p) = \mathsf{T} \qquad [\text{for } p \in \mathcal{L}_p].$$

The semantics of the standard propositional connectives is as in classical logic, e.g.,

$$\langle \sigma, i \rangle \models A \vee B \quad \text{iff} \quad \langle \sigma, i \rangle \models A \quad \text{or} \quad \langle \sigma, i \rangle \models B .$$

The semantics of the unary future-time temporal operators is defined as follows.

$$\langle \sigma, i \rangle \models \bigcirc A \quad \text{iff} \quad \langle \sigma, i + 1 \rangle \models A$$
$$\langle \sigma, i \rangle \models \Diamond A \quad \text{iff} \quad \text{there exists } j \in \mathbb{N} \text{ such that } j \geq i \text{ and } \langle \sigma, j \rangle \models A$$
$$\langle \sigma, i \rangle \models \Box A \quad \text{iff} \quad \text{for all } j \in \mathbb{N}, \text{ if } j \geq i \text{ then } \langle \sigma, j \rangle \models A$$

Additionally, the syntax includes two binary future-time temporal operators, interpreted as follows.

$$\langle \sigma, i \rangle \models A \mathcal{U} B \quad \text{iff} \quad \text{there exists } k \in \mathbb{N}, \text{ such that } k \geq i \text{ and } \langle \sigma, k \rangle \models B$$
$$\text{and for all } j \in \mathbb{N}, \text{ if } i \leq j < k \text{ then } \langle \sigma, j \rangle \models A$$
$$\langle \sigma, i \rangle \models A \mathcal{W} B \quad \text{iff} \quad \langle \sigma, i \rangle \models A \mathcal{U} B \quad \text{or} \quad \langle \sigma, i \rangle \models \Box A$$

As temporal formulae are interpreted at a particular state-index, i, then indices less than i represent states that are 'in the past' with respect to state s_i. The semantics of the unary past-time operators is given as follows.

$$\langle \sigma, i \rangle \models \bullet A \quad \text{iff} \quad \langle \sigma, i - 1 \rangle \models A \quad \text{and} \quad i > 0$$
$$\langle \sigma, i \rangle \models \blacklozenge A \quad \text{iff} \quad \text{there exists } j \in \mathbb{N}, \text{ s.t. } 0 \leq j < i \text{ and } \langle \sigma, j \rangle \models A$$
$$\langle \sigma, i \rangle \models \blacksquare A \quad \text{iff} \quad \text{for all } j \in \mathbb{N}, \text{ if } 0 \leq j < i \text{ then } \langle \sigma, j \rangle \models A$$

Note that, in contrast to the future-time operators, the '$\blacklozenge$' ("sometime in the past") and '$\blacksquare$' ("always in the past") operators are interpreted as being *strict*, i.e., the current index is not included in their definition. Apart from their strictness, the binary past-time operators are similar to their future-time counterparts; their semantics is defined as follows.

$$\langle \sigma, i \rangle \models A \mathcal{S} B \quad \text{iff} \quad \text{there exists } k \in \mathbb{N}, \text{ s.t. } 0 \leq k < i \text{ and } \langle \sigma, k \rangle \models B$$
$$\text{and for all } j \in \mathbb{N}, \text{ if } k < j < i \text{ then } \langle \sigma, j \rangle \models A$$
$$\langle \sigma, i \rangle \models A \mathcal{Z} B \quad \text{iff} \quad \langle \sigma, i \rangle \models A \mathcal{S} B \quad \text{or} \quad \langle \sigma, i \rangle \models \blacksquare A .$$

Finally, the '**start**' operator is defined such that it can only be satisfied at the beginning of time, i.e. where $i = 0$.

2.3   Separated Normal Form

As an object's behaviour is represented by a temporal formula, we can transform this formula into Separated Normal Form (SNF) (Fisher 1992, Fisher 1997*a*). This not only removes the majority of the temporal operators, but also translates the formula into a set of *rules* suitable for direct execution (see §3). Each of these rules is of one of the following forms.

$$\mathbf{start} \quad \Rightarrow \quad \bigvee_{j=1}^{r} m_j \qquad \text{(an } \textit{initial} \; \square\text{-rule)}$$

$$\mathbf{\bullet} \bigwedge_{i=1}^{q} k_i \quad \Rightarrow \quad \bigvee_{j=1}^{r} m_j \qquad \text{(a } \textit{global} \; \square\text{-rule)}$$

$$\mathbf{start} \quad \Rightarrow \quad \Diamond l \qquad \text{(an } \textit{initial} \; \Diamond\text{-rule)}$$

$$\mathbf{\bullet} \bigwedge_{i=1}^{q} k_i \quad \Rightarrow \quad \Diamond l \qquad \text{(a } \textit{global} \; \Diamond\text{-rule)}$$

where each k_i, m_j or l is a literal. Note that the left-hand side of each *initial* rule is a constraint only on the *first* state, while the left-hand side of each *global* rule represents a constraint upon the previous state. The right-hand side of each $\square$-rule is simply a disjunction of literals referring to the current state, while the right-hand side of each $\Diamond$-rule is a single eventuality (i.e., '$\Diamond$' applied to a literal).

While the details of the transformation process will not be given here, it is important to note that Concurrent METATEM programs are represented as sets of rules (i.e. implications) where the left-hand side of each rule is a past-time formula, while the right-hand side of each rule is a present or future-time formula. This simple form leads naturally on to an operational model for the execution of such rules, as described in the next section.

3 CONCURRENT METATEM

The motivation for the development of Concurrent METATEM (Fisher 1993) has been provided from many areas. Being based upon executable logic, it can be utilised as part of the formal specification and prototyping of reactive systems. In addition, as it uses *temporal*, rather than classical, logic the language provides a high-level programming notation in which the dynamic attributes of individual components can be concisely represented (Barringer, Fisher, Gabbay, Gough & Owens 1995). This, together with its use of a novel model of concurrent computation, ensures that it has a range of applications in distributed and concurrent systems (Fisher 1994).

Concurrent METATEM is an object-based programming language comprising two distinct aspects:

1. the fundamental behaviour of a single object is represented as a temporal formula and animation of this behaviour is achieved through the direct execution of the formula (Fisher 1996);

2. objects are placed within an operational framework providing both asynchronous concurrency and broadcast message-passing.

While these aspects are, to a large extent, independent, the use of *broadcast* communication provides a natural link between them as it represents both a flexible communication model for concurrent objects (Birman 1991) and a natural interpretation of distributed deduction (Fisher 1997*b*). Thus, these features together provide an coherent and consistent programming model within which a variety of reactive systems can be represented and implemented.

3.1   Objects

The basic elements of Concurrent METATEM are objects. These are considered to be encapsulated entities, executing independently, and having complete control over their own internal behaviour. There are two elements to each object: its *interface definition* and its *internal definition*. The definition of which messages an object recognises, together with a definition of the messages that an object may itself produce, is provided by the interface definition for that particular object. The internal definition of each object is provided by a temporal specification.

An object's interface consists of three components, namely a unique *identifier*, which names the object, a set of symbols defining what messages will be accepted by the object (these are called *environment* propositions) and a set of symbols defining messages that the object may send (these are called *component* propositions). For example, the interface definition of a 'car' object might be:

car(go,stop,turn)[fuel,overheat]

Here, car is the identifier that names the object, {go,stop,turn} are the environment propositions, and {fuel,overheat} are the component propositions.

In order to animate the behaviour of an object, we choose to execute its temporal specification directly (Fisher 1996). Execution of a temporal formula corresponds to the construction of a model for that formula and, in order to execute a set of SNF rules representing the behaviour of a Concurrent METATEM object, we utilise the *imperative future* (Barringer, Fisher, Gabbay, Owens & Reynolds 1996) approach. This evaluates the SNF rules at every moment in time, using information about the history of the object in order to constrain future execution.

The operator used to represent the basic temporal indeterminacy within the SNF rules is the *sometime* operator, '$\Diamond$'. When $\Diamond\varphi$ is executed, the system must try to ensure that φ *eventually* becomes true. As such eventualities might not be able to be satisfied immediately, we must keep a record of the unsatisfied eventualities, retrying them as execution proceeds. It should be noted that the use of temporal logic as the basis for the computation rules gives an extra level of expressive power over the corresponding classical logics. In particular, operators such as '$\Diamond$' give the opportunity to specify future-time (temporal) indeterminacy.

As an example of a simple set of rules which form a fragment of an object's description, consider the following.

$$
\begin{array}{rcl}
\textbf{start} & \Rightarrow & \neg\mathsf{moving} \\
\bullet\,\mathsf{go} & \Rightarrow & \Diamond\mathsf{moving} \\
\bullet\,(\mathsf{moving} \wedge \mathsf{go}) & \Rightarrow & \mathsf{overheat} \vee \mathsf{fuel}
\end{array}
$$

Here, we see that moving is false at the start of execution and, whenever go is satisfied in the last moment in time, a commitment to eventually satisfy moving is made. Similarly, whenever both go and moving are satisfied in the last moment in time, then either overheat or fuel must be satisfied.

3.2 Concurrency and Communication

It is fundamental to our approach that all objects are (potentially) concurrently active. In particular, they may be asynchronously executing. Each object, in executing its temporal formula, independently constructs its own temporal model. Within Concurrent METATEM, a mechanism is provided for communication between separate objects which simply consists of partitioning each object's propositions into those controlled by the object and those controlled by its environment. As above, the former are termed either *component* or *internal* propositions while the latter are termed *environment* propositions. Within the individual object's execution, if a component proposition is satisfied, this has the side-effect of *broadcasting* the value of that proposition to all other objects. If a particular message is received, a corresponding environment proposition is satisfied in the object's execution. If an internal proposition is satisfied, this has no external effect.

To fit in with this logical view of communication, whilst also providing a flexible and powerful message-passing mechanism, *broadcast* message-passing is used to pass information between objects. Here, when an object sends a message it does not send it to a specified *destination*, it merely sends it to its environment where it can be received by *all* other objects. Although broadcast is the basic mechanism, both multicast and point-to-point message-passing can be defined on top of this (Fisher 1994). Finally, the default behaviour for a message is that if it is broadcast, then it will *eventually* be received at all possible receivers. Also note that, by default, the order of messages is not preserved, though such a constraint can be added, if required.

3.3 Applications and Implementation

The combination of executable temporal logic, asynchronous message-passing and broadcast communication provides a powerful and flexible basis for the development of reactive systems. Concurrent METATEM is being utilised in the development of a range of applications in areas from distributed artificial intelligence (Fisher & Wooldridge 1993), agent societies (Fisher & Wooldridge 1995), concurrent theorem-proving (Fisher 1997b), and systems simulation (Finger, Fisher & Owens 1993). A survey of some of the potential applications of the language is given in (Fisher 1994).

4　PRINCIPLES OF REFINEMENT

Given that Concurrent METATEM objects can be defined and executed, we now consider the refinement of their representations. In particular, we examine the refinement of an object's internal behaviour (via manipulation of its SNF rules), refinement of an object's interface (thus affecting how it interacts with the environment), and refinement of a single object into a *set* of objects exhibiting equivalent behaviour.

First, we will provide a definition of specifications for Concurrent METATEM objects. Rather than consisting solely of the appropriate temporal rules, an object's specification must also provide the partition of propositions into component, environment and internal sets. Note that, in the following definition, props is a function that extracts all proposition symbols from a given set of SNF rules.

Definition 1 (Specification) *A specification of a Concurrent* METATEM *object is given as a tuple* $\langle R, P_E, P_C, P_I \rangle$, *where*

- R *is the set of SNF rules comprising the object,*
- P_E *is the object's set of* environment *propositions,*
- P_C *is the object's set of* component *propositions, and*
- P_I *is the object's set of* internal *propositions,*

and where both $\mathsf{props}(R) \subseteq P_E \cup P_C \cup P_I$ *and* $P_C \cap P_I = \emptyset$.

We now consider a variety of different classes of refinement, beginning with standard refinement of temporal specifications. The notation we use for refinement of specifications is $S_1 \longrightarrow S_2$, meaning that S_2 is a refinement of S_1.

4.1　Refining an Object's Internal Behaviour

Standard refinement of temporal specifications (Manna & Pnueli 1992) can be applied to specifications of Concurrent METATEM objects. Such refinement can be carried out in the following circumstances:

$\langle R, P_E, P_C, P_I \rangle \longrightarrow \langle R', P_E, P_C, P_I \rangle$ if, and only if,

$$\vdash \left(\Box \bigwedge_{r' \in R'} r' \right) \Rightarrow \left(\Box \bigwedge_{r \in R} r \right)$$

Example 1   This first (simple) example involves the removal of eventualities. The original object defined by

$$\text{ex1(announce)[give,receive]:}$$
$$\textbf{start} \;\Rightarrow\; \Diamond\text{give}$$

can be refined to

$$\text{ex1(announce)[give,receive]:}$$
$$\textbf{start} \;\Rightarrow\; \text{x}$$
$$\text{\textbf{O}x} \;\Rightarrow\; \text{give}$$

showing that a give message will be produced on the second execution step of object ex1. This follows since

$$\vdash \; (\,\Box(\textbf{start} \Rightarrow \text{x}) \,\wedge\, \Box(\textbf{O}\text{x} \Rightarrow \text{give})) \;\Rightarrow\; \Box(\textbf{start} \Rightarrow \Diamond\text{give})$$

Example 2   Another simple example of the reduction of non-determinism concerns the removal of disjunctions. Here, the object defined by

$$\text{ex2(ins)[outs]:}$$
$$\textbf{O}\text{p} \;\Rightarrow\; \text{q} \vee \text{r}$$

can be refined to

$$\text{ex2(ins)[outs]:}$$
$$\textbf{O}\text{p} \;\Rightarrow\; \text{q}$$

Thus, as in standard formal development, non-determinism within the temporal specification is reduced by such refinement steps.

4.2   Interface Refinement

In addition to the refinement of an object's rule set, as above, we can apply refinement to the object's interface. However, certain restrictions on this are enforced. Thus,

$$\langle R, P_E, P_C, P_I \rangle \;\longrightarrow\; \langle R, P'_E, P'_C, P'_I \rangle$$

just as long as the following constraints are observed.

1. Any increase in component, environment or internal sets involves propositions not already present in *any* of those sets, e.g. for environment propositions

$$(P'_E - P_E) \cap (P_E \cup P_C \cup P_I) = \emptyset$$

2. Any decrease in component, environment or internal sets does not involve propositions which appear in the original rule set, e.g., again for environment propositions

$$(P_E - P_E') \cap \mathsf{props}(R) = \emptyset$$

Example 3   To provide a simple example showing why such restrictions are necessary, we give an *illegal* refinment below which may produce unwelcome behaviour. Thus, if we refine the interface of

ex3(ins)[outs,p]:
  O true   ⇒   ◇p

by removing **p** from its component set, we produce

ex3(ins)[outs]:
  O true   ⇒   ◇p

However, while **p** messages are broadcast from the original object infinitely often, no **p** messages are ever broadcast from the refined object. If any other object is dependent upon these messages, then this can obviously lead to radically different behaviour across the system.

4.3   Object Decomposition

Next, we consider refining a single object into a set of objects that together implement the required behaviour. For example, in Fig. 1, an object **A** is refined into three objects **B1**, **B2** and **B3**, which communicate together to provide **A**'s behaviour.

This decomposition can take place as long as

$$\vdash \left(\bigwedge_{i=1}^{n} [\![\langle R^i, P_E^i, P_C^i, P_I^i \rangle]\!] \right) \Rightarrow [\![\langle R, P_E, P_C, P_I \rangle]\!]$$

where $[\![\;]\!]$ provides the temporal semantics of each object (Pnueli 1981), under appropriate communication constraints (Fisher 1995*b*). Note that this validation also checks that the sets of environment, internal and component predicates are consistent across the distributed system.

Unfortunately, simple (yet non-trivial) examples of this type of transformation are difficult to find. The problem is that, even for relatively small specifications, the temporal formula representing the semantics of an object tends to be large. Lack of space precludes the inclusion of such examples.

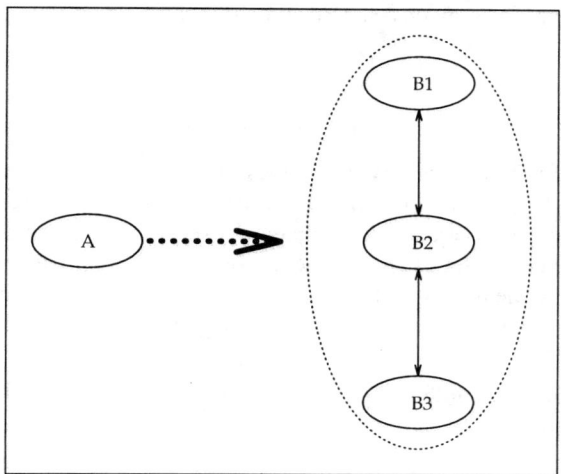

Figure 1 Decomposing an Object

4.4 Fixed Transformations

In an effort to simplify the refinement procedure for the system developer, particularly in the case where an object is decomposed into a new set of objects, we introduce fixed transformations which preserve the behaviour of the systems. While
many of the simpler transformations concern the single object refinements described
above (in §4.1 and §4.2), the ones we describe here relate to object decomposition
(as in §4.3) and have been found to be very useful in simplifying the derivation of
refinement.

We label the first transformation, $T1$; the second one, called $T2$ is essentially a
generalisation of the first.

$\underline{T1}$: Given a specification containing the rules

$$\begin{array}{ccc} \bullet p & \Rightarrow & q \\ \bullet q & \Rightarrow & r \end{array}$$

where r and q do not occur on the left-hand side of any other rules, apart from
the above, then the specification of the new object is $\langle \{ \bullet q \Rightarrow r \}, \{q\}, \{r\}, \emptyset \rangle$
where $\bullet q \Rightarrow r$ is removed from the original object's ruleset and q is added to
its set of component propositions. Note that, if r does not occur in *any* other rule
in the original object, this proposition can be removed from its set of component
propositions.

$\underline{T2}$: This is similar to $T1$, except that the r message is recognised by the original
object, hence providing a mechanism for passing information between the new

and original object. Thus, given a specification containing the rules

$$
\begin{aligned}
\bullet\, p &\Rightarrow q \\
\bullet\, q &\Rightarrow r \\
\bullet\, r &\Rightarrow s
\end{aligned}
$$

where r and q do not occur in any other rules, then the specification of the new object is again $\langle\{\,\bullet\, q \Rightarrow r\}, \{q\}, \{r\}, \emptyset\rangle$. As before, $\bullet\, q \Rightarrow r$ is removed from the original object's ruleset and q is added to its set of component propositions, but now r is added to its set of environment propositions. While this transformation produces the same new object as in $T1$, the difference is that, using $T2$, the r message will effectively be passed back to the original object.

Other transformations follow this pattern. The important property of these transformations is as follows.

Theorem 1 *If a fixed transformation is applied to specification S to give S', then $[\![S]\!] \Leftrightarrow [\![S']\!]$ under the standard communication constraints.*

The standard communication constraints are that if a message is broadcast it will eventually arrive at all other objects. Given this, the above theorem can be established by appealing to the temporal semantics of Concurrent METATEM (Fisher 1995b).

Example 4   As a simple example of the use of $T2$, we can transform the object

ex4(a)[s,b]:
$$
\begin{aligned}
\bullet\, a &\Rightarrow p \\
\bullet\, p &\Rightarrow q \\
\bullet\, q &\Rightarrow r \\
\bullet\, q &\Rightarrow b \\
\bullet\, r &\Rightarrow s \\
\bullet\, p &\Rightarrow s
\end{aligned}
$$

to a system comprising two objects, i.e.

ex4(a,r)[s,q]:
$$
\begin{aligned}
\bullet\, a &\Rightarrow p \\
\bullet\, p &\Rightarrow q \\
\bullet\, r &\Rightarrow s \\
\bullet\, p &\Rightarrow s
\end{aligned}
$$

new4(q)[r,b]:
$$
\begin{aligned}
\bullet\, q &\Rightarrow r \\
\bullet\, q &\Rightarrow b
\end{aligned}
$$

Thus, the sub-computation concerning the q proposition (message) can be isolated within the newly spawned object.

4.5 Development Example

In this section, we provide a simple example exhibiting the varieties of refinment steps described above, particularly with respect to decomposing an object into a new set of communicating objects. This example, which consists of a very basic planning system, not only has obvious relevance to the domain of multi-agent systems but also shows how specifications incorporating general liveness constraints may be refined. The original object is specified simply by

planner(goal)[plan]:

$$
\begin{array}{rcl}
\bullet\, goal & \Rightarrow & \Diamond plan \\
\neg subplan1\; \mathcal{S}\; goal & \Rightarrow & \neg plan \\
\neg subplan2\; \mathcal{S}\; goal & \Rightarrow & \neg plan \\
\neg subplan3\; \mathcal{S}\; goal & \Rightarrow & \neg plan \\
\bullet\, goal & \Rightarrow & \Diamond subplan1 \\
\bullet\, goal & \Rightarrow & \Diamond subplan2 \\
\bullet\, goal & \Rightarrow & \Diamond subplan3
\end{array}
$$

Here, once a goal message is received by the object, it guarantees to eventually produce a plan that achieves that goal. However, the plan can not be produced until all of its three subplans have been completed. Thus, the object also undertakes to generate these subplans.

An obvious structural refinement for this object is to decompose it into four objects, one effectively coordinating the planning activity, the other three producing the three subplans. This refined system can be represented as

planner(goal,subplan1,subplan2,subplan3)[plan,goal1,goal2,goal3]:

$$
\begin{array}{rcl}
\bullet\, goal & \Rightarrow & \Diamond plan \\
(\neg subplan1\; \vee\; \neg subplan2\; \vee\; \neg subplan3)\; \mathcal{S}\; goal & \Rightarrow & \neg plan \\
\bullet\, goal & \Rightarrow & goal1 \\
\bullet\, goal & \Rightarrow & goal2 \\
\bullet\, goal & \Rightarrow & goal3
\end{array}
$$

p1(goal1)[subplan1]:

$$\bullet\, goal1 \;\Rightarrow\; \Diamond subplan1$$

p2(goal2)[subplan2]:

$$\bullet\, goal2 \;\Rightarrow\; \Diamond subplan2$$

p3(goal3)[subplan3]:

$$\bullet\, goal3 \;\Rightarrow\; \Diamond subplan3$$

To see how the original specification is transformed into this, we will now consider the refinement steps used in a little more detail.

1. First, we can merge the three rules utilising the '$\mathcal{S}$' operator into one, i.e.

$$(\neg\mathsf{subplan1} \vee \neg\mathsf{subplan2} \vee \neg\mathsf{subplan3})\,\mathcal{S}\,\mathsf{goal} \;\Rightarrow\; \neg\mathsf{plan}$$

   using standard temporal refinement.
2. Next, we refine each of the '$\bullet\mathsf{goal} \;\Rightarrow\; \Diamond\mathsf{subplan}_i$' rules in order to introduce intermediate steps characterised by new variables. Thus, '$\bullet\mathsf{goal} \;\Rightarrow\; \Diamond\mathsf{subplan1}$' becomes

$$\begin{array}{rcl} \bullet\mathsf{goal} & \Rightarrow & \mathsf{goal1} \\ \bullet\mathsf{goal1} & \Rightarrow & \Diamond\mathsf{subplan1} \end{array}$$

   again using standard temporal refinement. N.B., goal1 will be used as the coordinating message between the planner and p1.
3. Finally, this specification is then distributed amongst four objects, with the planner object retaining all, and only, those SNF rules that contain the goal proposition. This is achieved by fixed transformations based upon $T2$ being applied to rules of the form $\bullet\mathsf{goal}_i \;\Rightarrow\; \mathsf{subplan}_i$ generated in step (2) above.

While this example is relatively simple, similar refinements occur in many systems. Since all temporal specifications are translated into SNF, wherein the main temporal operator is '$\Diamond$', and since the global structures within the application are often provided by grouping, which itself is based upon the decomposition of objects into appropriate sets of objects, it is perhaps not surprising that many applications require transformation steps of this form.

5   CONCLUSIONS AND FUTURE WORK

We have provided a basic framework for refinement in Concurrent METATEM, thus allowing the principled development of a range of concurrent object-based systems. While refinement proofs remain, in general, difficult, the use of fixed transformations provides the system developer with a toolbox of fast (no verification required), simple and safe (behaviour preserving) refinements.

   These refinement techniques are powerful, yet relatively simple, primarily because of the fit between the computational model and the temporal execution mechanism. It is important to note that the simplicity of both the refinement conditions and of the fixed transformations derived is a consequence of the easy match between temporal specification, temporal execution and the particular model of computation used. Specifically, utilising objects makes the general refinement of temporal specifications simpler (Barringer, Kuiper & Pnueli 1984), using broadcast messages obviates the need to keep track of senders and receivers of messages, and using a simple communication constraint, such as a message broadcast will eventually arrive at all objects, avoids tight coupling of objects. Although simple, this model of object-based computation can be used to represent applications in a number of areas. In particular, in the multi-agent systems area agents are often quite simply specified, yet it is the patterns of communication and grouping between them that gives these systems power.

While there is much work in the areas of refinement of temporal specifications (Manna & Pnueli 1992), formal methods for object-oriented systems (Buchs & Guelfi 1993), and the development of concurrent object-based systems (Agha 1986), there is little research directly relevant to that presented here. This is mainly because the object model we use is much simpler than most formal models of object-oriented systems, the operational model is unusual compared with the majority of work on concurrent object-based systems, and previous work on refinement of temporal specifications has neither consider the refinement of executable temporal specifications, nor the refinement of specifications under such a model of computation.

There are three directions for future work.

1. The extension of these refinements to first-order temporal logics.
2. The derivation of slightly more complex fixed transformations concerning cases where the objects produced need to communicate together in order to arrive at a common view and where true *grouping* (Birman 1991) is employed (again these are useful for the multi-agent systems area).
3. The development of a toolbox of fixed transformations has led us to consider producing a visual environment for object decomposition. Here, users select the rule(s) they wish to move to new objects, the system checks that this can be achieved and then generates a template for the new objects.

REFERENCES

Agha, G. (1986), *Actors - A Model for Concurrent Computation in Distributed Systems*, MIT Press.

Barringer, H., Fisher, M., Gabbay, D., Gough, G. & Owens, R. (1995), 'METATEM: An Introduction', *Formal Aspects of Computing* 7(5), 533–549.

Barringer, H., Fisher, M., Gabbay, D., Owens, R. & Reynolds, M., eds (1996), *The Imperative Future: Principles of Executable Temporal Logics*, Research Studies Press, Chichester, United Kingdom.

Barringer, H., Kuiper, R. & Pnueli, A. (1984), Now You May Compose Temporal Logic Specifications, *in* 'Proceedings of the Sixteenth ACM Symposium on the Theory of Computing'.

Birman, K. P. (1991), The Process Group Approach to Reliable Distributed Computing, Techanical Report TR91-1216, Department of Computer Science, Cornell University.

Buchs, D. & Guelfi, N. (1993), Formal Development of Actor Programs using Structured Algebraic Petri Nets, *in* 'Parallel Architectures and Languages, Europe (PARLE)', Munich, Germany. (Published in *Lecture Notes in Computer Science*, volume 694, Springer-Verlag).

Emerson, E. A. (1990), Temporal and Modal Logic, *in* J. van Leeuwen, ed., 'Handbook of Theoretical Computer Science', Elsevier, pp. 996–1072.

Finger, M., Fisher, M. & Owens, R. (1993), METATEM at Work: Modelling Reactive Systems Using Executable Temporal Logic, *in* 'Sixth International Conference on Industrial and Engineering Applications of Artificial Intelligence and Expert Systems', Gordon and Breach Publishers, Edinburgh, U.K.

Fisher, M. (1992), A Normal Form for First-Order Temporal Formulae, *in* 'Proceedings of Eleventh International Conference on Automated Deduction (CADE)', Saratoga Springs, New York. (Published in *Lecture Notes in Computer Science*, volume 607, Springer-Verlag).

Fisher, M. (1993), Concurrent METATEM — A Language for Modeling Reactive Systems, *in* 'Parallel Architectures and Languages, Europe (PARLE)', Munich, Germany. (Published in *Lecture Notes in Computer Science*, volume 694, Springer-Verlag).

Fisher, M. (1994), A Survey of Concurrent METATEM — The Language and its Applications, *in* 'First International Conference on Temporal Logic (ICTL)', Bonn, Germany. (Published in *Lecture Notes in Computer Science*, volume 827, Springer-Verlag).

Fisher, M. (1995*a*), Representing and Executing Agent-Based Systems, *in* M. Wooldridge & N. R. Jennings, eds, 'Intelligent Agents', Springer-Verlag.

Fisher, M. (1995*b*), Towards a Semantics for Concurrent METATEM, *in* M. Fisher & R. Owens, eds, 'Executable Modal and Temporal Logics', Springer-Verlag.

Fisher, M. (1996), 'An Introduction to Executable Temporal Logics', *Knowledge Engineering Review* **11**(1), 43–56.

Fisher, M. (1997*a*), 'A Normal Form for Temporal Logic and its Application in Theorem-Proving and Execution', *Journal of Logic and Computation* **7**(4).

Fisher, M. (1997*b*), An Open Approach to Concurrent Theorem-Proving, *in* 'Parallel Processing for Artificial Intelligence III', Elsevier Science B.V.

Fisher, M. & Wooldridge, M. (1993), Executable Temporal Logic for Distributed A.I., *in* 'Twelfth International Workshop on Distributed A.I.', Hidden Valley Resort, Pennsylvania.

Fisher, M. & Wooldridge, M. (1995), A Logical Approach to the Representation of Societies of Agents, *in* N. Gilbert & R. Conte, eds, 'Artificial Societies', UCL Press.

Gabbay, D., Pnueli, A., Shelah, S. & Stavi, J. (1980), The Temporal Analysis of Fairness, *in* 'Proceedings of the Seventh ACM Symposium on the Principles of Programming Languages', Las Vegas, Nevada, pp. 163–173.

Lichtenstein, O., Pnueli, A. & Zuck, L. (1985), 'The Glory of the Past', *Lecture Notes in Computer Science* **193**, 196–218.

Manna, Z. & Pnueli, A. (1992), *The Temporal Logic of Reactive and Concurrent Systems: Specification*, Springer-Verlag, New York.

Pnueli, A. (1981), 'The Temporal Semantics of Concurrent Programs', *Theoretical Computer Science* **13**, 45–60.

31

Applying LOTOS to the Design of TINA Applications

E. Koerner
Université de Liège
Institut d'Electricité Montefiore B28, B- 4000 Liège, Belgium
Email: koerner@montefiore.ulg.ac.be

L. Strick
GMD FOKUS
Hardenbergplatz 2, D- 10623 Berlin, Germany
Email: Strick@fokus.gmd.de

Abstract

We integrate the formal language LOTOS into a design framework for TINA-based applications. The behaviour model of TINA service components in the ODP computational viewpoint can be formally specified and verified. Compliance checks with the enterprise model can then be carried out by comparing requirements against the computational specification. Aspects of specifying computational objects with multiple interfaces are also pointed out. All discussions are based on a case study of a floor control component designed for a TINA multimedia collaboration service.

Keywords

TINA, LOTOS, ODP viewpoints, verification, viewpoint compliance

1 Introduction

The TINA architecture promises the fast and cost effective introduction of a wide range of telecommunications applications with advanced multimedia, multi-party and mobility features [TN-ARC]. The basic structure of TINA systems consists of an application layer providing telecommunications services, that run on top of a distributed processing environment (DPE) hiding the underlying heterogeneous communications and computing environment.

The project TANGRAM [DEE+96] at GMD Fokus in Berlin studies the applicability of the TINA concepts. One of its main objectives is to provide support for the development of distributed telecommunications applications. TINA proposes design guidelines [TN-SDG] that use the RM-ODP viewpoints [ITU-X901] as a main thrust to outline a development process. However, the TINA-C work on design guidelines has been discontinued. In the TANGRAM project the TINA design process has been enhanced with methods that have originally been used for the design of management services [BK-DEI]. Further, it has been possible to reuse the corresponding CASE-tool called COMPASS (Composition of Management Applications and Services) as an efficient support of the development process. Currently, formal description techniques are not integrated into this process. In this paper, we show how LOTOS [BoB87] [ISO8807] can fill this gap.

In section 2, an overview of the notations being used in the ODP viewpoints is given. The possible use of LOTOS is described in section 3. Then we present the design of a component for floor control in CSCW applications. The formal specification of this component in a TINA environment is presented in section 5. The sixth section deals with the verification activities that can be undertaken with such a formal specification. Section 7 concludes this article.

2 Design Framework

To guide the design of TINA services a corresponding framework has to be defined. It provides graphical and syntactical notations that appropriately concretise the abstract language constructs for the RM-ODP viewpoints.

The enterprise specification is broken up into two parts. The Business Model captures the business context of a TINA system in terms of stakeholders, business roles, administrative domains and reference points [TN-REF]. It provides structure for the TINA system and drives the modification of specifications as a result of business evolution. The means to describe the Business Model are a (non-standard) graphical representation and natural language. The Service Model is the complete description of the actual service and whatever requirements are related to the service from the stakeholders' point of view. The concept of Use Cases [Jac92] is used as the support for the Service Model. It supports the specification of interactions between roles, played by stakeholders and subsystems, and a system in terms of activities. Use Cases help to break down the complexity of a whole TINA service into treatable scenarios. The specification of Use Cases consists of graphical schemata together with a natural language description of activities.

Use Cases are functional in nature. The object orientation is introduced in the information specification. The Static Object Model defines the static (invariant) information structure in terms of information object types with their attributes and the relationships among them. A Dynamic Object Model is used to specify how information objects change over time. It captures dynamic behaviour and dynamic constraints of the information objects. OMT's graphical notations for object and dynamic modelling are used [Rum91]. The semi-formal syntax qGDMO/qGRM [TN-IMC] is proposed by TINA-C and allows to do in-depth information modelling.

For the computational specification an object model is used to describe the object classes from which the necessary object instances can be created at runtime. It includes the relations at the conceptual level, such as object type, interface type and inheritance. For the graphical specification of computational object classes a non-standard extension of OMT is used that specifically supports the TINA computational modelling concepts [TN-CMC]. As such, it can be related to TINA's semi-formal Object Definition Language (ODL) [TN-ODL]. The activity model concentrates on the modelling of typical behaviour patterns performed by a set of interworking computational object instances (COs). The main purpose of the interaction model is to derive behaviour requirements for a single CO from the global system behaviour. An interaction diagram grammar is used to model the interworking of object instances.

In the engineering viewpoint notations are defined for an object model, interaction model and a configuration model. They are not further described in this paper.

The ODP viewpoint specifications of a telecommunications service can be related to the construction stage of the TINA service lifecycle model [TN-SA]. This service lifecycle model identifies the set of processes that are required to support the development and operation of a service. The order in which the activities take place is not described in TINA. A suitable development process has, for instance, been described in [SRP96].

3 Usage of LOTOS in the Design Framework

There are several reasons to include a formal description technique into the design framework:

In the computational viewpoint TINA-ODL only allows to describe object behaviour through natural language strings. A behaviour description using an appropriate formal language is thus useful. It could be attached to each computational object class. This corresponds to adding a behaviour model as a third submodel to the computational viewpoint.

TINA promotes a component approach to service construction. Specifications of reusable service components should be verified before they are released.

The object model of TINA allows objects with multiple interfaces. It should be noted that in this model interfaces do not have independent state and behaviour. The state and behaviour are actually in the object that the interface belongs to. Consequently, interactions with an object via one interface may influence interactions via other interfaces of the object. Further, invocations at a server interface of an object may result in that object invoking an operation through a client interface associated to that object. When a computational object is constructed with multiple interfaces it has to be assured that the resulting behaviour is the desired one. A formal specification can help to identify additional behaviour required when interface specifications are reused in an object definition. Further, conflicts among operations at different interfaces of the same object may be isolated.

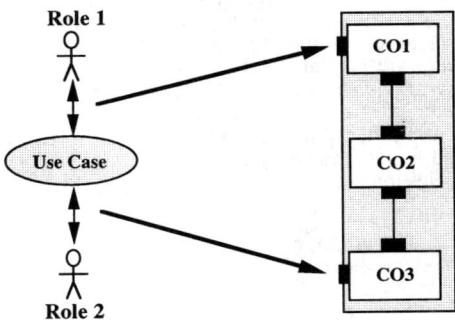

Figure 1 Relation between enterprise and computational model

The use of a formal language may also allow one to rigidly check for relations between viewpoint specifications. For instance, requirements tracing demands for the compliance[*] of the computational specification with the enterprise specification. Once a computational specification exists, one must be able to define a computational configuration that provides a solution to the requirements identified in the enterprise viewpoint. The compliance can be checked by extracting and formalising properties in the enterprise viewpoint, projecting them into the computational viewpoint and then performing appropriate model checking of the computational configuration.

---

[*] When considering correspondences between viewpoint specifications several terms come into play like consistency, conformance or compliance. Here the following definition of compliance is assumed: "Compliance is a relation between two specifications, A and B, that holds when specification A makes requirements which are fulfilled by specification B (when B complies with A)" [ITU-X901].

In the concrete case of the design architecture a requirement stated in a Use Case has to be satisfied by a configuration of ODL-defined object instances. More precisely, roles defined in the use cases have an interface to the system. These interfaces are mapped to computational interfaces at objects representing the role (for example, user agents or session managers). The compliance check is carried out with respect to these interfaces (see figure 1).

LOTOS has been intensively studied and applied in the ODP community. It is used to give formal semantics to the ODP viewpoint languages as part of the RM-ODP [ITU-X904]. The application of LOTOS for ODP viewpoint specifications has been further concretised by the idea of LOTOS sub-languages for the information, computational and engineering viewpoint [Vog95]. In this approach the syntax of LOTOS is restricted, in order to enforce an object-based specification style respecting the concepts, rules and structures for each viewpoint.

The sublanguage LOTOScomp for the computational viewpoint is thus readily applicable for our purposes. Its grammar is given in extended Backus-Naur-Form. Its main constituents are process templates for computational objects, interfaces and operations. Operational and stream interfaces are distinguished. Within the former an operation process template is defined, within the latter a flow process template is used.

In the remainder of this paper a case study of a floor control component is discussed in order to investigate the different aspects of integration of LOTOS into the design architecture.

4 The Floor Control Component

Floor control is a component for CSCW applications like multimedia collaboration [Koe96]. It is used to attribute the right to perform a particular activity in a multimedia collaboration session. For instance, in an audio-visual conference a token owner is attributed the right to provide audio-visual input.

In figure 2 the decomposition of floor control into use cases is shown. Floor control involves requesting a token, assigning it, and at a later point releasing it again. Assignment and release are part of the *Token Passing* activity. Requesting a token is described in the *Request Token* use case. The requirement to have the possibility to withdraw a token request is captured in the *Revoke Token* activity. The extends-associations have been introduced here to indicate that the order in which the extending use cases are executed is determined by the external stimuli given to the extended use case by the different roles interacting with it.

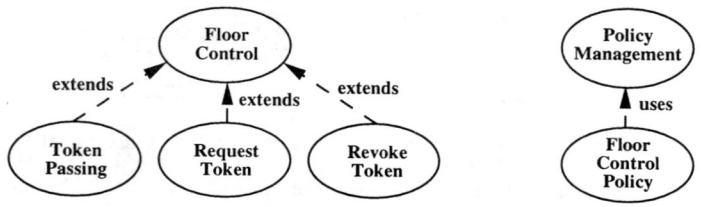

Figure 2 Enterprise service model for floor control

The floor control component should be highly configurable. The use case *Floor Control Policy* describes the corresponding configuration management activity. Three basic floor control policies are defined:
- In the *Automatic* mode the floor is passed automatically among parties according to one of several possible automatic assignment policies.
- In the *Master* mode the decision about the next token owner is taken by a master. If the master releases the token, automatic assignment is enabled until the next token assignment.

- In the *Tokenholder* mode the current token owner designates the next token owner. If the tokenholder releases the token, automatic assignment is enabled until the next token assignment.

Further, different rules may be set for token release:
- *Explicit*: An explicit token release requires the token owner to free the token himself.
- *Autopreemptive*: An automatic preemption takes the token away from the token owner after a specified time. Explicit release is still possible.
- *Masterpreemptive*: A preemption takes the token away from the owner upon request by the master. Explicit release is still possible.

The use case *Policy Management* is defined as an abstract use case (uses-association) and serves as a container for configuration management activities. Such an abstraction may lead to the identification of a policy management function as an engineering support.

Telelecturing is a typical service that may include floor control as a feature. For that application, the master mode with masterpreemptive token release would be a typical floor control policy. The lecturer would take on the master role and own the token as long as he is addressing the audience. The students may have the opportunity to intervene in the lecture by requesting the floor. The lecturer may then decide to assign the token to one of his students. As long as the student owns the token, his associated audio output stream is distributed to the audience. If the teacher requests the token it is immediately assigned back to him. Otherwise, the student may finish his intervention and subsequently release the token which will then be passed back to the teacher.

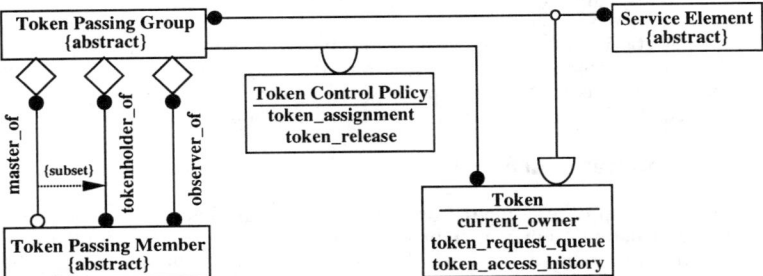

Figure 3 Static information model for floor control

A possible information model for floor control is shown in figure 3. A *Token* is modelled as a relationship that controls the association between a *Token Passing Group* and a *Service Element* such as a videoconference or a shared application. A *Token* is characterised by the current owner, the request queue and an access history holding the list of recent token owners. Three roles are distinguished for the token passing interaction. The *master* role is optional as it may be unassigned in the automatic and tokenholder mode. A *tokenholder* is entitled to become owner of the token. Observers are merely monitoring the token passing. The interaction between the *Token Passing Group* and a *Token* object is governed by the *Token Control Policy*. The attribute *token_assignment* determines the basic floor control policy while *token_release* determines the policy rules for token release.

In the computational viewpoint four interfaces are defined for floor control. A token passing interface provides operations for assigning, releasing and requesting the token, as well as for cancelling pending token requests. A policy management interface provides an operation for the modification of floor control policy. Both interfaces have a corresponding notification interface that defines announcements to be sent out to the parties. Thus, an announcement can be sent when a token has been passed, requested, a token request has been revoked or the floor control policy has been modified.

In a TINA application design these interfaces will be attached to objects. Figure 4 shows a typical object configuration of service session related objects [TN-SA]. A party has an associated user application (UAP) as a session endpoint in the user domain and a user service session manager (USM) in the provider domain that manages the service aspects particular to that party. The service session manager (SSM) coordinates all service capabilities that are shared among the users in the service session. The floor control interfaces may consequently be instantiated as UAP and SSM interfaces. A UAP will be a client for token passing and token policy management and a server for the corresponding notification interfaces. The SSM has to instantiate these interfaces with opposite causality.

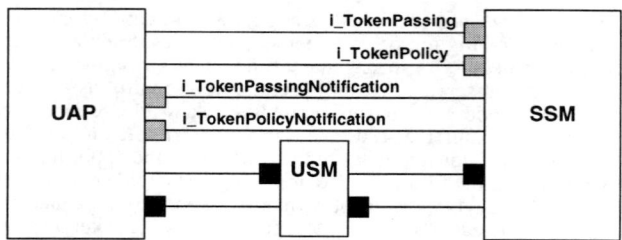

Legend: UAP: User Application, USM: User Service Session Manager, SSM: Service Session Manager

Figure 4 Computational object configuration

Of course, other configurations are possible. For instance, a concurrency access control server in the provider domain may coordinate token passing. The floor control operations may also need to be passed through the USM. Moreover, the token passing interfaces may be included in a base interface type by interface inheritance.

5 Formal Computational Specification

Applying LOTOScomp the computational object configuration in figure 4 yields the global process structure shown in figure 5. This structure is quite general for TINA service sessions and can be reused as a skeleton to plug in different behaviours. There is thus one process definition per computational object (CO) type. These types are instantiated in an object space process. The process DPE supports the modelling of the remote method invocation paradigm in LOTOS. The COs communicate through their interfaces via three gates available at the DPE:
- *announcement* carries announcement type operational interactions,
- *invocation* and *termination* allow for interrogation type operational interaction.

Interrogations terminating with a failure are signalled at the UAP gate *exception*.

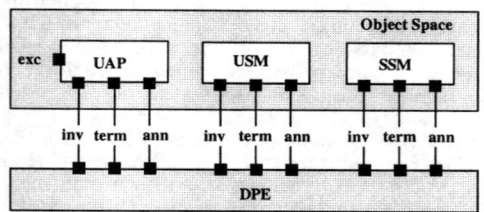

Legend: inv: invocation, term: termination, ann: announcement, exc: exception

Figure 5 Process structure for TINA service session

The internal structure of a CO process definition is exemplified in figure 6. It is composed of several client interfaces, several server interfaces and a core process definition. Each interface process may inherit multiply. In the floor control case study a SSM server interface inherits from the token passing and floor control policy management interfaces, while the client interface inherits from the corresponding notification interfaces. The interface processes simply specify ordering constraints on invocation and termination signals. Besides, they maintain knowledge about their operational binding. In our case study three parties are foreseen and all interface bindings are assumed to be established. Referencing of interfaces is achieved through multiplexing at the DPE gates. In other scenarios, dynamic instantiation of objects and interfaces may be needed. This can be achieved through the object space process.

The core capabilities are specified in the core component processes. We consider floor control as a single component. When a service is constructed out of several service components the core processes may need to synchronise on an internal gate. For the floor control component a typical case is the composition with an audio-visual conferencing or a role management component.

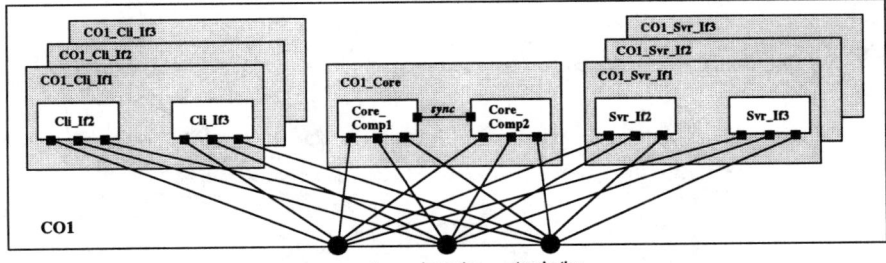

Legend: Cli_If: client interface, Svr_If: server interface, Core_Comp: core component

Figure 6 Process structure for a TINA computational object

The data type part of the specification has been written using data type language extensions, as offered by the APERO tools [Pec96] included in the Eucalyptus toolbox [Gar96]. The original text has to be pre-processed by the APERO translator to get a valid LOTOS specification. This provides for a smaller and more readable specification and for some level of immunity w.r.t. underlying processing tools. In order to have a chance to generate a graph for the system, all the base data sets have been defined as small sets of constants. Further, queues have to be bounded to a minimal depth. For instance, it is sufficient to define the token request queue with a depth of one.

The data type part of the specification is composed of:
• base values: interface references, operation types, exception codes, defined as enumeration,
• information object types and identifiers: token, token policy, party and party session group, service element,
• operation parameter type: this type wraps a list of operation parameters into one parameter to be passed at a gate between a CO interface and the DPE.

6 Verification of the Floor Control Component

The CADP package [FGK+96] included in the Eucalyptus toolbox has been used to carry out the verification. The Cæsar tool serves to generate a Labelled Transition System (LTS) from the LOTOS specification. The Aldebaran tool provides the facilities to minimise the resulting graph and to perform model-based verification.

6.1 Verification Examples

The first verification activity concerns the SSM as the most critical element in our configuration. It exemplifies the verification of a service component. The UAP processes and the DPE then constitute the environment. Three UAP processes are instantiated running the master, tokenholder and observer role, respectively. The UAP processes keep track of the state of token passing by observing incoming notifications, i.e. incorrect behaviour has not been voluntarily introduced. All invocations of interrogations are blocking. Further, the DPE process can buffer two operation invocations such that race conditions can occur. For example, the master may invoke the assignment of the token to a tokenholder concurrently with this tokenholder cancelling his token request. If the revoke token operation is processed first by the SSM, then the assignment operation of the master has to terminate with an exception stating that the token has not been requested by that tokenholder. Announcements are sent from the SSM to the three UAPs via a multicast binding. Besides, the DPE is idealised in that it does not cause any losses, duplication or reordering of signals. USM processes do not participate in the token passing interaction.

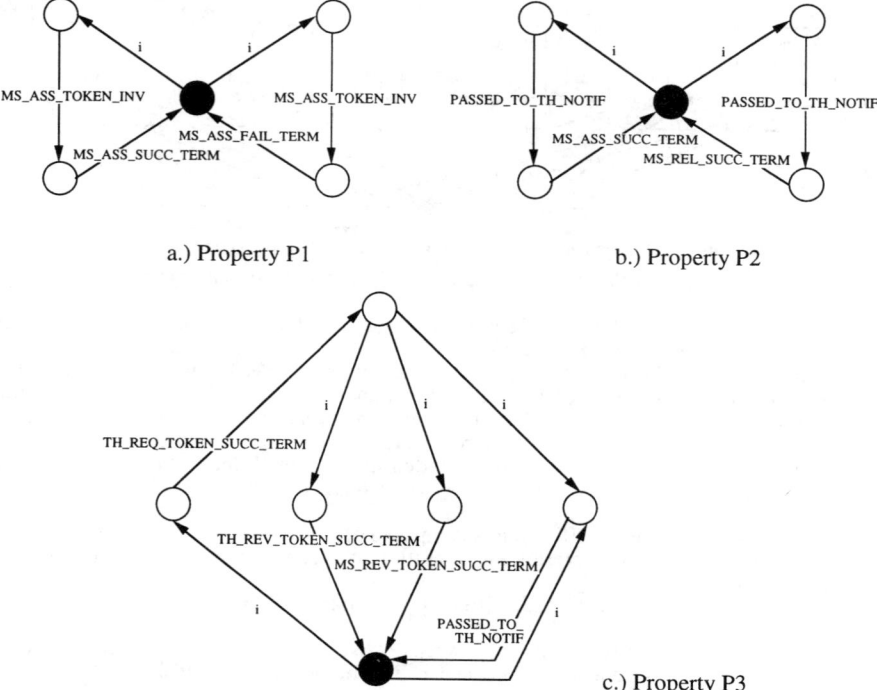

a.) Property P1

b.) Property P2

c.) Property P3

Legend: ● : initial state, **MS**: master, **TH**: tokenholder, **INV**: invocation, **TERM**: termination, **SUCC**: success, **FAIL**: failure (exception), **NOTIF**: notification (announcement), **ASS**: assign token, **REL**: release token, **REQ_TOKEN**: request token, **REV_TOKEN**: revoke token request

Figure 7 Behavioural properties at the SSM interfaces

Interesting behavioural properties of that specification can be formulated as follows:
* P1: The invocation of a token assignment operation will always terminate with either success or an exception.

- P2: Each announcement that a token has been passed to a tokenholder always engenders a successful termination of a token assignment or token release by the master.
- P3: A successful token request by a tokenholder will always be followed by the successful cancellation of that request by the tokenholder or by the master, or be followed by the assignment of the token to that tokenholder.

All properties are liveness properties. The LTSs in figure 7 have thus been obtained by reducing the original graph modulo the branching equivalence, renaming of the relevant actions and hiding of the irrelevant actions. The three properties are particular for the master mode. Other properties have to be formulated when the floor control policy is changed. Property P1 can be reformulated for all operations and assures that there exist no states where an operation invocation will always fail. Regarding property 3 one can note that a tokenholder may receive a token without a preceding request. This typically occurs when the master initiates the token release operation and the token request queue is empty. In that case the automatic mode becomes active and the next token owner is chosen randomly.

A second aspect of verification is the validation of compliance with the use cases. The use cases have been defined from the end-user role point of view. That means that we have to include the UAP processes into the verification, as the end-users interact with the UAPs via a human computer interface. Further, the major source of inspiration to device the use cases has been the users' expectations of the system expressed in the requirements. Hence, it has to be considered that they usually do not capture error situations and their recovery.

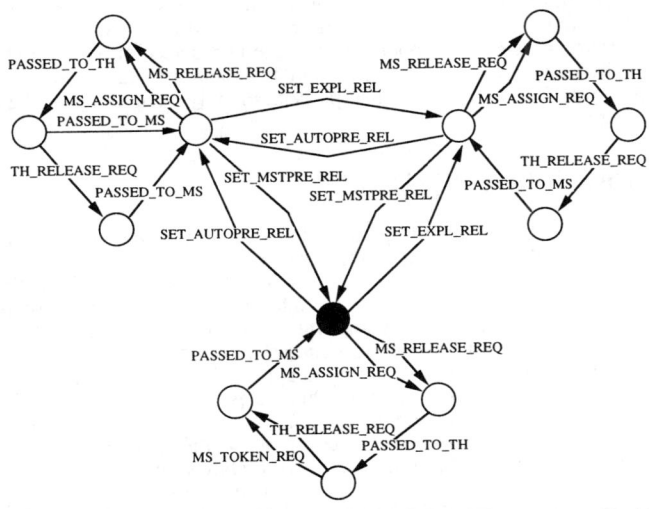

Legend: **SET_EXPL_REL: explicit token release, SET_AUTOPRE_REL: autopreemptive token release, SET_MASTERPRE_REL: masterpreemptive token release**

Figure 8 LTS of the use case "token passing"

A procedure can then be advised as follows:
- formulate a LOTOS specification P_{UC} of the use case. A LOTOS process for the master mode of the use case "token passing" (see figure 2) generates the LTS shown in figure 8,
- project this specification into the computational viewpoint by associating the transactions in the use case to one or a sequence of signals at the UAP. For example, a transaction for token assignment can be associated with the invocation signal of the

assign token operation at the client UAP. As a result of this projection one obtains a specification P'_{UC},

- relate the behaviour of P'_{UC} and the computational behaviour P_{comp} by the safety preorder ($\leq_s$) [BFG+91], i.e. $P'_{UC} \leq_s P_{comp}$. Informally, "$P \leq_s Q$" is a simulation relation meaning that the behaviour exhibited by P is possible in Q (i.e. Q can simulate P).

This procedure has been readily applicable to our specification by using the model checking feature for the safety preorder in the Aldebaran tool. As a limitation of this procedure, one can note that most use cases have an implicit notion of liveness. In particular, one execution of a use case is eventually followed by another execution of that use case. If one insists to demonstrate this property, we currently do not see another recourse than formulating several liveness properties (like in figure 7) based on the knowledge of error handling introduced in the computational model. This approach is, however, much less appealing as it demands much more effort.

A third aspect of verification of TINA components has been approached by adding the role attribution part of a role management component to our specification. Such a component has been specified independently of the floor control component. It supplies, among others, a role management interface with an operation to dynamically modify roles and an interface to notify parties of role changes. A second functionality of role management which is the association of access control policy to roles has not been considered in the specification.We have integrated this component into our computational objects as shown in figure 6. Putting the floor control and the role management core components together requires their synchronisation at an internal gate to keep the roles consistent. It is then important to verify that the SSM will keep the global state of token passing consistent. Race conditions may occur when operations are invoked at UAPs that have not yet received or processed notifications of role changes. Further, if the token passing component has not been prepared for dynamic role changes, modifications in its behaviour are needed. For instance, an entry in the token request queue has to be removed if the concerned party has newly been assigned the master role in token passing.

6.2 Experiences with the Tools

The complete specification including the role management part needs 1329 lines of extended LOTOS code, excluding comments and blank lines. After translation to standard LOTOS by APERO the specification is 1982 lines long. Seen that size, the major concern during the tool-based verification has been the state explosion problem.

We have used a Sun Ultra-2 workstation running Solaris 2.5 with 128 Mbytes of RAM for the processing. It was not possible to generate a graph for the whole specification. Each time a particular aspect of verification is approached, one has to take decisions to simplify or remove parts of the specification that are estimated irrelevant to this aspect. If this is still not sufficient to achieve a graph, the specification has to be suitably decomposed and the verification has to be carried out on the different parts.

A tool specific problem is that the specification style imposed by the LOTOScomp sublanguage obviously causes the Cæsar tool to generate excessively huge graphs. The largest LTS has been generated for the case including the role management part and comprised 345601 states and 710840 transitions. After minimisation by the strong equivalence using Aldebaran this graph "implodes" to 1161 states and 2365 transitions. Formerly, we had made the similar experience with a specification of the ODP trader using LOTOScomp. As a counterexample, the verification activities of the Equicrypt protocol show that much different ratios are possible: a LTS of 786681 and 4161795 transitions reduced to 69754 states and 520633 transitions after minimisation by strong equivalence [LBK+96]. The latter specification was not based on ODP architectural concepts.

Nevertheless, through the verification activities shortcomings in the initial behaviour specification of the floor control component could be identified and eliminated. Race conditions were detected that require handling and the definition of additional exception codes.

Another problem for the tools is related to the large number of process definitions required to write a specification based on LOTOScomp. This does not only slow down the LTS generation, but also the simulator Xeludo [STS94]. To specify the case study 50 process definitions were finally needed. The simulator could then no longer be used at a decent speed. 20 of these processes have been defined from LOTOScomp templates for computational operations. The problem can thus be alleviated by removing the restriction to use these templates and by writing the operations directly into the interface process definitions.

7 Conclusion

The integration of LOTOS into the design framework for TINA systems has been discussed in this article. LOTOScomp has allowed us to express the required computational viewpoint concepts. Formal specifications of TINA service components will surely be useful in the future, as the prescriptive part of the TINA service architecture with the associated definition of reference points is beginning to mature.

LOTOS specifications could be attached to the behaviour clauses of TINA ODL specifications. A compiler from CORBA IDL or TINA ODL to LOTOScomp would be a useful support to speed up the process of formalisation. It can be estimated that more than one third of the case study could have been generated automatically by feeding TINA ODL into such a compiler.

TINA applications exhibit an enormous complexity and the application of verification tools currently remains restricted to parts of a specification. We are nevertheless convinced that even a limited verification helps to detect specification flaws early. Improvements in the algorithms implanted in the tools can also be expected in the near future such that the verification activities could be pushed further.

Through the use of LOTOS we can also relate the specifications of the enterprise and computational viewpoint in a uniform manner. As a further step, the sublanguage LOTOSeng could be used for the engineering viewpoint specification [Vog95]. The engineering viewpoint can be seen as a refinement of the computational viewpoint in which transparency support functions and deployment aspects become visible. Each computational object then corresponds to one (possibly replicated) basic engineering object (BEO). A LOTOScomp and a LOTOSeng specification should then be related by observation equivalence with respect to operation invocation at COs and BEOs.

8 References

[BFG+91]   A. Bouajjani, J.-C. Fernandez, S. Graf, C. Rodriguez and J. Sifakis, *Safety for Branching Time Semantics*, In: 18th ICALP, Berlin, Springer Verlag, July 1991

[BK-DEI]   BERKOM Project DEIMOS, *The COMPASS Tool - Architecture and Specification*, GMD FOKUS, September 1995

[BoB87]   T. Bolognesi, E. Brinksma, *Introduction to the ISO Specification Language LOTOS*, Computer Networks and ISDN Systems 14 (1) 25-59, 1987

[DEE+96]   M. Khayrat Durmosch, Christian Egelhaaf, Klaus-Dietrich Engel, Peter Schoo *Design and Implementation of a Multimedia Communication Service in a Distributed Environment based on the TINA-C Architecture*, Aachen Workshop on Trends in Distributed Systems, October 1996

[FGK+96]   J.-C. Fernandez, H. Garavel, A. Kerbrat, R. Mateescu, L. Mounier and M. Sighireanu, *CADP (CAESAR/ALDEBARAN Development Package): A Protocol Validation and Verification Toolbox*, Proc. of the 8th Conference on Computer-Aided Verification, New Brunswick, August 1996

[Gar96]   H. Garavel, *An Overview of the Eucalyptus Toolbox*, Proc. of the COST247 Workshop, June 1996

[ISO8807]   ISO IS 8807, *Open Systems Interconnection - LOTOS - A formal description technique based on the temporal ordering of observational behaviour*, 1989

[ITU-X901] ISO IS 10746-1 | ITU-T Rec. X.901, *ODP Reference Model: Overview*, June 1995

[ITU-X904] ISO IS 10746-4 | ITU-T Rec. X.904, *ODP Reference Model: Architectural Semantics*, 1996

[Jac92]   Ivar Jacobsen et al., *Object Oriented Software Engineering- A Use Case Driven Approach*, Addison Wesley/ACM-Press, 1992

[Koe96]   Eckhart Koerner, *Group Management for a Multimedia Collaboration Service*, EUNICE96 Summer School on Telecommunication Services, September 1996

[LBK+96]   G.Leduc, O. Bonaventure, E.Koerner, L.Leonard, C.Pecheur, D. Zanetti, *Specification and Verification of a TTP protocol for the conditional access to services*, Proc. of 12th J. Cartier Workshop on "Formal Methods and their Applications: Telecommunications, VLSI and Real-Time Computerized Control Systems", Montreal, October 1996

[Pec96]   Charles Pecheur, *Improving the Specification of Data Types in LOTOS*, Doctoral dissertation, University of Liege, July 1996

[Rum91]   J. Rumbaugh et al., *Object-Oriented Modelling and Design*, Prentice Hall, 1991

[SRP96]   Linda Strick, Jenia Rosa, Stephan Paschke, *COMPASS: A CASE development tool for distributed systems*, OOPSLA96 Workshop on Methodologies for Distributed Objects, October 1996

[STS94]   B. Stepien, J. Tourrilhes and J.Sincennes, *ELUDO: The University of Ottawa Toolkit*, Technical Report, University of Ottawa, 1994, available by FTP at lotos.csi.uottawa.ca

[TN-ARC]   TINA-C, *Overall Concepts and Principles of TINA*, Archiving Label: TB_MDC.018_1.0_94, February 1995

[TN-CMC]   TINA-C, *Computational Modelling Concepts*, Archiving Label: TB_A2.HC.012_1.2_94, February 1995

[TN-IMC]   TINA-C, *Information Modelling Concepts*, Archiving Label: TB_EAC.001_1.2_94, April 1995

[TN-ODL]   TINA-C, *TINA Object Definition Language (TINA-ODL) Manual Version 2.3*, Archiving Label: TR_NM.002_2.2_96, July 1996

[TN-REF]   TINA-C, *TINA Reference Points Version 3.1*, Archiving Label: EN_RCJ.030_3.1_96, December 1996

[TN-SA]   TINA-C, *Definition of Service Architecture Version 4.0*, Archiving Label: TB_RM.001_4.0_96, December 1996

[TN-SDG]   TINA-C, *Service Design Guidelines*, Archiving Label: TP_JS_001_0.1_95, March 1995

[Vog95]   Andreas Vogel, *Towards a Formal Computational Model for Distributed Multimedia Applications*, Formal Description Techniques VII, Chapman & Hall, 1995

INDEX OF CONTRIBUTORS

KEYWORD INDEX